AF606380

Progress in Botany 64

Springer

Berlin
Heidelberg
New York
Hong Kong
London
Milan
Paris
Tokyo

64 PROGRESS IN BOTANY

Genetics
Physiology
Systematics
Ecology

Edited by

K. Esser, Bochum
U. Lüttge, Darmstadt
W. Beyschlag, Bielefeld
F. Hellwig, Jena

With 40 Figures

ISSN 0340-4773
ISBN 3-540-43620-0 Springer-Verlag Berlin Heidelberg New York

The Library of Congress Card Number 33-15850

Springer-Verlag Berlin Heidelberg New York
a member of BertelsmannSpringer Science+Business Media GmbH
http://www.springer.de

Cover design: Design & Production, Heidelberg
Typesetting: M. Masson-Scheurer, Neckargemünd
SPIN 10833552 31/3150 - 5 4 3 2 1 0 - Printed on acid-free paper

Contents

Review

Genetics

Physiology

Systematics

Ecology

List of Editors

Professor Dr. Dr. h. c. mult. K. Esser
Lehrstuhl für Allgemeine Botanik, Ruhr Universität
Postfach 10 21 48
44780 Bochum, Germany

Phone: +49-234-32-22211; Fax: +49-234-32-14211
e-mail: karl.esser@ruhr-uni-bochum.de

Professor Dr. U. Lüttge
TU Darmstadt, Institut für Botanik, FB Biologie (10)
Schnittspahnstraße 3-5
64287 Darmstadt, Germany

Phone: +49-6151-163200; Fax: +49-6151-164630
e-mail: luettge@bio.tu-darmstadt.de

Professor Dr. W. Beyschlag
Fakultät für Biologie, Lehrstuhl für Experimentelle Ökologie und Ökosystembiologie
Universität Bielefeld, Universitätsstraße 25
33615 Bielefeld, Germany

Phone: +49-521-106-5573; Fax: +49-521-106-6038
e-mail: w.beyschlag@biologie.uni-bielefeld.de

Professor Dr. F. Hellwig
Friedrich-Schiller-Universität Jena
Biologisch-Pharmazeutische Fakultät
Institut für Spezielle Botanik
Philosophenweg 16
07743 Jena, Germany

Phone +49-3641-949250; Fax +49-3641-949252
e-mail: hellwig@otto.biologie.uni-jena.de

Wilhelm Nultsch was born on 20 March 1927 in Magdeburg, Germany. From 1946 to 1953 he studied biology, chemistry, physics, philosophy and theory and methodology of education at Martin-Luther-Universität Halle-Wittenberg. In 1951 he sat his final university examination and in 1953 he received his doctoral degree (Dr. rer. nat.) in botany, supervised by Johannes Buder. In the following year, he was Buder's assistant lecturer. During this time, he developed new methods for embedding plant and animal materials in paraffin and celloidin using tetrahydrofuran as intermedium. In 1954, he left the university and for 6 years he conducted research in a chemical factory in Magdeburg, where he took part in the development of bactericides, fungicides and herbicides, resulting in two patents. In 1959, he habilitated at Universität Halle in botany and went to Universität Tübingen, where he became Privatdozent. In 1966, he was appointed full professor of botany and director of the Botanical Institute and Botanical Garden of Universität Marburg. In 1994, just one year before his retirement as university professor, he became director of the Biologische Anstalt Helgoland with the Central Institute Hamburg, the Marine Station Helgoland and the Littoral Station List/Sylt. In 1998, after the Biologische Anstalt Helgoland had been incorporated into Stiftung Alfred Wegener Institute Bremerhaven, he retired.

His main field of research, besides the phytopathological studies mentioned above, was photobiology, especially photobiology of microorganisms: phototaxis, photophobic responses, photokinesis of cyanobacteria, diatoms, flagellates and the unicellular red alga *Porphyridium cruentum*. Other topics were the photodynamic effects of dyes on microorganisms, light-induced chromatophore displacements in macroalgae, photoinhibition in and effects of UV on marine algae. The field experiments necessary for the studies with marine organisms were carried out on the island of Helgoland, on the island of Hainan (South China Sea), on King George Island (Antarctica) and on the arctic coast of northern Norway (Kårvika Marine Station). A large part of the work was carried out in cooperation with scientists from Chile, China, Greece, India, Italy, Poland, Russia, Sweden, and the USA. The resulting papers were published in various journals, NATO-ASI books and proceedings. Numerous papers appeared in *Archives of Microbiology*, *Marine Biology*, *Marine Ecology Progress Series*, *Botanica Acta*, *Cell Motility* and *Cytoskeleton*, among others.

W. Nultsch published two textbooks: *Allgemeine Botanik*, Thieme, Stuttgart, first published 1964, 11th edition 2001, translated into six languages (Dutch, English, French, Polish, Portuguese, Spanish) and *Mikroskopisch-Botanisches Praktikum*, Thieme, Stuttgart, first published 1968, 11th edition 2001. In addition, he has written chapters for several textbooks. Finally, he was co-editor of the book, *Environmental UV Photobiology*, Plenum Press, New York, London, 1993.

In 1975, W. Nultsch was elected corresponding member of the Wissenschaftliche Gesellschaft at Johann-Wolfgang-Goethe-Universität, Frankfurt. From 1980 to 1988 he was referee for botany for the Deutsche Forschungsgemeinschaft and for 16 years he was a member of the scientific committee *Antarktisforschung* of the same institution. In addition, he acted as referee for the Alexander-von-Humboldt-Stiftung and the Stiftung Volkswagenwerk. After the reunification of East and West Germany, he was appointed member of the committee for the reorganization of the (East)German Academy of Sciences and member of the committee for reorganization of the Faculty of Natural Sciences of the Martin-Luther-Universität Halle-Wittenberg. He was president of the Deutsche Botanische Gesellschaft (1985–1994) and later became an honorary member, president of the European Society for Photobiology (1991–1993) and president of the Union Deutscher Biologischer Gesellschaften (1994–1995). From 1981 to 1988 he acted as associate editor of the journal, *Photochemistry and Photobiology*. In 1996 he was awarded the Order of Merit of the Federal Republic of Germany.

He has been married to Dorothea Nultsch (née Simon) since 1950. They have two children. W. Nultsch is grateful to his wife for her patience and her help when they were experiencing hard times.

Chromatophore Displacements in Marine Macroalgae: Physiology and Ecological Relevance

Wilhelm Nultsch

1 Introduction

In Volume 60 of this series, Wolfgang Haupt published a review on chloroplast movements. Thus it may seem surprising that the present review in principle deals with the same topic, as chloroplasts belong of course to the chromatophores. I think, however, that the reader of this report will soon recognize that both reviews differ considerably with regard to the aim of the investigations, the methodology used and the organisms investigated.

The first question may be, why are seaweeds used? There are several good reasons for this choice:

1. Many flat thalli of green, brown and red algae consist of only one, two or three cell layers without intercellular spaces which would scatter light. Thus they have the great advantage that they allow the measurement of transmittance changes.
2. During the experiments, they live in a seawater flow-through system like their natural environment and survive for many days or even weeks without any damage.
3. The different pigment content of chloroplasts, phaeoplasts and rhodoplasts (therefore chromatophores) allows an approach to the question whether and to what extent photosynthetic pigments influence the displacement directly or indirectly.
4. As the macroalgae chosen live in the intertidal belt where they are covered by seawater layers of changing thickness, or even become uncovered during low tide, the comparison of the light sensitivity of samples from different depths will show whether the ability to displace the chromatophores from the so-called low-intensity arrangement to the high-intensity arrangement, and vice versa, is an important factor for the zonation of the algae in the intertidal belt.
5. Field experiments (which are indispensable to answer ecological questions) can be carried out easily with macroalgae, e.g., exposure to light at different depths.

Progress in Botany, Vol. 64

A frequently asked question is: What is the advantage of chromatophore displacements for the algae when they are floating in the sea? This question will be answered in Section 7. However, during low tide, the algae, at least those in the upper and middle eulittoral, often become completely uncovered and occupy a fixed position in which they are exposed to direct sunlight.

2 Measurement of Transmittance Changes

As described already by Senn (1908, 1919), in algae displaying light-induced chromatophore displacements three different arrangements of chromatophores occur under natural conditions. The arrangement can be different in different species, e.g., depending on the shape and the number of chromatophores. In the seaweeds investigated in dim light the chromatophores occupy preferably the cell walls which are perpendicular to the light beam or, if the cell contains only one chromatophore, it moves in a position facing the light. This is the so-called low-intensity arrangement (LIA) or face position. In bright light the chromatophores move to the walls parallel to the light beam and a single chromatophore moves to a side wall, showing its profile. This is the high-intensity arrangement (HIA) or profile position. In darkness the chromatophores of several algae occupy positions that are different from LIA and HIA (dark arrangement, DA). Accordingly, one has to distinguish a high-intensity movement from LIA to HIA, a low-intensity movement from HIA to LIA, and dark movements from either LIA or HIA to DA. These movements are accompanied by transmittance changes which can be used to measure the response photometrically, as has been done to some extent by several authors with higher plants and mosses (Biebl 1954, 1955; Seybold 1955, 1956; Zurzycki 1961; Lechowski 1974).

All the methods used in the past to measure chloroplast displacements microscopically (Zurzycki 1953, 1961; Zurzycka and Zurzycki 1957) are time-consuming and unsuitable for carrying out long-term experiments. Therefore, an automated recording microphotometer system was devised by Pfau et al. (1974) to determine quantitatively the amplitudes and kinetics of light-induced transmittance changes as a measure of chromatophore displacements under continuous registration. Low-intensity irradiation was achieved by inserting a neutral density filter between the light source and the object. Simultaneously, a filter of the same transmittance mounted between the sample and the photomultiplier was automatically removed so that the photon fluence rate impinging on the photomultiplier remained constant. The inverse procedure was used to induce high-intensity movement. The resulting signals were recorded. For more information, see Pfau et al. (1974). In order to make this system usable for long-term experiments a flow-through cuvette was constructed in which the alga was fixed with a perforated foil. The cuvette was coupled with a recycling system for nutrient medium supply and gas exchange. Thus it was possible to investigate the same area of an algal thallus under physiological conditions over several weeks. Long-term experiments demonstrated that the test objects still gave good reactions after about one month of experimental treatment. The level of transmittance remained constant during this time. Misleading results due to different thallus thickness and pigmentation could be excluded in this way. As light-induced changes in the shape and/or size of chromatophores can also cause transmittance changes (Zurzycki 1966), one has to ensure that the transmittance changes measured are really due to the movement of chromatophores.

Therefore, a fast scanning microspectrophotometer with an electrodynamic moving condenser was used (Nultsch and Benedetti 1978); this is described in detail by Benedetti et al. (1976). It allows a positional accuracy of 0.15 µm within a scanned area of 500×500 µm. The scanning area in the *Dictyota* experiment was 50×50 µm with a resolution of 62×62 measuring points. If one starts with HIA and decreases the light intensity, low-intensity movement is induced. For 5 h the *Dictyota* cell was scanned every 25 min. The corresponding pictures of the cell were played back in color on a video screen so that the changes of the chromatophore positions could be observed. When the transmittance integrals measured during this experiment are plotted versus time, the time course of low-intensity movement is obtained which is essentially identical to the one measured by Pfau et al. (1974). These and corresponding experiments with high-intensity movement demonstrate that the changes in transmittance are actually the result of the movement of the chromatophores and, hence, are a reliable measure of light-induced chromatophore displacements.

However, a single-beam photometer is not the appropriate instrument to measure action spectra, because the measured transmittance changes caused by different stimulus wavelengths are comparable only when the corresponding chromatophore arrangements are measured at a constant wavelength. Therefore, a dual-beam microscope photometer was devised in which stimulus and measuring light can be varied independently of each other (Pfau et al. 1988). This way light-induced transmittance changes can be measured during simultaneous stimulus irradiation. The measuring light is set to a precisely stabilized constant fluence rate and a fixed wavelength of 672 nm, while the stimulus light can be varied in wavelength and fluence rate. The measuring beam is chopped by means of a synchronous motor drive attached to a rotor blade with four symmetric segments. Thus, at a frequency of 50 cycles/s, the beam is modulated with 200 counts per second (cps). As shown in Fig. 1, both beams are mixed at the 10% transmitting mirror and focused

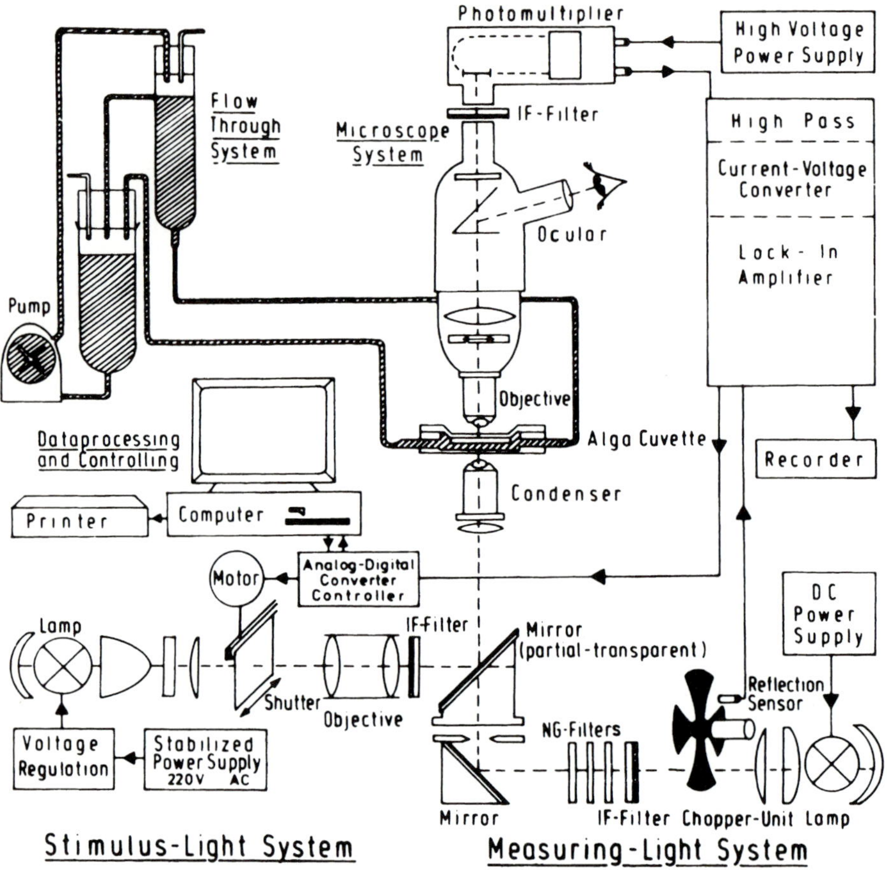

Fig. 1. Computer-controlled dual-beam microphotometer system. *Lower left* Stimulus-light system with variable wavelength and variable radiant intensity. *Lower right* Measuring-light system with a fixed wavelength of 672 nm, modulated by a chopper disk. *Middle* Microscope system with the alga mounted in a flow-through cuvette. *Upper right* Electronic instrumentation to convert the measuring light signal into a DC signal. *Middle left* Computer instrumentation for controlling the system and processing the data. *Upper left* Flow-through system to supply the alga with a constant stream of seawater. (Reprinted from Hanelt and Nultsch 1989, with permission from Elsevier Science)

on the object plane of a microscope in which the condenser has been replaced by a "Neofluar"type objective (6.3×). Before separation of the stimulus and the measuring beam, light passes an analyzer combined with a variable polarizer which allows a fine tuning of the measuring light. The signal output of the multiplier is connected with a high pass circuit where any DC components of the signal are filtered out. The residual AC signal is amplified in a current-voltage converter and then fed into a Lock-In amplifier, which is tuned to the modulated measuring signal and kept in phase by a reference signal of the reflex sensor in the chopper unit. The resulting DC-signal is filtered and plotted on a recorder, giving a measure of thallus transmittance. The samples are placed into a flow-through cuvette and kept in place with a foil that is perforated to allow a fast exchange of ions and gas. The bottom of the cuvette consists of Pyrex glass that does not absorb UV-A light. The cuvette is coupled with a flow-through system that has an overflow in order to maintain a constant hydrostatic pressure.

This highly sophisticated system was modified further by Hanelt and Nultsch (1989) in order to measure transient transmittance changes caused by phaeoplast movements. As shown in Fig. 1, it is a computer-controlled microphotometer. A wavelength of 673 nm and a fluence rate of 2.5×10^{-3} W m^{-2} were chosen as measuring light and the measuring beam was modulated by a chopper unit. After passing the algal thallus the measuring light was attenuated to different extents, depending on the phaeoplast arrangement. The photomultiplier on the top of the microscope converts the attenuated measuring light into a current signal. This is amplified, filtered and then converted into a DC voltage. The analog voltage signal is digitized by an analog/digital converter. This is connected to a computer that stores and processes the data. In addition, the computer controls a shutter in the stimulus light beam, to produce definite stimulus pulses. Thus, by continuously computing the thallus transmittance, it is possible to apply stimulus light pulses at definite transmittance levels for definite periods of time.

3 Occurrence of Light-Induced Transmittance Changes in Marine Algae

As it has been shown that chromatophore displacements can easily be detected by measuring transmittance changes with the aid of a microphotometer, the occurrence of light-induced transmittance changes in many species of green, brown and red marine algae was investigated (Nultsch and Pfau 1981). The results are summarized in Table 1.

As seen in Table 1, the green algae investigated show no transmittance changes and, hence, no chromatophore displacements with the exception of *Ulva lactuca* which, however, follow circadian rhythms. The light-induced transmittance changes measured in some red algae could not be correlated with chromatophore displacements because changes in the positions of rhodoplasts have not been observed unequivocally. Moreover, comparison of recordings of the transmittance changes measured in thalli of red algae, e.g., *Chondrus crispus*, with those of brown algae (see below) reveals striking differences in their kinetics (Nultsch and Pfau 1979). This suggests that the transmittance changes measured in red algae are due to different processes, e.g., to conforma-

Table 1. Occurrence of light-induced transmittance changes in marine algae. + Transmittance changes were measured; – transmittance changes not detectable; (+) transmittance changes follow circadian rhythms. (Modified after Nultsch and Pfau 1979)

Brown algae		Red algae		Green algae	
Ascophyllum nodosum	+	*Ahnfeltia plicata*	–	*Bryopsis plumosa*	–
Alaria esculenta	–	*Bonnemaisonia hamifera*	–	*Enteromorpha linza*	–
Desmarestia aculeata	–	*Ceramium deslongchampsii*	–	*Enteromorpha compressa*	–
Dictyopteris polypodioides	–	*Ceramium rubrum*	–	*Enteromorpha intestinais*	–
Dictyota dichotoma	+	*Chondrus crispus*	+	*Monostroma grevillei*	–
Ectocarpus siliculosus	–	*Cystoclonium purpureum*	+	*Prasiola stipitata*	–
Fucus spiralis	+	*Delesseria sanguinea*	–	*Ulva lactuca*	(+)
Fucus serratus	+	*Dumontia incrassata*	+		
Fucus vesiculosus	+	*Furcellaria fastigiata*	–		
Halidrys siliquosa	+	*Halarachnion ligulatum*	–		
Laminaria digitata	+	*Membranoptera alata*	–		
Laminaria hyperborea	+	*Phycodrys rubens*	–		
Laminaria saccharina	+	*Phyllophora brodiaei*	+		
Petalonia fascia	+	*Phyllophora membranifolia*	+		
Petalonia zosterifolia	+	*Plocamium cartilagineum*	–		
Punctaria plantaginea	+	*Polyides rotundus*	–		
Pilayella littoralis	–	*Polysiphonia urceolata*	–		
Scytosiphon lomentaria	+	*Pophyra purpurea*	+		
		Porphyra umbilicalis	+		
		Rhodomela confervoides	–		

tional and configurational changes in the thylakoid membranes, as reported by Murakami and Packer (1970) for *Porphyra*.

Chromatophore displacements and concomitant transmittance changes were observed in most of the brown algae investigated. Exceptions are *Ectocarpus siliculosus* and *Pilayella littoralis* which have no flat thallus and *Alaria esculenta* and *Desmarestia aculeata* which live in the lower eulittoral and sublittoral, respectively. Thus, they occur preferentially in those species that have a more-or-less flat thallus and live in the intertidal belt, where they are often fixed during low tide so that they are exposed to direct sunlight. The unexpected behavior of *Dictyopteris*, which has also a flat monolayered thallus, will be discussed later. The adult *Fucus* thalli which are multilayered and differentiated into epidermis, cortex and medulla show different phaeoplast arrangements in dim and strong light as well as in darkness (Rüffer et al. 1978). However, as the phaeoplast arrangements in the epidermal, cortical and medullary cells are also different from each other, *Fucus* is apparently not a suitable object for physiological studies. Therefore, most of the investigations were carried out with *Dictyota dichotoma* (Huds.) Lamour whose three-layered thallus consists of one layer of large, phaeoplast-free medullary cells, covered on both sides by smaller cortical cells which contain numerous phaeoplasts. Thus, the optical properties of the thallus are suitable for transmittance measurement. In addition, *Dictyota* can be cultivated and easily handled in field experiments.

As in other organisms three different arrangements of phaeoplasts occur. In dim light they occupy preferentially the cell walls that are perpendicular to the light beam (LIA). In bright light they move to the walls parallel to the light beam (HIA). In darkness they occupy the anticlinal and the inner periclinal cell walls (DA). It must be mentioned, however, that the phaeoplasts in the cortical cells covering the centers of the neighboring medullary cells do not respond to changes in irradiance (so-called medullary cell pattern), probably as a result of "cytotaxis" caused by chemical gradients (Senn 1908). As shown in Fig. 2, high-intensity movement results in an increase and low-intensity movement in a corre-

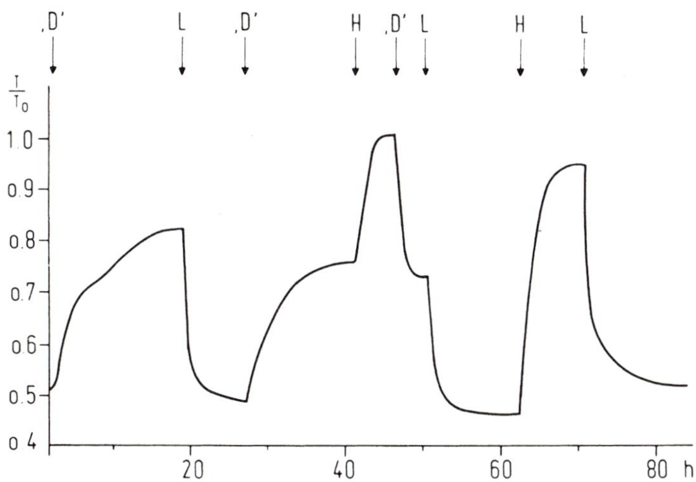

Fig. 2. Low-intensity, high-intensity and dark movements of phaeoplasts, recorded as transmittance changes vs. time following radiation with blue light (439 nm) of different intensities. Low intensity (L) = 1 W m^{-2}, high intensity (H) = 25 W m^{-2}, 'D' = "relative" darkness under the experimental conditions (red measuring light 10^{-4} W m^{-2}). *Abscissa* Time in hours; *ordinate* thallus transmittance (T/To) in relative units. (Reprinted from Rüffer et al. 1981, with permission from Urban & Fischer)

sponding decrease of transmittance. Since DA looks like an intermediate state between LIA and HIA, its transmittance value lies between those of LIA and HIA (Rüffer et al. 1981).

4 Action Spectrum and Photoreceptor Problems

As comprehensively reported in the recent review by Haupt (1999), in most green plants chloroplast displacements are caused by radiation of short wavelengths. The action spectra measured in different laboratories show maxima around 370 and 450 nm, suggesting a flavin as photoreceptor pigment. As other UV-A- and blue-light absorbing pigments cannot be ruled out, these pigments are usually summarized as cryptochromes. The only exceptions among the algae are the chloroplast movements of *Mougeotia* and *Mesotaenium*, which are governed by red light, indicating phytochrome as photoreceptor pigment. This is particularly supported by the reversible red/far-red antagonism.

Since even carotenoids have been suspected to act as photoreceptors in blue-light-induced responses (De Fabo 1980; Shropshire 1980), it was of interest to measure the action spectrum of phaeoplast movements in *Dictyota*, because brown algae contain a relatively large amount of fucoxanthin, a carotenoid absorbing in vivo up to 590 nm (Goedheer 1970). Thus, if these carotenoids were involved in photoperception, one should expect an extension of the action spectrum to longer wavelengths.

The transmittance changes caused by monochromatic light were measured with the computer-controlled dual-beam microphotometer system described above (Fig. 1). If the high-intensity (LIA→HIA) or the low-intensity (HIA→LIA) movements were measured, the algae would have been exposed repeatedly to high fluence rates for a relatively long time, so that they were light stressed or even damaged. Therefore, the action spectrum of the movement from the dark to the low-intensity arrangement (DA→LIA) was measured. As in DA the phaeoplasts occupy also the inner periclinal walls in addition to the anticlinal ones, the transmittance changes measured this way are smaller, but sufficient to obtain significant results. Moreover, as it takes a relatively long time to achieve steady-state transmittance levels, the changes in transmittance in response to short light pulses were used to shorten the measuring time, so that more experiments could be carried out with the same thallus. Monochromatic light was produced by inserting interference filters. Different fluences were achieved by opening the shutter in the stimulus light beam for different periods of time at a constant fluence rate. For each wavelength, fluence-response curves were measured in the range 358–743 nm. As a measure the standardized slopes of the fluence-response curves were used. As the action spectrum (Fig. 3) shows, the

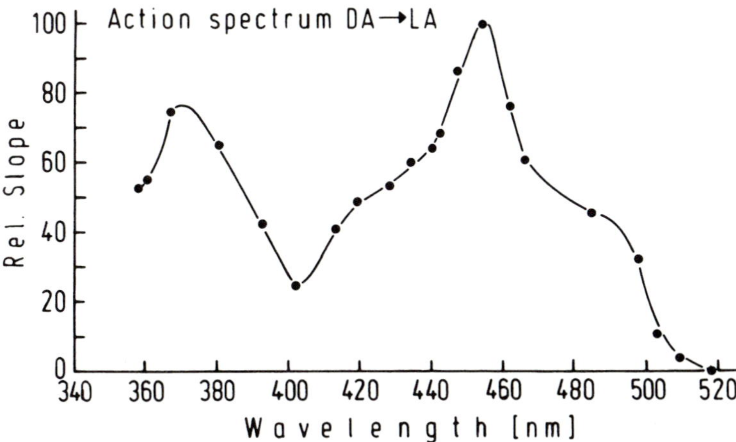

Fig. 3. Action spectrum of phaeoplast movements from dark to low-intensity arrangement of *Dictyota dichotoma*. As a measure the standardized slopes of the fluence-response curves were used. *Abscissa* Wavelengths in nm. *Ordinate* Slopes of the fluence-response curves in relation to the standard of the averaged slopes of the curves at 454 nm. (Reprinted from Hanelt and Nultsch 1989, with permission from Elsevier Science)

whole range of wavelengths between 358 and 518 nm is active. Clear maxima were found at 370 and 454 nm, corresponding to those in the absorption spectrum of a flavoprotein (Galland and Senger 1988). Because of the ineffectiveness of wavelengths above 518 nm, fucoxanthin can with certainty be excluded as possible photoreceptor. Thus the photoreceptor of the phaeoplast displacement in *Dictyota dichotoma* is a flavoprotein rather than a carotenoid.

The ineffectiveness of fucoxanthin is not surprising because this pigment is located within the phaeoplasts, whereas all relevant photoreceptor studies carried out in the past indicate that the photoreceptor molecules are located outside the plastids in the cortical cytoplasm parallel to the cell surface. If this were true also for *Dictyota*, an action dichroism should be demonstrable in this alga. In fact, irradiation with weak polarized white or blue (442 nm) light causes a large part of the phaeoplasts to migrate to the anticlinal cell walls parallel to the E-vector of polarization (Pfau et al. 1979), similar to the arrangements described by Zurzycki (1967) for the moss *Funaria*. The action spectrum of the dichroic arrangement is largely consistent with the one of *Funaria* with the exception that, unlike *Funaria*, wavelengths between 365 and 400 nm also caused action dichroism, while light above 500 nm was ineffective. Another difference to Zurzycki's results is that in strong light no dichroism effect is observed. Under those conditions the phaeoplasts occupy a typical unpolarized high-intensity arrangement. As in *Funaria*, in the active spectral range no significant differences in the effectiveness of the different wavelengths could be detected. These results indicate that also in *Dictyota* the photoreceptor molecules are orientated parallel to the cell surface as in other cryptochrome-mediated chloroplast movements.

5 Circadian Rhythms in Chromatophore Movements

In contrast to the chromatophore movements in *Dictyota* and other brown algae, the chloroplast movement in the green alga *Ulva lactuca* is governed by circadian rhythms (Senn 1908; Britz and Briggs 1976). The cells of the bilayered thallus have a single, large, cup-shaped chloroplast. During the day, the chloroplast occupies the outer face of the cell (face position), while at night it is along the side (profile position). These positional changes result in considerable transmittance changes which can easily be measured with a microphotometer while the disks cut out of the thallus are mounted in a flow-through cuvette allowing long-term measurements. The measuring wavelength in these experiments was 475 nm (Britz et al. 1976).

After the samples had become arrhythmic, rhythms could be reinitiated with 12-h white light irradiation (915 lx). With the continuously recording microphotometer rhythms could be monitored for up to 10 days. The free running period in "relative" darkness (=in red measuring light) at 20 °C was 24–25 h. However, depending on the measuring beam intensity, the free running periods varied between 21 and about 26 h. In addition, the end position assumed by the chloroplasts after the rhythms have dampened out is also influenced by environmental conditions. Thus, the rhythmic chloroplast movement is sensitive to modification, especially by light.

As shown, the phaeoplast movements of *Dityota dichotoma* from profile to face position and vice versa are caused by changes from low- to high-intensity irradiation, and vice versa. Nevertheless, circadian oscillations of the amplitudes and kinetics of the resulting transmittance changes are initiated when the algae are grown in light:dark cycles of about natural day lengths (Nultsch et al. 1984). If the thalli were transferred in continuous light or darkness, the circadian rhythmic oscillations continued. The lengths of the free running periods varied between 24 and 29 h. As circadian transmittance changes can also be initiated in isolated cortical cell layers, monolayers mounted upside down were used for microscopic counting of the phaeoplast numbers at the inner periclines. This would not be possible with intact thalli because of difficulties in focusing. These experiments have shown that the low-intensity arrangement is the most stable one, i.e., the number of phaeoplasts at the walls perpendicular to the light beam is relatively constant. If the thalli were exposed to photon fluence rates above or below the low-intensity level, rhythmic changes in the phaeoplast arrangements occurred. The unstable high-intensity arrangement is due to the varying numbers of phaeoplasts at the periclinal cell walls, as the circadian clock causes variations of HIA→LIA as well as LIA→HIA movement and, hence, corresponding oscillations of the transmittance amplitudes. The dark arrangement is apparently the most unstable, as the recordings of transmittance in darkness show the most striking circadian oscillations. Accordingly, the number of phaeoplasts at the periclinal walls displays considerable variations, if maxima and minima of circadian transmittance oscillations are compared. This may explain the different results of the investigations by Senn (1908, 1919) and Rüffer et al. (1981). Simultaneously, the experiments with isolated cortical layers demonstrate that the above-mentioned "medullary cell pattern" of the phaeoplast arrangement at the inner periclines of the cortical cells has no influence on the circadian rhythms of phaeoplast movements as they are not altered by the destruction of the medullary cells by bisecting the thallus. Moreover, the experiments show that the site of the physiological clock is in the cortical cells themselves.

It cannot be excluded with certainty that other phenomena causing transmittance changes, such as changes in the shape of chromatophores, changes in their pigment content or in their ultrastructure, are also subject to circadian oscillations. However, the observed changes in the phaeoplast arrangements are certainly sufficient to explain the recorded transmittance changes.

The occurrence of circadian rhythms in light-induced chromatophore displacements was controversial in the past. Zurzycka and Zurzycki (1953) did not find circadian rhythmic oscillations in the chloroplast movements of *Lemna trisulca*. Later, Flohrs and Haupt (1971) observed daily fluctuations in the response of *Mougeotia* chloroplasts to red and far-red light stimuli. Thus, the investigations with *Dictyota dichotoma* clearly indicate that light-induced chromatophore movements can also be influenced by the physiological clock.

6 Field Experiments

As all the experiments reported so far were carried out in the laboratory, i.e., with parallel light of definite fluence rates and wavelengths, impinging perpendicularly on the thalli, at constant temperature and controlled hydrostatic pressure, it was of greatest interest to see what happens under natural conditions. Therefore, field experiments were carried out at the rocky shore of the island Helgoland (Hanelt and Nultsch 1990). The ability to displace their chromatophores should be particularly important for algae living in the intertidal belt where their exposure to light varies during the day according to the position of the sun, cloud cover and tide level. In particular, the algae in the upper and middle eulittoral are frequently uncovered by seawater during low tide and consequently fixed in their position to sunlight. Indeed, if thalli of species which display light-induced chromatophore displacements are collected at low tide, their chromatophores are in HIA.

Dictyota dichotoma, however, lives in the lower eulittoral where it rarely becomes completely uncovered so that its thalli are floating and do not occupy a fixed position with respect to sunlight. In order to find out what position the chromatophores occupy in floating thalli the algae were clamped between two microscopic slides and mounted on a carrier especially devised to submerge algae in the sea. After LIA was induced the transmittance of the thalli was measured as a reference value. During the night, chromatophores were in DA. Next morning the experiment was started by submerging the algae in the sea, e.g., 10 cm below the water surface. The device was fastened to a pontoon so that the depth was constant and independent of the tidal level. Every hour one thallus was removed from the device and fixed in a solution of 1% glutaraldehyde in seawater in order to prevent changes in the phaeoplast arrangement during transport and measurement. The pigmentation was not influenced by this treatment. Then the thallus transmittance was measured again and compared with the reference value. In addition, the chromatophore arrangement was controlled microscopically and photomicrographs were taken. The experiment was finished in the evening. During the whole experiment the attenuation of the daylight by the seawater was measured with a waterproof photodetector in the vicinity of the device.

Immediately after sunrise the phaeoplasts began to migrate from DA to LIA, but only for about 1 h. Then they moved with increasing fluence rate to HIA (which was complete around noon) and maintained for several hours. Finally, with decreasing fluence rate, the phaeoplasts moved to LIA. These results indicate that the phaeoplast arrangement of floating thalli under natural conditions depends much more on the fluence rate than on the light direction. This is consistent with Schönbohm's (1966, 1980) observation in *Mougeotia* that we have to distinguish between an orienting light signal (which in *Mougetia* is perceived by phytochrome) and an intensity-dependent signal which is perceived by a blue-light receptor and determines the chromatophore arrangement, with the difference that in *Dictyota* both signals are mediated by the same blue-light photoreceptor, but the intensity-dependent signal dominates the directional cue.

In a second series of experiments, the dependence of phaeoplast arrangement on water depth was measured. In this case thalli were exposed to sunlight in different depths of seawater. All thalli were removed from the device at the same time (14:00 h) and fixed. The transmittance values were determined and the light attenuation by the water body was measured as described above. Again photomicrographs were taken. Both transmittance measurements and photomicrographs show that, even at a depth of 1 m, the phaeoplasts were in HIA. Between 1 and 3 m they occupied intermediate positions between HIA and LIA. Finally, at 4 m, an almost complete LIA was reached. The phaeoplasts of thalli floating at or near the surface of the waterbody are in HIA almost all of the day and move to LIA transiently in the early morning and in the later afternoon, after and before they migrate to DA. A transient cloud cover of a bright sky does not significantly influence the arrangement, as the overall fluence rate remains above the high-intensity level. The duration of LIA becomes the longer the deeper the thalli are submerged and, hence, the lower the overall fluence rate is.

7 Ecological Role of Chromatophore Displacement in Algae

Zurzycki (1955, 1975), who worked with the moss *Funaria hygrometrica* and the seed plant *Lemna trisulca*, suggested that photosynthetic activity could be regulated by the chloroplast arrangement, as the rate of photosynthesis increased after irradiance had been changed from high to low intensity. He concluded that this increase was due to the absorbance increase caused by the movement of chloroplasts from HIA to LIA. This seems not to be true for *Mougeotia* in which the rate of photosynthesis reached a new level before the chloroplasts had occupied the face position (Zurzycki 1955). Lechowski (1974) reported that in the seed plant *Ajuga reptans* the rate of photosynthesis in HIA was reduced to about 50% compared to that in LIA, although the change in light absorption caused by this movement was only about 6%.

Nultsch et al. (1981) found that in three seaweeds the rate of oxygen production was strongly decreased by pre-irradiation with white, blue and red light of high fluence rates: in *Dictyota dichotoma*, which shows phaeoplast displacements, in *Alaria esculenta*, which does not show phaeoplast displacements, and in *Ulva lactuca*, in which chloroplast displacement is governed by the physiological clock. Thus in these algae the decrease of the PS rate is independent of the respective chromatophore arrangement. Consequently, the function of chromatophore displacements is apparently not the regulation of photosynthetic activity. This conclusion is strongly supported by the results of the field experiments, because the phaeoplast displacement in *Dictyota* is a relatively sluggish process that requires some hours to become complete and is therefore not able to follow fast changes of the fluence rates.

These findings are in good agreement with the observation by Titlyanov et al. (1978) that in *Ulva fenestrata* changes in photosynthetic capacity were correlated with changes in chlorophyll content but not with the chloroplast position. Witztum et al. (1979) reported that in a mutant of *Lemna paucicostata* light-induced chloroplast displacements occurred although the photosynthetic electron transport was blocked. However, high-intensity movement can be regarded as a slow adaptation to gradually increasing fluence rates caused by the decreasing sea level during low tide as well as by the diurnal course of the overall fluence rate. As an adaptation to transient alterations in the fluence rate photo-inhibition and its recovery are more suitable, because they have faster kinetics and are more effective in regulating photosynthetic activity (Nultsch et al. 1987; Hanelt et al.1995).

The movement from LIA to HIA might be rather a light-protective mechanism. In order to find out whether this is true, the algae were exposed to fluences that are high enough to cause photodamage and correspond roughly to the light conditions in the natural environment when the algae are uncovered by seawater or just below the surface during low tide (Hanelt and Nultsch 1991). For these experiments a dual-beam irradiation device was used (Fig. 4). Disks 4 mm in diameter were cut out of the *Dictyota* thalli and put into equally sized holes of black masks. The masks were placed in quartz cuvettes filled with seawater and inserted into the light beam of a Shimadzu UV-VIS 2000 recording spectrophotometer equipped with an integrating sphere. The thallus absorption was measured in LIA, induced by 0.5 W m^{-2} 450 nm. Then, the cuvette was placed between two light sources under constant conditions. One side of the alga was irradiated with blue light (450 nm) to induce or maintain the desired phaeoplast arrangement (0.5 W m^{-2} for LIA, 40 W m^{-2} for HIA). The other side of the thallus was irradiated with either visible radiation >495 nm (GG 495, Schott) or UV-B (band pass 290–310 nm) in order to check the photodamaging effectiveness. For UV-B a Xenon lamp and a grating-monochromator were used; for visible light a halogen lamp. Two thalli were irradiated simultaneously, but exposed to different irradiation programs. Wavelengths and total fluences

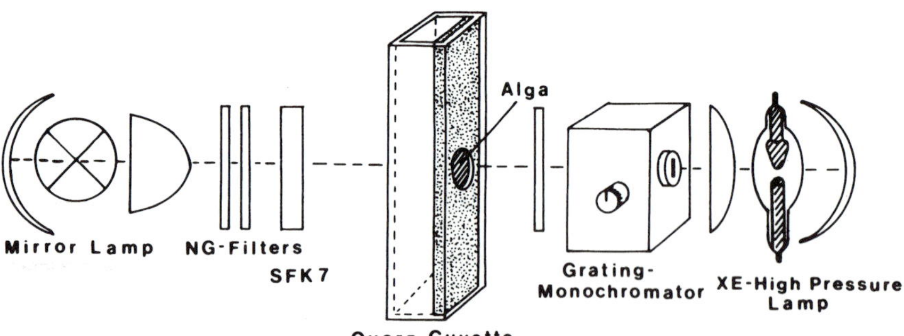

Fig. 4. Dual beam irradiation device. *Left* Light source for induction of the respective chromatophore arrangement. *Middle* Photometer cuvette with a black mask for keeping the alga in place. *Right* Light source for production of UV-B radiation, which is replaced by a halogen lamp and a GG495 filter in the GG495 nm experiments. (Reprinted from Hanelt and Nultsch 1991, with permission from Urban & Fischer)

were equal in both programs, but in program (1) the thallus was irradiated with phaeoplasts in LIA, in (2) with phaeoplasts in HIA. As an example, the GG 495 program is shown:

1. LIA
   [450 nm 0.5 W m^{-2}]→[2 h 450 nm 0.5 W m^{-2}+GG 495 400 W m^{-2}]→[1 h 30 min 450 nm 40 W m^{-2}]→[2 h 450 nm 0.5 W m^{-2}]→[17 h recovery at 450 nm 0.5 W m^{-2}].
2. HIA
   [450 nm 0.5 W m^{-2}]→[2 h 450 nm 0.5 W m^{-2}]→[1 h 30 min 450 nm 40 W m^{-2}]→[2 h 450 nm 0.5 W m^{-2}+GG 495 400 W m^{-2}]→[17 h recovery at 450 nm 0.5 W m^{-2}].

The corresponding programs for UV-B are not shown. At the end of the experiments the absorption of thalli in LIA, induced with 450 nm 0.5 W m^{-2}, was measured again. The difference spectrum of absorbance before and after irradiation indicates the extent of photodamage, either due to pigment bleaching and/or to possible changes in shape, size and arrangement of phaeoplasts. In addition, the thalli were inspected by microscope.

UV-B (300 nm, 0.5 W m^{-2}) does not induce movement to HIA (Pfau et al. 1988). The absorption of UV-B is not clearly influenced by the changes in phaeoplast arrangement. Consequently, the photodamage caused by UV-B is not significantly correlated to the phaeoplast arrangement. However, UV-B affects the mechanism of phaeoplast movement and, hence, their arrangement. In some cortical cells the phaeoplasts do not return to the periclinal walls in LIA, but form clusters. In addition, a slight destruction of pigmentation was observed.

Contrary to UV-B, the absorption of GG495 light depends considerably on the phaeoplast arrangement. The difference spectra, calculated from ten thalli, show striking differences in the absorbance decrease caused by irradiation of thalli with phaeoplasts being in LIA and HIA, respectively (Fig. 5). The absorption change in the blue (Soret band) is about 50% smaller when the phaeoplasts were in HIA during irradiation. A strong effect is also seen in the red. In general, the absorption decrease is apparently mainly restricted to the wavelength ranges absorbed by the chlorophylls *a* and *c*, whereas the changes in the range absorbed by fucoxanthin are much smaller. If relative difference spectra (difference spectra shown in Fig. 5 divided by the absorption spectra measured before the experiment) are calculated it results that the absorption by chlorophyll *a* decreased by about 6%, but the absorption by fucoxanthin only by 2%, when phaeoplasts were irradiated in LIA. It must be emphasized that the ability to return to LIA was not impaired by the first irradiation with strong light. Therefore, the effects of repeated strong light irradiation could be investigated. After the first repetition of the irradiation program the additional photodamage was twice as large as after the first

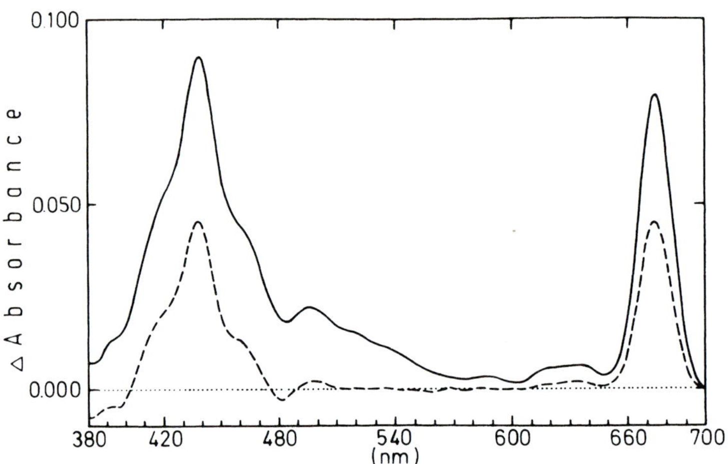

Fig. 5. Difference spectra of thallus absorption before and after irradiation with visible light >495 nm. The *solide line* shows the difference spectrum when the phaeoplasts were irradiated in LIA, the *dashed line* the difference spectrum when they were irradiated in HIA. (Reprinted from Hanelt and Nultsch 1991, with permission from Urban & Fischer)

one. The relative difference spectrum revealed an absorbance decrease of 11% in the whole wavelength range (also absorbed by fucoxanthin). Observations by microscope have shown that the ability of the phaeoplasts to return to LIA was now significantly impaired and clusters were formed as after UV-B treatment. This damage was more striking in thalli irradiated in LIA than in HIA. After a third strong light treatment the phaeoplasts were damaged and the normal LIA was disturbed.

The fluences of UV-B and of light >495 nm were of the same order of magnitude as those to which the algae are exposed in their natural environment. Thus, the fluences applied were not nonphysiological. Zurzycki (1957) observed strong photodamage (changes in shape and/or arrangement of chloroplasts, chlorophyll destruction and inactivation of photosynthesis) in *Lemna* after irradiation with strong white light (about 200 W m^{-2}). Photoinhibition of photosynthesis commences simultaneously with movement to HIA, but it reaches its final level much earlier because of its faster kinetics (Nultsch et al. 1981, 1987; Hanelt et al.1995). Apparently, the destruction of photosynthetic pigments occurs only if photoinhibition and movement to HIA are not longer sufficient to protect the photosynthetic apparatus from an excess of impinging quanta.

Another hypothesis concerning the ecological function of chromatophore arrangement has been suggested by Walczak and Gabrys (1981), Psaras (1986) and Seitz (1987). These authors suppose that a CO_2 gradient in the cell may be responsible for chromatophore movement. This is improbable in the case of *Dictyota* because an increase in HCO_3^- concentration neither prevents nor delays movement to HIA (Hanelt and Nultsch 1991). Thus, the photoprotective role of HIA is out of question.

Of ecological importance is also the question, whether correlations exist between chromatophore displacement and algal zonation in the phytal. Therefore, dose-response curves were measured with *Fucus spiralis* var. *platycarpus* Thur. which is found in the upper littoral and supralittoral, so that it is often exposed to direct sunlight, *Fucus vesiculosus* L. which lives in the middle and upper eulittoral, and *Fucus serratus* L. which grows in the lower and middle eulittoral (Nultsch and Pfau 1979). The results suggest that those correlations could exist, as the more exposed the localities of the algae are the lower is the irradiance at which minimal absorption is reached. No significant differences were detected in *Laminaria saccharina* (L.) Lamour, *Laminaria digitata* (Huds.) Lamour and *Laminaria hyperborea* (Gunn.) Fosl. all of which live in the sublittoral, where they are, if at all, scarcely exposed to direct sunlight. However, these data are insufficient to conclude whether correlations between chromatophore displacement and zonation exist.

8 Concluding Remarks

As shown in this report, measurements of transmittance changes combined with observations by microscope are reliable means to measure chromatophore displacements quantitatively, especially because the same part of the same thallus can be measured over weeks up to a month in the flow-through cuvette. Several questions could be answered in this way.

Light-induced chromatophore displacements occur in many brown algae, but not in the green and red algae investigated. The pigments of the chromatophores are not involved in photoperception. The photoreceptor, at least in *Dictyota* and some other brown algae, is apparently a flavoprotein (cryptochrome) and, as action dichroism indicates, its molecules are oriented parallel to the surface as in other cryptochrome-mediated chloroplast movements. Although the chromatophore displacements are induced by light and not governed by the physiological clock as in *Ulva*, circadian oscillations of amplitudes and kinetics of transmittance changes were initiated when the algae were grown in light:dark cycles of about natural day lengths. Of especial interest were the results of field experiments. They suggest that the normal position of the phaeoplasts is the low-intensity arrangement, because, in this position, the phaeoplasts are able to capture a maximum of photons. However, if the fluence rate exceeds a distinct threshold value, which may vary from thallus to thallus, the chromatophores begin to move to the high-intensity arrangement the more, the higher the photon-fluence rate is. Of course, the absorption cross-section decreases this way. Nevertheless, the photosynthetic activity is not significantly reduced because of the higher photon-flux density, unless photoinhibition occurs. This is true also for thalli floating in the sea. The phaeoplasts are in high-intensity arrangement if the total fluence rate is high but they are in low-intensity arrangement if the total fluence rate is low. This demonstrates that the phaeoplast arrangement depends much more on the fluence rate than on the light direction which cannot be perceived by floating thalli that steadily change their position and are also exposed to light scattered and reflected by particles suspended in the seawater. Thus, the ecological function of chromatophore displacement is photoprotection from high irradiances, whereas photosynthetic activity at supraoptimal fluence rates is regulated by photoinhibition.

Nevertheless, several questions remain open:

1. Why do green algae not change their chloroplast arrangements, though many of them (*Ulva lactuca*, several *Enteromorpha* and *Monostroma* species) live in the upper eulittoral where they are often exposed to direct sunlight? Does the rapid adaptation of photosynthetic capacity to changing light conditions as observed by Häder et al. (2001) with *Enteromorpha linza* make it unnecessary to decrease the absorprtion cross section?
2. What is the function of the circadian chloroplast movement in the bilayered *Ulva* in which the chloroplasts occupy the face position during the day even if they are exposed to direct sunlight but occupy the profile position during the night? Moreover, why do the chloroplasts of the related, also bilayered *Enteromorpha linza* not show the same phenomenon?
3. Why do the threelayered *Dictyota* and the monolayered *Laminaria* germlings (of course also the multilayered adult *Laminaria*) perform light-induced phaeoplast displacement but not the monolayered *Dictyopteris*?
4. What is the function of dark arrangement? Do the chromatophores leave the outer periclinal walls (in *Ulva* as in *Dictyota*) in order to facilitate the uptake of oxygen for respiration?
5. What is the function of the so-called medullary cell pattern in *Dictyota*?

The chromatophores covering the center of the medullary cells do not protect their nuclei from photodamage, as the nuclei are often found near the anticlinal cell walls and are consequently not protected from irradiation. Moreover, this pattern also occurs in darkness.

References

Benedetti PA, Bianchini G, Chiti G (1976) Fast scanning microspectroscopy: an electrodynamic moving condenser method. Appl Optics 15:2554–2558

Biebl R (1954) Lichttransmission und Chloroplastenbewegung. Flora 141:163–177

Biebl R (1955) Tagesgänge der Lichttransmission verschiedener Blätter. Flora 142:280–294

Britz SJ, Briggs WR (1976) Circadian rhythms of chloroplast orientation and photosynthetic capacity in *Ulva*. Plant Physiol 58:22–27

Britz SJ, Pfau J, Nultsch W, Briggs WR (1976) Automatic monitoring of a circadian rhythm of change in light transmittance in *Ulva*. Plant Physiol 58:17–21

De Fabo E (1980) On the nature of the blue light photoreceptor: still an open question. In: Senger H (ed) The blue light syndrome. Springer, Berlin Heidelberg New York, pp 187–197

Flohrs H, Haupt W (1971) Tagesperiodische Empfindlichkeitsschwankungen der lichtinduzierten Chloroplastenbewegung von *Mougeotia*. Z Pflanzenphysiol 65:65–69

Galland P, Senger H (1988) The role of flavins as photoreceptors. J Photochem Photobiol B Biol 1:277–294

Goedheer JC (1970) On the pigment system of brown algae. Photosynthetica 4:97–106

Häder DP, Lebert M, Helbling EW (2001) Effects of solar radiation on the Patagonian macroalga *Enteromorpha linza* (L) J. Agardh – Chlorophyceae. J Photochem Photobiol B Biol 62:43–54

Hanelt D, Nultsch W (1989) Action spectrum of phaeoplast displacements from the dark to the low intensity arrangement in the brown alga *Dictyota dichotoma*. J Photochem Photobiol B Biol 4:111–121

Hanelt D, Nultsch W (1990) Daily changes of the phaeoplast arrangement in the brown alga *Dictyota dichotoma* as studied in field experiments. Mar Ecol Prog Ser 61:273–279

Hanelt D, Nultsch W (1991) The role of chromatophore arrangement in protecting the chromatophores of the brown alga *Dictyota dichotoma* against photodamage. J Plant Physiol 138:470–475

Hanelt D, Uhrmacher S, Nultsch W (1995) The effect of photoinhibition on photosynthetic oxygen production in the brown alga *Dictyota dichotoma*. Bot Acta 108:99–106

Haupt W (1999) Chloroplast movement: from phenomenology to molecular biology. In: Esser K, Kadereit JW, Lüttge U, Runge M (eds) Progress in Botany, vol 60. Springer, Berlin Heidelberg New York, pp 3–36

Lechowski Z (1974) Chloroplast arrangement as a factor of photosynthesis in multilayered leaves. Acta Soc Bot Pol 43:531–540

Murakami S, Packer W (1970) Light-induced changes in conformation and configuration of the thylakoid membranes of *Ulva* and *Porphyra* chloroplasts in vivo. Plant Physiol 45:289–299

Nultsch W, Benedetti PA (1978) Microspectrophotometric measurements of light-induced chromatophore movements in a single cell of the brown alga *Dictyota dichotoma*. Z Pflanzenphysiol 87:173–180

Nultsch W, Pfau J (1979) Occurrence and biological role of light-induced chromatophore displacements in seaweeds. Mar Biol 51:77–82

Nultsch W, Pfau J (1981) Investigations on light-induced plastid movement in seaweeds. In: Fogg GE, Jones WE (eds) Proc 8th Int Seaweed Symp Bangor. The Mar Sci Lab, Menai Bridge, pp 209–216

Nultsch W, Pfau J, Rüffer U (1981) Do correlations exist between chromatophore arrangement and photosynthetic activity in seaweeds? Mar Biol 62:111–117

Nultsch W, Rüffer U, Pfau J (1984) Circadian rhythms in the chromatophore movement of *Dictyota dichotoma*. Mar Biol 81:217–222

Nultsch W, Pfau J, Materna-Weide M (1987) Fluence and wavelength dependence of photoinhibition in the brown alga *Dictyota dichotoma*. Mar Ecol Prog Ser 41:93–97

Pfau J, Throm G, Nultsch W (1974) Recording microphotometer for determination of light-induced chromatophore movements in brown algae. Z Pflanzenphysiol 71:242–260

Pfau J, Rüffer U, Nultsch W (1979) Der Einfluß polarisierten Lichtes auf die Chromatophorenanordnung von *Dictyota dichotoma*. Ber Dtsch Bot Ges 92:695–715

Pfau J, Hanelt D, Nultsch W (1988) A new dual beam microphotometer for determination of action spectra of light-induced phaeoplast movements in *Dictyota dichotoma*. J Plant Physiol 133:572–579

Psaras GK (1986) Chloroplast arrangement along intercellular spaces in the leaves of Mediterranean shrubs. J Plant Physiol 126:189–193

Rüffer U, Nultsch W, Pfau J (1978) Untersuchungen zur lichtinduzierten Chromatophorenverlagerung bei *Fucus vesiculosus*. Helgol Wiss Meeresunters 31:333–346

Rüffer U, Pfau J, Nultsch W (1981) Movements and arrangements of *Dictyota* phaeoplasts in light and darkness. Z Pflanzenphysiol 101:283–293

Schönbohm E (1966) Der Einfluß von Rotlicht auf die negative Phototaxis des *Mougeotia*-Chloroplasten: die Bedeutung eines Gradienten von P730 für die Orientierung. Z Pflanzenphysiol 55:278–286

Schönbohm E (1980) Phytochrome and non-phytochrome dependent blue light effects on intracellular movement in fresh water algae. In: Senger H (ed) The blue light syndrome. Springer, Berlin Heidelberg New York, pp 69–96

Seitz K (1987) Light-dependent movement of chloroplasts in higher plant cells. Acta Physiol Plant 9:137–148

Senn G (1908) Die Gestalts- und Lageveränderungen der Pflanzen-Chromatophoren. Engelmann, Leipzig

Senn G (1919) Weitere Untersuchungen über Gestalts- und Lageveränderungen der Chromatophoren. Z Bot 11:81–141

Seybold A (1955) Beiträge zur Optik der Laubblätter. Beitr Biol Pfl 31:499–513

Seybold A (1956) Hat die Chromatophorenverlagerung in Laubblättern eine Bedeutung? Naturwissenschaften 43:90–91

Shropshire W Jr (1980) Carotenoids as primary photoreceptors in blue light responses. In: Senger H (ed) The blue light syndrome. Springer, Berlin Heidelberg New York, pp 172–186

Titlyanov EA, Kolmakow PV, Lee BD, Horvath I (1978) Functional states of the photosynthetic apparatus of the marine green alga *Ulva fenestrata*. Acta Bot Acad Sci Hung 24:167–177

Walczak T, Gabrys H (1981) The CO_2 effect on light-induced chloroplast translocations in higher plant leaves. Z Pflanzenphysiol 101:367–375

Witztum A, Posner HB, Gower RA (1979) Phototactic chloroplast displacement in the photosynthetic mutant *Lemna paucicostata* Strain 1073. Ann Bot 44:1–4

Zurzycka A, Zurzycki J (1953) Studies on phototactic movements of chloroplasts I. Acta Soc Bot Pol 22:667–678

Zurzycka A, Zurzycki J (1957) Cinematographic studies on phototactic movements of chloroplasts. Acta Soc Bot Pol 26:177–206

Zurzycki J (1953) Arrangement of chloroplasts and light absorption in plant cell. Acta Soc Bot Pol 22:299–320

Zurzycki J (1955) Chloroplast arrangement as a factor in photosynthesis. Acta Soc Bot Pol 24:27–63

Zurzycki J (1957) The destructive effect of intense light on the photosynthetic apparatus. Acta Soc Bot Pol 26:157–175

Zurzycki J (1961) The influence of chloroplast displacement on the optical properties of leaves. Acta Soc Bot Pol 30:503–527

Zurzycki J (1966) Investigation on the contraction of chloroplasts in *Mnium undulatum* (L.) Weis. II. Studies on isolated chloroplasts. Acta Soc Bot Pol 35:281–291

Zurzycki J (1967) Properties and localization of the photoreceptor active in displacements of chloroplasts in *Funaria hygrometrica*. I. Action spectrum. Acta Soc Bot Pol 36:133–142

Zurzycki J (1975) Adjustment processes of the photosynthetic apparatus to light conditions, their mechanism and biological significance. Pol Ecol Stud 1:41–49

Professor Dr. W. Nultsch
Hasenmoor 6
25462 Rellingen, Germany

Genetics

Alternative Respiration in Plants and Fungi: Some Aspects of Its Biological Role

Heike Röhr and Ulf Stahl

1 Introduction

Electron transfer from NADH to molecular oxygen in animals within the mitochondrion proceeds via at least three respiratory complexes (Fig. 1): NADH:ubiquinone oxidoreductase (complex I), ubiquinol:cytochrome *c* oxidoreductase (complex III) and cytochrome *c* oxidase (complex IV). These "standard" respiratory enzymes form a linear respiratory chain, and electron transport of each is coupled to proton translocation out of the mitochondrion. In this way, a proton motive force is generated which is subsequently used for ATP synthesis by the ATPase/ATP synthase. In addition, electron transfer also occurs from succinate dehydrogenase (complex II) to ubiquinone (Q), but it is not coupled to proton translocation.

In contrast, plants and many fungi may use a branched respiratory chain with additional alternative components (Fig. 1; for reviews, see

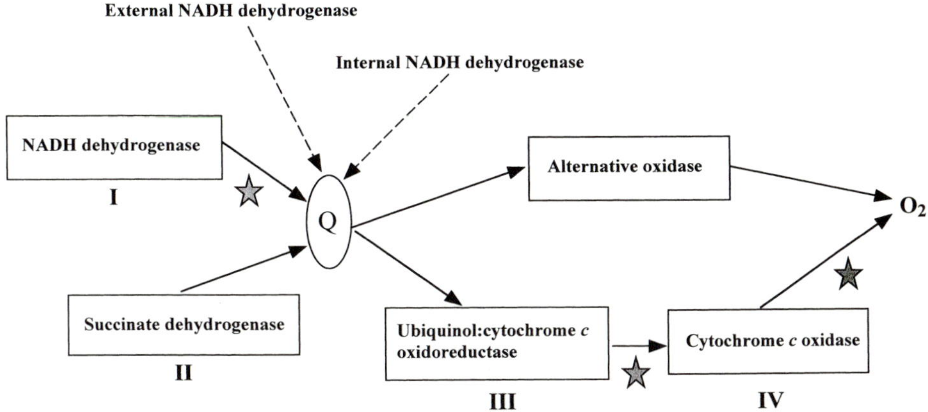

Fig. 1. The branched respiratory chain in plants and fungi. Electrons are transferred from the ubiquinone pool either to the alternative oxidase or to the ubiquinol:cytochrome *c* oxidoreductase and cytochrome *c* oxidase, both ways resulting in reducing molecular oxygen to water. *Gray stars* indicate where enzyme activity is linked to proton translocation. *Q* Ubiquinone pool; *I* NADH dehydrogenase; *II* succinate dehydrogenase; *III* ubiquinol:cytochrome *c* oxidoreductase; *IV* cytochrome *c* oxidase

Progress in Botany, Vol. 64

Lambers 1982; Siedow and Berthold 1986; Moore and Siedow 1991; McIntosh 1994; Joseph-Horne et al. 2001).

1. There are at least two alternative NADH dehydrogenases: one faces the intermembrane space oxidizing cytoplasmic (external) NADH, the other one faces the matrix space of mitochondria and catalyzes oxidation of endogenously (internally) generated NADH (Prömper et al. 1993). The latter is thought to be the counterpart of complex I (De Vries et al. 1992).
2. Apart from the cytochrome *c* oxidase, the mitochondria of plants and fungi also contain an alternative oxidase that directly accepts electrons from ubiquinol and catalyzes the reduction of molecular oxygen to water.

Nevertheless, the participation of alternative components in the respiratory system, which are all non-energy conserving, seems to ensure proton translocation through at least one site during electron transfer from NADH to oxygen (Joseph-Horne et al. 2001).

The biological function of alternative respiration, particularly the mode of action of the alternative oxidase, is still an open question. Since alternative respiratory enzymes are known to have a significantly lower affinity to NADH and ubiquinol than their counterparts in the cytochrome *c* pathway, earlier investigations hypothesized that alternative respiration may act as an "overflow mechanism", reducing equivalents which regulate the redox state of the cell in the case of catabolic overflow (Lambowitz et al. 1989; Lambers 1982). Recent results have indicated that alternative respiration may have more differentiated functions, according to the kind of organism.

The function of alternative respiration seems to be clear in the thermogenic tissue of plants where energy is released as heat in germinating seeds or attracts insects for pollination (Meeuse 1975). Moreover, Maxwell et al. (1999) found that alternative respiration in plants lowers oxidative stress because generation of reactive oxygen species (ROS) is limited by action of the alternative pathway. Other results have indicated that alternative respiration is induced by ROS in the pathogenic fungus *Magnaporthe grisea* (Yukioka et al. 1998) but not in the ascomycete *Podospora anserina* (Borghouts et al. 2001). Furthermore, the role of the alternative oxidase in fungicide resistance has been discussed in the case of phytopathogenic fungi such as *Septoria tritici* and *Magnaporthe grisea* (Joseph-Horne et al. 1998; Yukioka et al. 1998; Affourtit et al. 2000).

This report will focus on exceptional features of alternative respiration. Furthermore, current knowledge of the biological function of the alternative respiratory pathway in plants and fungi will be summarized.

2 Alternative Respiration

a) Alternative NADH Dehydrogenases

The transfer of electrons from NADH into the respiratory chain in animals and humans is only based on complex I. As a consequence, NADH oxidation in the mitochondria of these organisms is totally blocked by inhibitors such as rotenone or piericidin A. However, most fungal and plant mitochondria are also capable of oxidizing internal and external NADH when complex I is blocked by inhibitors or mutation (for review, see Kerscher 2000).

The first evidence of alternative NADH dehydrogenase activity was found by Bonner and Voss (1961) who observed that plant mitochondria, in contrast to mammalian mitochondria, were also able to oxidize externally added NAD(P)H. Similar results were obtained for *Saccharomyces carlsbergensis*, although EPR studies revealed the absence of complex I in mitochondria (Ohnishi et al. 1966; Joseph-Horne et al. 2001). In contrast, mitochondria of the fungus *N. crassa* were also found to possess both internal (facing the mitochondrial matrix) and external (facing the intermembrane space) alternative NADH dehydrogenases in addition to complex I (Weiss et al. 1970).

Both external and internal NADH dehydrogenases in plant and fungal mitochondria are encoded by a single nuclear gene. Their only prosthetic group is represented by FAD, in contrast to the FMN and multiple FeS centers detected in complex I (Joseph-Horne 2001). The isoenzymes exhibit insensitivity to inhibitors of complex I like rotenone and piericidin A, and their action is always non-proton translocating (de Vries and Grivell 1988). Little is known about the metabolic function and regulation of alternative NADH dehydrogenases. Some are suggested to also accept NADPH as substrate; some require calcium ions for activity (Kerscher 2000).

Internal alternative NADH:ubiquinone oxidoreductases may compete with complex I for the substrates NADH and ubiquinone. The affinity of internal isoenzymes for these substrates is apparently lower than that of their counterpart (Joseph-Horne et al. 2001). Nevertheless, it might be an advantage for organisms to have the possibility to express one or the other protein under certain conditions (Kerscher 2000). This hypothesis is consistent with results reported with mitochondria of *N. crassa*, where the internal isoenzyme acts preferentially during the exponential and complex I during the stationary growth phase (Schwitzguebel and Palmer 1982). Moreover, *N. crassa* is also viable in the absence of complex I because the internal isoenzyme is able to substitute for this respiratory complex (Videira 1998). Similar results were obtained from studies with *A. niger* (Wallrath et al. 1991).

External NADH dehydrogenases obviously enable the feeding of the respiratory chain with electrons from NADH generated in the cytoplasm. Interestingly, recent results from yeast indicate that the external enzyme is able to fully substitute the glycerol-3-phosphate dehydrogenase system which is known to have a shuttle function in this organism (Larsson et al. 1998).

Melo et al. (1999) cloned a cDNA of *N. crassa* encoding a putative NADH dehydrogenase with striking homology to NDI (encoding for an internal NADH dehydrogenase), NDE1 and NDE2 (encoding for putative external NADH dehydrogenases) of *Saccharomyces cerevisiae* on the amino acid level. Apart from the expected NADH and FAD binding motifs, an additional Ca^{2+}-binding motif was predicted from the sequence. This motif is absent in yeast sequences but found to be present in the external NADH dehydrogenase of plant mitochondria (Rasmusson et al. 1999), suggesting that Ca^{2+} plays a regulatory role, at least for external enzymes (Ramusson and Møller 1991; Møller et al. 1993). As recently confirmed, in addition to complex I, plant mitochondria may contain up to four alternative NAD(P)H dehydrogenases. They are associated with both sides of the inner mitochondrial membrane and differ in their induction kinetics and Ca^{2+} requirements (Rasmusson et al. 1999).

b) Alternative NADH Dehydrogenases in Yeast

Complex I of the respiratory chain has been found in nearly all fungi investigated. Interestingly, the most prominent example lacking this enzyme is the baker's yeast *Saccharomyces cerevisiae*. Complex I has also not been found in *S. carlsbergensis* and *Kluyveromyces lactis* (Joseph-Horne et al. 2001). Lacking complex I, mitochondria of *S. cerevisiae* contain three NADH dehydrogenases which enable NADH to be oxidized and the reducing equivalents to be transported into the respiratory chain: NDE1, NDE2 (external) and NDI1 (internal).

NDI1 consists of only one 53-kDa subunit and contains noncovalently bound FAD (de Vries and Grivell 1988). The enzyme, localized in the inner mitochondrial membrane and facing the matrix site, oxidizes NADH produced by the tricarboxylic acid (TCA) cycle and the mitochondrial alcohol dehydrogenase. NDI1 is thought to be the physiological counterpart of complex I lacking in *S. cerevisiae*. Biochemical studies of *ndi1* null mutants revealed the existence of (additional) external NADH dehydrogenase(s), as these mutants are unable to oxidize mitochondrial (internal) NADH generated by the TCA cycle (Marres et al. 1991). Cytosolic (external) NADH is produced by the oxidation of ethanol to acetate and the operation of the glyoxylate cycle (Marres et al. 1991). Putative external NADH dehydrogenases in *S. cerevisiae*, NDE1 and NDE2, have been identified in two independent studies (Luttik et al. 1996; Small and McAllister-Henn 1998). The authors found that two open reading frames (ORFs; *YMR145c* and *YDL085w*) of the *S. cerevisiae* genome show high homology to the NDI1 gene. Both groups could demonstrate that disruption of NDE1 led to a significant (three-fold) reduction of total NADH dehydrogenase activity compared with the wild type in cells cultivated in glucose and, moreover, to four-fold reduction of oxygen uptake of isolated mitochondria. However, these physiological effects could not be shown when NDE2 was deleted. Disruption resulted in an inconspicuous phenotype, indicating that NDE1 is the more important of the two external NADH dehydrogenases (Luttik et al. 1996). Nevertheless, NADH-dependent mitochondrial respiration was totally abolished in a nde1/nde2 double null mutant, whereas respiration of other substrates was not affected at all. The physiological role of NDE2 thus remains unclear.

c) The Alternative Oxidase (AOX)

The alternative oxidase is an ubiquinol oxidase found in mitochondria of a wide range of organisms. It acts parallel to the cytochrome *c* oxidase (complex IV) as the terminal oxidase in the electron transfer chain, also catalyzing the reduction of molecular oxygen to water. It is currently generally accepted that the electron flow through this alternative pathway branches from the conventional, cytochrome pathway at the level of the ubiquinone pool (see Fig. 1; Lambers 1982; Moore and Siedow 1991). Therefore, the alternative pathway consists of a single enzyme that functions as a quinol oxidase. In contrast to mitochondrial cytochrome *c* oxidase, the action of the alternative oxidase is not linked to proton translocation. Moreover, two of the three sides of energy conservation during the respiratory chain are bypassed and, as a consequence, free energy is dissipated as heat (Siedow and Berthold 1986; Berthold et al. 2000; Siedow and Umbach 2000). The alternative oxidase is resistant to inhibitors that are known to block complex III, the cytochrome bc_1 complex (antimycin A), and complex IV, the cytochrome *c* oxidase (cyanide, azide), but is inhibited specifically by several substances such as salicylhydroxamic acid (SHAM) and *n*-propyl gallate (Schonbaum et al. 1971; Moore and Siedow 1991).

α) Structure and Diversity

The alternative oxidase has been found in all plants, in many, but not all, fungi tested and in some protozoa. Interestingly, no evidence has been found which proves the existence of an alternative oxidase in metazoan mitochondria or any prokaryote (Siedow and Umbach 2000). The most striking structural difference between alternative oxidases isolated from plants and non-plants is that the plant oxidase is present as a dimer in the inner mitochondrial membrane (Umbach and Siedow 1993). The monomeric subunits are capable of either being covalently linked by a disulfide bond or not, but no mechanism is currently known that could catalyze oxidation or reduction of the disulfide bond (Umbach and Siedow 1993). Further results indicate that the enzyme in its dimeric form is essentially inactive (Siedow and Umbach 2000). Therefore, the significance of the dimeric structure of the alternative oxidase in plants is still not clear. The alternative oxidase originated from other sources than plant mitochondria is a monomer, due to the absence of the critical Cys residues responsible for disulfide bond formation in the plant enzyme (Umbach and Siedow 1993, 2000; Joseph-Horne et al. 2000).

The nature of the alternative oxidase catalytic site has been the object of extensive study for many years. Obviously, the alternative oxidase was able to catalyze oxygen reduction to water, despite the fact that the enzyme only comprises one single polypeptide of about 32 kDa (Siedow and Umbach 2000).

Recently, two models of the structure of the active site of the alternative oxidase have been proposed. Siedow and coworkers (1995) predicted two transmembrane helices exhibiting both N- and C-termini on the one side of the inner mitochondrial membrane. In addition, three conserved E-X-X-H motifs have been found within the sequence. Based on this observation, Siedow et al. (1995) proposed that the catalytic site of the alternative oxidase contains a di-iron center, similar to those found in a subgroup of di-iron carboxylate proteins, the ribonucleotide reductases (RNR) R2-like proteins. The active site of these proteins consists of a binuclear iron center coordinated by two histidines and four carboxylate residues (Andersson and Nordlund 1999). With the rising number of known sequences of alternative oxidases, it became apparent that the histidine residue of one of the E-X-X-H motifs, which is supposed to form the di-iron center, is not as conserved as previously assumed. Therefore, Andersson and Nordlund (1999) presented a revised model for the active site of the alternative oxidase.

β) Regulation

The regulation of the alternative oxidase has been studied extensively in the last decade, particularly in plant mitochondria (for reviews, see Moore and Siedow 1991; McIntosh 1994; Siedow and Umbach 2000; Affourtit et al. 2001). First results clearly indicate that alternative oxidase enzyme activity depends on the concentration of reduced ubiquinone (Moore and Siedow 1991; Millenaar et al. 1998), and, in addition, on the amount of the alternative oxidase protein present in mitochondria (Elton and McIntosh 1987; Hiser et al 1996). If the level of alternative oxidase mRNA and protein is high, for example, in the case of overexpressing the enzyme, the capacity of the alternative pathway increases (Hiser et al. 1996).

The alternative pathway only appeared to be engaged by increasing the concentration of reduced ubiquinone (Moore and Siedow 1991). Several kinetic approaches were presented to describe the experimental data concerning the behavior of the alternative oxidase with respect to the reduction levels of the ubiquinone pool (Siedow and Moore 1993; Wagner and Wagner 1995). However, it was suggested that high levels of reduced ubiquinone result in an increase in the formation of free radicals. Furthermore, engagement of the alternative pathway may obviously decrease the amount of reduced ubiquinones in mitochondria. Thus, alternative respiration was shown to stabilize the reduction state of the ubiquinone pool in vivo (Millenaar et al. 1998). In recent years, much more information about the regulation of the alternative oxidase has been obtained. Additional potential mechanisms of post-transcriptional regulation have been characterized in plants in particular. Millar et al. (1993) demonstrated that certain α-keto acids, mainly pyruvate (see also Day et al. 1994), stimulate the alternative oxidase activity even when the

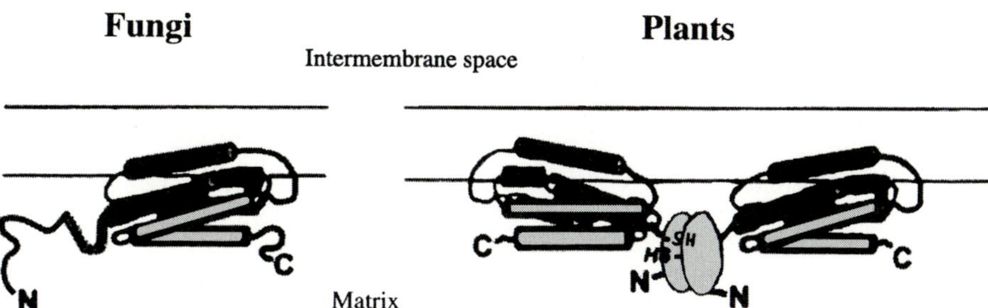

Fig. 2. Topology of the alternative oxidase in the inner mitochondrial membrane. The enzyme is present as a dimer in the case of plants or as a monomer in fungi. The *darker segments* represent the two hydrophobic domains that earlier were thought to form a membrane-spanning domain. The figure is in accordance with the model proposed by Andersson and Nordlund (1999) that did not predict any transmembrane domain. (Adapted from Siedow and Umbach 2000)

level of reduced ubiquinone in the mitochondrial membrane is low (Umbach et al. 1994). Moreover, as noted above, the plant alternative oxidase exists as a dimer in mitochondria (Umbach and Siedow 1993; Fig. 2). The dimeric structure of the enzyme may be present either in a disulfide-linked, less active state, or in a more active state when the disulfide bond has been reduced. A relationship between these two regulatory mechanisms was based on the observation that stimulation by pyruvate principally occurs when the sulfhydryl/disulfide system is in the reduced state (Umbach et al. 1994). Pyruvate has little or no effect on alternative oxidase activity when the disulfide bond is formed.

Interestingly, it has been found that only two cysteine residues are conserved in the amino acid sequences of higher plants (Vanlerberghe and McIntosh 1997). Considering the results achieved so far, it was suggested that these two cysteines, both located in the N-terminal region of the protein, are involved in both formation of the intersubunit disulfide bond and site of pyruvate stimulation. Consequently, Rhoads et al. (1998) used site-directed mutagenesis to change the two conserved Cys residues, Cys-78 and Cys-128, in *Arabidopsis thaliana* sequence numbering, to alanine. The results indicated that one single Cys residue, Cys-78, serves as both the regulatory sulfhydryl/disulfide and site of activation by α-keto acids. However, other not yet characterized protein–protein interactions may also be important for regulation, as pyruvate binding does not prevent dimerization (Joseph-Horne et al. 2001).

The alternative pathway is induced in fungi, as in some plants, when the cytochrome *c* pathway is blocked by respiratory inhibitors like antimycin A, cyanide or azide (Lambowitz et al 1989; Sakajo et al. 1991; Li et al. 1996) or, alternatively, when this pathway is affected by mutation (Schulte et al. 1989). Whilst inhibitory profiles and structural models of the alternative oxidase appear to be very similar in fungi and plants, recent data indicate some apparent differences in regulation and function. The most striking difference is the observation that fungal alternative oxidases only seem to exist in monomeric form (Umbach and Siedow 2000; Fig. 2). Moreover, an alignment of all available fungal alternative oxidase amino acid sequences revealed that the cysteine residue associated with pyruvate regulation and disulfide bond formation in plants is completely absent in fungal oxidases. Only the yeast *Hansenula anomala* has a cysteine residue close to Cys-78 in *A. thaliana* (Joseph-Horne and al. 2000). As a consequence, no influence of pyruvate on alternative oxidase activity could be observed in fungi. These results suggest that the role of the alternative oxidase in fungi might be different from that in plants.

3 The Putative Role of Alternative Respiration in Plants and Fungi

The alternative oxidase ubiquitous in all plant and many fungal mitochondria tested is also present in some protists (e.g., *Acanthamoeba castellanii*, Jarmuskiewicz et al. 1997; *Plasmodium falciparum*, Murphy and Lang-Unnasch 1999; *Trypanosoma brucei*, Helfert et al. 2001). The results obtained so far suggest that the alternative oxidase may have different functions in different organisms. The first conclusive function of this enzyme came from aroid plants where alternative respiration was found to be involved in thermogenesis (Meeuse 1975; Rhoads and McIntosh 1991). Uncoupled respiration and, as a consequence, heat production is a desired outcome in thermogenic tissue for releasing volatile compounds to attract insects for pollination. Since alternative respiration has also been detected in the mitochondria of non-aroid plants (e.g., *Arabidopsis thaliana*; Kumar and Söll 1992) and fungi (e.g., *N. crassa*; Lambowitz et al. 1989), it became evident that the cyanide-resistant respiratory pathway provides an additional function. Considering the fact that alternative respiration is not coupled to energy conservation, Lambers (1982) emphasized that this pathway acts as an "energy overflow" when the cytochrome *c* pathway is saturated or limited (Moore and Siedow 1991). In this way, oxidation of NADH and FADH and electron transport is guaranteed, and the metabolism is thus maintained. This observation was consistent with the regulation of the alternative oxidase by high levels of reduced ubiquinone (see Sect. 2.c.β) and may generally apply to all organisms possessing an alternative respiratory pathway.

Recent results have indicated that the alternative oxidase possibly contributes to a defense system against reactive oxygen species (ROS) in transgenic tobacco mitochondria (Maxwell et al. 1999). The authors showed that the level of ROS in mitochondria is significantly lower when the alternative pathway is active in comparison to levels during operation of the cytochrome *c* pathway. This finding is of biological importance as ROS are potentially toxic.

ROS may cause oxidative damage to cells, affecting macromolecules such as polypeptides and DNA, probably leading to cell death.

Apart from its uncontested significance in thermogenesis in aroid species, Maxwell et al. (1999) demonstrated that alternative respiration in plant cells has another conclusive and experimentally sustained function.

Investigations of great importance regarding regulation and function of the alternative oxidase have been mainly carried out on the plant enzyme. However, knowledge about the function of alternative respiration in fungi is very rare and diffuse. Similar to results obtained from *Petunia hybridia* cells (Wagner 1995), gene expression of the alternative oxidase is induced by active oxygen species as second messengers in some fungi. Superoxide anion is suggested to induce alternative respiration in the yeast *H. anomala* (Minagawa et al. 1992). In addition, both superoxide and hydrogen peroxide (H_2O_2) have been shown to significantly enhance transcription levels of the alternative oxidase in the phytopathogen *Magnaporthe grisea* (Yukioka et al. 1998). However, the contrary effect was observed for *Podospora anserina* where transcript levels of the alternative oxidase even decreased in the presence of H_2O_2 (Borghouts et al. 2001).

Recent results obtained from phytopathogenic fungi such as *Magnaporthe grisea* (Yukioka et al. 1998), *Septoria tritici* (Affourtit et al. 2000) and *Gaeumannomyces graminis* var. *tritici* (Joseph-Horne et al. 1998) have indicated a possible role of alternative respiration in fungicide resistance. Most commercial fungicides interact with complex III, the cytochrome bc_1 complex, of the respiratory chain. Interestingly, inhibition of complex III by antimycin A resulted in an apparent increase in superoxide formation (Maxwell et al. 1999). Similar observations were obtained when examining the effect of the fungicide SSF-126, a potent inhibitor of the cytochrome bc_1 complex, on *Magnaporthe grisea* mitochondria (Yukioka et al. 1998). The authors detected an induction of superoxide generation after treatment of *M. grisea* mitochondria with SSF-126 and observed, as an apparent response, increased levels of the alternative oxidase transcript. Together with the similar induction of alternative respiration by hydrogen peroxide, they thus suggested active oxygen species to be signal mediators to activate expression of the alternative oxidase in *M. grisea*. Moreover, alternative respiration was thought to contribute a certain resistance to fungicides.

4 Uncoupling Proteins (UCPs)

As mentioned above, it is generally accepted that the electrochemical proton potential gradient ($\Delta\mu H^+$) generated by respiratory chain and subsequent proton translocation is used by the F_1F_0-ATP synthase to produce ATP. Alternatively, this proton gradient can be dissipated as heat by alternative respiration or the uncoupling proteins (UCPs). UCPs allow protons released by the respiratory chain to re-enter the matrix and, therefore, bypass the ATP synthase, thus permitting dissipation of $\Delta\mu H^+$ as heat, a process called (transient) thermogenesis (Vercesi 2001). These proteins (found in mammalian and plant mitochondria) belong to a distinct cluster of mitochondrial anion-carrier proteins.

A 32-kDa uncoupling protein was found in the brown adipose tissue mitochondria (BATM) of newborn and hibernating or cold-adapted mammals (BAT-specific UCP1; for review, see Nicholls 2001). Furthermore, in addition to UCP1, four other mammalian UCPs have been recognized: the ubiquitous UCP2 (Fleury et al. 1997), skeletal muscle specific UCP3 (Sanchis et al. 1998) and brain-specific UCP4 (Mao et al. 1999). Their role remains largely unclear. Verseci et al. (1995) discovered the first uncoupling protein in plant mitochondria, namely PUMP (plant uncoupling mitochondrial protein) or pUCP (plant uncoupling protein). Other genes encoding pUCPs have been identified in recent years, e.g., StUCP from potato (*Solanum tuberosum*; Laloi et al. 1997), AtUCP1 (Maia et al. 1998) and AtUCP2 (Watanabe et al. 1999). In addition, partial sequences from other species are also known (Jezek et al. 2001).

As plant uncoupling proteins (pUCPs) have been found in both thermogenic and non-thermogenic plants, a strictly thermogenic role of pUCPs has not yet been proven (Jezek et al. 2001).

The process of uncoupling proton translocation from ATP synthesis by the action of UCPs only occurs in the presence of fatty acids (FAs), acting as cofactors for their activity. The transport of protons is achieved by some kind of free fatty acid cycling (see Skulachev 1991).

Both uncoupling proteins and alternative oxidase are common features of plant mitochondria. The action of both enzymes is energy-dissipating and, moreover, appears to prevent ROS formation (Casolo et al. 2000). Interestingly, recent results have shown that an increase in free fatty acid (FFA) concentration in mitochondria may simultaneously enhance UCP activity and switch off the alternative oxidase. These results indicate that UPCs and alternative oxidase never work together at their maximal activity, and it could be hypothesized that both energy-dissipating enzymes act sequentially during the cell life of plants (Jarmuskiewicz et al. 2001).

5 Conclusions

Although much information has been obtained on the regulation, nature and function of alternative respiration, many questions remained unresolved. Apart from engagement in thermogenesis in aroid plants, alternative respiration has recently been found to play a role in cellular defense against oxidative stress in plants (Maxwell et al. 1999). Active oxygen species are found to enhance transcription of the alternative oxidase in some but not all fungi. As relevant reports concerning the function of alternative respiration in fungi are very rare, all suggestions could only be rather speculative. Considering recent results obtained from both phytopathogenic fungi where active oxygen species may act as second messengers in fungicide resistance, and from the ascomycete *Podospora anserina* where oxidative stress seem to negatively affect transcription of the alternative oxidase, one could suppose that the function is not conserved among fungi. Moreover, the role of alternative respiration in the aging process in *P. anserina* has been controversially discussed for many years. Some data indicate that alternative respiration is increased in aged cultures while respiration via cytochrome *c* oxidase decreases and is thus correlated with the aging process (Frese 1993; Koll et al. 2001). Other results have shown that mRNA levels of the alternative oxidase do not increase during the aging process (Borghouts et al. 2001), and, additionally, that elimination of the gene has no effect on the onset and progress of aging (Lorin et al. 2001). However, further investigations should shed more light on this still unanswered question.

One very interesting evolutionary aspect is the fact that alternative respiration is not common to all organisms. Until now, the alternative oxidase has been detected in mitochondria of plants and many fungi, and in some protists. However, some yeasts (*S. cerevisiae*, *S. pombe*) as well as animals and humans obviously lack this enzyme. In addition, alternative respiration has never been reported for bacteria. On the contrary, uncoupling proteins (UCPs) have been found in a wide range of organisms, including humans, animals, plants and yeast. Both UCPs and the alternative oxidase represent an energy-dissipating system, but only UCPs appear to have been conserved during evolution. This observation suggests that the presence of UCPs in mitochondria may provide an evolutionary advantage also for mammals and the alternative oxidase does not.

Acknowledgements. We thank Dr. Markus Wedde for critical comments on the manuscript and Roslin Bensmann for her support and correction of the English.

References

Affourtit C, Heaney SP, Moore AL (2000) Mitochondrial electron transfer in the heat pathogenic fungus *Septoria tritici*: on the role of alternative respiratory enzymes in fungicide resistance. Biochim Biophys Acta 1459:291–298

Affourtit C, Krab K, Moore AL (2001) Control of plant mitochondria. Biochim Biophys Acta 1504:58–69

Andersson ME, Nordlund P (1999) A revised model of the active site of the alternative oxidase. FEBS Lett 449:17–22

Berthold DA, Andersson ME, Nordlund P (2000) New insight into the structure and function of the alternative oxidase. Biochim Biophys Acta 1460:241–254

Bonner WD, Voss DO (1961) Nature 191:682–684

Borghouts C, Werner A, Elthon T, Osiewacz HD (2001) Copper-modulated gene expression and senescence in the filamentous fungus *Podospora anserina*. Mol Cell Biol 21:390–399

Casolo V, Braidot E, Chiandussi E, Macri F, Vianello A (2000) The role of mild uncoupling and non-coupled respiration in the regulation of hydrogen peroxide generation by plant mitochondria. FEBS Lett 474:53–57

Day DA, Millar AH, Wiskich JT, Whelan J (1994) Regulation of alternative oxidase activity by pyruvate in soybean mitochondria. Plant Physiol 106:1421–1427

De Vries S, Grivell LA (1988) Purification and characterization of a rotenone-insensitive NADH:Q6 oxidoreductase from mitochondria of *Saccharomyces cerevisiae*. Eur J Biochem 176:377–384

De Vries S, van Witzenburg R, Grivell LA, Marres CAM (1992) Primary structure and import pathway of the rotenone-insensitive NADH-ubiquinone oxidoreductase of mitochondria from *Saccharomyces cerevisiae*. Eur J Biochem 203:587–592

Elthon TE, McIntosh L (1987) Identification of the alternative terminal oxidase of higher plant mitochondria. Proc Natl Acad Sci USA 84:8399–8403

Fleury C, Neverova M, Collins S, Raimbault S, Champigny O, Bouillaud F, Seldin MF, Surwit RS, Ricquier D, Warden CH (1997) Uncoupling protein-2: a novel gene linked to obesity and hyperinsulinemia. Nat Genet 15:269–272

Frese D (1993) Die molekularbiologische Analyse der physiologischen Phänomene des Seneszenzsyndroms bei dem Ascomyceten *Podospora anserina*. Dissertation, Bibliotheca Mycologica, Band 149. J Cramer, Berlin

Helfert S, Estévez AM, Bakker B, Michels P, Clayton C (2001) Roles of triosephosphate isomerase and aerobic metabolism in *Trypanosoma brucei*. Biochem J 357:117–125

Hiser C, Kapranov P, McIntosh L (1996) Genetic modification of respiratory capacity in potato. Plant Physiol 110:277–286

Jarmuskiewicz W, Wagner AM, Wagner MJ, Hryniewiecka L (1997) Immunological identification of the alternative oxidase of *Acanthamoeba castellanii* mitochondria. FEBS Lett 411:110–114

Jarmuskiewicz W, Sluse-Goffart CM, Vercesi AE, Sluse FE (2001) Alternative oxidase and uncoupling protein: thermogenesis versus energy balance. Biosci Rep 21:213–222

Jezek P, Borecky J, Zácková M, Costa ADT, Arruda P (2001) Possible basic and specific functions of plant uncoupling proteins (pUCP). Biosci Rep 21:237–245

Joseph-Horne T, Wood P, Wood CK, Moore AL, Headrick J, Hollomon D (1998) Characterization of a split respiratory pathway in the wheat "take-all" fungus, *Gaeumannomyces graminis* var. *tritici*. J Biol Chem 273:11127–11133

Joseph-Horne T, Babij J, Wood IM, Hollomon D, Sessions RB (2000) New sequence data enable modelling of the fungal alternative oxidase and explain the absence of regulation of pyruvate. FEBS Lett 481:141–146

Joseph-Horne T, Hollomon D, Wood PM (2001) Fungal respiration: a fusion of standard and alternative components. Biochim Biophys Acta 1504:179–195

Kerscher SJ (2000) Diversity and origin of alternative NADH:ubiquinone oxidoreductases. Biochim Biophys Acta 1459:274–283

Koll F, Sidoti C, Rincheval V, Lecellier G (2001) Mitochondrial membrane potential and ageing in *Podospora anserina*. Mech Ageing Dev 122:205–217

Kumar AM, Söll D (1992) Arabidopsis alternative oxidase sustains *Escherichia coli* respiration. Proc Natl Acad Sci USA 89:10842–10846

Laloi M, Klein M, Riesmeier JW, Muller-Rober B, Fleury C, Bouillaud F, Ricquier D (1997) A plant cold-induced uncoupling protein. Nature 389:135–136

Lambers H (1982) Cyanide-resistant respiration: a non-phosphorylating electron transport pathway acting as an energy overflow. Physiol Plant 55:478–485

Lambowitz AM, Sabourin JR, Bertrand H, Nickels R, McIntosh L (1989) Immunological identification of the alternative oxidase of *Neurospora crassa* mitochondria. Mol Cell Biol 9:1362–1364

Larsson C, Pahlman IL, Ansell R, Rigoulet M, Adler L (1998) The importance of glycerol-3-phosphate shuttle during aerobic growth of *Saccharomyces cerevisiae*. Yeast 14:347–357

Li Q, Ritzel RG, McLean LLT, McIntosh L, Ko T, Bertrand H, Nargang FE (1996) Cloning and analysis of the alternative oxidase gene of *Neurospora crassa*. Genetics 142:129–140

Lorin S, Dufour E, Boulay J, Begel O, Marsy S, Sainsard-Chanet A (2001) Overexpression of the alternative oxidase restores senescence and fertility in a long-lived respiration-deficient mutant of *Podospora anserina*. Mol Microbiol 42:1259–1267

Luttik MAH, Overkamp KM, Kötter P, de Vries S, van Dijken JP, Pronk JT (1998) The *Saccharomyces cerevisiae NDE1* and *NDE2* genes encode separate mitochondrial NADH dehydrogenases catalyzing the oxidation of cytosolic NADH. J Biol Chem 273:24529–24534

Maia IG, Benedetti CE, Leite A, Turcinelli SR, Vercesi AE, Arruda P (1998) *AtPUMP*: an *Arabidopsis* gene encoding a plant uncoupling mitochondrial protein. FEBS Lett 429:403–406

Mao W, Yu XX, Zhong A, Li W, Brush J, Sherwood SW, Adams SH, Pan G (1999) UCP4, a novel brain-specific mitochondrial protein that reduces membrane potential in mammalian cells. FEBS Lett 443:326–330, Erratum in: FEBS Lett 449:293

Marres CAM, de Vries S, Grivell LA (1991) Isolation and inactivation of the nuclear gene encoding the rotenone-insensitive internal NADH:ubiquinone oxidoreductase of mitochondria from *Saccharomyces cerevisiae*. Eur J Biochem 195:857–862

Maxwell DP, Wang Y, McIntosh L (1999) The alternative oxidase lowers mitochondrial reactive oxygen production in plant cells. Proc Natl Acad Sci USA 96:8271–8276

McIntosh L (1994) Molecular biology of the alternative oxidase. Plant Physiol. 105:781–786

Meeuse BD (1975) Thermogenic respiration in aroids. Annu Rev Plant Physiol 26:117–126

Melo AM, Duarte M, Videira A (1999) Primary structure and characterisation of a 64 kDa NADH dehydrogenase from the inner mitochondrial membrane of *Neurospora crassa* mitochondria. Biochim Biophys Acta 1412:282–287

Millar AH, Wiskich JT, Whelan J, Day DA (1993) Organic acid activation of the alternative oxidase of plant mitochondria. FEBS Lett 329:259–262

Millenaar FF, Benschop JJ, Wagner AM, Lambers H (1998) The role of the alternative oxidase in stabilizing the in vivo reduction state of the ubiquinone pool and the activation state of the alternative oxidase. Plant Physiol 118:599–607

Minagawa N, Koga S, Nakano M, Sakajo S, Yoshimoto A (1992) Possible involvement of superoxide anion in the induction of cyanide-resistant respiration in *Hansenula anomala*. FEBS Lett 302:217–219

Møller IM, Rasmusson AG, Fredlund KM (1993) NAD(P)H-ubiquinone oxidoreductases in plant mitochondria. J Bioenerg Biomembr 25:377–384

Moore AL, Siedow JN (1991) The regulation and nature of the cyanide-resistant alternative oxidase of plant mitochondria. Biochim Biophys Acta 1059:121–140

Murphy AD, Lang-Unnasch N (1999) Alternative oxidase inhibitors potentiate the activity of atovaquone against *Plasmodium falciparum*. Antimicrob Agents Chemother 43:651–654

Nicholls DG (2001) A history of UCP1. Biochem Soc Trans 29:751–7555

Ohnishi T, Kawaguchi K, Hagihara B (1966) Preparation and some properties of yeast mitochondria. J Biol Chem 241:1797–1806

Prömper C, Schneider R, Weiss H (1993) The role of the proton-pumping and alternative respiratory chain NADH:ubiquinone oxidoreductases in overflow catabolism of *Aspergillus niger*. Eur J Biochem 1993 Aug 15 216:223–230

Rasmusson AG, Møller IM (1991) Effect of calcium ions and inhibitors on internal NAD(P)H dehydrogenases in plant mitochondria. Eur J Biochem 202:617–623

Rasmusson AG, Svensson AS, Knoop V, Grohmann L, Brennicke A (1999) Homologues of yeast and bacterial rotenone-insensitive NADH dehydrogenases in higher eukaryotes: two enzymes are present in potato mitochondria. Plant J 20:79–87

Rhoads DM, McIntosh L (1991) Isolation and identification of a cDNA clone encoding an alternative oxidase protein of *Sauromatum guttatum* (Schott). Proc Natl Acad Sci USA 88:2122–2126

Rhoads DM, Umbach AL, Sweet CR, Lennon AM, Rauch GS, Siedow JN (1998) Regulation of the cyanide-resistant alternative oxidase of plant mitochondria. Identification of the cysteine residue involved in α-keto acid stimulation and intersubunit disulfide bond formation. J Biol Chem 273:30750–30756

Sakajo S, Minagawa N, Komiyama T, Yoshimoto A (1991) Molecular cloning of cDNA for antimycin A-inducible mRNA and its role in cyanide-resistant respiration in *Hansenula anomala*. Biochim Biophys Acta 1090:102–108

Sanchis D, Busquets S, Alvarez B, Ricquier D, Lopez-Soriano FJ, Argiles JM (1998) Skeletal muscle UCP2 and UCP3 gene expression in a rat cancer cachexia model. FEBS Lett 436:415–418

Schonbaum GR, Bonner WD Jr, Storey BT, Bahr JT (1971) Specific inhibition of the cyanide-resistant respiratory pathway in plant mitochondria by hydroxamic acids. Plant Physiol 47:124–128

Schulte E, Kück U, Esser K (1989) Multipartite structure of mitochondrial DNA in a fungal longlife mutant. Plasmid 21:79–84

Schwitzguebel JP, Palmer JM (1982) Properties of mitochondria as a function of growth stages of *Neurospora crassa*. J Bacteriol 149:612–619

Siedow JN, Berthold DA (1986) The alternative oxidase: a cyanide-resistant respiratory pathway in higher plants. Physiol Plant 66:569–573

Siedow JN, Moore AL (1993) A kinetic model for the regulation of electron transfer through the cyanide-resistant pathway in plant mitochondria. Biochim Biophys Acta 1142:165–174

Siedow JN, Umbach AL (2000) The mitochondrial cyanide-resistant oxidase: structural conservation and regulatory diversity. Biochim Biophys Acta 1459:432–439

Siedow JN, Umbach AL, Moore AL (1995) The active site of the cyanide-resistant oxidase from plant mitochondria contains a binuclear center. FEBS Lett 362:10–14

Skulachev VP (1991) Fatty acid circuit as a physiological mechanism of uncoupling of oxidative phosphorylation. FEBS Lett 294:158–162

Small WC, McAlister-Henn (1998) Identification of a cytosolically directed NADH dehydrogenase in mitochondria of *Saccharomyces cerevisiae*. J Bacteriol 180:4051–4055

Umbach AL, Siedow JN (1993) Covalent and noncovalent dimers of the cyanide-resistant alternative oxidase protein in higher plant mitochondria and their relationship to enzyme activity. Plant Physiol 103:845–854

Umbach AL, Siedow JN (2000) The cyanide-resistant alternative oxidases from the fungi *Pichia stipitis* and *Neurospora crassa* are monomeric and lack regulatory features of the plant enzyme. Arch Biochem Biophys 378:234–245

Umbach AL, Wiskich JT, Siedow JN (1994) Regulation of alternative oxidase kinetics by pyruvate and intermolecular disulfide bond redox status in soybean seedling mitochondria. FEBS Lett 348:181–184

Vanlerberghe GC, McIntosh L (1997) Alternative oxidase: from gene to function. Annu Rev Plant Physiol Plant Mol Biol 48:703–734

Vercesi AE (2001) The discovery of an uncoupling mitochondrial protein in plants. Biosci Rep 21:195–200

Vercesi AE, Martins IS, Silva MAP, Leite HMF, Midea Cuccovia I, Chaimovich H (1995) PUMPing plants. Nature 375:24

Videira A (1998) Complex I from the fungus *Neurospora crassa*. Biochim Biophys Acta 1364:89–100

Wagner AM (1995) A role for active oxygen species as second messengers in the induction of alternative oxidase gene expression in *Petunia hybrida* cells. FEBS Lett 368:339–342

Wagner AM, Wagner MJ (1995) Measurements of in vivo ubiquinone reduction levels in plant cells. Plant Physiol 108:277–283

Wallrath J, Schmidt M, Weiss H (1991) Concomitant loss of respiratory chain NADH:ubiquinone reductase (complex I) and citric acid accumulation in *Aspergillus niger*. Appl Microbiol Biotechnol 36:76–81

Watanabe A, Nakazono M, Tsutsumi N, Hirai A (1999) AtUCP2: a novel isoform of the mitochondrial uncoupling protein of *Arabidopsis thaliana*. Plant Cell Physiol 40:1160–6

Weiss H, von Jagow G, Klingenberg ME, Bücher T (1970) Characterization of *Neurospora crassa* mitochondria prepared with a grind-mill. Eur J Biochem 14:75–82

Yukioka H, Inagaki S, Tanaka R, Katoh K, Miki N, Mizutani A, Masuko M (1998) Transcriptional activation of the alternative oxidase gene of the fungus *Magnaporthe grisea* by respiratory-inhibiting fungicide and hydrogen peroxide. Biochim Biophys Acta 1442:161–169

Prof. Dr. Ulf Stahl
Heike Röhr
University of Technology Berlin
Dept. of Microbiology and Genetics
Gustav-Meyer-Allee 25
13355 Berlin, Germany

e-mail: ulf-stahl@lb.tu-berlin.de

Mutants and Transgenics – a Comparison of Barley Resources in Crop Breeding

Christer Jansson and Hilde-Gunn Opsahl Ferstad

1 Introduction

Barley is the most cultivated crop in Sweden, Norway, Denmark and Finland. Globally, barley is the most cultivated cereal and the forth most important cereal crop (after wheat, maize and rice) and is grown on 70 Mha with an annual yield of 160 Mt. Today, barley grain is mostly used as animal feed and malt and, to a lesser extent, in human food.

The origin of barley is much debated and has been suggested to be either the Nile Valley in Egypt, China, or the Near East. Possibly in a crop cultivated for so long the origin can never be resolved with certainty. In southern Egypt, barley was a staple food as far back as 18,000 years ago (Wendorf et al. 1979). Barley is an excellent energy source since the grain consists of 80% carbohydrates, mostly starch, and in many countries in Africa and Asia barley is an important part of the diet, either for bread making or for specific recipes. Barley was considered a high-energy food already during the Roman times, when the gladiators were called "hordeari" or "barley men" because they were fed a barley diet before going to the Circus. In northern Europe, barley was the most common cereal in cooking well into the 16th century. The largest use of food barley is found in regions where other cereals do not grow well due to altitude, low rainfall, or soil salinity. Barley can be grown in hostile, arid, salty, cool environments. In Tibet, Nepal, Ethiopia, and the Andes, farmers cultivate barley on the mountain slopes at elevations higher than other cereals. In many regions in North Africa, the Middle East, Afghanistan, Pakistan, Eritrea, and Yemen, it is often the only possible rain-fed crop. Consequently, barley is considered to be the most drought- and salinity-tolerant of the cereals.

Due to its remarkable capacity to adapt to marginal areas and different environmental conditions, barley has the widest geographical distribution of all crops. For this reason, barley has accumulated a vast array of genetic variability. This is manifested in the extensive amount of barley land races around the world. Together with a vast material of barley mutants, these land races are deposited in different gene banks. The total number of barley accessions in these gene banks is estimated to 485,000. Of these, 55,000 accessions are found in Europe, distributed among 35 gene banks, e.g., the Nordic gene bank in Alnarp, Sweden, and the gene bank at IPK in Gatersleben, Germany. Other large barley germplasm centers are maintained by the Oregon State University Barley Project, USA, and the Barley Germplasm Center, Okayama, Japan. The largest barley collection is located at PGRC in Saskatoon, Canada, with around 43,000 accessions. The large numbers of barley genotypes represented in the various gene banks are publicly available and constitute an extremely important genetic resource for barley research and breeding. Furthermore, since barley is a true diploid, self-compatible species, it has come to serve as a model plant for wheat and other members of the Triticeae family.

Progress in Botany, Vol. 64

Both forward (mutant studies) and reverse genetics (isolating genes directly) are valuable tools for identifying genes for breeding. The defective kernel 1 (Dek1) mutant was described decades ago, but the corresponding gene has only recently been cloned, sequenced and described (Lid SE, Gruis D, Jung R, Lorentzen JA, Ananiev E, Chamberlin M, Niu X,Meeley R, Nichols S, Olsen O-A, unpubl. results). Other genes isolated from barley have been identified by different gene screening techniques and studied in transgenic plants. Such genes are the Ltp1, Ltp2, ESR (embryo surrounding region) and AGPase (Kall et al. 1994; Thorbjornsen et al. 1996; Opsahl-Ferstad et al. 1997; Bonello et al. 2000).

2 Barley Mutagenesis

A mutation is a sudden, random heritable change in a genetic material. Mutations are the source of all genetic variation and hence the basis of evolution. A gene mutation results in a new allele, whereas a chromosome mutation causes changes in a segment of a chromosome, a whole chromosome, or a set of chromosomes. Transposable elements, which are DNA sequences with the capacity to move about in the genome, can also be a source of mutation as they insert to or depart from the genome. Mutations can be induced or they can occur naturally, in which case they are said to be spontaneous. Spontaneous mutations occur as a result of errors during cell division and DNA repair. Some of these "spontaneous mutations" are the result of naturally occurring mutagens in the environment, e.g., solar radiation. Others, on the other hand, are bona fide spontaneous, such as those caused by DNA replication errors. Spontaneous mutations are usually quite rare; average mutation rates are 10^{-5}–10^{-6} events per locus per generation (in both prokaryotes and eukaryotes). Mutagenesis as a term refers to the deliberate production of induced mutations through the use of different treatments, such as exposure to radiation – neutrons, gamma-rays, X-rays, UV – or chemicals – sodium azide, ethyl methanesulfonate, mustard gas. With plants, passage through tissue-culture steps also causes similar effects, so-called somaclonal variation.

Barley mutagenesis has a long history in plant research and breeding. Barley was one of the first plants to be subjected to radiation-induced mutations, shortly after its introduction in the studies on *Drosophila* in the mid-1920s. In the past 80 years, barley is the crop species that has been studied most exhaustively in respect to induced mutations. Much of the knowledge of efficient methods of mutagen application and handling of the progeny after mutagen treatment has been gained from studies with barley (Ramage 1987). Due to the large number of developmental mutants, barley has been an important model plant for studies on plant development. Indeed, genes of importance in seed development were first described from barley mutants (Felker et al. 1985). The application of different mutagenic agents in crop breeding has increased crop biodiversity and productivity all over the world. A large proportion of these varieties are food crops released in developing countries. Some of them were obtained as infrequent mutation of specific genes responsible for agronomically important plant characters. This has resulted in the widespread use of these mutated genes in plant breeding programs throughout the world and has brought about an enormous economic impact, e.g., in barley, sunflower, soybean, rice and many other crops. Usually, seeds are exposed to the treatment and the effects will not be displayed until the second generation after treatment. Most of the mutations are lethal or deleterious, or at best neutral, and only a minor proportion of them will turn out to be beneficial and desirable. Because of this, a mutagenesis program will usually screen very large populations (10,000+) for the best plants. Two barley cultivars that are in the pedigrees of many barley cultivars grown today (Diamant and Golden Promise) derive from mutation programs.

Anther and microspore cultures for formation of haploid cells have become very important in vitro culture methods in conjunction with radiation- or chemical-induced mutations to improve seed propagated crops, such as barley, rice, wheat and maize (Ahloowalia 1998; Castillo et al. 2001). Mutagenesis of haploid cells allows a rapid detection of the mutated phenotype. It also offers the possibility of screening for recessive mutations in the first generation. The production of haploid mutants and the subsequent production of dihaploids with mutated genes constitute a rapid and powerful method for mutational breeding of barley and other cereals. Today, new barley varieties are continuously being released from doubled-haploid barley plants following mutagenesis (Kasha et al. 1997).

a) Radiation Mutagenesis

Radiation was the first mutagenic agent known. In 1903, the botanist de Vries discovered mutations in plants and the following year proposed the use of radiation for induction of mutations. Subsequently, employment of induced mutations in plant breeding was suggested by many workers after detection of X-ray mutagenesis in the late 1920s (Gustafsson 1969). The most common sources of radiation in induction of mutants are gamma-rays and X-rays. Both are energetic enough to produce reactive ions and are, therefore, referred to as ionizing radiation. The genetic effects of ionizing radiation on DNA include rearrangements and deletions. A less exploited source of ionizing radiation is neutron bombardment. Neutron radiation has proven a very powerful approach for the production of deletion libraries in barley (Falk et al.

2001). With this method, deletions in the range of 100–10,000 bp can be generated. Based on the known number of neutron-induced mutations in barley, one can expect that between 10,000 and 20,000 mutagenized plants are required in order to achieve a reasonable (>90%) probability to identify at least one deletion mutant per gene.

The number of officially released crop mutant varieties obtained through irradiation is listed in the FAO/IAEA database (http://www-infocris.iaea.org/MVD/). To date, the list exceeds 2200 entries (Maluszynski et al. 2000). Almost half of these have been released during the last 15 years. Barley is second on the list (269 mutant varieties) after rice (434 mutant varieties). Included in these records are some outstanding examples of cultivars, e.g., Diamant and Trumpf in barley (Micke 1999). In addition, many of the mutated genes from gamma or X-ray irradiation have been used in cross-breeding programs, often without indicating the nature of the desired genes. This is exemplified in barley by the extensive use of the *denso* gene as a source for semi-dwarfness. The *denso* gene was induced by X-ray irradiation in the tall Moravian variety Valticki (Bouma and Ohniutka 1991). It has been suggested that more than 150 malting barley varieties in all continents carry the *denso* gene (Maluszynski et al. 2000). Overall, the semi-dwarfness trait obtained by radiation breeding in barley varieties Pallas in North Europe, Diamant in central Europe and Golden Promise in the UK and USA have significantly contributed to the economy in many countries (Castillo et al. 2001).

b) Chemical Mutagenesis

The first report of a mutagenic action of a chemical was in 1942 by Charlotte Auerbach, who demonstrated that nitrogen mustard, an ingredient in the poisonous mustard gas used in World Wars I and II, could induce mutations. Since then, many mutagenic chemicals have been identified. These various chemicals react with bases of the DNA and cause gene or chromosome mutations. The chemical mutagens can be categorized by their mode of action in (1) base analogs (bromouracil, aminopurine); (2) alkylating agents and others that affect structure and pairing properties of bases (ethyl methanesulfonate, methanesulfonate, nitrous acid, nitrosoguanidine, sodium azide); (3) intercalating agents (acridine orange, proflavin, ethidium bromide); and (4) agents altering DNA structure (psoralens, peroxides).

The most common chemical agents in plant mutagenesis are ethyl methanesulfonate (EMS) and sodium azide (N_3Na). EMS is an alkylating compound. The most frequent product of EMS-mediated alkylation of DNA is 7-alkylguanine, which can base-pair with thymine yielding a G→A substitution. Another product is 4-ethylthymine, which can base-pair with G resulting in a T→C substitution. N_3Na is converted in cells to o-azidoalanine, which is the mutagenic agent. N_3Na is a very potent seed mutagen and in barley it is one of the most efficient (Konzak et al. 1972; Nilan et al. 1973; Bosnes et al. 1987; Castillo et al. 2001). This is probably explained by a substantial formation of o-azidoalanine in the barley embryo (Lundqvist 1992). N_3Na has been successfully employed in barley breeding, e.g., for studies on plant development (Olsen et al. 1998; Becraft et al. 2000; Olsen 2001) and for improving malting varieties (Molina-Cano et al. 1999) and nutritional qualities (Rasmussen and Hatzack 1998).

c) Transposon Mutagenesis

Whereas "normal" genes are arranged in a linear and stable order along the chromosomes, some genes, called transposable elements or transposons, are mobile (for a review, see Falk et al. 2001). Transposable elements comprise two classes: the class I elements or retrotransposons, which replicate via an RNA intermediate; and the class II or DNA transposons, which move as DNA by a cut-and-paste mechanism (Finnegan 1990). The class II elements or transposons were the first to be actively studied in plants. They were discovered in maize by the geneticist Barbara McClintock in the 1930s during her work on chromosome breakage (McClintock 1939, 1956, 1984). She named the chromosome-breaking site the dissociation (Ds) locus. She also soon identified a second locus that was required to activate chromosome breakage, and she designated this the activator (Ac) locus. Barbara McClintock was awarded the Nobel Prize for Physiology or Medicine in 1983 for her work on mobile genetic elements. The maize Ac/Ds elements were the first transposons to be discovered. Since then, transposons have been found in the genome of most, if not all, organisms studied (for a recent review on the history of the Ac/Ds transposons, see Federoff 2001). Other common class II transposons in maize are the *En/Spm* and *Mutator* systems (Döring and Starlinger 1986; Gierl and Saedler 1986). Maize transposons have been cloned by Federoff and coworkers (Fedoroff 1983) and are now widely used in transposon mutagenesis for gene tagging and subsequent gene discovery. The first successful transposon tagging was of the *bronze* gene in maize, on the anthocyanin biosynthetic pathway (Fedoroff et al. 1984).

The maize two-element transposon systems Ac/Ds and En/Spm have been successfully used for gene tagging in heterologous dicotyledonous species, and later also in the heterologous monocots rice, wheat and barley (reviewed by Falk et al. 2001). An efficient method for the utilization of the Ac/Ds system in barley has recently been described by Koprek et al. (2000).

Unlike the class II DNA transposable elements, integrated copies of retrotransposons are not excised as a part of transposition. Transposition is instead a replicative process and would be better described as propagation. The nature of retrotransposons and their applications in mutagenesis have recently been reviewed by Falk et al. (2001). A prominent family of retrotransposons in barley is BARE-1, which comprises 3% of the genome (Vicient et al. 1999). The genetics and activities of the BARE-1 retrotransposons in barley have been elegantly studied by Schulman and coworkers (Jääskeläinen 1999; Vicient et al. 1999; Kalendar et al. 2000).

3 Transgenic Barley

The two most common methods for production of transgenic cereals are the biolistic approach and transformation mediated by the soil bacterium *Agrobacterium tumefaciens.* Grasses were for a long time considered to be outside the host range of *Agrobacterium* and thus the first transgenic cereals were produced using biolistic transformation. However, during the last 10 years, protocols for *Agrobacterium*-based transformation have been established also for maize, rice, wheat and barley (for reviews, see Lemaux et al. 1999; Vasil 1999; Jacobsen et al. 2000). Barley transformation has been extensively reviewed by Jacobsen et al. (2000) and Lemaux et al. (1999).

a) Biolistic Transformation

Biolistic transformation, also referred to as particle or microprojectile bombardment, is carried out with a so-called DNA gun, or gene gun. The gene gun was conceived by John Stanford at Cornell University in 1984. In biolistic transformation, DNA-coated gold microprojectiles (microcarriers) are placed on a plastic sheet (macrocarrier). A helium pressure is built up in a gas acceleration tube. At a certain helium pressure, a disk (rupture disk) at the bottom of the tube ruptures. The shock wave generated by the compressed helium propels the macrocarrier into a stopping screen. The microcarriers continue through a partial vacuum until impact with the target tissue. The most common commercial gene gun design is that of Bio-Rad's PDS-1000/He. For reviews of biolistic transformation and the construction and operation of the gene gun, see Klein and Jones (1999) and Mäenpää et al. (1999).

A major benefit of biolistic transformation is that genes can be delivered directly into intact tissue. Thus, the method can be used for any organism, even those that are recalcitrant or poorly susceptible to *Agrobacterium* infection. Since reliable protocols for *Agrobacterium*-

mediated transformation of monocots have been slow in coming, the biolistic approach has been the method of choice for the construction of transgenic cereals. Another powerful feature of particle bombardment is that there is no biological constraint to the size of the DNA fragments that can be introduced into the plant genome. On the other hand, a disadvantage of the biolistic approach is that it often leads to the integration of several copies of the transgene into the host genome. This, in turn, can result in cosuppression mechanisms (Matzke and Matzke 1995).

b) *Agrobacterium*-Mediated Transformation

Agrobacterium tumefaciens is a soil bacterium that genetically can transform plant cells with a segment of plasmid DNA, called transfer DNA (T-DNA). The plasmid is called tumor-inducing (Ti plasmid) since it contains oncogenes that result in production of plant tumors called crown galls. Following wounding of a plant, bacteria are attracted to the wounded site in response to phenolic compounds released by the plant cells. Replacing the oncogenes with genes of interest allows the introduction of desirable genes into the plant chromosomes by *Agrobacterium* infection. In general, *Agrobacterium*-mediated transformation does not seem to suffer from cosuppression to the same extent as biolistic transformation does.

c) Transformation of Barley

The first efficient transformation of barley was reported in 1994 using the gene gun (Wan and Lemaux 1994). It was then the last of the world's major cereals for which transformation protocols were established (Jacobsen et al. 2000). *Agrobacterium*-mediated gene transfer was not adopted in cereals until recently. Success was reported first in maize (Gould et al. 1991) and rice (Chan et al. 1992), and later in wheat (Cheng et al. 1997) and barley (Tingay et al. 1997). Transformation of barley with either the gene gun or *Agrobacterium* exhibits a strong genotype dependency and the most successful results have been obtained with a small number of cultivars, primarily Golden Promise and Igri (Jacobsen et al. 2000). Thus, transformation protocols developed for a model cultivar such as Golden Promise need to be adjusted and optimized for commercial cultivars and breeding lines before use (Roussy et al. 2001). An alternative approach is to transform Golden Promise and then introduce the transformant into a breeding program and to backcross to local varieties or breeding lines of choice.

Transformation of barley has been utilized in both transient assays and in stabile transformed plants. Applications range from fundamental studies to improvement of agronomic traits, such as pest resistance (Wan and Lemaux 1994; McGrath et al. 1997; Leckband and Lörz 1998) and nutritional quality (Brinch-Pedersen et al. 1996).

4 Future Prospects

Barley mutants have contributed greatly to crop breeding during the last 80 years. They will undoubtedly continue to do so also after the introduction of transgenic cereals. In fact, the vast number of barley mutants and varieties present in different gene bank depositories constitutes an enormous potential and, hitherto almost unexploited, resource for gene mining. In the era of post-genomics, tools such as DNA microarray analyses and proteomics will allow high-throughout transcript and protein profiling of the mutants and offer unprecedented possibilities to identify genes responsible for interesting phenotypes. In this way, genes conferring desirable agronomic traits can be introduced into barley breeding programs, both for so-called conventional breeding and for production of transgenic barley varieties. At the same time, transgenic barley will become increasingly more important in crop breeding. As methods for transformation are refined, production and utilization of transgenic cereals will be routine in plant breeding projects.

The *Arabidopsis* and rice genomes (120 and 430 Mbp, respectively) have now been fully sequenced. By contrast, complete sequencing of large genomes such as that of barley (5000 Mbp) is not yet practical. A more affordable solution for gene mining in such large genomes is to utilize EST (expressed sequence tag) databases. Some 150,000 barley ESTs are now released in the public domain, and another 100,000 ESTs are to be submitted in 2002 (http://harvest.ucr.edu). These EST libraries will serve as a very valuable source in the production of barley DNA microarrays.

In addition to its use for food, malt and feed, transgenic barley and barley mutants will be used for production of renewable raw materials, such as starch for the manufacture of bioplastics. Transgenic crops will also be designed as bioreactors for the production of pharmaceuticals and industrial enzymes.

References

Ahloowalia BS (1998) Radiation techniques in crop and plant breeding – multiplying the benefits. IAEA Bull 40:37–39

Becraft PW, Brown RC, Lemmon BE, Olsen O-A, Opsahl-Ferstad H-G (2000) Developmental biology of endosperm development. Kluwer, Dordrecht

Bonello J-F, Opsahl-Ferstad H-G, Perez P, Dumas C, Rogowsky PM (2000) *Esr* genes show different levels of expression in the same region of maize endosperm. Gene 246:219–227

Bosnes M, Harris E, Aigeltinger L, Olsen O-A (1987) Morphology and ultrastructure of 11 barley shrunken endosperm mutants. TAG 74:177–187

Bouma J, Ohnoutka Z (1991) Importance and application of the mutant "Diamant" in spring barley breeding. Plant mutation breeding for crop improvement, vol 1. IAEA, Vienna, pp 127–133

Brinch-Pedersen H, Galili G, Knudsen S, Holm PB (1996) Engineering of the aspartate family biosynthetic pathway in barley (*Hordeum vulgare* L.) by transformation with heterologous genes encoding feedback-insensitive aspartate kinase and dihydrodipicolinate synthase. Plant Mol Biol 32:611–620

Castillo AM, Cistué L, Vallés MP, Sanz JM, Romagosa I, Molina-Cano JL (2001) Efficient production of androgenic doubled-haploid mutants in barley by the application of sodium azide to anther and microspore cultures. Plant Cell Rep 20:105–111

Chan M, Lee MT, Chang H (1992) Transformation of indica rice (*Oryza sativa* L.) mediated by *Agrobacterium*. Plant Cell Physiol 33:577–583

Cheng M, Fry JE, Pang S, Zhou H, Hironaka CM, Duncan DR, Conner TW, Wan Y (1997) Genetic transformation of wheat mediated by *Agrobacterium tumefaciens*. Plant Physiol 115:971–980

Döring H-P, Starlinger P (1986) Molecular genetics of transposable elements in plants. Annu Rev Genet 20:175–200

Falk A, Rassmusen S, Schulman A, Jansson C (2001) Barley mutagenesis. Progress in Botany 62. Springer, Berlin Heidelberg New York, pp 34–50

Federoff N (2001) How jumping genes were discovered. Nat Struct Biol 8:300–301

Fedoroff NV (1983) Controlling elements in maize. Mobile genetic elements. Academic Press, New York, pp 1–63

Fedoroff NV, Furtek DB, Nelson OE (1984) Cloning of the *bronze* locus in maize by a simple and generalizable procedure using the transposable element *Activator* (*Ac*). Proc Natl Acad Sci USA 81:3825–3829

Felker FC, Peterson DM, Nelson OE (1985) Anatomy of immature grains of eight maternal effect shrunken endosperm barley mutants. Am J Bot 72:248–256

Finnegan DJ (1990) Transposable elements and DNA transposition in eukaryotes. Curr Opin Cell Biol 2:471–477

Gierl A, Saedler H (1986) The *En/Spm* transposable element of *Zea mays*. Plant Mol Biol 13:261–266

Gould J, Devey M, Hasegawa O, Ulian EC, Peterson G, Smith RH (1991) Transformation of *Zea mays* L. using *Agrobacterium tumefaciens* and the shoot apex. Plant Physiol 95:426–432

Gustafsson A (1969) A study on induced mutations in plants. Induced mutations in plants. IAEE, Vienna, pp 9–31

Jacobsen JV, Matthews PM, Abbott DC, Wang M-B, ad Waterhouse PM (2000) Transgenic barley. Am Assoc Cereal Chem, pp 88–114

Jääskeläinen M, Mykkänen A-H, Arna T, Vicient CM, Suoniemi A, Kalendar R, Savilahti H, Schulman AH (1999) Retrotransposon BARE-1: expression of encoded proteins and formation of virus-like particles in barley cells. Plant J 20:413–422

Kalendar R, Tanskanen J, Immonen S, Nevo E, Schulman AH (2000) Genome evolution of wild barley (*Hordeum spontaneum*) by BARE-1 retrotransposon dynamics in response to sharp microclimatic divergence. Proc Natl Acad Sci USA 97:6603–6607

Kalla, R, Shimamoto K, Potter R, Nielsen PS, Linnestad C, Olsen O-A (1994) The promoter of the barley aleurone-specific gene encoding a putative 7 kDa lipid transfer protein confers aleurone cell-specific expression in transgenic rice. Plant J 6:849–860

Kasha KJ, Ziauddin A, Cho U-H (1997) Haploids in cereal improvement: anther and microspore culture. Gene manipulation in plant improvement II. Plenum Press, New York, pp 213–235

Klein TM, Jones TJ (1999) Methods of genetic transformation: the gene gun. Molecular improvement of cereal crops. Kluwer, Dordrecht, pp 21–42

Konzak CF, Wickman IM, Dekock MJ (1972) Advances in methods of mutagen treatment. IAEA, Vienna, Proc FAO/IEAE Int Symp, pp 95–119

Koprek T, McElroy D, Louwerse J, Williams-Carrier R, Lemaux PG (2000) An efficient method for dispersing *Ds* elements in the barley genome as a tool for determining gene function. Plant J 24:243–363

Leckband G, Lörz H (1998) Transformation and expression of a stilbene synthase gene of Vitis vinifera L. in barley and wheat for increased fungal resistance. Theor Appl Genet 96:1004–1012

Lemaux PG, Cho M-J, Zhang S, Bregitzer P (1999) Transient cereals: *Hordeum vulgare* L. (barley). Molecular improvement of cereal crops. Kluwer, Dordrecht, pp 255–316

Lundqvist U (1992) Mutation research in barley. Thesis, The Swedish University of Agricultural Sciences, Svalöv, Sweden

Mäenpää P, Baena-Conzáles E, Ahlandsberg S, Jansson C (1999) Transformation of nuclear and plastomic plant genomes by biolistic particle bombardment. Mol Biotechnol 13:67–72

Maluszynski M, Nichterlein K, van Zanten L, Ahloowalia BS (2000) Officially released mutant varieties – The FAO/IAEA database. Mutation breeding review 12. IAEA, Vienna

Matzke MA, Matzke AJM (1995) How and why do plants inactivate homologous (trans)genes? Plant Physiol 107:679–685

McClintock B (1939) The behavior in successive nuclear divisions of a chromosome broken at meiosis. Proc Natl Acad Sci USA 25:405–416

McClintock B (1956) Controlling elements and the gene. Cold Spring Harbor Symp Quant Biol 21:197–216

McClintock B (1984) The significance of responses of the genome to challenge. Science 226:792–801

McGrath PF, Vincent JR, Lei C, Pawlowski WP, Torbert KA, Gu W, Kaeppler HF, Wan Y, Lemaux PG, Rines HR, Somers DA, Larkins BA, Lister RA (1997) Coat protein-mediated resistance to isolates of barley yellow dwarf in oats and barley. Eur J Plant Pathol 103:695–710

Micke A (1999) Mutations and in vitro mutation breeding. Breeding in crop plants. Kalyani Publishers, Ludhiana, India, pp 1–19

Molina-Cano JL, Sopena A, Swantson JS, Casas AM, Moralejo MA, Ubieto A, Lara I, Pérez-Vendrell AM, Romagosa I (1999) A mutant induced in the malting barley cv. Triumph with reduced dormancy and ABA response. Theor Appl Genet 98:347–355

Nilan RA, Sideris EG, Kleinhofs A, Sander C, Konzak CF (1973) Azide – a potent mutagen. Mutat Res 17:142–144

Olsen O-A (2001) Endosperm development: cellularization and cell fate specification. Annu Rev Plant Physiol Plant Mol Biol 52:233–267

Olsen O-A, Lemmon BE, Brown RC (1998) A model for aleurone cell development. Trends Plant Sci 3:168–169

Opsahl-Ferstad H-G, Le Deunff E, Dumas C, Rogowsky P (1997) *ZmEsr*, a novel endosperm specific gene expressed in a restricted region around the maize embryo. Plant J 12:235–246

Ramage RT (1987) A history of barley breeding methods. Plant breeding reviews 5. Van Nostrand Reinhold Company, New York, pp 95–138

Rasmussen SK, ad Hatzack F (1998) Identification of two low-phytate barley (*Hordeum vulgare* L.) grain mutants by TLC and genetic analysis. Hereditas 129:107–113

Roussy I, Ahlandsberg S, Jansson C (2001) Transformation and generation capacities for five Nordic barley elite cultivars - evaluation of tissue culture response and transient expression. Hereditas 134:97–101

Thorbjornsen T, Villand P, Kleczkowski LA, Olsen O-A (1996) A single gene encodes two different transcripts for the ADP-glucose pyrophosphorylase small subunit from barley (*Hordeum vulgare*). Biochem J 313:149–154

Tingay S, McElroy D, Kalla R, Fieg S, Wang M, Thornton S, Bretell R (1997) *Agrobacterium tumefaciens*-mediated barley transformation. Plant J 11:1369–1376

Vasil IK (1999) Molecular improvement of cereal crops - an introduction. Molecular improvement of cereal crops. Kluwer, Dordrecht, pp 1–8

Vicient CM, Suoniemei A, Anamthawat-Jonsson K, Tanskanen J, Beharav A, Nevo E, Schulman AH (1999) Retrotransposon *BARE-1* and its role in genome evolution in the genus *Hordeum*. Plant Cell 11:1769–1784

Wan Y, Lemaux PG (1994) Generation of large numbers of independently transformed fertile barley plants. Plant Physiol 104:37–48

Wendorf F, Schild R, El Hadidi N, Close AE, Kobisiewicz M, Wieckowska H, Issawi B, Hass H (1979) Use of barley in the Egyptian late paleotithic. Science 205:1341–1347

Christer Jansson
Department of Plant Biology
The Swedish University of Agricultural Sciences
P.O. Box 7080,
75007 Uppsala, Sweden

e-mail: christer.jansson@vbiol.slu.se

Hilde-Gunn Opsahl-Ferstad
Department of Chemistry and Biotechnology
Agricultural University of Norway
P.O. Box 5040,
1432 Aas, Norway

Molecular Genetics of Chloroplast Biogenesis

Jörg Nickelsen

1 Introduction

During evolution, the eukaryotic plant cell that we see today is thought to have developed by fusion of former free-living prokaryotes, a process referred to as endosymbiosis. This widely accepted theory predicts mitochondria to have originated from α-proteobacteria, which were engulfed in a first endosymbiotic step by an ancient host – probably an archaebacterium (Martin and Müller 1998; Brocks et al. 1999). Then, this new, heterotrophically growing organism gained the capability for photoautotrophic growth by the endosymbiotic uptake of a cyanobacteria-like prokaryote which, later, became the chloroplast.

In addition to others, one essential piece of evidence supporting this idea is the fact that both mitochondria and chloroplasts contain their own genetic material. These extrachromosomal genomes usually consist of circular, autonomously replicating DNA molecules, which are present in multiple copy numbers.

To date, complete nucleotide sequences of 23 different plastid genomes are known (http://www.ncbi.nlm.nih.gov/PMGifs/Genomes/plastids_tax. html). Generally, only 100 protein-coding genes, a complete set of tRNAs and the ribosomal rRNA genes are encoded by the plastome (Sugiura 1992). The complete genome of the recent cyanobacterium *Synechocystsis* sp. PCC 6803, however, contains approximately 3200 genes. Thus, the former cyanobacterial endosymbiont must have lost most of its genes during evolution, especially those with functions required for a free-living lifestyle. Additionally, approximately 1400 genes migrated to the nucleus (Abdallah et al. 2000), where they gained nuclear elements of gene expression, such as promoters, as well as translation and polyadenylation signals. In addition, more than 50% of these genes obtained a chloroplast signal sequence, which allowed them to be reimported by the organelle once they had been synthesized on cytoplasmic ribosomes (Schleiff and Soll 2000). Furthermore, the plastid proteome appears to comprise about 1500 additional proteins with genes originating from the former mitochondrial host cell. This suggests a significant change in the proteome composition during the formation of chloroplasts (Abdallah et al. 2000).

An important consequence of the extensive gene transfer following endosymbiosis is that most of the multimeric plastid protein complexes are assembled from subunits, some of which are encoded by the nuclear and

Progress in Botany, Vol. 64

others by the plastid genome. These can, for example, include ribosomes, the Rubisco enzyme, or the photosynthetic complexes of the thylakoid membrane. Thus, the cell faces the problem of coordinating gene expression within its different DNA-containing compartments in order to assure the synthesis of these subunits in stoichiometric amounts for their correct assembly. This situation is further complicated by the fact that the respective gene expression machineries differ considerably due to their different phylogenetic origins. Accumulating genetic and biochemical evidence now indicates that the coordination between these machineries is mediated by an intracellular communication system, which is based on nucleus-encoded factors controlling chloroplast biogenesis. Together with the substantial transfer of genes originating from the endosymbiotic cyanobacterium to the nucleus, this reflects a coherent principle for the evolution of eukaryotic plant cells, demonstrating that the nuclear genome gained genetic dominance over the endosymbiotic genome during its phylogenetic development.

The aim of this review is to summarize our current knowledge of the genetic background of this regulatory communication system by selectively focusing on some recent advancements in the identification of its constituent nucleus-encoding factors.

2 Genetic Model Systems and Molecular Tools

In recent years, the development of several genetic and molecular tools has substantially improved the design of approaches for the analysis of interactions between nucleus and chloroplast. Current genetic model systems include the green alga *Chlamydomonas reinhardtii*, the dicot *Arabidopsis thaliana* and the monocot *Zea mays*. Each of these organisms offers distinct advantages regarding their experimental accessibility for both genetic and molecular analyses, and these have recently been reviewed (Barkan 1998; Barkan and Goldschmidt-Clermont 2000; Nickelsen and Kück 2000; Dent et al. 2001). In particular, the isolation and characterization of many photosynthetic nuclear mutants, which are affected in chloroplast function, now provide a source for the identification of regulatory factors involved in chloroplast biogenesis. In maize, cloning of nuclear genes defined by mutations has been accomplished by transposon tagging. In *A. thaliana*, both map-based cloning and insertional mutagenesis led to the isolation of such regulatory genes, whereas, in *C. reinhardtii*, similar loci can be cloned by complementation of mutants through rapid and efficient nuclear transformation techniques, which are available for this haploid organism.

In contrast to higher plants, however, *C. reinhardtii* mutants are fertile and viable in the dark, enabling their detailed analysis by means of classical genetic and biochemical approaches. In particular, like yeast,

C. reinhardtii is amenable to the analysis of suppressor mutations, which affect non-photosynthetic mutants. This provides a selection strategy not only for 'loss-of-function' but also for 'gain-of-function' mutants (Nickelsen 2000; Wostrikoff et al. 2001). As non-photosynthetic mutants of higher plants can only be propagated as heterozygotes, a similar suppressor analysis is currently difficult to perform with these systems.

Moreover, *C. reinhardtii* is the only one of these organisms in which both the nuclear genome and the chloroplast genome can be transformed (Boynton et al 1988). By using the biolistic particle bombardment system, DNA can be delivered to the chloroplast and is integrated into the genome via homologous recombination, thus allowing site-directed mutagenesis of endogenous plastid genes and their regulatory *cis*-acting sequence elements and/or the introduction of chimeric reporter genes. The combined application of chloroplast transformation and classical genetics enables the analysis of chloroplast mutants in different nuclear backgrounds. This has led to the precise identification of chloroplast target sequences for nucleus-encoded functions (Nickelsen et al. 1994; Vaistij et al. 2000a).

3 Plastid Development

To date, only a limited number of mutants have been isolated in which the initial steps of chloroplast development are affected. This may suggest the requirement for an intact plastid compartment not only for photosynthesis, but also for the production of other essential compounds like hormones, lipids and cofactors, which are essential for plant function in general. For instance, the mutants *dcl* from tomato, *dag* from *Antirrhinum majus* or *dal* from *A. thaliana* exhibit abnormal mesophyll cell development and contain small plastids nearly devoid of thylakoid membranes and, thus, resembling proplastids (Chatterjee et al. 1996; Keddie et al. 1996; Babiychuk et al. 1997). The respective genes affected in these mutants have been cloned and were shown to encode novel polypeptides with DAG and DAL proteins sharing 40% amino acid identity, but it is not yet known whether the *DAG* and *DAL* genes are orthologs.

In *dag* mutant cells, the mRNA for *rpoB* (a subunit of the plastid-encoded RNA polymerase) is not detectable, although it is expressed in wild-type plants at the same developmental stage. Hence, it has been suggested that a block in the synthesis of the plastid RNA polymerase prevents the production of plastid components necessary for the formation of the photosynthetic apparatus (Chatterjee et al. 1996).

Recently, the tobacco *VDL* gene has been cloned and shown to encode a 53-kDa DEAD box RNA helicase, which is imported into the chloroplast where it is suggested to control early plastid differentiation (Wang et al.

2000). Furthermore, several additional mutations affecting chloroplast morphology, division, and pigmentation have been described most of which cause pleiotropic effects (for reviews, see Goldschmidt-Clermont 1998; Leon et al. 1998).

A particular interesting aspect of chloroplast biogenesis concerns the formation of the lipid bilayer of the thylakoid membrane. While profound knowledge of the organization and function of this essential chloroplast subcompartment has been accumulated throughout the last decades, relatively little is known about from where these membranes originate and how they develop. Since thylakoid galactolipids are synthesized at the inner envelope membrane of the chloroplast, they must be transported in some way to reach their final destination within the thylakoid membrane system (Andersson et al. 2001). Several models have been proposed to explain how this lipid transfer is accomplished. These include transient fusions of the inner envelope with the thylakoid, specific lipid transfer proteins or a vesicular transport system resembling the well-characterized cytosolic one. The latter model is supported by electron microscopy studies (Hoober et al. 1991; Morre et al. 1991) and by the finding that a dynamin-like protein from *A. thaliana*, which is encoded by the nuclear *ADL-1* gene, is required for thylakoid membrane biogenesis.

Dynamins are members of a family of high molecular weight GTP-binding proteins, which can be found in a variety of eukaryotic systems ranging from yeast to humans. Although these proteins are involved in different biological processes, i.e., neurotransmitter recycling in animals, vacuole protein sorting in yeast or cell plate formation during plant cell division, a common feature shared by almost all dynamins is their participation in trafficking of cellular material.

The ADL-1 protein is localized at the outward-facing side of thylakoid membranes and a minor fraction is present at the inner envelope membrane. It has been suggested that this factor is directly involved in vesicle formation playing a similar role to that of dynamin 1 during endocytosis for neurotransmitter recycling in the brain (Park et al. 1998).

Further evidence pointing to a chloroplast vesicular transport system was obtained with the isolation of a NSF-like protein, Pftf, from pepper chromoplasts (Hugueney et al. 1995) and the recent identification of the nuclear *VIPP1* locus in *A. thaliana*. When wild-type *Arabidopsis* plants are exposed to low temperature (4 °C), budding of vesicles from the inner envelope membrane can be observed in electron micrographs. In contrast, no thylakoid membranes form in *VIPP1* deletion mutants and budding of vesicles is abolished. This deficiency can be complemented by the wild-type *VIPP1* gene, confirming its essential role for vesicle trafficking in chloroplasts. Additionally, a finding consistent with this idea is that the VIPP1 protein is located at both the inner envelope membrane and the thylakoids (Kroll et al. 2001).

Genes encoding VIPP1-like proteins are present in several plants and cyanobacteria that carry out oxygenic photosynthesis. Similar to higher plants, when the *VIPP1* gene from *Synechocystis* is inactivated, the evolution of the thylakoid membrane system is prevented, suggesting an ancient function of this protein during the genesis of photosynthetic activity. However, in contrast to *A. thaliana*, the VIPP1 protein is exclusively located to the plasma membrane and no immunoreactive material is detected in fractions containing thylakoid membranes (Westphal et al. 2001). Whether this reflects any differences in the molecular function of both VIPP1 proteins remains to be elucidated.

When viewed holistically, accumulating genetic and biochemical evidence supports the idea of a dynamic membrane trafficking system in chloroplasts, which may have facilitated the establishment of the complex thylakoid membrane system with its differentiation in grana and stroma regions during the evolution of oxygenic photosynthesis.

4 Nuclear Control of Chloroplast Gene Expression

The ongoing analysis of several photosynthetic mutants has revealed that, similar to the early stages of chloroplast development, later steps are also tightly controlled by the nuclear genome via regulatory factors affecting almost all steps of chloroplast gene expression, which include transcription, RNA metabolism, translation and assembly of multimeric complexes. This allows the plant cell to adequately react to endogenous and/or external signals, for example, to developmental-dependent and tissue-specific signals or varying light conditions.

a) Transcription

Plastid genes are transcribed by at least two completely different RNA-polymerases, the bacteria-like, so-called PEP (plastid-encoded polymerase) enzyme, whose core-subunits are encoded by the plastid genome, and the NEP (nuclear-encoded polymerase) enzyme, which is encoded by the nuclear *RpoZ* gene (Hedtke et al. 1997). Both enzymes unite with additional nucleus-encoded accessory subunits, which confer regulative capacities to the core complexes (for reviews, see Allison 2000; Lerbs-Mache 2000). Detailed mapping of plastid promoters revealed that photosynthetic genes are transcribed by the PEP, while non-photosynthetic genes contain promoters for both enzymes (Silhavy and Maliga 1998). Interestingly, the plastid *rpo* operon encoding the PEP core subunits is transcribed from an NEP-dependent promoter; therefore, the PEP enzyme is synthesized under direct control of the nuclear genome.

This probably represents an essential step for the cellular integration of the chloroplast during its endosymbiotic development.

So far, genetic screens have not succeeded in revealing any nuclear genes involved in plastid transcriptional control with the exception of the *76-5EN* mutant in *C. reinhardtii* (Hong and Spreitzer 1994). However, its role as a transcriptional mutant is still a matter of debate (Nickelsen 1998). Therefore, the question remains, why such screenings have failed to detect nuclear mutants affected in plastid transcription. One possible explanation is that transcriptional mutations might escape the usual screening strategies, which are mainly based on gene inactivation in the nucleus, because the genes concerned tend to play a more general role in cell viability.

b) RNA Metabolism

In contrast to the nucleus, chloroplast gene expression is mainly regulated at the post-transcriptional level. In particular, the steps of RNA maturation and stabilization appear to represent key control points for this regulation, as documented by many nuclear mutants that have been isolated and most of which show defects in the processing/stability of distinct chloroplast transcripts. Some of the corresponding genes have recently been cloned (Table 1), applying different techniques depending on the model organism used.

α) Intron Splicing

Several plastid genes are disrupted by introns, which are classified into the main groups I and II based on their conserved primary and secondary structures. Although some of these introns show autocatalytic self-splicing activity in vitro, genetic and biochemical data suggest that splicing in general requires the assistance of *trans*-acting factors in vivo.

Interestingly, splicing of the four group I introns of the *psbA* RNA in *C. reinhardtii* is regulated by light and depends on active photosynthetic electron transport. This finding suggests that synthesis of the *psbA* gene product, the D1 protein of photosystem II, is regulated, at least in part, at the level of RNA maturation (Deshpande et al. 1997).

Consequently, one main group of cloned genes for chloroplast gene expression comprises those required for splicing of introns in plastid precursor transcripts. To date, five nuclear genes have been isolated, all of which are involved in plastid group II intron splicing.

In maize, the CRS1 protein specifically mediates the intron removal from the *atpF* precursor RNA. This factor is part of a ribonucleoprotein complex, which appears to contain *atpF* intron RNA, suggesting that

CRS1 interacts directly with the target transcript. Sequence analysis of the CRS1 protein reveals highly basic and repeated amino acid motifs, which might represent a novel RNA-binding domain (Till et al. 2001). A second mutation at the *crs2* locus has more pleitropic effects since the splicing of a whole subgroup of introns, namely group IIB, is defective (Vogel et al. 1999). Similar to CRS1, the CRS2 protein was suggested to facilitate splicing via direct interaction with plastid intron RNA.

Table 1. Cloning of genetically defined nuclear loci involved in chloroplast gene expression

Mutant	Homology of gene product	cp-target gene expression step	Organism	Reference
raa1	No	*psaA* RNA *trans*-splicing	*C. reinhardtii*	Barkan and Goldschmidt-Clermont (2000)
raa2	Pseudouridine synthetase	*psaA* RNA *trans*-splicing	*C. reinhardtii*	Perron et al. (1999)
raa3	Pyridoxamine 5′ phosphate oxidase	*psaA* RNA *trans*-splicing	*C. reinhardtii*	Rivier et al. (2001)
crs1	No	*atpF* splicing	*Z. mays*	Till et al. (2001)
crs2	Peptydil-tRNA hydrolase	Splicing of group IIB introns	*Z. mays*	Jenkins and Barkan (2001)
mbb1	TPR-protein	*psbB* RNA stabilization	*C. reinhardtii*	Vaistij et al. (2000b)
hcf107	TPR-protein	*psbH* RNA stabilization	*A. thaliana*	Felder et al. (2001)
nac2	TPR-protein	*psbD* RNA stabilization	*C. reinhardtii*	Boudreau et al. (2000)
crp1	PPR-protein	*petA*, *petD* RNA processing	*Z. mays*	Fisk et al. (1999)
tr72	Poly(ADP) ribose polymerase	*tscA* RNA 3′ end processing	*C. reinhardtii*	Nickelsen and Kück (unpubl.)
ac115	No	*psbD* RNA translation	*C. reinhardtii*	Rattanachaikunsopon et al. (1999)
tbc2	CRP1	*psbC* RNA translation	*C. reinhardtii*	Zerges (2000)
hcf136	Synechocystis ORF slr2034	PSII assembly	*A. thaliana*	Meurer et al. (1998)
hcf164	Thioredoxin	$cytb_6f$ assembly	*A. thaliana*	Lennartz et al. (2001)

CRS2 is related to peptidyl-tRNA hydrolase enzymes based on homology searches, but several conserved amino acids, which are critical for enzyme activity in *E. coli*, are absent from the maize factor indicating that it may lack peptidyl-tRNA hydrolase activity. Consistent with this notion is that the *Crs2* gene cannot complement an *E. coli* peptidyl-tRNA hydrolase mutant (Jenkins and Barkan 2001).

Another group of splicing factors is represented by those that function in *trans*-splicing of *psaA* precursor transcripts in chloroplasts of *C. reinhardtii*. This complex RNA maturation process requires at least 14 nuclear loci (for a review, see Nickelsen and Kück 2000), three of which have now been identified. While *Raa1* and *Raa3* encode novel proteins, the *Raa2* gene product shares homologies with pseudouridine synthetases (Barkan and Goldschmidt-Clermont 2000). However, similar to the situation found for CRS2 from maize, the enzymatic synthetase activity is not required for the splicing function of Raa2, as demonstrated by the analysis of site-directed *raa2* mutants (Perron et al. 1999).

The Raa3 protein is part of a chloroplast high molecular weight complex, which contains both the precursor transcript of *psaA* exon1 and *tscA* RNA, a small choroplast-encoded RNA co-factor required for the *trans*-splicing of *psaA* exons 1 and 2 (Rivier et al. 2001). This resembles the organization of snRNP complexes, which are involved in nuclear splicing events and, accordingly, it has been hypothesized that the case of *psaA trans*-splicing reflects an intermediate form in the evolution of nuclear introns from organellar group II introns (Sharp 1991).

In conclusion, the picture emerging from the identification of the first chloroplast splicing factors is that, apparently, both novel proteins and enzymes, which lost their ancestral activity, were recruited as facilitators of plastid intron splicing during evolution. It seems likely that this process was promoted by the ability of these factors to interact with RNA, which made them well suited to develop into RNA chaperones.

β) RNA Stabilization and Processing

In the past, comparison of plastid transcription rates with the steady-state levels of the corresponding transcripts has revealed that light or cell cycle dependent changes in RNA accumulation are mainly determined by post-transcriptional processes (Gruissem and Tonkyn 1993; Salvador et al. 1993). Based on in vitro experiments, it was hypothesized that so-called stem-loop structures within the 3′ untranslated regions (UTRs) of chloroplast transcripts together with bound protein factors represent the main determinants for individual transcript half-lives by protecting upstream sequences against exonucleolytic attack by a plastid PNPase (Baginsky et al. 2001). However, more recent genetic data indicate that 5′ UTRs of at least some chloroplast mRNAs are essential for their stable accumulation (for a review, see Nickelsen 1998). Moreover, the combined application of chloroplast transformation and genetic crossing in *C. reinhardtii* enabled the identification of 5′ UTRs as target sequences of nucleus-encoded factors, which are required for mRNA stabilization of either *psbD*, *petD* or the *psbB* mRNA (Nickelsen et al. 1994; Drager et al. 1998; Vaistij et al. 2000a). Two corresponding nuclear genes, *Nac2* and *Mbb1*, which affect either *psbD* or *psbB/psbT/psbH* transcripts, have been cloned. Both factors share a protein–protein interaction motif called the tetratricopeptide (TPR) domain.

The TPR repeat is a degenerate 34 amino acid motif present in tandem arrays in proteins fulfilling a wide range of different functions including, e.g., the control of cell division, DNA replication, transcription, splicing and protein transport. Each TPR repeat consists of two anti-parallel α-helices (subdomains A and B) and multiple TPRs can form a right-handed superhelical structure, which serves as a surface for mediating inter- and intra-molecular protein–protein interactions (Das et al. 1998; Blatch and Lässle 1999).

Interestingly, similar TPR motifs have recently been identified in the nuclear *Hcf107* gene from *A. thaliana*, which is involved in the processing and stabilization of translation-competent *psbH* transcripts in chloroplasts (Felder et al. 2001), thus affecting the expression of the same chloroplast gene cluster as Mbb1 in *C. reinhardtii*. Phylogenetic analysis revealed that the two genes are evolutionary orthologs indicating that the molecular mechanisms of chloroplast gene expression are closely related between algae and higher plants.

A variation of the TPR motif, the so-called pentatricopeptide repeat (PPR) domain, is formed by Crp1, a factor required for processing of *petA* and *petD* precursor RNA in maize (Fisk et al. 1999; Small and Peeters 2000). Further *trans*-acting proteins containing TPR units include the chloroplast-encoded photosystem I assembly factor Ycf3 (Boudreau et al. 1997) and the *ycf37* gene product, which also has been shown to be involved in photosystem I accumulation in *Synechocystis* (Wilde et al. 2001). Moreover, a thorough computer-assisted search for TPR-containing open reading frames in the complete genome sequence of this former endosymbiont revealed the presence of a total of 22 putative TPR proteins, for most of which no function has yet been assigned (Ossenbühl and Nickelsen, unpubl. results). Thus, it appears that the ancient TPR motif has been conserved throughout evolution suggesting that it fulfils an essential role for the function/organization of regulatory units within chloroplasts. However, the precise molecular working mode of these factors is still to be established.

The presence of a TPR domain in the different proteins suggests that they form parts of multi-subunit complexes and, indeed, sedimentation analysis and size exclusion chromatography demonstrate that Nac2, Mbb1 and Crp1 co-fractionate with high molecular weight material. In Nac2, a single conserved amino acid of the TPR domain was mutated leading to the inactivation of the Nac2 function. In these site-directed mutant strains, the size of the Nac2 complex increased dramatically, indicating that the mutation caused abnormal complex aggregation (Boudreau et al. 2000).

To date, one of the most far-reaching analyses has been performed for the case of the chloroplast *psbD* gene in *C. reinhardtii* encoding the D2 protein of the photosystem II reaction center. Based on the identification of the *Nac2* gene as a *psbD* gene regulator and the concomitant application of in vitro RNA binding assays, a model for the post-transcriptional control of *psbD* gene expression has been proposed, which is depicted in Fig. 1. Probably, Nac2 interacts with a *cis*-acting element within the *psbD* 5′ UTR called PRB2 (Nickelsen et al. 1999). As a consequence, the *psbD* mRNA is protected against exonucleolytic attack from the 5′ end of the message. Subsequently, Nac2 guides a 40-kDa protein (RBP40) to its cognate binding site, which is located immediately

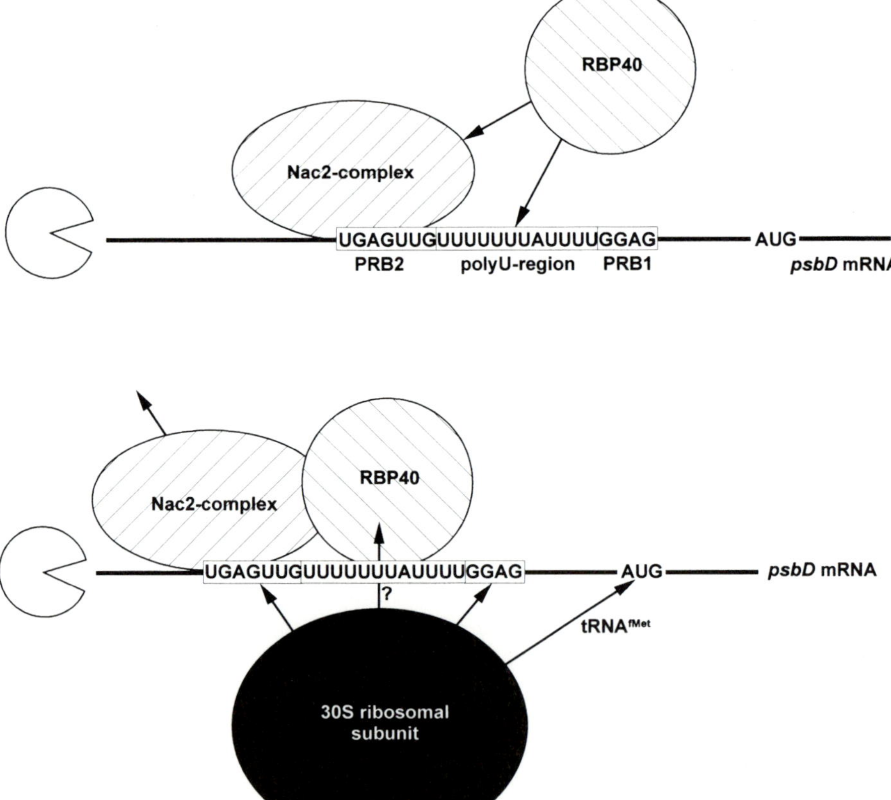

Fig. 1. Posttranscriptional regulation of *psbD* gene expression in *Chlamydomonas*. For explanation, see text

downstream of PRB2. This U-rich element is required for D2 synthesis, strongly suggesting that RBP40 is involved in the initial steps of *psbD* mRNA translation (Ossenbühl and Nickelsen 2000). Once RBP40 has bound to the 5′ UTR, Nac2 is released and ribosomal assembly can take place. Thus, Nac2 might connect processes of RNA stabilization and protein synthesis. A similar dual function has been proposed for the Hcf107 factor of *A. thaliana*, which affects both *psbH* RNA processing and *psbB* mRNA translation (Felder et al. 2001), and for Crp1 from maize, which is required for the processing of *petD* RNA, and also for the synthesis of both *petA* and *petD* gene products (Fisk et al. 1999).

γ) RNA Editing

RNA editing represents another plastid RNA maturation mechanism, which results in the modification of single nucleotide positions. Since editing cannot be detected in green algae, it has probably arisen later on during the phylogenetic development of chloroplasts. Genetic data obtained from tobacco indicate that nucleus-encoded functions are involved in this process (Bock and Koop 1997; Schmitz-Linneweber et al. 2001). Furthermore, the analysis of site-directed chloroplast mutants in tobacco demonstrated that the specificity of the editing reaction is determined by single nucleotide positions in the vicinity of the editing site on chloroplast RNAs (Hermann and Bock 1999). Hopefully, the recent development of an in vitro editing system for tobacco chloroplasts will provide the appropriate tool for the identification and characterization of the *trans*-acting factors involved (Hirose and Sugiura 2001).

c) Translation and Assembly of Complexes

The basic plastid protein synthesis apparatus, for instance, the ribosomes, stayed unaffected by processes of endosymbiosis and, hence, is of procaryotic origin. Similar to the situation found for the metabolism of chloroplast RNAs, this basic machinery is regulated by nucleus-encoded factors. These allow the gene-specific control of protein synthesis rates as documented by the characterization of corresponding mutants also affecting this level of gene expression. The process of polypeptide production includes the initiation of translation, which is usually mediated by the 5′ UTRs of plastid mRNAs and is considered as the rate-limiting step for synthesis. In addition, especially in chloroplasts, downstream events such as translation elongation and protein complex assembly appear to be extensively controlled in a gene-specific manner.

α) Translation Initiation

In addition to the basic *cis*-acting elements specifying translation initiation, e.g., the Shine-Dalgarno motif and the AUG start codon, plastid 5′ UTRs have been shown to contain crucial gene-specific regulatory translational elements. In a few cases, however, the 3′ RNA regions have also been suggested to be involved in protein synthesis, which is similar to the situation found in the cytoplasm (Rott et al. 1998). Several site-directed chloroplast mutants and the subsequent analysis of second-site suppressors in *C. reinhardtii* support this idea and, apparently, both essential secondary structures and distinct sequence elements constitute the sites, which are recognized by translational regulators (Staub and Maliga 1994; Bruick and Mayfield 1999; Fargo et al. 1999; Higgs et al. 1999; Nickelsen et al. 1999).

Biochemical approaches have now led to the identification of several plastid proteins, which represent good candidates for such regulators since they exhibit high affinity binding to the RNA elements in vitro. RNA binding proteins include the above-mentioned RBP40 factor interacting with a translational U-rich element within the *psbD* 5′ UTR *of C. reinhardtii* (Fig. 1) and a 47-kDa protein (RB47), which is part of a complex concerned with the regulation of *psbA* mRNA translation.

The *psbA* 5′ UTR binding complex has been investigated in greater detail. In addition to RB47, this complex consists of three other subunits that are 60, 55, and 38 kDa in size (RB60, RP55 and RB38). The genes for RB47 and RB60 have been cloned and shown to encode a poly(A) binding protein and a protein disulfide isomerase, respectively (Kim and Mayfield 1997; Yohn et al. 1998). It has been postulated that by redox and phosphorylation-dependent regulation of this complex, *psbA* translation is directly coupled to photosynthetic electron flow, which can be sensed via the redox state of the ferredoxin-thioredoxin system and/or the plastoquinone pool (Trebitsh and Danon 2001). In contrast, regulation of *psbA* mRNA translation in cyanobacteria appears not to be controlled at the level of translation initiation, but instead on the level of translation elongation, suggesting that, during evolution, eukaryotes developed some novel molecular strategies for gene regulation (Tyystjärvi et al. 2001).

However, recent data indicate that RB47 also interacts with the chloroplast *rps7* 5′ UTR together with the *rps7* gene product, the chloroplast ribosomal protein S7, which might auto-regulate its expression similar to ribosomal proteins in prokaryotes (Fargo et al. 2001). This mode of regulation might also apply to photosynthetic chloroplast genes, as exemplified by the synthesis of cytochrome f subunits of the cytb$_6$f complex in *C. reinhardtii*. The translation of the corresponding *petA* mRNA is repressed through its 5′ UTR when free, non-assembled cytochrome f accumulates in the thylakoid membrane (Choquet et al. 1998).

The translation of the *psbC* mRNA in chloroplasts of *C. reinhardtii* depends on at least three nuclear loci, one of which, called *Tbc2*, has recently been isolated (Zerges 2000). *Tbc2* encodes a novel protein of approximately 140 kDa that, interestingly, shares some homology with the Crp1 factor from maize chloroplasts controlling RNA processing and translation of *petA* and *petD* mRNAs (Table 1). This finding suggests that at least some aspects of the function of these two proteins are conserved between *C. reinhardtii* and higher plants.

β) Translation Elongation and Post-translational Steps

The significance of regulatory processes during the elongation of polypeptide chains in plastids is documented by the observation that, in higher plants, mRNAs for light-regulated genes such as *psbA* or *rbcL* are already associated with polysomes in the dark, although only after light treatment efficient protein synthesis occurs (Klein et al. 1988; Edhofer et al. 1998), which apparently depends on the photosynthetic proton gradient across the thylakoid membrane (Mühlbauer and Eichacker 1998). In *C. reinhardtii*, two nuclear mutants, *nac1* and *ac115*, were isolated both of which affect the synthesis of the D2 protein of photosystem II at a step following translation initiation. The *Ac115* gene has been cloned by complementation and encodes a novel small protein of 113 amino acids; however, its precise function is still unclear (Rattanachaikunsopon et al. 1999). The characterization of a non-allelic suppressor of *ac115* indicates that additional nuclear gene products are involved in this process (Wu and Kuchka 1995).

One main focus of current research work is directed to the assembly of complexes of the thylakoid membrane, which mediate photosynthetic electron transfer. By applying both biochemical and genetic approaches, the involved steps are in the process of clarification (for reviews, see Wollman et al. 1999; Choquet and Vallon 2000). Similar to other gene expression steps, the assembly appears to be controlled by regulatory factors, which do not form part of the complexes themselves, but instead, transiently interact with certain subunits during the assembly process. In *A. thaliana*, the *Hcf136* and *Hcf164* genes encode such factors, which are required for photosystem II assembly and the biogenesis of the cytb$_6$f complex, respectively (Meurer et al. 1998; Lennartz et al. 2001).

Hcf136 homologues have been found in cyanobacteria and in the cyanelle genome of *Cyanophora paradoxa*, suggesting that these represent an ancient member of regulatory factors for photosynthetic activity. The same applies to Hcf164, which is closely related to the thioredoxin-like protein TxlA from *Synechocystis* sp. PCC 6803. Both factors are localized to the thylakoid lumen; however, their precise molecular function is still unknown.

Two other plastid-encoded factors, which are involved in the assembly of photosystem I, are represented by the *ycf3* and *ycf4* gene products.

The targeted disruption of both genes did not allow the accumulation of photosystem I, although its different subunits are synthesized at normal rates (Boudreau et al. 1997). Both factors cannot be found associated with the purified complexes suggesting that they might function as chaperones for the correct assembly of photosystem I.

4 Subcompartmentalization of Chloroplast Gene Expression Steps

Whilst the identification of several nuclear genes which are involved in chloroplast biogenesis now facilitates the piecing together of the underlying molecular mechanisms, the subcompartmentalization of chloroplast gene expression remains an intriguing, unresolved matter as far as the localization of the different regulatory factors is concerned. It may be confidently assumed that distinct factors associate with different plastid subfractions that include the soluble stroma or the membrane phases of thylakoids and the inner envelope. Another membranous subcompartment, the so-called 'low density membrane' (LDM) system, has recently been described (Zerges and Rochaix 1998).

LDMs are non-green chloroplast membranes of low buoyant density, which resemble the inner envelope membrane with regard to their acyl lipid composition, i.e., the high content of monogalactosyldiaglycerol and digalactosyldiaglycerol and the reduced content of phospholipids when compared with the outer envelope. Apparently, this membranous subfraction is associated at least in part with thylakoids in a magnesium-dependent manner (Zerges and Rochaix 1998).

Considering the above-mentioned hypothesis of a vesicular transport system within chloroplasts, it is tempting to speculate that LDMs represent an intermediate state during thylakoid membrane biogenesis, which originates from the inner chloroplast envelope.

During plant development, initially, chloroplast DNA, which is organized in distinct nucleoids, associates with the inner proplastid envelope through a factor called the PEND protein. Upon full chloroplast maturation, the cpDNA is dispersed and found at the thylakoid membrane (Sato et al. 1999). Consequently, the plastid transcription machinery and probably other factors acting further downstream, like those involved in RNA metabolism and/or translation, should be distributed in a similar way. Indeed, association with the inner chloroplast envelope membrane has been reported for the translation termination factor RF4 from spinach (Rolland et al. 1999). Co-fractionation with the LDM system has been observed for the RNA splicing factor *Raa2* (Table 1) and several RNA binding activities, which have been proposed to be involved in various post-transcriptional gene expression steps in *C. reinhardtii* (Zerges and Rochaix 1999) or in pea (Ossenbühl and Nickelsen, unpubl.

results). Conversely, many other factors, e.g., the above-mentioned Crp1 and Crs2 from maize or Raa3, Mbb1 and Nac2 from *C. reinhardtii* (Table 1), were detected in the chloroplast stroma. Nevertheless, it has been hypothesized that the inner envelope/LDMs represent the sites of thylakoid membrane protein synthesis and that de novo formed photosynthetic complexes are transported via vesicles to thylakoids (Zerges 2000). This idea is now strongly supported by the recent observation that, in cyanobacteria, the initial steps of photosystem assembly occur at the plasma membrane and not at the thylakoids (Zak et al. 2001).

By using a two-dimensional separation procedure, thylakoid and plasma membrane proteins were separated and subsequent immunodetection revealed that many proteins of the photosystem I and II reaction centers were present in the plasma membrane. Moreover, these subunits appeared to form chlorophyll-containing complexes, which were able to undergo light-induced charge separation (Zak et al. 2001).

In contrast, a number of both genetic and biochemical data indicate that the synthesis and integration of photosynthetic membrane proteins take place at the thylakoids. For instance, the application of an in vitro run on translation system indicates a co-translational insertion of thylakoid membrane proteins (for a review, see Wollman et al. 1999) and, consistent with this, *psbA* and *psbD* mRNAs were found to be partially associated with thylakoids (Herrin and Michaels 1985). A maize mutant lacking the chloroplast SecY homologue of the thylakoid protein translocation apparatus is defective in plastid translation, providing further evidence for an essential role of thylakoids in chloroplast gene expression (Roy and Barkan 1998). Thus, it appears that different pathways, which are spatially separated and which might partly overlap in time, exist in plastids.

This can be illustrated by the case of the widely studied *psbA* gene. As already mentioned, the translation of the *psbA* message is regulated by a multiprotein complex containing RB47 and RB60 in *C. reinhardtii*. RB47 has been localized to LDMs and it has been shown that RB60 is partitioned between the stroma and the membrane phase following chloroplast fractionation experiments. Interestingly, recent RNA binding studies identified a novel protein, RBP63, which binds to a translational element within the *psbA* 5′ UTR with high affinity but, in contrast to RB40 and RB60, was found to be associated exclusively with stromal thylakoid membranes (Ossenbühl et al. 2002). This spatial distribution of *trans*-acting factors for D1 synthesis might reflect two different processes of photosystem II generation. One is the de novo assembly in premature developing chloroplasts at LDMs and the other its maintenance in mature chloroplasts at thylakoids.

The repair of photosystem II mainly involves the exchange of photo-damaged D1 protein for newly synthesized protein, a process that occurs at the stroma lamellae of thylakoid membranes. The repaired photosystem II is subsequently retranslocated to thylakoid grana regions (Adir et al. 1990). Assuming a co-translational insertion of D1 during photosystem II repair (Zhang et al. 2001), it is likely that *psbA* mRNA translation is restricted to stromal thylakoid membranes. This leads to the prediction of a molecular tether, which targets chloroplast transcripts to stromal thylakoids and might be represented by RBP63 (Ossenbühl et al. 2002).

5 Conclusions and Perspectives

The establishment of chloroplasts as an organellar structure typical for plant cells represents one of the most significant steps during the evolution of eukaryotism, which involved the transformation of a free-living cyanobacterium into an autonomy-lacking 'cellular slave'. The recent cloning of genetically defined loci required for chloroplast biogenesis now provides the first insights into how this 'slave' is controlled. Apparently, different strategies were applied to regulate almost all steps of plastid development. Initially, former cyanobacterial regulatory genes were maintained by the endosymbiotic host within its nuclear genome, e.g., the *VIPP1* gene involved in vesicular membrane transport or the complex assembly factor genes *Hcf136* and *Hcf164*. Then, novel factors, which probably originated from the former host genome, gained control of distinct chloroplast gene expression steps, which include the splicing factors Raa1 and Crs1 or the *Ac115* gene product. Another group of factors comprises enzymes, which were recruited as RNA chaperones but lost their initial activity such as Raa2, Raa3 and Crs2 (Table 1). Finally, four proteins, namely Mbb1, Nac2, Hcf107 and Crp1, form TPR or TPR-like domains suggesting that this domain might play an important role in the organization of regulatory units within chloroplasts.

An additional important outcome from the data available to date is that many of the molecular principles, which were realized during the phylogenetic development of the plastid compartment, are conserved between lower eukaryotes like algae and higher plants. For example, the post-transcriptional regulation of the plastid *psbB/psbT/psbH* gene cluster is mediated by the TPR-containing orthologs *Mbb1* and *Hcf107*.

In the future, the cloning of additional regulatory genes, which nowadays is a routine procedure, will help to further explain the intracellular communication system. However, only complementing biochemical approaches will enable one to dissect the precise molecular details of its working mode.

Other interesting aspects of chloroplast biogenesis, which have not been addressed in this review due to size limitations, concern the signal(s) that are generated by the chloroplast, resulting in a feedback regulation of nuclear gene expression (Kropat et al. 1997) and the genetic interactions between mitochondria and chloroplasts (Bennoun and Delosme 1999). The analysis of these processes is likely to provide an even more complex picture of the intracellular network that underlies the development of the eukaryotic cell.

Acknowledgements. I thank U. Kück for providing basic support. The experimental work of the author is funded by the Deutsche Forschungsgemeinschaft (SFB480-B8).

References

Abdallah F, Salamini F, Leister D (2000) A prediction of the size and evolutionary origin of the proteome of chloroplasts of *Arabidopsis*. Trends Plant Sci 5:141–142

Adir N, Shochat S, Ohad I (1990) Light-dependent D1 protein synthesis and translocation is regulated by reaction centre II. Reaction centre II serves as an acceptor for the D1 precursor. J Biol Chem 265:12563–12568

Allison LA (2000) The role of sigma factors in plastid transcription. Biochimie 82:537–548

Andersson MX, Kjellberg JM, Sandelius AS (2001) Chloroplast biogenesis. Regulation of lipid transport to the thylakoid in chloroplasts isolated from expanding and fully expanded leaves of pea. Plant Physiol 127:184–193

Babiychuk E, Fuangthong M, Van Montagu M, Inze D, Kushnir S (1997) Efficient gene tagging in *Arabidopsis thaliana* using a gene trap approach. Proc Natl Acad Sci USA 94:12722–12727

Baginsky S, Shteiman-Kotler A, Liveanu V, Yehudai-Resheff S, Bellaoui M, Settlage RE, Shabanowitz J, Hunt DF, Schuster G, Gruissem W (2001) Chloroplast PNPase exists as a homo-multimer enzyme complex that is distinct from the *Escherichia coli* degradosome. RNA 7:1464–1475

Barkan A (1998) Approaches to investigating nuclear genes that function in chloroplast biogenesis in land plants. Methods Enzymol 297:293–310

Barkan A, Goldschmidt-Clermont (2000) Participation of nuclear genes in chloroplast gene expression. Biochimie 82:559–572

Bennoun P, Delosme M (1999) Chloroplast suppressors that act on a mitochondrial mutation in *Chlamydomonas reinhardtii*. Mol Gen Genet 262:85–89

Blatch GL, Lässle M (1999) The tetratricopeptide repeat: a structural motif mediating protein-protein interactions. Bioessays 21:932–939

Bock R, Koop H (1997) Extraplastidic site-specific factors mediate RNA editing in chloroplasts. EMBO J 16:3282–3288

Boudreau E, Takahashi Y, Lemieux C, Turmel M, Rochaix J-D (1997) The chloroplast *ycf3* and *ycf4* open reading frames of *Chlamydomonas reinhardtii* are required for the accumulation of the photosystem I complex. EMBO J 15:6095–6104

Boudreau E, Nickelsen J, Lemaire S, Ossenbühl F, Rochaix J-D (2000) The *Nac2* gene of *Chlamydomonas reinhardtii* encodes a chloroplast TPR protein involved in *psbD* mRNA stability, processing and/or translation. EMBO J 19:3366–3376

Boynton JE, Gillham NW, Harris EH, Hosler JP, Johnston AR, Jones BL, Randolph-Anderson D, Robertson TM, Klein KB, Shark B, Sanford J (1988) Chloroplast transformation in *Chlamydomonas* with high velocity microprojectiles. Science 240:1534–1538

Brocks JJ, Logan GA, Buick R, Summons RE (1999) Archean molecular fossils and the early rise of eukaryotes. Science 285:1033–1036

Bruick RK, Mayfield SP (1999) Light-activated translation of chloroplast mRNAs. Trends Plant Sci 4:190–195

Chatterjee M, Sparvoli S, Edmunds C, Garosi P, Findlay K, Martin C (1996) *DAG*, a gene required for chloroplast differentiation and palisade development in *Antirrhinum majus*. EMBO J 15:4194–4207

Choquet Y, Vallon O (2000) Synthesis, assembly, and degradation of thylakoid membrane proteins. Biochimie 82:615–634

Choquet Y, Stern DB, Wostrikoff K, Kuras R, Girard-Bascou J, Wollman FA (1998) Translation of cytochrome f is autoregulated through the 5′ untranslated region of *petA* mRNA in *Chlamydomonas* chloroplasts. Proc Natl Acad Sci USA 95:4380–4385

Das AK, Cohen PW, Barford D (1998) The structure of the tetratricopeptide repeats of protein phosphatase 5: implications for TPR-mediated protein-protein interactions. EMBO J 17:1192–1199

Dent RM, Han M, Niyogi KK (2001) Functional genomics of plant photosynthesis in the fast lane using *Chlamydomonas reinhardtii.* Trends Plant Sci 6:364–371

Deshpande NN, Bao Y, Herrin DL (1997) Evidence for light/redox regulated splicing of the *psbA* introns in *Chlamydomonas* chloroplasts. RNA 3:37–48

Drager RG, Girard-Bascou J, Choquet Y, Kindle KL, Stern DB (1998) In vivo evidence for 5′-3′ exoribonuclease degradation of an unstable chloroplast mRNA. Plant J 13:85–96

Edhofer I, Mühlbauer SK, Eichacker LA (1998) Light regulates the rate of translation elongation of chloroplast reaction centre protein D1. Eur J Biochem 257:78–84

Fargo DC, Boynton JE, Gillham NW (1999) Mutations altering the predicted secondary structure of a chloroplast 5′ untranslated region affect its physical and biochemical properties as well as its ability to promote translation of reporter mRNAs both in the *Chlamydomonas reinhardtii* chloroplast and in *Escherichia coli.* Mol Cell Biol 19:6980–6990

Fargo DC, Boynton JE, Gillham NW (2001) Chloroplast ribosomal protein S7 of *Chlamydomonas* binds to chloroplast mRNA leader sequences and may be involved in translation initiation. Plant Cell 13:207–218

Felder S, Meierhoff K, Sane AP, Meurer J, Driemel C, Plücken H, Klaff P, Stein B, Bechtold N, Westhoff P (2001) The nucleus-encoded *HCF107* gene of *Arabidopsis* provides a link between intercistronic RNA processing and the accumulation of translation-competent *psbH* transcripts in chloroplasts. Plant Cell 13:2127–2141

Fisk DG, Walker MB, Barkan A (1999) Molecular cloning of the maize gene *crp1* reveals similarity between regulators of mitochondrial and chloroplast gene expression. EMBO J 18:2621–2630

Goldschmidt-Clermont M (1998) Coordination of nuclear and chloroplast gene expression in plant cells. Int Rev Cytol 177:115–180

Gruissem W, Tonkyn JC (1993) Control mechanisms of plastid gene expression. Crit Rev Plant Sci 12:19–55

Hedtke B, Börner T, Weihe A (1997) Mitochondrial and chloroplast phage-type RNA polymerases in *Arabidopsis.* Science 277:809–811

Hermann M, Bock R (1999) Transfer of plastid RNA-editing activity to novel sites suggests a critical role for spacing in editing-site recognition. Proc Natl Acad Sci USA 96:4856–4861

Herrin D, Michaels A (1985) The chloroplast 32 kDa protein is synthesized on thylakoid-bound ribosomes in *Chlamydomonas reinhardtii.* FEBS Lett 184:90–95

Higgs DC, Shapiro RS, Kindle KL, Stern DB (1999) Small *cis*-acting sequences that specify secondary structures in a chloroplast mRNA are essential for RNA stability and translation. Mol Cell Biol 19:8479–8491

Hirose T, Sugiura M (2001) Involvement of a site-specific *trans*-acting factor and a common RNA-binding protein in the editing of chloroplast mRNAs: development of a chloroplast in vitro RNA editing system. EMBO J 20:1144–1152

Hoober JK, Boyd CO, Paavola LG (1991) Origin of the thylakoid membranes in *Chlamydomonas reinhardtii y-1* in the light or dark. Plant Physiol 96:1321–1328

Hong S, Spreitzer RJ (1994) Nuclear mutation inhibits expression of the chloroplast gene that encodes the large subunit of ribulose-1,5-bisphosphate carboxylase/oxygenase. Plant Physiol 106:673–678

Hugueney P, Bouvier F, Badillo A, d'Harlingue A, Kuntz M, Camara B (1995) Identification of a plastid protein involved in vesicle fusion and/or membrane protein translocation. Proc Natl Acad Sci USA 92:5630–5634

Jenkins BD, Barkan A (2001) Recruitment of a peptidyl-tRNA hydrolase as a facilitator of group II intron splicing in chloroplasts. EMBO J 20:872–879

Keddie JS, Carroll B, Jones JDG, Gruissem W (1996) The *DCL* gene of tomato is required for chloroplast development and palisade morphogenesis in leaves. EMBO J 15:4208–4217

Kim J, Mayfield SP (1997) Protein disulphide isomerase as a regulator of chloroplast translational activation. Science 278:1954–1957

Klein RR, Mason HS, Mullet JE (1988) Light-regulated translation of chloroplast proteins. I. Transcripts of *psaA*-psaB, *psbA* and *rbcL* are associated with polysomes in dark-grown and illuminated barley seedlings. J Cell Biol 106:289–301

Kroll D, Meierhoff K, Bechtold N, Kinoshita M, Westphal S, Vothknecht UC, Soll J, Westhoff P (2001) *VIPP1*, a nuclear gene of *Arabidopsis thaliana* essential for thylakoid membrane formation. Proc Natl Acad Sci USA 98:4238–4242

Kropat J, Oster U, Rüdiger W, Beck CF (1997) Chlorophyll precursors are signals of chloroplast origin involved in light induction of nuclear heat-shock genes. Proc Natl Acad Sci USA 94:14168–14172

Lennartz K, Plücken H, Seidler A, Westhoff P, Bechthold N, Meierhoff K (2001) *HCF164* encodes a thioredoxin-like protein involved in the biogenesis of the cytochrome b_6f complex in *Arabidopsis*. Plant Cell 13:2539–2551

Leon P, Arroyo A, Mackenzie S (1998) Nuclear control of plastid and mitochondrial development in higher plants. Annu Rev Plant Physiol Plant Mol Biol 49:453–480

Lerbs-Mache S (2000) Regulation of rDNA transcription in plastids of higher plants. Biochimie 82:525–535

Martin W, Müller M (1998) The hydrogen hypothesis for the first eukaryote. Nature 392:37–41

Meurer J, Plücken H, Kowallik KV, Westhoff P (1998) A nuclear-encoded protein of prokaryotic origin is essential for the stability of photosystem II in *Arabidopsis thaliana*. EMBO J 17:5286–5297

Morre DJ, Sellden G, Sundquist C, Sandelius AS (1991) Stromal low temperature compartment derived from the inner membrane of the chloroplast envelope. Plant Physiol 97:1558–1564

Mühlbauer SK, Eichacker LA (1998) Light-dependent formation of the photosynthetic proton gradient regulates translation elongation in chloroplasts. J Biol Chem 273:20935–20940

Nickelsen J (1998) Chloroplast RNA stability. In: Rochaix J-D, Goldschmidt-Clermont M, Merchant S (eds) The molecular biology of chloroplasts and mitochondria in *Chlamydomonas*. Kluwer, Dordrecht, pp 151–163

Nickelsen J (2000) Mutations at three different nuclear loci of *Chlamydomonas* suppress a defect in chloroplast *psbD* mRNA accumulation. Curr Genet 37:136–142

Nickelsen J, Kück U (2000) The unicellular green alga *Chlamydomonas reinhardtii* – an experimental system to study chloroplast RNA metabolism. Naturwissenschaften 87:97–107

Nickelsen J, van Dillewijn J, Rahire M, Rochaix J-D (1994) Determinants for stability of the chloroplast *psbD* RNA are located within its short leader region in *Chlamydomonas reinhardtii*. EMBO J 13:3182–3191

Nickelsen J, Fleischmann M, Boudreau E, Rahire M, Rochaix J-D (1999) Identification of *cis*-acting RNA leader elements required for chloroplast *psbD* gene expression in *Chlamydomonas*. Plant Cell 11:957–970

Ossenbühl F, Nickelsen J (2000) *Cis*- and *trans*-acting determinants for translation of *psbD* mRNA in *Chlamydomonas reinhardtii*. Mol Cell Biol 20:8134–8142

Ossenbühl F, Hartmann K, Nickelsen J (2002) A chloroplast RNA binding protein from stromal thylakoid membranes specifically binds to the 5' untranslated region of the psbA mRNA. Eur J Biochem, in press

Park JM, Cho JH, Kang SG, Jang HJ, Pih KT, Piao HL, Cho MJ, Hwang I (1998) A dynamin-like protein in *Arabidopsis thaliana* is involved in biogenesis of thylakoid membranes. EMBO J 17:859–867

Perron K, Goldschmidt-Clermont M, Rochaix J-D (1999) A factor related to pseudouridine synthases is required for chloroplast group II intron *trans*-splicing in *Chlamydomonas reinhardtii.* EMBO J 15:6481–6490

Rattanachaikunsopon P, Rosch C, Kuchka MR (1999) Cloning and characterization of the nuclear *AC115* gene of *Chlamydomonas reinhardtii.* Plant Mol Biol 39:1–10

Rivier C, Goldschmidt-Clermont M, Rochaix J-D (2001) Identification of an RNA-protein complex involved in chloroplast group II intron *trans*-splicing in *Chlamydomonas reinhardtii.* EMBO J 20:1765–1773

Rolland N, Janosi L, Block MA, Shuda M, Teyssier E, Miege C, Cheniclet C, Carde JP, Kaji A, Joyard J (1999) Plant ribosome recycling factor homologue is a chloroplastic protein and is bactericidal in *Escherichia coli* carrying temperature-sensitive ribosome recycling factor. Proc Natl Acad Sci USA 96:5464–5469

Rott R, Levy H, Drager RG, Stern DB, Schuster G (1998) 3'-Processed mRNA is preferentially translated in *Chlamydomonas reinhardtii* chloroplasts. Mol Cell Biol 18:4605–4611

Roy LM, Barkan A (1998) A SecY homologue is required for the elaboration of the chloroplast thylakoid membrane and for normal chloroplast gene expression. J Cell Biol 141:385–395

Salvador ML, Klein U, Bogorad L (1993) Light regulated and endogenous fluctuations of chloroplast transcript levels in *Chlamydomonas.* Regulation by transcription and RNA degradation. Plant J 3:213–219

Sato N, Rolland N, Block MA, Joyard J (1999) Do plastid envelope membranes play a role in the expression of the plastid genome? Biochimie 81:619–629

Schleiff E, Soll J (2000) Travelling of proteins through membranes: translocation into chloroplasts. Planta 211:449–456

Schmitz-Linneweber C, Tillich M, Herrmann RG, Maier RM (2001) Heterologous, splicing-dependent RNA editing in chloroplasts: allotetraploidy provides *trans*-factors. EMBO J 20:4874–4883

Sharp PA (1991) Five easy pieces. Science 254:663

Silhavy D, Maliga P (1998) Mapping of the promoters for the nucleus-encoded plastid RNA polymerase (NEP) in the *iojap* maize mutant. Curr Genet 33:340–344

Small ID, Peeters N (2000) The PPR motif – a TPR-related motif prevalent in plant organellar proteins. Trends Biochem Sci 25:46–47

Staub JM, Maliga P (1994) Translation of *psbA* mRNA is regulated by light via the 5' untranslated region in tobacco plastids. Plant J 6:547–553

Sugiura M (1992) The chloroplast genome. Plant Mol Biol 19:149–168

Till B, Schmitz-Linneweber C, Williams-Carrier R, Barkan A (2001) CRS1 is a novel group II intron splicing factor that was derived from a domain of ancient origin. RNA 7:1227–1238

Trebitsh T, Danon A (2001) Translation of chloroplast *psbA* mRNA is regulated by signals initiated by both photosystems II and I. Proc Natl Acad Sci USA 98:12289–12294

Tyystjärvi T, Herranen M, Aro E-M (2001) Regulation of translation elongation in cyanobacteria: membrane targeting of the ribosome nascent-chain complexes controls the synthesis of D1 protein. Mol Microbiol 40:476–484

Vaistij F, Goldschmidt-Clermont M, Rochaix J-D (2000a) Stability determinants of the chloroplast *psbB/T/H* mRNAs of *Chlamydomonas reinhardtii.* Plant J 21:469–482

Vaistij F, Boudreau E, Lemaire SD, Goldschmidt-Clermont M, Rochaix JD (2000b) Characterization of Mbb1, a nucleus-encoded tetratricopeptide-like repeat protein required for expression of the chloroplast *psbB/psbT/psbH* gene cluster in *Chlamydomonas reinhardtii*. Proc Natl Acad Sci USA 97:14813–14818

Vogel J, Börner T, Hess WR (1999) Comparative analysis of splicing of the complete set of chloroplast group II introns in three higher plant mutants. Nucleic Acids Res 27:3866–3873

Wang Y, Duby G, Purnelle B, Boutry M (2000) Tobacco *VDL* gene encodes a plastid DEAD box RNA helicase and is involved in chloroplast differentiation and plant morphogenesis. Plant Cell 12:2129–2142

Westphal S, Heins L, Soll J, Vothknecht UC (2001) *VIPP1* deletion mutant of *Synechocystis*: a connection between bacterial phage shock and thylakoid biogenesis? Proc Natl Acad Sci USA 98:4243–4248

Wilde A, Lünser K, Ossenbühl F, Nickelsen J, Börner T (2001) Characterization of the cyanobacterial *ycf37*: mutation decreases the photosystem I content. Biochem J 357:211–216

Wollman FA, Minai L, Nechushtai R (1999) The biogenesis and assembly of photosynthetic proteins in thylakoid membranes. Biochim Biophys Acta 1411:21–85

Wostrikoff K, Choquet Y, Wollman FA, Girard-Bascou J (2001) TCA1, a single nuclear-encoded translational activator specific for *petA* mRNA in *Chlamydomonas reinhardtii* chloroplast. Genetics 159:119–132

Wu HY, Kuchka MR (1995) A nuclear suppressor overcomes defects in the synthesis of the chloroplast *psbD* gene product caused by mutations in two distinct nuclear genes of *Chlamydomonas*. Curr Genet 27:263–269

Yohn CB, Cohen A, Danon A, Mayfield SP (1998) A poly(A) binding protein functions in the chloroplast as a message-specific translation factor. Proc Natl Acad Sci USA 95:2238–2243

Zak E, Norling B, Maitra R, Huang F, Andersson B, Pakrasi HB (2001) The initial steps of biogenesis of cyanobacterial photosystems occur in plasma membranes. Proc Natl Acad Sci USA 98:13443–13448

Zerges W (2000) Translation in chloroplasts. Biochimie 82:583–601

Zerges W, Rochaix J-D (1998) Low density membranes are associated with RNA-binding proteins and thylakoids in the chloroplast of *Chlamydomonas reinhardtii*. J Cell Biol 140:101–110

Zerges W, Girard-Bascou J, Rochaix J-D (1997) Translation of the *psbC* mRNA and incorporation of its polypeptide product into photosystem II is controlled by interactions between the *psbC* 5′ leader and the *NCT* loci in *Chlamydomonas*. Mol Cell Biol 17:3440–3448

Zhang L, Paakkarinen V, Suorsa M, Aro EM (2001) A SecY homologue is involved in chloroplast-encoded D1 protein biogenesis. J Biol Chem 276:37809–37814

Jörg Nickelsen
Lehrstuhl Allgemeine und Molekulare Botanik
Ruhr-Universität Bochum,
Universitätsstraße 150
44780 Bochum, Germany

e-mail: joerg.nickelsen@ruhr-uni-bochum.de

Extranuclear Inheritance: Genetics and Biogenesis of Mitochondria

Thomas Lisowsky, Karlheinz Esser, Martin Ingenhoven, Elke Pratje, and Georg Michaelis

1 Introduction

In our previous article in Volume 62 of this series (Esser et al. 2001), we discussed new results obtained for mitochondrial genomes, mitochondrial evolution, the cross-talk between mitochondria, cytosol and the nucleus, and finally, for the mitochondrial protein transport in higher plants. In the present review, different topics have been selected: mitochondrial genomes (Sect. 2); apoptosis in plants (Sect. 3); and formation of disulfide bridges in organellar proteins (Sect. 4).

In human developmental biology apoptosis is a very dynamic research field that stimulates work on other organisms like higher plants or even fungi. Mitochondria and cytochrome c release play a major role in this process. Our last section is devoted to the new protein family of sulfhydryl oxidases, that catalyze the introduction of disulfide bonds into proteins. Recent studies have characterized these enzymes as essential components for the biogenesis of functional mitochondria.

For additional aspects of mitochondria the reader is referred to recent reviews on mitochondrial inheritance and division (Berger and Yaffe 2000; Boldogh et al. 2001; Gilson and Beech 2001; Yoon and McNiven 2001), evolution of the mitochondrial genetic system (Saccone et al. 2000), gene expression in higher plant mitochondria (Giegé and Brennicke 2001), protein import (Pfanner and Geissler 2001; Truscott et al. 2001), RNA import (Schneider and Marechal-Drouard 2000; Entelis et al. 2001), protein unfolding (Matouschek et al. 2000), ATP-dependent proteases (Van Dyck and Langer 1999), and plant mitochondrial carriers (Laloi 1999).

2 Mitochondrial Genomes

a) Higher Plants

In recent years, several complete mitochondrial genomes have been sequenced, mostly from animals, protists, fungi and algae. The mitochondrial sequences from higher plants are derived from the two dicotyledonous angiosperms *Arabidopsis thaliana* (Unseld et al. 1997; see Prog-

Progress in Botany, Vol. 64

ress in Botany 2001) and *Beta vulgaris* (Kubo et al. 2000). Both mitochondrial genomes are similar in size (*A. thaliana* 367 kb, *B. vulgaris* 369 kb) and gene content. However, four genes were identified in *B. vulgaris* (rps13, trnF-GAA, ccb577 and trnC2-GCA) that do not occur in *A. thaliana*. In contrast, four genes found in *A. thaliana* (ccb228, rpl2, rpl16, trnY2-GUA) are absent or represented by pseudogenes in *B. vulgaris*. Sequences shared by the two plant species represent about 21% of their mitochondrial genomes.

Kubo et al. (2000) predicted at least 370 C to U editing sites. This number is somewhat less than the reported 441 editing sites in *A. thaliana* (Giegé and Brennicke 1999). The genes for the large rRNA (rrn26) and the tRNAfmet-CAU of *B. vulgaris* are arranged in a three-copy repeat of 6222 bp. The 100% sequence identity of the three repeated elements indicates that corrections of changes in each of the copies are ongoing processes. Sugar beet mitochondrial rRNA and tRNA genes lack introns, but, in seven protein-coding genes, 20 group II introns have been identified and six of them are trans-splicing introns. A novel tRNA-Cys gene (trnC2-GCA) has been found that shows no sequence homology with any tRNA-Cys gene sequenced so far in plant mitochondria. This tRNA gene is transcribed into a mature cysteine tRNA.

On the other hand, 559 open reading frames (ORFs; comprising more than 60 codons) exhibited no homology to any reported gene. Therefore, it was suggested that it is unlikely that these ORFs code for important protein information. In *B. vulgaris*, like in *A. thaliana*, about 56% of the mitochondrial genome shows no similarity to any sequence in databases and probably carries no obvious informational features.

The phylogenetic relationship of the five groups of extant seed plants was investigated by sequencing the mitochondrial cox1, atpA and small ribosomal RNA genes (Bowe et al. 2000; Chaw et al. 2000). These analyses included the plastid rbcI and the nuclear 18S rRNA genes. The mitochondrial and plastid sequences place the *Gnetales* within the conifers as a sister group of the *Pinaceae*. Thus, the morphological similarities between angiosperms and *Gnetales*, like the reproductive structures and double fertilization, arose independently. A monophyletic origin of extant gymnosperms is strongly supported, with cycads separated from the other gymnosperms first, followed by *Ginkgo*.

b) Algae

The complete mitochondrial DNA sequences from several green algae reveal remarkable differences in size, gene content and gene organization (Table 1). Six of the seven green algal mitochondrial genomes completely sequenced so far show a circular map. The gene number varies

Table 1. Characteristic features of mitochondrial genomes from green algae. The listed number of genes excludes duplications; the listed number of ORFs includes intronic ORFs. Data were taken from the following references: *Chlamydomonas reinhardtii* (Michaelis et al. 1990; Vahrenholz et al. 1993), *Chlamydomonas eugametos* (Denovan-Wright et al. 1998), *Chlorogonium elongatum* (Kroymann and Zetsche 1998), *Scenedesmus obliquus* (Kück et al. 2000; Nedelcu et al. 2000), *Pedinomonas minor* (Turmel et al. 1999), *Prototheca wickerhamii* (Wolff et al. 1994), *Nephroselmis olivacea* (Turmel et al. 1999)

	Chlamydomonas reinhardtii	*Chlamydomonas eugametos*	*Chlorogonium elongatum*	*Scenedesmus obliquus*	*Pedinomonas minor*	*Prototheca wickerhamii*	*Nephroselmis olivacea*
Genome size (kb)	15.8	22.9	22.7	42.9	25.1	55.3	45.2
Genome structure	Linear	Circle	Circle	Circle	Circle	Circle	Circle
Number of genes	13	19	18	47	21	65	69
Protein genes	7	7	7	13	11	30	33
rRNA	2	2	2	2	2	3	3
tRNA	3	3	3	27	8	26	26
ORFs	1	7	6	5	0	6	7
Number of introns	0	9	6	4	1	5	4

between 69 in *Nephroselmis olivacea* (Turmel et al. 1999) and 13 in *Chlamydomonas reinhardtii* (Michaelis et al. 1990), reflecting the evolution of the green algal mitochondrial genomes. The gene density of *Scenedesmus obliquus* is the lowest (60.6%) and non-coding sequences are dispersed over the mitochondrial genome. A similar low gene density is described for *Pedinomonas minor*, but in this alga non-coding sequences are arranged in only one defined region of the circle (Turmel et al. 1999). The 86% coding information reported for *Chlorogonium elongatum* (Kroymann and Zetsche 1998) is the highest value of all reported algal mtDNAs. *Nephroselmis olivacea* seems to have the most ancestral algal mt genome and the genes are organized into at least 15 transcriptional units (Turmel et al. 1999). A gene arrangement in transcriptional units of one (*Chlamydomonas eugametos, Chlorogonium elongatum, Pedinomonas minor*), two (*Chlamydomonas reinhardtii, Prototheca wickerhamii*) or three (*Scenedesmus obliquus)* seems to be characteristic for the more advanced green algae (Nedelcu et al. 2000).

The mitochondrial genes for tRNAs differ in the various green algae, even if their number is similar. For example, six tRNA genes are unique to *Scenedesmus* (Kück et al. 2000; Nedelcu et al. 2000) and the tmW-UCA is so far unique to *Pedinomonas*. Several deviations from the standard genetic code have been reported (Hayashi-Ishimaru et al. 1996). The termination codon UGA is decoded as tryptophan in *Pedinomonas minor* (Turmel et al. 1999), while UAG codes for leucine in *Scenedesmus obliquus* (Kück et al. 2000; Nedelcu et al. 2000). In *Scenedesmus*, the UCA codon is used as a stop codon. Codon changes by RNA editing have been excluded by RT-PCR analysis (Kück et al. 2000).

Fragmented and scrambled rRNA genes are present in *Chlamydomonas, Chlorogonium, Scenedesmus* and *Pedinomonas*, but the numbers and positions of the breakpoints are variable. The rRNA genes are continuous in *Prototheca* and *Nephroselmis.*

In summary, the size of the mitochondrial DNA, the gene content and the fragmentation pattern of the rRNA genes suggest a gradual evolution in green algae from the ancestral mtDNA of *Nephroselmis* to the intermediate mitochondrial genome of *Scenedesmus* and finally to the reduced advanced types of *Chlamydomonas.*

A relationship between red algae and green plants has been discussed for more than a century. Recently, Stiller et al. (2001) asked for caution to consider a sister relationship between red algae and green plants. In contrast, a comparison of the mtDNA sequences of five rhodophytes gave strong evidence that red algae share a common ancestry with green algae and higher plants (Burger et al. 1999). Since 1995, when the first complete mtDNA from a red alga, *Chondrus crispus*, was published (Leblanc et al. 1995, see Progress in Botany 1997), the sequences of *Cyanidioschyzon merolae* (Ohta et al. 1998) and *Porphyra purpurea* (Burger et al. 1999) have been completed. In addition, partial sequences from

Cyanidium caldarium (Viehmann et al. 1996) and *Gracillariopsis lemaneiformis* have been published. The mitochondrial genomes of these red algae are similar in size (25–37 kb) and gene content and exhibit a derived pattern. Approximately half of the mitochondrial genes have apparently been transferred to the nucleus after separation of the rhodophytes from the common ancestor shared with chlorophytes (Burger et al. 1999).

3 Apoptosis in Plants

Apoptosis, a form of programmed cell death (PCD), has been extensively characterized in mammalian cells in the last 10 years. It is a genetically defined program that involves a cascade of zymogen-activated specific cysteine proteases, designated caspases (cysteine aspases). Activation is achieved by receptor systems, cytochrome c (Cyt c) release and reactive oxygen species (ROS) from mitochondria (for reviews, see Green and Reed 1998; Mignotte and Vayssiere 1998).

Once started, apoptosis causes characteristic alterations in the cell: a reduced volume and partition of the cytoplasm especially by altered microtubule organization, proteolytic degradation of specific proteins, endonucleolytic fragmentation as well as condensation of chromatin, and formation of membrane bound vesicles (apoptotic bodies) are observed. Different forms of PCD that are not always accompanied by all characteristics of apoptosis are described in the literature. PCD can also be independent of caspases, as shown in studies with caspase inhibitors and caspase knockout cells (del Pozo and Lam 1998; Kawahara et al. 1998; McCarthy et al. 1997; Vercammen et al. 1998; Yoshida et al. 1998; Chautan et al. 1999). Two PCD extremes are distinguished: (1) apoptosis, an active death program requiring ATP and activation of protease mechanisms, and (2) oncosis, the cell death after the inability of the cell to repair damage.

In plants, much less is known about PCD, and here it seems to display more the characteristics of a programmed form of oncosis. Nevertheless, recent studies show that some characteristics and important mechanisms are ubiquitous in different types of PCD in all multicellular organisms including plants. These mechanisms are always associated with altered mitochondrial functions (Jones 2000), and certain components of these mechanisms exist in unicellular eukaryotes and even in bacteria (Engelberg-Kulka and Glaser 1999; Madeo et al. 1999; Ligr et al. 2001).

a) Examples of Programmed Cell Death in Plants

Like in other multicellular organisms, PCD can be developmentally determined in plants: (1) in tissues or cells with temporary function, e.g., plasmodial tapetum cells, that are important for the secretion of substances during pollen formation, (2) in tissues exerting functions after cell death, as, e.g., xylem or sclerenchym tissues, and (3) during senescence of tissues or whole organs, as, e.g., leaves at the end of their life span. In addition, PCD can also be exerted as a reaction to the damage of cells by stress, or by infectious pathogens, in order to prevent their spreading over the entire organism by rapid death of the infected and neighboring host cells.

In plants, special cellular processes and organelles exist that may be linked to PCD. One example is cytoplasmic male sterility (CMS). This defect prevents the production of functional pollen, and is often caused by the expression of chimeric mitochondrial genes (Conley and Hanson 1995; Wise et al. 1996; Lisowsky et al. 1999). The precise action of the mitochondrial gene products and of respective nuclear restorer genes is not understood. Analyzing S-type CMS of maize, Wen and Chase (1999) observed the characteristics of PCD in collapsing pollen and they speculate that the CMS gene triggers apoptosis. If this observation can be confirmed, it is conceivable that respective restorer genes might regulate or even encode anti-apoptotic factors. Balk and Leaver (2001) describe the premature death of tapetum cells in PET1-CMS of sunflower that is caused by another mitochondrial mutation. This defect is associated with Cyt c release and other characteristics of PCD.

Ricinosomes are plant-specific organelles that are exclusively formed in senescent tissues by budding from the ER after induction of PCD. They deliver certain peptidases into the cytosol, while mitochondria and glyoxysomes are degraded (Gietl and Schmid 2001; Schmid et al. 2001). Although the involved papain-type cysteine endopeptidases (CysEP) do not belong to the caspase family, they are also activated by removal of amino-terminal sequences. Thereby they exert caspase-similar functions. A transport of certain stress-inducible cysteine peptidases from the ER into lytic vacuoles is mediated by ER bodies that form a proteinase-storing system in epidermal cells of *Arabidopsis* preparing for PCD (Hayashi et al. 2001).

b) Differences in PCD in Mammalian and Plant Cells

It is characteristic for plants that a large number of secondary metabolites are likely to regulate or induce PCD of different cell types or tissues. Examples are the involvement of jasmonate, ethylene or salicylate (Asai et al. 2000; Xie and Chen 1999, 2000). In spite of similar effects of PCD in animal and plant cells, studies on plant apoptosis and the identification of plant factors revealed surprising differences. It seems that in plants functions similar to caspases can be exerted by caspase-like proteases. For example, the papain-type cysteine peptidases have been identified in several inhibitor studies (del Pozo and Lam 1998; Lam and del Pozo 2000). In contrast to animal cells, plant proteases, regulator proteins, and signaling pathways involved in PCD are still mostly unidentified or clearly differ from those of mammalian cells (Fig. 1). Genes encoding members of the most important protein families involved in PCD (such as the caspases and the regulators Bcl2 and Bax) are absent from the genomes of plants and yeasts. One reason might be that in animal cells different types of PCD can be carried out that are controlled by negative and positive regulators. In contrast, PCD in plants seems to follow a rather simple mechanism that is always executed when initiated by metabolic changes, signals from mitochondria or release of hydrolases from the vacuole.

Therefore, two models are discussed for how PCD in plants can be exerted: (1) by the action of proteases with caspase-similar functions, or (2) by the direct effects of ROS generated after Cyt c release from mitochondria. The second model could represent a primordial mechanism, by which the early mitochondrial endosymbiont was capable of killing the host cell. In addition, plants contain plastids, a second organelle, that was taken up during evolution and that might have retained mechanisms for PCD. Signals from the chloroplasts seem to play a role at least in leaf senescence (Mehta et al. 1992; Ochs et al. 1999; Binyamin et al. 2001).

c) Conserved Mechanisms in Mammals and Plants

Obviously, there are many similarities between PCD of animal and plant cells. Cell death can be triggered by different internal and external signals, similar morphological alterations of cells undergoing PCD are observed, and, most importantly, mitochondria and Cyt c release play a key role in both systems. As an example, caspase-dependent PCD with the characteristic morphological alterations as well as other types of caspase-independent PCD have been observed in animal and in plants cells (Lam et al. 1999; Navarre and Wolpert 1999; Heath 2000; Sugiyama et al. 2000).

In addition, it has been shown that certain positive or negative regulator proteins from animal cells (Bax and Bcl2 homologues) result in similar effects when expressed in plant cells (Lacomme and Santa Cruz 1999; Mitsuhara et al. 1999), indicating that similar regulators, or at least homologues of their potential target proteins, must exist in plants, too.

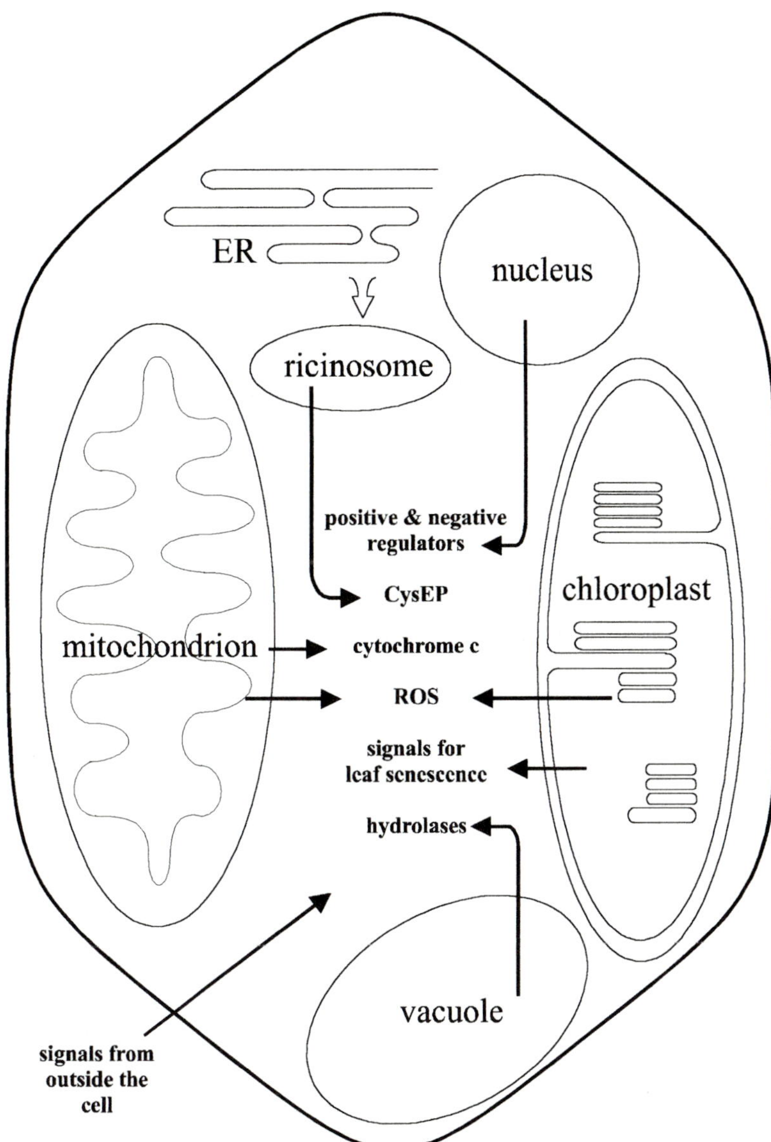

Fig. 1. Model of a typical plant cell displaying various organelles and their known functions in programmed cell death (PCD). *ROS* Reactive oxygen species; *Cys EP* cysteine endopeptidases; *ricinosome* new organelle generated by the endoplasmic reticulum (*ER*)

Numerous recent reports in the literature refer to the identification of plant substances that induce apoptosis or growth arrest in human cells and that possibly have cancer-protective effects. Examples are the recently described flavone eupatilin from *Artemisia* plants, arecoline from *Areca*, *Szygium aromaticum* flower bud substances, resveratrol from grapes, green tea extract, and others (Li et al. 2000; Chang et al. 2001; Kennedy et al. 2001; Park et al. 2001; Seo and Surh 2001; Tinhofer et al. 2001). Some of these substances may be of pharmacological value. These substances were also found to be involved in plant apoptosis, and several of them were capable of inducing cytochrome c dependent apoptosis.

d) A Key Role of Mitochondria is Common in PCD of Animal and Plant Cells

Most key events of the early stages of PCD focus on mitochondria, including (1) the release of Cyt c and the generation of ROS, (2) participation of mitochondrial proteins (such as Bcl2 in animal cells or the components of pore-forming proteins), (3) altered mitochondrial functions as electron transport and transmembrane potential, (4) alternative pathway respiration by induced expression of the alternative oxidase, and (5) diverse effects of signals and substances triggering or inhibiting these events. Two similar mechanisms are discussed for release of the death signal Cyt c by mitochondria (Fig. 2):

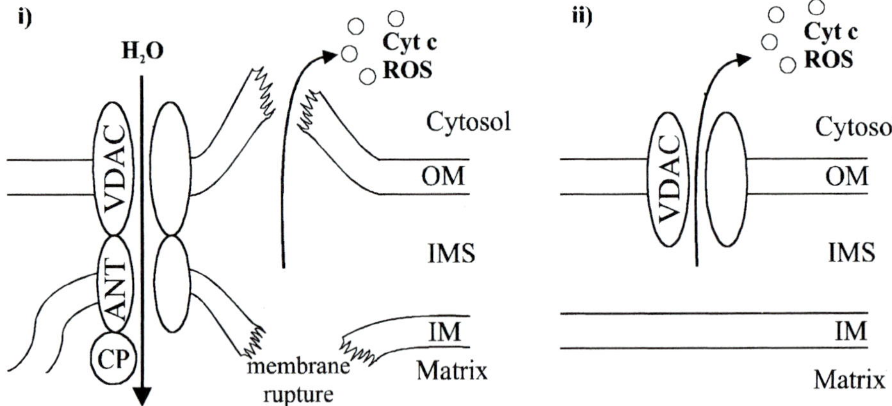

Fig. 2. Two possible mechanisms for the release of cytochrome c and ROS by mitochondria. **i** The formation of a permeability transition pore (PT), consisting of the voltage-dependent anion channel porin (*VDAC*), the adenine nucleotide translocator (*ANT*) and cyclophilin D (*CP*), may cause water influx, swelling and rupture of the mitochondrial membranes. **ii** Cytochrome c (*Cyt c*) and reactive oxygen species (*ROS*) are directly released from the intermembrane space via the *VDAC* channel. *IM* Inner membrane; *OM* outer membrane; *IMS* intermembrane space

1. Formation of the permeability transition pore (PT). This is a transient complex consisting of the voltage-dependent anion channel porin (VDAC) in the outer membrane, the adenine nucleotide translocator (ANT) in the inner membrane, and cyclophilin D in the matrix. This pore probably allows water influx and thereby swelling and rupture of the mitochondrial membranes.
2. Direct release of Cyt c from the intermembrane space via the VDAC channel, without membrane rupture.

The pore forming/opening is induced by oxidative stress, electron transport stress, and enhanced calcium ion levels, especially at low ATP levels. Nitric oxide (NO; Bosca and Hortelano 1999) or heat (Balk et al. 1999) exert similar effects.

The Bcl2 protein represents the most important inhibitor of apoptosis in animal cells. It is part of a protein complex in the inner membrane of mitochondria, where it interacts with VDAC and prevents Cyt c release. From heterologous gene expression studies it is suggested that the permeability of the transition pore (PT) is regulated by binding of Bcl-like proteins in plants, too. The function of such regulators is not only restricted to the control of membrane permeability, because the Bcl2 protein family is assumed to interact with apoptotic proteins in order to colocalize them to the mitochondrial membranes, preventing activation of proteases in the cytosol. The mitochondrial alternative oxidase (AOX) possibly represents a plant-specific regulator of PCD, limiting the release of ROS upon inhibition of oxidative phosphorylation (Maxwell et al. 1999). In addition, mitochondrial redox-sensitive factors containing sulfhydryl groups possibly can shift the redox potential to a more reduced state, protecting the cell from ROS and preventing induction of cell death (Jones 2000; Moller 2001).

e) The Role of ROS as Signal for PCD

All conditions that stress cells beyond the metabolic balance can trigger release of the death signal Cyt c from mitochondria. Independently of caspase activation, that is lacking in plant cells, removal of Cyt c from mitochondria blocks electron transport, thereby generating reactive oxygen species (ROS) levels that may be lethal (Jones 2000; Moller 2001). This is supported by the observation of Mittler and Lam (1996) that PCD in tobacco plants is suppressed by low oxygen pressure. Normally, limited amounts of ROS are continuously produced by mitochondria and other organelles, but their negative effects are prevented by the presence of diverse protecting mechanisms and substances such as ascorbate and reduced glutathione (Noctor and Foyer 1998) and the ROS-homeostasis enzymes catalase or superoxide dismutase (Longo et al. 1997). Fröhlich and Madeo (2000) report that regulation of PCD by oxygen radicals occurs in the unicellular yeast *Saccharomyces cerevisiae*, too, and suggest that a former response to oxidative cell damage developed to ROS producing/releasing mechanisms for PCD. In summary, ROS seem to have a regulatory function used by mitochondria to control and execute cell death.

f) Conclusions

The highly regulated cellular events of apoptosis are best analyzed in mammalian cells, where a number of characteristics of apoptotic processes can be identified. The ongoing research on apoptosis factors in connection with the sequencing of whole plant genomes possibly will identify the respective and probably distinct factors in plants. The recent studies in plants indicate that the role of ROS and mitochondria as stress sensors, regulators and/or initiators of cell death may represent the only factors that are common to all eukaryotic cells. Therefore, these factors are likely to define the basic principles of programmed cell death.

4 Genetic Basis of Disulfide Bridge Formation in Mitochondria and Plastids

The genetic basis of the introduction of functional thiol bonds into organellar proteins is a very new research topic (Lisowsky 2001). The first nuclear genes that perform important functions in this process have only recently been identified (Lee et al. 2000; Nakamoto et al. 2000; Lisowsky et al. 2001). Functional disulfide bridges in organellar proteins create a special problem, since the transport of proteins across membranes, and thereby also their import into organelles, only takes place if these polypeptides have been completely unfolded for the translocation process (Lithgow et al. 1995; Soll and Tien 1998). The active unfolding at the import complexes of the membranes removes all disulfide bridges from the translocated proteins (Wachter et al. 1994; Lill and Neupert 1996). For example, it is known that during the transfer of mitochondrial precursor proteins from the outer side to the inner side of the outer membrane, these proteins are actively unfolded in an ATP-independent manner strictly dependent on the TOM complex (Wachter et al. 1994; Pfanner and Meijer 1995). An example for the import of the rieske/sulfur protein of the cytochrome bc_1 complex illustrates this problem (for details, see Fig. 3).

The rieske/sulfur protein is of special interest as it contains a functionally important internal disulfide bond (Tian et al. 1999). Therefore, this example stresses the problem that functional disulfide bridges will have to be newly formed in organelles after the import process. Cur-

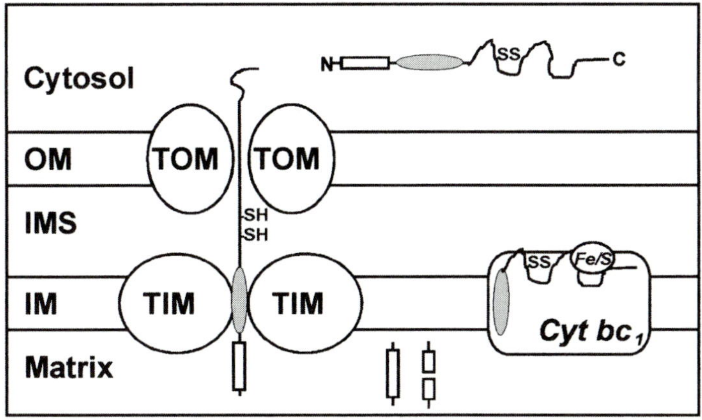

Fig. 3. Unfolding of proteins upon translocation across the outer mitochondrial membrane. The import of the nuclear encoded mitochondrial rieske/sulfur protein is illustrated. This protein is translated in the *cytosol* and then binds to the *TOM* complex. The *TOM* complex actively unfolds proteins and transfers them via the intermembrane space (*IMS*) to the *TIM* complex. The import complexes *TIM* and *TOM* and the import mechanism for the rieske/sulfur protein are actually more complex than shown. The central point of this schematic illustration is the finding that during import the internal disulfide bridge of the rieske/sulfur protein is removed. After cleavage of the leader peptide the protein is integrated into the cytochrome bc_1 complex (*Cyt* bc_1). Upon refolding of the protein the internal disulfide bridge is reconstituted. Furthermore, during this process, an Fe/S cluster (*Fe/S*) is attached to the protein

rently, little is known about the mechanisms and the components that take part in this reaction inside organelles. Today, in biochemistry textbooks, the general mechanism for the introduction of disulfide bonds into eukaryotic proteins is described for the endoplasmic reticulum (Woycechowsky and Raines 2000), but the respective enzymes are restricted to this cellular compartment (Pollard et al. 1998; Gerber et al. 2001) and are not found in organelles. So far, the problem of disulfide bridge formation in proteins after translocation across membranes has not been investigated systematically for organelles like mitochondria and plastids.

The evolutionary relationship of mitochondria and plastids to eubacteria (Gray et al. 1999) makes it worthwhile to take a preliminary look at the model organism *E. coli*. For this eubacterium the most detailed studies have been conducted for the periplasm that, in many aspects, serves as an analogous model for the intermembrane space of mitochondria. No counterparts of the sec-dependent major bacterial export system have been identified in mitochondria, but the yeast mitochondrial inner membrane peptidase (Imp), consisting of the two subunits Imp1 and Imp2, is homologous to the *E. coli* signal peptidase (Lep), an essential component of the bacterial protein export pathway (Pratje et al. 1994). Export of proteins from the cytoplasm of *E. coli* into

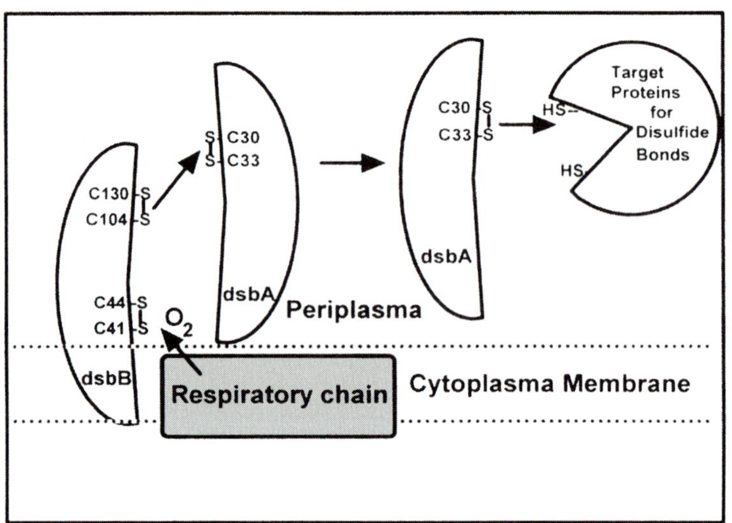

Fig. 4. Model for disulfide bond formation in the periplasm of *E. coli*. The export of proteins into the periplasm of *E. coli* serves as a model for the re-export of proteins from the matrix into the intermembrane space of mitochondria. In *E. coli* periplasmic proteins with functional disulfide bridges are modified by the dsbA/dsbB proteins. Target proteins with free cysteine residues that have to be transformed into a disulfide bridge interact with dsbA. The internal disulfide bridge between the cysteine residues *C30* and *C33* of *dsbA* is transferred to the target protein. Regeneration of the internal disulfide bond of dsbA is facilitated by close interaction with the C104–C130 cysteine pair of dsbB that is integrated into the cytoplasm membrane. Intramolecular transfer reactions in *dsbB* between C41–C44 and C104–C130 regenerate the C104–C130 disulfide bond. Latest research identifies a close physical interaction of dsbB with the respiratory chain. Free oxygen (O_2) from the respiratory chain keeps the C41–C44 cysteine pair in the oxidized state

the periplasm is also coupled with unfolding of the secreted proteins (Raina and Missiakas 1997; Bader et al. 1999; Kobayashi and Ito 1999). The identification of functional disulfide bridges in secreted bacterial proteins is important to address this problem. In a genetic approach, new mutants for disulfide bridge formation in the periplasm have been isolated (Raina and Missiakas 1997). The analysis of these mutants resulted in the identification of the dsbA and dsbB proteins (dsb: disulfide bond forming enzymes). The current model for the *E. coli* dsb system in the periplasm is illustrated in Fig. 4.

One interesting point is the close association of this redox system with the respiratory chain. The accumulation of oxygen in the neighborhood of the cytochrome oxidase is essential to keep the internal disulfide bridge of dsbB in a permanently oxidized state (Bader et al. 1999; Kobayashi and Ito 1999). In contrast to the bacterial lep signal peptidase, it has not been possible to identify homologous proteins for dsbA or dsbB in the organelles of eukaryotes. The complete molecular data currently available from the genome projects of model organisms like yeast, humans and *Arabidopsis thaliana* demonstrate that in lower and higher eukaryotes other components must have been adapted to this function.

The first important point for the investigation of this problem in eukaryotes is therefore the identification of possible target proteins for disulfide bridge formation in organelles and to determine what kind of different functional disulfide bridges are found in proteins of organellar compartments. Today, we are just at the beginning of the research for identification of newly formed disulfide bridges after protein import into organelles, but some functionally important disulfide bonds have already been described. For plant mitochondria, a first systematic approach for the identification of important functional sulfhydryl groups for the protein import into mitochondria has been conducted (von Stedingk et al. 1997). These studies identified the importance of thiol groups for the import machinery of the outer mitochondrial membrane. In addition, selected data from other organellar proteins are available. Because these data are still very incomplete, we have summarized the identified organellar proteins with functional disulfide bonds from all

Table 2. Examples for diverse functions of disulfide bridges in organellar proteins

Protein	Localization	Function of disulfide bond	Reference(s)
Rieske/sulfur protein	Cytochrome bc_1 complex	Internal disulfide bond	Tian et al. (1999)
Subunit 4+6 of ATPase	Inner membrane of mitochondria	Linkage of subunits	Velours et al. (1998)
Erv1p	Intermembrane space of mitochondria	Dimerization	Lee et al. (2000)
Ubiquitin tag of sperm mitochondria	Outer membrane of sperm mitochondria	Masking of the ubiquitin tag for degradation	Sutovsky et al. (1999, 2000)
Alternative oxidase	Inner membrane of mitochondria	Dimerization, regulation of activity	Hoefnagel and Wiskich (1998); McCabe et al. (1998)
γ-Subunit of ATPase	Thylakoid membrane of plastids	Internal bond, thiol modulation of activity	Evron and McCarty (2000); Konno et al. (2000)
Apo Cytf	Thylakoid membrane of plastids	Attachment of cytochrome	Fabianek et al. (1997, 1998); Page and Ferguson (1997)
urf13p	Inner membrane of plant CMS mitochondria	Oligomerization of pore complex in CMS plants	Rhoads et al. (1998)

available eukaryotic organisms in Table 2. This table gives a short summary of the respective proteins, their localization, and the different types of functional disulfide bridges found in mitochondria or plastids.

For plastids, the comparison with the genetic data from bacteria was most helpful to identify the first specific components for disulfide bridge formation in these organelles. It turned out that thioredoxin molecules are of special importance. A major advancement was the finding that the eubacterial system for the attachment of cytochrome c to proteins (Page and Ferguson 1997) has been adapted to the introduction of cytochromes for the formation of apo-cytochrome f of plastids (Fabianek et al. 1997, 1998; Monika et al. 1997). Therefore, today, the best characterized example in plants is the introduction of a functional disulfide bridge into apo-cytochrome f. The formation of this sulfur bond is coupled with the attachment of cytochrome to the protein (Nakamoto et al. 2000). Four protein components participate in the disulfide bridge transfer reaction (see Fig. 5).

The four plastid components are homologous to the bacterial Ccmg system for cytochrome c assembly (Nakamoto et al. 2000). On the stromal side of the thylakoid membrane a thioredoxin transfers a disulfide bridge to the CcdA protein. CcdA interacts with Hcf164 (Lennartz et al. 2001) and Hcf164 finally transfers the sulfur bond to apo-cytochrome f (for details, see Fig. 5). For other enzymatic pathways, neither the re-

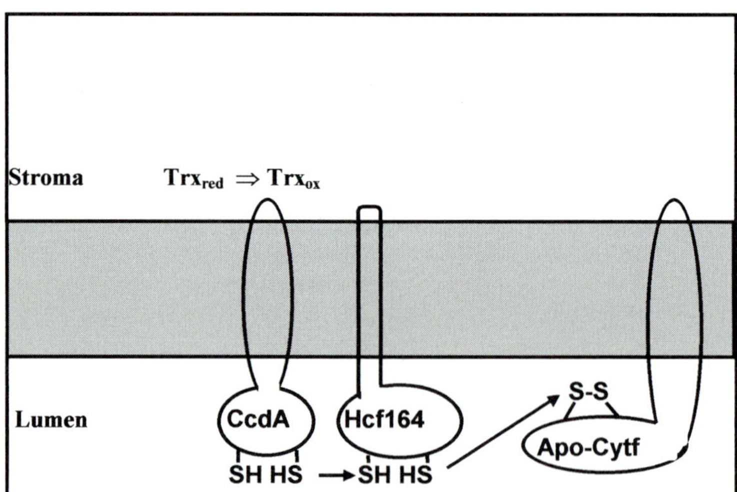

Fig. 5. Model for the introduction of a disulfide bridge into apo-cytf in plastids. A specialized thioredoxin (*Trx*) transfers a disulfide bridge to the *CcdA* protein. Thereby this thioredoxin changes from the reduced (red) form into the oxidized (ox) form. Later thioredoxin is regenerated in the stroma. The closed disulfide bridge of the *CcdA* protein interacts with *Hcf164* (Lennartz et al. 2001) and Hcf164 finally transfers the disulfide bridge to *apo-cytf*. This mechanism is essential for the integration of heme into the apo-cytf protein

spective enzymes nor the precise enzymatic mechanisms are currently known, although other plastid proteins have already been described to harbor functional disulfide bonds. The ATP synthase complex of plastids contains several functionally important disulfide bridges for the regulation of its activity (Evron and McCarty 2000; Konno et al. 2000); the alternative oxidase seems to be regulated by linkage of an internal cysteine pair (Hoefnagel and Wiskich 1998; McCarty et al. 1998), and the mitochondrial urf13 protein from cytoplasmic sterile maize mitochondria forms pore complexes via disulfide bridges (Rhoads et al. 1998). For mitochondria, it is remarkable that so far only proteins of the intermembrane space or the inner membrane have been found to contain functional sulfur bonds. This finding is in agreement with the current dogma that the highly reducing conditions in the mitochondrial matrix do not allow the formation of stable disulfide bridges between or in proteins. Consequently, it has not been possible to identify any mitochondrial matrix protein with functional disulfide bonds and therefore the current research concentrates on the investigation of mitochondrial intermembrane space proteins.

A major breakthrough for mitochondria was a genetic approach in the model organism yeast (Grivell et al. 1999) that identified a new redox protein localized in the mitochondrial intermembrane space (Lisowsky 1992; Lange et al. 2001). The nuclear gene *ERV1* (Essential for Respiration and Vegetative growth) was identified by complementation of a

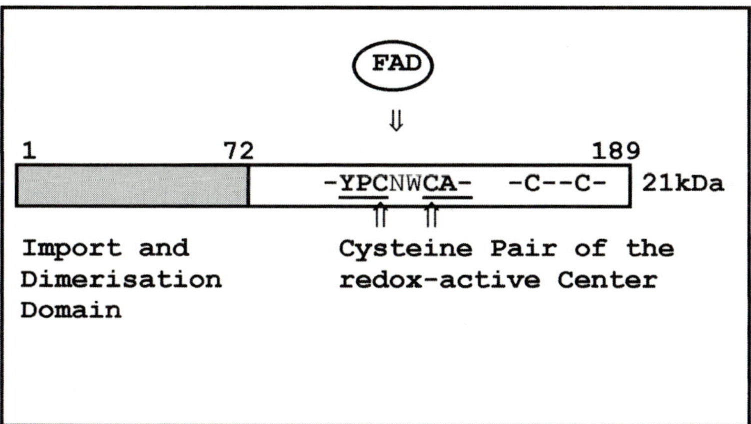

Fig. 6. Functional protein domains of the yeast Erv1p sulfhydryl oxidase. The small protein of 21 kDa contains 189 amino acids. The first 72 amino acids are essential for import into the mitochondrial intermembrane space and for dimerization. Covalent linkage of the dimers is facilitated by the two cysteines in this domain. Another cysteine pair in the carboxyl-terminal part of the protein forms the redox-active center. The consensus motif for Erv1p/Alrp enzymes is *underlined*. Each monomer of Erv1p binds one molecule of FAD next to the redox-active center. Another conserved cysteine pair with possible functional implications is located near the carboxyl terminus of the protein

conditional yeast mutant with a complex defect in oxidative phosphorylation and normal cell growth (Lisowsky 1992, 1994). Yeast Erv1p was found to be essential for the membrane morphology of mitochondria and the maintenance of intact organelles in the cell (Becher et al. 1999). Erv1p is a new sulfhydryl oxidase consisting of 189 amino acids that form characteristic functional domains (see Fig. 6; Lee et al. 2000).

Our latest research identifies Erv1p as a FAD-linked sulfhydryl oxidase of the new Erv1p/Alrp protein family (Lee et al. 2000; Lisowsky et al. 2001). These enzymes use FAD, O_2 and an internal redox-active cysteine pair (Chivers et al. 1997) to generate disulfide bridges in protein substrates. Sulfhydryl oxidases act as facile catalysts for the introduction of disulfide bonds into protein substrates outside the endoplasmic reticulum (Lisowsky 2001). Of general importance is the finding that very specialized sulfhydryl oxidases for mitochondrial biogenesis are present in all eukaryotes (Lisowsky et al. 1995; Polimeno et al. 1999; Lisowsky et al. 2001). The reversible introduction of disulfide bonds into mitochondrial proteins probably serves many different functions. According to the latest model (represented in Fig. 7), sulfhydryl oxidases could participate in shaping the membrane morphology of mitochondria (Yaffe 1999), the formation of contact sites between inner and outer membrane (Perkins et al. 1997), and in the modification of enzyme activities or receptor specificities (Nakamura et al. 1997). In addition, Erv1p was found to be essential for the export of Fe/S clusters from mitochondria to the cytosol (Lill et al. 1999; Lange et al. 2001). It has been reported that Erv1p attaches Fe/S clusters to special shuttle proteins for this transport.

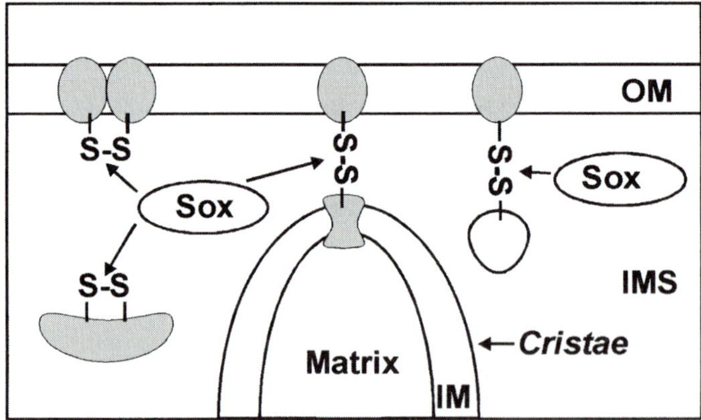

Fig. 7. Possible targets of a sulfhydryl oxidase (*Sox*) in the intermembrane space (*IMS*) of mitochondria. The formation of *cristae* is supported by defined contact points between the inner membrane (*IM*) and the outer membrane (*OM*). Modifications of the outer membrane can be facilitated by the attachment of other molecules or by the linkage of integral membrane proteins. The activity of soluble enzymes can be regulated by internal disulfide bonds

This demonstrates that sulfhydryl oxidases can fulfill a set of distinct functions in the mitochondrial intermembrane space and that mitochondria perform a lot more functions than the production of ATP (Zorov et al. 1997). The identification of additional target molecules for this enzymatic activity is a major goal of future research.

5 Summary and New Aspects

The identification of the importance of disulfide bridges in mitochondrial and plastid proteins adds new regulatory means to the nuclear control of organellar biogenesis and development. First genetic approaches characterized new nuclear genes whose gene products perform important functions in this process. It transpires that mitochondria and plastids adapted different specific mechanisms for this problem. In plastids the use of specialized thioredoxins seems to be the dominant mechanism whereas inside mitochondria FAD-linked sulfhydryl oxidases perform the most important tasks. The analysis of the diverse mechanisms for the formation of disulfide bonds in organelles is far from complete, and it can be expected that many additional diverse mechanisms and components have been recruited for this problem during evolution. A common feature for all organelles is the finding that the eubacterial dsbA/dsbB components have not been recruited for this function.

The reversible modification of thiol groups is especially attractive to explain complex developmental problems like the interactions of organellar membranes with the cytoskeleton or other membranes (Perkins et al. 1992; Yaffe 1999), the regulation of enzyme activities, the attachment of marker molecules and conformational changes of receptors (Lisowsky 2001). Therefore, much future work will be necessary to elucidate the precise molecular targets and mechanisms of the involved enzymes.

References

Asai T, Stone JM, Heard JE, Kovtun Y, Yorgey P, Sheen J, Ausubel FM (2000) Fumonisin B1-induced cell death in *Arabidopsis* protoplasts requires jasmonate-, ethylene-, and salicylate-dependent signaling pathways. Plant Cell 12:1823–1836

Bader M, Muse W, Ballou DP, Gassner C, Bardwell JC (1999) Oxidative protein folding is driven by the electron transport system. Cell 98:21727

Balk J, Leaver CJ (2001) The PET1-CMS mitochondrial mutation in sunflower is associated with premature programmed cell death and cytochrome c release. Plant Cell 8:1803–1818

Balk J, Leaver CJ, McCabe PF (1999) Translocation of cytochrome c from the mitochondria to the cytosol occurs during heat-induced programmed cell death in cucumber plants. FEBS Lett 463:151–154

Becher D, Kricke J, Stein G, Lisowsky T (1999) A mutant for the yeast *scERV1* gene displays a new defect in mitochondrial morphology and distribution. Yeast 15:1171–1181

Berger KH, Yaffe MP (2000) Mitochondrial DNA inheritance in *Saccharomyces cerevisiae.* Trends Microbiol 8:508–513

Binyamin L, Falah M, Portnoy V, Soudry E, Gepstein S (2001) The early light-induced protein is also produced during leaf senescence of *Nicotiana tabacum.* Planta 212:591–597

Boldogh IR, Yang HC, Pon LA (2001) Mitochondrial inheritance in budding yeast. Traffic 2:368–374

Bosca L, Hortelano S (1999) Mechanisms of nitric oxide-dependent apoptosis: involvement of mitochondrial mediators. Cell Signal 11:239–244

Bowe LM, Coat G, dePamphilis CW (2000) Phylogeny of seed plants based on all three genomic compartments: extant gymnosperms are monophyletic and Gnetales' closest relatives are conifers. Proc Natl Acad Sci USA 97:4092–4097

Burger G, Saint-Louis D, Gray MW, Lang BF (1999) Complete sequence of the mitochondrial DNA of the red alga *Porphyra purpurea.* Cyanobacterial introns and shared ancestry of red and green algae. Plant Cell 11:1675–1694

Chang MC, Ho YS, Lee PH, Chan CP, Lee JJ, Hahn LJ, Wang YJ, Jeng JH (2001) Areca nut extract and arecoline induced the cell cycle arrest but not apoptosis of cultured oral KB epithelial cells: association of glutathione, reactive oxygen species and mitochondrial membrane potential. Carcinogenesis 22:1527–1535

Chautan M, Chazal G, Cecconi F, Gruss P, Golstein P (1999) Interdigital cell death can occur through a necrotic and caspase-independent pathway. Curr Biol 9:967–970

Chaw SM, Parkinson CL, Cheng Y, Vincent TM, Palmer JD (2000) Seed plant phylogeny inferred from all three plant genomes: monophyly of extant gymnosperms and origin of *Gnetales* from conifers. Proc Natl Acad Sci USA 97:4086–4091

Chivers PT, Prehoda KE, Raines RT (1997) The CXXC motif: a rheostat in the active site. Biochemistry 36:4061–4066

Conley CA, Hanson MR (1995) How do alterations in plant mitochondrial genomes disrupt pollen development? J Bioenerg Biomembr 27:447–457

del Pozo O, Lam E (1998) Caspases and programmed cell death in the hypersensitive response of plants to pathogens. Curr Biol 8:1129–1132

Denovan-Wright EM, Nedelcu AM, Lee RW (1998) Complete sequence of the mitochondrial DNA of *Chlamydomonas eugametos.* Plant Mol Biol 36:285–295

Engelberg-Kulka H, Glaser G (1999) Addiction modules and programmed cell death and antideath in bacterial cultures. Annu Rev Microbiol 53:43–70

Entelis NS, Kolesnikova OA, Martin RP, Tarassov IA (2001) RNA delivery into mitochondria. Adv Drug Deliv Rev 49:199–215

Esser K, Lisowsky T, Michaelis G, Pratje E (2001) Extranuclear inheritance: genetics and biogenesis of mitochondria. In: Behnke H-D, Lüttge U, Esser K, Kadereit JW, Runge M (eds) Progress in botany, vol 62. Springer, Berlin Heidelberg New York, pp 71–97

Evron Y, McCarty-RE (2000) Simultaneous measurement of delta pH and electron transport in chloroplast thylakoids by 9-aminoacridine fluorescence. Plant Physiol 124:407–414

Fabianek RA, Huber-Wunderlich M, Glockshuber R, Künzler P, Hennecke H, Thöny-Meyer L (1997) Characterization of the *Bradyrhizobium japanicum* CycY protein, a membrane-anchored periplasmic thioredoxin that may play a role as a reductant in the biogenesis of c-type cytochromes. J Biol Chem 272:4467–4473

Fabianek RA, Hennecke H, Thöny-Meyer L (1998) The active-site cysteines of the periplasmic thioredoxin-like protein Ccmg of *Escherichia coli* are important but not essential for cytochrome c maturation in vivo. J Bacteriol 180:1947–1950

Fröhlich KU, Madeo F (2000) Apoptosis in yeast – a monocellular organism exhibits altruistic behaviour. FEBS Lett 473:6–9

Gerber J, Mühlenhoff U, Hofhaus G, Lill R, Lisowsky T (2001) Yeast Erv2p is the first microsomal FAD-linked sulfhydryl oxidase of the Ervlp/Alrp protein family. J Biol Chem 276:23486–23491

Giegé P, Brennicke A (1999) RNA editing in *Arabidopsis* mitochondria effects 441 C to U changes in ORFs. Proc Natl Acad Sci USA 96:15324–15329

Giegé P, Brennicke A (2001) From gene to protein in higher plant mitochondria. CR Acad Sci III 324:209–217

Gietl C, Schmid M (2001) Ricinosomes: an organelle for developmentally regulated programmed cell death in senescing plant tissues. Naturwissenschaften 88:49–58

Gilson PR, Beech PL (2001) Cell division protein FtsZ: running rings around bacteria, chloroplasts and mitochondria. Res Microbiol 152:3–10

Gray MW, Burger G, Lang BF (1999) Mitochondrial evolution. Science 283:1476–1481

Green DR, Reed JC (1998) Mitochondria and apoptosis. Science 281:1309–1312

Grivell LA, Artal-Sanz M, Hakkaart G, de-Jong L, Nijtmans LGJ, van-Oosterum K, Siep M, van-der-Spek H (1999) Mitochondrial assembly in yeast. FEBS Lett 452:57–60

Hayashi Y, Yamada K, Shimada T, Matsushima R, Nishizawa NK, Nishimura M, Hara-Nishimura I (2001) A proteinase-storing body that prepares for cell death or stresses in the epidermal cells of *Arabidopsis*. Plant Cell Physiol 42:894–899

Hayashi-Ishimaru Y, Ohama T, Kawatsu Y, Nakamura K, Osawa S (1996) UAG is a sense codon in several chlorophycean mitochondria. Curr Genet 30:29–33

Heath MC (2000) Hypersensitive response-related death. Plant Mol Biol 44:321–334

Hoefnagel MHN, Wiskich JT (1998) Activation of the plant alternative oxidase by high reduction levels of the Q-pool and pyruvate. Arch Biochem Biophys 355:262–270

Jones A (2000) Does the plant mitochondrion integrate cellular stress and regulate programmed cell death? Trends Plant Sci 5:225–230

Kawahara A, Ohsawa Y, Matsumura H, Uchiyama Y, Nagata S (1998) Caspase-independent cell killing by Fas-associated protein with death domain. J Cell Biol 143:1353–1360

Kennedy DO, Kojima A, Yano Y, Hasuma T, Otani S, Matsui-Yuasa I (2001) Growth inhibitory effect of green tea extract in Ehrlich ascites tumor cells involves cytochrome c release and caspase activation. Cancer Lett 166:9–15

Kobayashi T, Ito K (1999) Respiratory chain strongly oxidizes the CXXC motif of DsbB in the *Escherichia coli* disulfide bond formation pathway. EMBO J 18:1192–1198

Konno H, Yodogawa M, Stumpp MT, Kroth P, Strotmann H, Motohashi K, Amano T, Hisabori T (2000) Inverse regulation of F-1-ATPase activity by a mutation at the regulatory region on the gamma subunit of chloroplast ATP synthase. Biochem J 352:783–788

Kroymann J, Zetsche K (1998) The mitochondrial genome of Chlorogonium elongatum inferred from the complete sequence. J Mol Evol 47:431–440

Kubo T, Nishizawa S, Sugawara A, Itchoda N, Estiati A, Mikami T (2000) The complete nucleotide sequence of the mitochondrial genome of sugar beet (Beta vulgaris L.) reveals a novel gene for tRNA(Cys)(GCA). Nucleic Acids Res 28:2571–2576

Kück U, Jekosch K, Holzhamer P (2000) DNA sequence analysis of the complete mitochondrial genome of the green alga *Scenedesmus obliquus*: evidence for UAG being a leucine and UCA being a non-sense codon Gene 253:13–18

Lacomme C, Santa Cruz S (1999) Bax-induced cell death in tobacco is similar to the hypersensitive response. Proc Natl Acad Sci USA 96:7956–7961

Laloi M (1999) Plant mitochondrial carriers: an overview. Cell Mol Life Sci 56:918–944

Lam E, del Pozo O (2000) Caspase-like protease involvement in the control of plant cell death. Plant Mol Biol 44:417–428

Lam E, Pontier D, del Pozo O (1999) Die and let live – programmed cell death in plants. Curr Opin Plant Biol 2:502–507

Lange H, Lisowsky T, Gerber J, Mühlenhoff U, Gyula K, Lill R (2001) Essential function of the mitochondrial sulfhydryl oxidase Erv1p/Alrp in the maturation of cytosolic Fe/S proteins. EMBO Rep 2:715–720

Leblanc C, Boyen C, Richard O, Bonnard G, Grienenberger JM, Kloareg B (1995) Complete sequence of the mitochondrial DNA of the rhodophyte *Chondrus crispus* (Gigartinales). Gene content and genome organization. J Mol Biol 250:484–495

Lee J-E, Hofhaus G, Lisowsky T (2000) Erv1p from *Saccharomyces cerevisiae* is a FAD-linked sulfhydryl oxidase. FEBS Lett 477:62–66

Lennartz K, Plücken H, Seidler A, Westhoff P, Bechtold N, Meierhoff K (2001) HCF164 encodes a thioredoxin-like protein involved in the biogenesis of the cytochrome b(6)f complex in *Arabidopsis*. Plant Cell 11:2539–2551

Li HC, Yashiki S, Sonoda J, Lou H, Ghosh SK, Byrnes JJ, Lema C, Fujiyoshi T, Karasuyama M, Sonoda S (2000) Green tea polyphenols induce apoptosis in vitro in peripheral blood T lymphocytes of adult T-cell leukemia patients. Jpn J Cancer Res 91:34–40

Ligr M, Velten I, Fröhlich E, Madeo F, Ledig M, Fröhlich KU, Wolf DH, Hilt W (2001) The proteasomal substrate stm1 participates in apoptosis-like cell death in yeast. Mol Biol Cell 12:2422–2432

Lill R, Neupert W (1996) Mechanisms of protein import across the mitochondrial outer membrane. Trends Cell Biol 6:56–61

Lill R, Diekert K, Kaut A, Lange H, Pelzer W, Prohl C, Kispal G (1999) The essential role of mitochondria in the biogenesis of cellular iron-sulfur proteins. Biol Chem 380:1157–1166

Lisowsky T (1992) Dual function of a new nuclear gene for oxidative phosphorylation and vegetative growth in yeast. Mol Gen Genet 232:58–64

Lisowsky T (1994) *ERV1* is involved in the cell-division cycle and the maintenance of mitochondrial genomes in *Saccharomyces cerevisiae.* Curr Genet 26:15–20

Lisowsky T (2001) Sulfhydryl oxidases and the genetics of mitochondrial biogenesis. Rec Res Dev Curr Genet 1:1–10

Lisowsky T, Weinstat Saslow DL, Barton N, Reeders ST, Schneider MC (1995) A new human gene located in the PKD1 region of chromosome 16 is a functional homologue to *scERV1* of yeast. Genomics 1995 29:690–697

Lisowsky T, Esser K, Stein T, Pratje E, Michaelis G (1999) Extranuclear inheritance: genetics and biogenesis of mitochondria. In: Progress in botany, vol 60. Springer, Berlin Heidelberg New York, pp 99–118

Lisowsky T, Lee J-E, Polimeno L, Francavilla A, Hofhaus G (2001) Mammalian augmenter of liver regeneration (ALR) protein is a sulfhydryl oxidase. Dig Liv Dis 33:173–180

Lithgow T, Glick BS, Schatz G (1995) The protein receptor of mitochondria. Trends Biochem Sci 20:98–101

Longo VD, Ellerby LM, Bredesen DE, Valentine JS, Gralla EB (1997) Human Bcl-2 reverses survival defects in yeast lacking superoxide dismutase and delays death of wild-type yeast. J Cell Biol 137:1581–1588

Madeo F, Fröhlich E, Ligr M, Grey M, Sigrist SJ, Wolf DH, Fröhlich KU (1999) Oxygen stress: a regulator of apoptosis in yeast. J Cell Biol 145:757–767

Matouschek A, Pfanner N, Voos W (2000) Protein unfolding by mitochondria. The Hsp70 import motor. EMBO Rep 1:404–410

Maxwell DP, Wang Y, McIntosh L (1999) The alternative oxidase lowers mitochondrial reactive oxygen production in plant cells. Proc Natl Acad Sci USA 96:8271–8276

McCabe TC, Finnegan PM, Harvey Millar A, Day DA, Whelan J (1998) Differential expression of alternative oxidase genes in soybean cotyledons during postgerminate development. Plant Physiol 118:675–682

McCarthy NJ, Whyte MK, Gilbert CS, Evan GI (1997) Inhibition of Ced-3/ICE-related proteases does not prevent cell death induced by oncogenes, DNA damage, or the Bcl-2 homologue Bak. J Cell Biol 136:215–227

Mehta RA, Fawcett TW, Porath D, Mattoo AK (1992) Oxidative stress causes rapid membrane translocation and in vivo degradation of ribulose-1,5-bisphosphate carboxylase/oxygenase. J Biol Chem 267:2810–2816

Michaelis G, Vahrenholz C, Pratje E (1990) Mitochondrial DNA of *Chlamydomonas reinhardtii*: the gene for apocytochrome b and the complete functional map of the 15.8 kb DNA. Mol Gen Genet 223:211–216

Mignotte B, Vayssiere JL (1998) Mitochondria and apoptosis. Eur J Biochem 252:1–15

Mitsuhara I, Malik KA, Miura M, Ohashi Y (1999) Animal cell-death suppressors Bcl-x(L) and Ced-9 inhibit cell death in tobacco plants. Curr Biol 9:775–778

Mittler R, Lam E (1996) Sacrifice in the face of foes: pathogen-induced programmed cell death in plants. Trends Microbiol 4:10–15

Moller IM (2001) Plant mitochondria and oxidative stress, electron transport, NADPH turnover, and metabolism of reactive oxygen species. Annu Rev Plant Physiol Plant Mol Biol 52:561–591

Monika EM, Goldman BS, Beckman DL, Kranz RG (1997) A thioreduction pathway tethered to the membrane for periplasmic cytochrome c biogenesis, in vitro and in vivo studies. J Mol Biol 271:679–692

Nakamoto SS, Hamel P, Merchant S (2000) Assembly of chloroplast cytochromes b and c. Biochimie 82:603–614

Nakamura H, Nakamura K, Yodi J (1997) Redox regulation of cellular activation. Ann Rev Immunol 15:351–369

Navarre DA, Wolpert TJ (1999) Victorin induction of an apoptotic/senescence-like response in oats. Plant Cell 11:237–249

Nedelcu AM, Lee RW, Lemieux C, Gray MW, Burger G (2000) The complete mitochondrial DNA sequence of *Scenedesmus obliquus* reflects an intermediate stage in the evolution of the green algal mitochondrial genome. Genome Res 10:819–831

Noctor G, Foyer CH (1998) Ascorbate and gluthathione: keeping active oxygen under control. Annu Rev Plant Physiol Plant Mol Biol 49:249–279

Ochs G, Schock G, Trischler M, Kosemund K, Wild A (1999) Complexity and expression of the glutamine synthetase multigene family in the amphidiploid crop *Brassica napus*. Plant Mol Biol 39:395–405

Otha N, Sato N, Kuroiwa T (1998) Structure and organization of the mitochondrial genome of the unicellular red alga (*Cyanidioschyzon merolae*) deduced from the complete nucleotide sequence. Nucleic Acids Res 26:5190–5298

Page MD, Ferguson SJ (1997) *Paracoccus denitrificans* CcmG is a periplasmic protein-disulfide oxidoreductase required for c- and aa3-type cytochrome biogenesis; evidence for reductase role in vivo. Mol Microbiol 24:977–990

Park HI, Jeong MH, Lim YJ, Park BS, Kim GC, Lee YM, Kim HM, Yoo KS, Yoo YH (2001) *Szygium aromaticum* (L.) Merr. et Perry (Myrtaceae) flower bud induces apoptosis of p815 mastocytoma cell line. Life Sci 69:553–566

Perkins G, Renken C, Martone ME, Young SJ, Ellisman M, Frey T (1997) Electron tomography of neuronal mitochondria: three-dimensional structure and organization of cristae and membrane contacts. Struct Biol 119:260–272

Pfanner N, Geissler A (2001) Versatility of the mitochondrial protein import machinery. Nat Rev Mol Cell Biol 2:339–349

Pfanner N, Meijer M (1995) Pulling in the proteins. Curr Biol 5:132–125

Polimeno L, Lisowsky T, Francavilla A (1999) The *ALR/ERV1* genes: from yeast to man – from mitochondria to liver regeneration: a new essential gene family. Ital J Gastro Hepatol 31:494–500

Pollard MG, Travers KJ, Weissman JS (1998) Ero1p: a novel and ubiquitous protein with an essential role in oxidative protein folding in the endoplasmic reticulum. Mol Cell 1:171–182

Pratje E, Esser K, Michaelis G (1994) The mitochondrial inner membrane peptidase. In: von Heijne G (ed) Signal peptidases. Landes Company, Austin, pp 105–112

Raina S, Missiakas D (1997) Making and breaking of disulfide bonds. Annu Rev Microbiol 51:179–202

Rhoads DM, Brunner-Neuenschwander B, Levings CS III, Siedow JN (1998) Cross-linking and disulfide bond formation of introduced cysteine residues suggest a modified model for the tertiary structure of URF13 in the pore-forming oligomers. Arch Biochem Biophys 354:158–164

Saccone C, Gissi C, Lanave C, Larizza A, Pesole G, Reyes A (2000) Evolution of the mitochondrial genetic system: an overview. Gene 261:153–159

Schmid M, Simpson DJ, Sarioglu H, Lottspeich F, Gietl C (2001) The ricinosomes of senescing plant tissue bud from the endoplasmic reticulum. Proc Natl Acad Sci USA 98:5353–8535

Schneider A, Marechal-Drouard L (2000) Mitochondrial tRNA import: are there distinct mechanisms? Trends Cell Biol 10:509–513

Seo HJ, Surh YJ (2001) Eupatilin, a pharmacologically active flavone derived from Artemisia plants, induces apoptosis in human promyelocytic leukemia cells. Mutat Res 496:191–198

Soll J, Tien R (1998) Protein transport into and across the chloroplastic envelope membranes. Plant Mol Biol 38:191–207

Stiller JW, Riley J, Hall BD (2001) Are red algae plants? A critical evaluation of three key molecular data sets. J Mol Evol 52:527–39

Sugiyama M, Ito J, Aoyagi S, Fukuda H (2000) Endonucleases. Plant Mol Biol 44:387–397

Sutovsky P, Moreno RD, Ramalho-Santos J, Dominko T, Simerly C, Schatten G (1999) Ubiquitin tag for sperm mitochondria. Nature 402:371

Sutovsky P, Moreno RD, Ramalho-Santos J, Dominko T, Simerly C, Schatten G (2000) Ubiquitinated sperm mitochondria, selective proteolysis, and the regulation of mitochondrial inheritance in mammalian embryos. Biol Reprod 63:582–590

Tian H, White S, Yu L, Yu CA (1999) Evidence for the head domain movement of the Rieske iron-sulfur protein in electron transfer reaction of the cytochrome bc(1) complex. J Biol Chem 274:7146–7152

Tinhofer I, Bernhard D, Senfter M, Anether G, Loeffler M, Kroemer G, Kofler R, Csordas A, Greil R (2001) Resveratrol, a tumor-suppressive compound from grapes, induces apoptosis via a novel mitochondrial pathway controlled by Bcl-2. FASEB J 15:1613–1615

Truscott KN, Pfanner N, Voos W (2001) Transport of proteins into mitochondria. Rev Physiol Biochem Pharmacol 143:81–136

Turmel M, Lemieux C, Burger G, Lang BF, Otis C, Plante I, Gray MW (1999) The complete mitochondrial DNA sequences of *Nephroselmis olivacea* and Pedinomonas minor. Two radically different evolutionary patterns within green algae. Plant Cell 11:1717–1730

Unseld M, Marienfeld JR, Brandt P, Brennicke A (1997) The mitochondrial genome of *Arabidopsis thaliana* contains 57 genes in 366,924 nucleotides. Nat Genet 15:57–61

Vahrenholz C, Riemen G, Pratje E, Dujon B, Michaelis G (1993) Mitochondrial DNA of *Chlamydomonas reinhardtii*: the structure of the ends of the linear 15.8-kb genome suggests mechanisms for DNA replication. Curr Genet 24:241–247

Van Dyck L, Langer T (1999) ATP-dependent proteases controlling mitochondrial function in the yeast *Saccharomyces cerevisiae*. Cell Mol Life Sci 56:825–842

Velours J, Spannagel C, Chaignepain S, Vaillier J, Arselin G, Graves PV, Velours G, Camougrand N (1998) Topography of the yeast ATP synthase F-0 sector. Biochimie 80:793–801

Vercammen D, Beyaert R, Denecker G, Goossens V, Van Loo G, Declercq W, Grooten J, Fiers W, Vandenabeele P (1998) Inhibition of caspases increases the sensitivity of L929 cells to necrosis mediated by tumor necrosis factor. J Exp Med 187:1477–1485

Viehmann S, Richard O, Boyen C, Zetsche K (1996) Genes for two subunits of succinate dehydrogenase from a cluster on the mitochondrial genome of Rhodophyta. Curr Genet 29:199–201

von Stedingk EM, Pavlov PF, Grinkevich VA, Glaser E (1997) Mitochondrial protein import: modification of sulfhydryl groups of the inner mitochondrial membrane import machinery in *Solanum tuberosum* inhibits protein import. Plant Mol Biol 35:809–820

Wachter C, Schatz G, Glick BS (1994) Protein import into mitochondria: the requirement for external ATP is precursor-specific whereas intramitochondrial ATP is universally needed for translocation into the matrix. Mol Biol Cell 5:465–474

Wen L-Y, Chase CD (1999) Mitochondrial gene expression in developing male gametophytes of male-fertile and S male-sterile maize. Sex Plant Reprod 11:323–330

Wise RP, Dill CL, Schnable PS (1996) Mutator-induced mutations of the rf1 nuclear fertility restorer of T-cytoplasm maize alter the accumulation of T-urf13 mitochondrial transcripts. Genetics 143:1383–1394

Wolff G, Plante I, Lang BF, Kück U, Burger G (1994) Complete sequence of the mitochondrial DNA of the chlorophyte alga *Prototheca wickerhamii*. Gene content and genome organization. J Mol Biol 237:75–86

Woycechowsky KJ, Raines RT (2000) Native disulfide bond formation in proteins. Curr Opin Chem Biol 4:533–539

Xie Z, Chen Z (1999) Salicylic acid induces rapid inhibition of mitochondrial electron transport and oxidative phosphorylation in tobacco cells. Plant Physiol 120:217–226

Xie Z, Chen Z (2000) Harpin-induced hypersensitive cell death is associated with altered mitochondrial functions in tobacco cells. Mol Plant Microbe Interact 13:183–190

Yaffe MP (1999) The machinery of mitochondrial inheritance and behavior. Science 283:1493–1497

Yoon Y, McNiven MA (2001) Mitochondrial division: new partners in membrane pinching. Curr Biol 11:67–70

Yoshida H, Kong YY, Yoshida R, Elia AJ, Hakem A, Hakem R, Penninger JM, Mak TW (1998) Apaf1 is required for mitochondrial pathways of apoptosis and brain development. Cell 94:739–750

Zorov DV, Krasnikov BF, Kuzminova AE, Vysokikh MYu, Zorova LD (1997) Mitochondria revisited. Alternative functions of mitochondria. Biosci Rep 17:507–520

Thomas Lisowsky
Karlheinz Esser
Martin Ingenhoven
Georg Michaelis
Botanisches Institut der Universität Düsseldorf
Universitätsstraße 1
40225 Düsseldorf, Germany

e-mail: michaelg@uni-duesseldorf.de

Elke Pratje
Institut für Allgemeine Botanik
Universität Hamburg,
Ohnhorststraße 18
22609 Hamburg, Germany

Genetics of Phytopathogenic Bacteria

Annette Burger and Rudolf Eichenlaub

1 Introduction

During the last 15 years, molecular genetics has generated a vast body of information on the mechanisms of the interactions between phytopathogenic bacteria and their host plants. In our last review on this subject (Ahlemeyer and Eichenlaub 2001), we focused on the Gram-negative bacteria where the understanding of the plant–microbe interaction has reached a relatively advanced stage. This is of course mainly due to the fact that Gram-negative bacteria have attracted the attention of numerous research groups, since they are generally easier to handle and genetic methods for an in-depth analysis are available. Also, there are more Gram-negative plant pathogenic bacteria known than Gram-positive. However, among the Gram-positive bacteria there are a number of agriculturally quite important plant pathogens some of which have been studied genetically to some extent.

2 Phytopathogenic Gram-Positive Bacteria

It is not the goal of this review to elaborate on all known Gram-positive phytopathogenic bacteria or to describe the various diseases they cause on different host plants. Instead, we will focus only on those bacteria for which genetic methods and tools, like transformation, cloning vectors, transposons etc., are available and for which pathogenicity genes can be identified and the mechanisms for induction of plant disease can be elucidated. This reduces the number of bacteria to *Clavibacter michiganensis*, *Leifsonia xyli*, *Streptomyces scabies*, and *Rhodococcus fascians*. All of these bacteria have a high G+C content of their DNA and belong to the *Actinomycetales*. The four genera have also been recently treated in some detail in the following review articles, which are recommended for further information: *Clavibacter* (Haapalainen et al. 1997; Metzler et al. 1997; Jahr et al. 1999), *Streptomyces* (Loria et al. 1997), *Rhodococcus fascians* (Goethals et al. 2001). *Leifsonia xyli*, formerly (until 1999)

Progress in Botany, Vol. 64

Clavibacter xyli, is treated in the reviews on *Clavibacter* (Haapalainen et al. 1997; Metzler et al. 1997).

3 *Clavibacter michiganensis*

Clavibacter is a genus of plant-pathogenic actinomycetes belonging to the family *Microbacteriaceae* (Stackebrand et al. 1997).

The former subspecies *Clavibacter xyli* ssp. *xyli* and *Clavibacter xyli* ssp. *cynodontis* are now reclassified as members of the new genus *Leifsonia* (Suzuki et al. 1999; Evtuschenko et al. 2000). *Rathayibacter tritici*, *R. iranicus* and *R. toxicus* were also formerly grouped in the genus *Clavibacter* but constitute a separate genus (Sasaki et al. 1998).

Currently, the genus *Clavibacter* is only represented by one species, *Clavibacter michiganensis*, with five subspecies that are all phytopathogenic. *Clavibacter michiganensis* ssp. *sepedonicus* induces ring rot of potato (*Solanum tuberosum*; Manzer and Genereux 1981), while *Clavibacter michiganensis* ssp. *nebraskensis* causes wilt and blight of maize (*Zea mays*; Schuster 1975). *Clavibacter michiganensis* ssp. *tesselarius* induces leaf freckles and spotting in wheat (*Triticum aestivum*; Carlson et al. 1982). *Clavibacter michiganensis* ssp. *insidiosus* is responsible for wilting and stunting of alfalfa (*Medicago sativa*; McCulloch 1925). Last, but not least, *Clavibacter michiganensis* ssp. *michiganensis* causes bacterial wilt and canker of tomato (*Lycopersicon esculentum*; Strider 1969; Davis et al. 1984).

Since the genetics of *Clavibacter michiganensis* ssp. *michiganensis* (*Cmm*) is more advanced than in the other subspecies, we will first describe recent findings on pathogenicity genes of this bacterium. As seen from Table 1, *Cmm* causes bacterial wilt and canker of tomato. This pathogen is a target of international quarantine regulations since bacterial canker of tomato causes severe crop failures and losses of up to 60% in all main cultivation areas (Strider 1969; Shirakawa et al. 1991), and neither resistant tomato cultivars nor effective chemical controls of this

Table 1. Phytopathogenic subspecies of *Clavibacter michiganensis*

Subspecies	Host plant	Disease
C. m. ssp. *michiganensis*	Tomato (*Lycopersicon esculentum*)	Wilt and canker
C. m. ssp. *sepedonicus*	Potato (*Solanum tuberosum*)	Wilt and ring rot of tuber
C. m. ssp. *nebraskensis*	Maize (*Zea mays*)	Wilt and blight
C.m. ssp. *insidiosus*	Alfalfa (*Medicago sativa*)	Wilt and stunting
C. m. ssp. *tesselarius*	Wheat (*Triticum aestivum*)	Leaf freckles and spots

pathogen are available (Thompson 1986). *Cmm* is generally transmitted by contaminated seeds or transplants (Fatmi and Schaad 1988; Gitaitis et al. 1991). Therefore, the only means for biocontrol is to prevent the spreading of the pathogen. This requires early detection and elimination of infected plant material. Consequently, there is great demand for reliable diagnostic methods to detect this important pathogen.

The bacteria infect the host plant via wounds and invade the xylem vessels, which is followed by systemic infection of the host. To explain the mechanisms for the induction of bacterial wilt several hypotheses have been presented. Exopolysaccharides (EPS) and glycoproteins produced by the bacteria were inferred to hinder water transport by plugging the xylem vessels in an in vitro assay on tomato seedlings (Rai and Strobel 1968a,b; Van den Bulk et al. 1989). However, analysis of the EPS of various *Cmm* strains differing in virulence and the pathogenic phenotype of EPS-negative mutants indicated that EPS is not a crucial pathogenicity factor in *Cmm* (Bermpohl et al. 1996).

It was also suggested that wilting might result from an enzymatic attack on xylem vessels and adjacent parenchymous cells (Wallis 1977; Benhamou 1991). Work from several laboratories has claimed that extracellular cell-wall-degrading enzymes have an important role in disease development. An endocellulase (Meletzus et al. 1993), polygalacturonase (Beimen et al. 1992), pectinmethylesterase (Strider 1969), and xylanase (Beimen et al. 1992) were identified in vitro and in vivo in *Cmm*. It is tempting to speculate that the macerating activity of such extracellular enzymes on plant cells is the reason for wilt development. Furthermore, a combined action with other virulence genes active on cell-wall components and/or genes involved in other mechanisms of pathogenic action also have to be considered.

An answer as to whether extracellular enzymes play a role in pathogenicity of *Cmm* was possible after the development of cloning vectors and a transformation system (Meletzus and Eichenlaub 1991; Laine et al. 1996). In the course of these experiments it was found that in *Cmm* strain NCPPB382 the genes responsible for wilting of infected tomato plants are located on two plasmids, pCM1 and pCM2 (Meletzus et al. 1993; Dreier et al. 1995, 1997). Curing of both plasmids converted the virulent strain NCPPB382 to an endophyte, designated CMM100, which effectively colonizes the host plant, but does not cause the development of disease symptoms. CMM100 was then used as a bacterial host strain for the cloning of the wilt-inducing genes mapping on plasmids pCM1 and pCM2. Analysis of plasmid pCM1 revealed that it carries the gene *cel*A, encoding an endo-β-1,4-glucanase, which constitutes a pathogenic determinant, responsible for wilt induction of infected tomato plants. This was confirmed by complementation of a *celA*$^-$ mutant, generated by insertion of an antibiotic cassette into the *celA* gene on pCM1, or strain CMM100 cured for pCM1. In both cases, transformation with vector pDM 302 carrying the wild-type *celA* gene restored endocellulase activity and the pathogenic phenotype.

The *celA* gene encodes a protein of 78 kDa and has a three-domain structure. The first 51 amino acids resemble a signal sequence with a putative processing site and thus the enzyme is most probably secreted. As typical for cellulases, the catalytic domain of about 350 amino acids is connected by a 27 amino acid linker region primarily consisting of serine and proline residues to a type IIa bacterial cellulose-binding domain of 84 amino acids. While the catalytic and the cellulose-binding domains exhibit extensive homologies to bacterial endocellulases, the third domain, which is connected to the cellulose-binding domain by a 28 amino acid linker region, is somehow unique. The only weak homology showing up in databank searches was to α-expansins (Jahr et al. 2000), plant proteins mediating cell-wall extension (Cosgrove 1998). It is entirely possible that the three-domain structure of *celA* of *Cmm* is the basis for the fact that here a hydrolytic enzyme is a primary pathogenicity factor responsible for the development of disease symptoms.

The second plasmid of strain NCPPB382, pCM2, also carries a gene locus related to pathogenicity, termed *pat*-1 (Meletzus et al. 1993; Dreier et al. 1997). The region of plasmid pCM2 encoding the pathogenicity locus *pat*-1 was mapped by deletion analysis and complementation studies to a 1.5-kb *Bgl*II/*Sma*I DNA fragment. Cloning of the *pat*-1 region in the endophytic, plasmid-free isolate CMM100 converted the recombinant strain into a virulent pathogen. Based on the nucleotide sequence of the *pat*-1 region, an open reading frame could be predicted, coding for a protein of 280 amino acids and 29.7 kDa. The first 33 amino acids resemble a signal sequence with a putative cleavage site between an alanine and a valine residue at positions 33 and 34, respectively. Furthermore, the Pat-1 protein shares homology with serine proteases from *Lysobacter enzymogenes* and *Streptomyces griseus.* At position 230 of the Pat-1 amino acid sequence, there exists a consensus G-D-S-G-G motif typical of the trypsin family of serine proteases, with serine at the active site. Therefore, it possible that the Pat-1 protein is a secreted protease. Although Northern hybridizations identified a 1.5-knt transcript of the *pat*-1 structural gene, the Pat-1 protein has not been purified yet nor has a proteolytic activity been demonstrated directly.

Downstream of the *pat*-1 structural gene, a peculiar repetitive sequence motif (*pat-1rep*) was found, consisting of 20 direct tandem repeats preceded by a run of 14 guanosine residues. DNA sequences homologous to *pat-1rep* were isolated and characterized from four other virulent *Cmm* strains, exhibiting a high extent of structural conservation. The deletion of this repetitive sequence reduced virulence significantly, but did not lead to a complete loss of the virulent phenotype. There is preliminary evidence that the *pat-1rep* repeats contribute to the stability of the *pat*-1 mRNA.

Recently, a modified transformation protocol and a transposon (Tn*1409* C) have been developed for *Cmm* (Gartemann and Eichenlaub 2001; Kirchner et al. 2001). These tools will allow gene replacement and transposon mutagenesis to saturation and thus open up the possibility to identify chromosomal loci that are essential for pathogenicity. Since it has been shown that Tn*1409* can also be used for transposon mutagenesis in other *Clavibacter michiganensis* subspecies, this new tool may also enhance the progress in genetic analysis of other members of the genus *Clavibacter.*

Closely related to *Cmm* is another quarantine organism, *Clavibacter michiganensis* ssp. *sepedonicus* (*Cms*). *Cms* causes the economically important disease bacterial ring rot of the potato tuber (*Solanum tuberosum*). Similar to the bacterial wilt of tomato, infection of potato plants by *Cms* is also accompanied by foliar symptoms, such as wilting, rolling and necrosis of the leaves. *Cms* is quite difficult to handle and therefore only very limited data are available on the molecular biology and genetics of this organism.

Most of the strains studied are reported to carry a plasmid of about 50 kb, pCS1 (Clark and Lawrence 1986; Mogen and Oleson 1987) which is frequently found to be integrated into the chromosome of *Cms* (Mogen et al. 1988). The insertion element IS*1121* present in pCS1 in two copies (Mogen et al. 1990) and in over 50 copies in the chromosome (Mogen and Oleson 1987) may serve as the site of integration of the plasmid. Due to the conservation of the plasmid in natural, virulent strains of *Cms* it was proposed that plasmid pCS1 may carry essential metabolic functions (Mogen et al. 1988, 1990).

Information on the function of this plasmid was recently obtained when the *celA* gene of *Cmm* was used as a probe in Southern hybridization showing that pCS1 carried a *celA* gene (Laine et al. 1996). After development of a transformation procedure by electroporation, cloning of *celA* of *Cms* was achieved by using vectors initially developed for *Cmm* (Laine et al. 1996). The sequence analysis showed that the *celA* nucleotide and cellulase amino acid sequences of *Cms* and *Cmm* are almost identical (Laine et al. 2000). Both enzymes have a three-domain structure and are encoded by a plasmid. The question as to whether *celA* is also a pathogenic determinant in *Cms* was answered by the naturally occurring plasmid-free strain P45 which does not produce a cellulase and is not virulent. Upon introduction of the *celA* gene carried by vector pHN216 into strain P45, cellulase production was restored and virulence was observed (Laine et al. 2000). However, virulence was lower than in other pathogenic strains of *Cms*, indicating that more genes than only *celA* may be required for full virulence. In this respect, *Cms* seems to resemble *Cmm*, where full virulence requires both *celA* and *pat*-1, while the presence of only one of the pathogenic factors results in a significant delay in symptom development (Meletzus et al. 1993).

Cms produces an extracellular protein of 40 kDa which elicits a hypersensitive response (HR) on the non-host plant tobacco (Nissinen 2001). It seems that this protein is also somehow involved in virulence since strain Cs4, which is *celA*$^+$ but lacks HR induction, is not able to efficiently colonize the host plant and thus is non-virulent. The strain P45 mentioned above, which is *celA*$^-$, nonvirulent, HR$^+$ and colonizes the host, was used in coinfections with strain Cs4. It was observed that in such a situation both strains now colonized the host and disease symptoms developed (Nissinen et al. 2001). This indicates that both cellulase production and the ability to induce a HR are necessary for virulence and, furthermore, that colonization of the host plant requires the gene for the 40-kDa elicitor protein whereas induction of disease symptoms seems to depend on the *celA* gene.

4 *Leifsonia xyli*

Leifsonia xyli ssp. *xyli* (*Lxx*) causes the ratoon stunting disease in sugarcane (Hughes 1974). Like *Clavibacter michiganensis*, *Lxx* colonizes the xylem vessels and causes stunting of infected plants possibly by impairing water transport. However, the involvement of other pathogenicity factors in the induction of disease symptoms cannot be excluded. *Lxx* is difficult to handle and grows extremely slow (Davis et al. 1980); however, very recently, a transformation procedure using broad host range vectors like pLAFR3 and transposon mutagenesis with Tn*4431* have been reported (Brumbley et al. 2002). Although the efficiency of transformation and transposition has to be improved, this is a very important step for further genetic analysis of this pathogen.

Closely related to *Lxx* is *Leifsonia xyli* ssp. *cynodontis* (*Lxc*) originally isolated from bermuda grass (*Cynodon dactylon*; Chen et al. 1977). Although *Lxc* is often isolated from diseased plants showing stunting and leaf chlorosis, it is not clear whether this bacterium is perhaps not a pathogen but rather an endophyte of the xylem causing only mild effects, like water stress. *Lxc* has been successfully transformed using IncP broad host range plasmids and vectors (Metzler et al. 1992), and a native plasmid has also been engineered into a cloning vector as well as a secretion vector (Taylor et al. 1993; Uratani et al. 1995). For the expression of cloned genes in *Lxc*, strong promoters were isolated (Haapalainen et al. 1996). However, due to the instability of cloning vectors in *Lxc*, integrative vectors were also constructed which integrate into the chromosome by a repetitive chromosomal element (Haapalainen et al. 1998). Using this integrative vector, a β-1,3-glucanase gene under the control of a previously isolated strong promoter of *Lxc* was stably expressed in *Lxc*, demonstrating that genetic engineering of this bacterium is possible.

A similar approach was taken in experiments to apply *Lxc* in crop protection: the biocontrol of the European corn borer (*Ostrinia nubilalis*). Besides bermuda grass, *Lxc* has been isolated from a variety of other monocotyledonous plants, like maize, rice, sorghum, and oats (Dimock et al. 1993). This broad range of host plants and the weak pathogenic effects caused by *Lxc* have furnished the idea of using this bacterium to deliver toxins into infected maize plants that are active against insects (Fahey et al. 1991). Thus, *Lxc* was engineered to express the insecticidal protein CryIA(c) from *Bacillus thuringiensis* ssp. *kurstaki* from a vector plasmid integrated into the chromosome of *Lxc* (Turner et al. 1991; Lampel et al. 1994). Recombinant strains were shown to produce the 133-kDa protoxin and to be toxic to insect larvae in bioassays, and they also exhibited significant insecticidal activity *in planta*. Although the constructs were somewhat unstable and frequently lost the toxin gene – this was discussed as acceptable in order to reduce environmental consequences of recombinant bacteria in agriculture – field tests showed that in maize the number of tunnels made by the corn borer larvae could be reduced up to about 80%. Despite such positive results, *Lxc* expressing the delta-endotoxin has never left the experimental stage in the biocontrol of the European corn borer, since it was already observed in control experiments in the field that the wild-type bacteria of *Lxc* reduced the yields of infected hybrid maize possibly due to physiological (water) stress caused by colonization of the xylem vessels. Thus it was obvious that the reduced yields to be expected more or less eliminated the benefits obtained by insect control. However, this example clearly shows that the use of bacterial endophytes for control of insects or other pathogens is plausible.

It has already been mentioned that plasmid-free strains of *Cmm*, such as strain CMM100, colonize tomato plants without causing disease symptoms and can be regarded as endophytes. Thus, a CMM100 hybrid strain was constructed carrying chitinase genes *chiA* or *chiC* from *Serratia marcescens* or *Streptomyces lividans*, respectively, expressed from a promoter of a chloramphenicol resistance gene (Bahro et al., in prep.). Under experimental conditions in a growth chamber, it was observed that a recombinant CMM100 strain expressing a bacterial chitinase protected infected tomato plants against the most prevalent and damaging vascular fungal pathogen of tomato, *Fusarium oxysporum* f. sp. *lycopersici*. However, it remains to be seen if such results are also obtained in field tests and how infection by the recombinant CMM100 strain affects the yield. Possibly, a similar outcome as for *Lxc*, expressing the delta-endotoxin, has to be anticipated.

5 *Streptomyces scabies*

Streptomycetes are Gram-positive high G+C soil bacteria that undergo a complex morphological and physiological differentiation process during their life-cycle. Their natural habitat is the soil where they live as a filamentous substrate mycelium that colonizes and surrounds organic compounds. Upon starvation streptomycetes start morphological and physiological differentiation. They begin to form aerial mycelium from which uninucleoidal spores can be dispersed in the soil, and each spore is capable of starting a new differentiation cycle (Chater 1998). During physiological differentiation, streptomycetes produce secondary active metabolites as antibiotics and extracellular hydrolytic enzymes which help to provide the bacteria with organic material from the soil. Among the more than 400 soil-saprophytic species that have been described to date, there are only a few which are plant-pathogenic. These species cause diseases of underground structures of different plants.

Streptomyces scabies (Lambert and Loria 1989b) induces an economically important disease of potato plants and other tap root crops, the common scab of potato. *Streptomyces acidiscabies* (Lambert and Loria 1989a), *S. turgidiscabies* (Miyajima et al 1998) and *S. ipomoeae* (Person and Martin 1940) cause similar symptoms to *S. scabies* and comprise the causal agents of different scab diseases of several root and tuber crops (Loria et al 1997). Taxonomic criteria indicate that the phytopathogenic streptomycetes really belong to different distinct species, but they all share one common characteristic: They all produce a phytotoxin of the thaxtomin family (Lawrence et al 1990; King et al 1991, 1994).

Thaxtomins [cyclo-(L-4-nitrotryptophyl-L-phenylalanyl) and related congeners] are a family of modified cyclic dipeptide phytotoxins. Thaxtomins A and B, the first of nine members of the family which have been described, were detected as the most abundant phytotoxins in *S. scabies* infected plant tissue. Purified thaxtomin A induced on potato tuber tissue necrosis and cell hypertrophy symptoms similar to those seen after infection by *S. scabies* (Lawrence et al. 1990). Chemical mutagenized *S. scabies* which were impaired or deficient in virulence produced significantly reduced or undetectable amounts of thaxtomin A in comparison to wild-type strains (Healy et al. 1997; Goyer et al. 1998). Furthermore, thaxtomin production perfectly correlated to pathogenicity in all strains tested. Thus, this compound is an important pathogenicity factor and it has been designated as a common pathogenicity determinant. The production of a very specific compound like thaxtomin by taxonomically different species implies that the structural genes for thaxtomin biosynthesis should be conserved among different species. The different strains could have acquired the genes by horizontal gene transfer and they might be organized similar to pathogenicity genes in Gram-negative bacteria on pathogenicity islands (Bukhalid et al. 1998; Healy et al. 1999).

The structure of the thaxtomins suggests that they are synthesized non-ribosomally by peptide synthetases. Thus, the corresponding genes for the thaxtomin biosynthesis of phytopathogenic *Streptomyces* strains were amplified by PCR using degenerate primers specific for conserved domains of peptide synthetases. As a nearly complete conservation of the thaxtomin biosynthesis genes was expected between different phytopathogenic *Streptomyces* species, Southern hybridizations of the amplificates were performed against the other strains to find the genes specific for phytopathogenicity and to exclude amplificates from strain-specific peptide synthetases involved in general metabolism. This strategy led to the identification of two peptide synthetase genes *txtA* and *txtB* from *S. acidiscabies*. They have a high G+C content of 71% and are organized probably as an operon without intergenic regions (Healy et al. 2000).

To date, there are no efficient tools available for the genetic manipulation of *S. scabies*, but DNA can be introduced into *S. acidiscabies* by conjugation. So a gene disruption mutant of *txtA* was constructed in this strain, and it was shown that this mutant could not produce thaxtomin any longer. Growth and morphological characteristics were not impaired but the mutant was avirulent on potato tubers. As complementation of the mutant restored pathogenicity, these experiments underline the importance of thaxtomin in the pathogenicity mechanism of streptomycetes (Healy et al. 2000).

In a further approach to understand pathogenicity mechanisms in phytopathogenic streptomycetes, a cosmid library from *S. scabies* was expressed in the non-pathogenic *Streptomyces* strain *S. lividans* 66 TK64, which is easy to handle with genetic methods and which comprises a model organism in *Streptomyces* research. Using an *S. scabies*-specific pathogenicity test (PTD-test, potato tuber disc), one recombinant *S. lividans* clone was detected which was converted to colonize potato tubers, to induce necrosis and to produce spores similar to the *S. scabies* spores. Furthermore, this recombinant strain could induce scab-like symptoms on immature potato tubers. By subcloning experiments a 9.4-kb *Bam*HI fragment was shown to be responsible for this conversion of non-pathogenic *S. lividans* to pathogenicity, and after mutagenization and deletion analysis of this region a 1.6-kb fragment was found to be responsible for the necrotizing activity. Sequence analysis of a 2.4-kb fragment including the necrotizing region indicated three open reading frames. One of the deduced proteins encodes a putative transposase (Orf_{tnp}), the other open reading frames did not find homologues in gene banks, but subcloning revealed that one of them (termed *nec1*) caused necrosis (Bukhalid and Loria 1997). Gene *nec1* (0.67 kb) encodes a water-soluble protein. This protein was able to convert *S. lividans* to pathogenicity although *S. lividans* TK64 recombinant for *nec1* did not produce thaxtomin.

To determine the significance of *nec1* the distribution of this gene and its surrounding region was analyzed in 43 different isolates of the species *S. scabies*, *S. acidiscabies and S. turgidiscabies* by Southern hybridization and PCR. All but one of the distinct phytopathogenic isolates contain this region in an identical organization with the genes *nec1* and orf_{tnp} linked together on one single *PvuII* fragment. This distribution and conservation of structure indicates that this pathogenicity-inducing region has also been acquired by the different unrelated strains by horizontal gene transfer (Bukhalid et al. 1998).

6 *Rhodococcus fascians*

Rhodococcus fascians is a Gram-positive bacterium that also belongs to the actinomycetes and is closely related to the corynebacteria (Goodfellow 1984, 1989). This species is phytopathogenic for a variety of plants and is characterized by a very wide host range. Infection of dicotyledonous plants often leads to symptoms as local proliferation of meristematic tissue. The plants are characterized by a loss of apical dominance and the development of multiple shoots giving rise to leafy galls (Brown 1927, Tilford 1936, Stapp 1961). Infection of monocotyledonous plants (lily, dahlia) is characterized by severe malformations of the bulbs and the formation of long side shoots (Miller et al. 1980) and, thus, such infected ornamental plants are not suitable for commercial use (Baker 1950). *Rhodococcus fascians* lives as a well-adapted epiphyte: The bacteria can colonize the plant surface as well as the interior of aerial parts of the plant (Cornelis et al. 2001). The symptoms of disease which are caused by *R. fascians* are similar to the symptoms which are caused by the Gram-negative species *Agrobacterium tumefaciens* or *Pseudomonas syringae* pv. *savastanoi*. The mechanisms of pathogenicity of these Gram-negative bacteria have been investigated for several years. Both species possess genes for enzymes to synthesize phytohormones (Morris 1986) and genes that derivatize these hormones (Roberto and Kosuge 1987).

In 1966, it was first shown that pathogenicity of *R. fascians* could also be attributed to the production of the phytohormone cytokinin (Thimann and Sachs 1966). Fasciation symptoms were triggered on sweet pea by the application of the cytokinin kinetin (Thimann and Sachs 1966) and since then different cytokinins have been found in the supernatants of different *R. fascians* strains (Klämbt et al. 1966; Scarbrough et al. 1973; Murai et al. 1980). Most work concerning the molecular basis of *Rhodococcus fascians* pathogenicity has been conducted in the last 10 years by a group in Belgium and has been summarized in a recent comprehensive review (Goethals et al. 2001). Thus, the following pages will only give a short overview of the genetic basis of *Rhodococcus* virulence.

The pathogenicity determinants of *R. fascians* D188 were detected in 1992. They are located on a linear conjugative plasmid pFiD188 (Fi for fasciation inducing) that is about 200 kb in size (Crespi et al. 1992). Strains that are cured for this plasmid become avirulent, whereas reintroduction of the plasmid into the curing derivatives restores virulence. Insertion mutagenesis via illegitimate recombination (Desomer et al. 1991) led to the identification of three loci on this plasmid that are involved in pathogenicity: the *att*-locus, the *hyp*-locus and the *fas*-locus (Crespi et al. 1992). Mutation of the *att*- or *hyp*-locus affects the degree of virulence of the corresponding mutants and leads to attenuated virulence or hypervirulence, respectively. Thus, these loci represent modulators of virulence. Molecular analysis of the *att*-locus showed that it contains nine open reading frames that are similar to genes from arginine and β-lactam biosynthesis. The expression of these genes is induced on the transcriptional level by leafy gall extracts, but not by extracts of uninfected plants. Expression of the *att* genes is mediated by AttR, a LysR family transcriptional activator. Furthermore, the *att*-locus seems to be involved in the formation of inducing factors (Ifs) which are found in gall extracts (Maes et al. 2001).

In contrast to mutations in the virulence modulating loci *att* and *hyp*, mutation of the *fas*-locus results in a complete loss of virulence (Crespi et al. 1992). By insertion mutagenesis this locus could be attributed to a 6.5-kb DNA fragment (Crespi et al. 1994) which carries six open reading frames. These open reading frames are orientated in the same direction and should comprise an operon. Their expression is specifically induced during the interaction with the host plant. The deduced proteins from *orf1* and the 5′ end of *orf2* show similarities to P-450 cytochromes and ferredoxins from actinomycetes. The 3′ end of *orf2* has regions with similarities with the deduced amino acids to transketolases from chemoautotrophic, photosynthetic or methylotrophic microorganisms (Crespi et al. 1994). Orf4 (*ipt*) shows homology to isopentenyltransferase genes from other phytopathogenic bacteria (Crespi et al. 1992, 1994) like *Agrobacterium tumefaciens*, *A. rhizogenes*, *Pseudomonas savastanoi* or *P. solanacearum*. Isopentenyltransferases usually catalyze the biosynthesis of isopentenyl-AMP which comprises a precursor of several cytokinins (Kaminek 1992). The specific activity of the *R. fascians* enzyme as an isopentenyltransferase was demonstrated after heterologous expression of the *ipt* gene in *E. coli* (Crespi et al. 1992). Therefore, it seems that this gene encodes a key enzyme for phytohormone synthesis that leads to gall formation on infected host plants. In *R. fascians* there is only expression of *ipt* in bacteria after induction by extracts from infected fasciated plant tissue indicating a tight regulation of pathogenicity.

Recent studies on the regulation of *Rhodococcus* pathogenicity show that there is a further gene (*fasR*) on the linear plasmid pFiD188 that is necessary for pathogenicity and for the expression of the genes in the *fas*-locus. The *fasR* gene is located between the two pathogenicity loci *fas* and *att* and encodes a regulatory protein belonging to the AraC family. A complex working model for the regulation of *fas* gene expression proposed by Temmermann et al. (2000) suggests that expression of the *fas* genes is controlled at the transcriptional as well as the translational level.

In a recent publication, Vereecke et al. (2002) describe the identification of the first chromosomal locus that affects virulence of *R. fascians*. This locus, *vicA*, encodes a malate synthase which is required for efficient *on planta* growth in fasciated, but not in normal plant tissue. This observation indicates that there is a change in metabolic requirements of the bacteria caused by the nutritional environment in the plant during the bacteria–plant interaction. Hyperplasia on plants thus could comprise a specific ecological niche for *R. fascians* and the interaction could be interpreted as a metabolic habitat modification (Vereecke et al. 2002).

7 Outlook

This overview has shown that, in most cases, the genetic study of the pathogenicity of Gram-positive phytopathogenic bacteria is still in its infancy. With the development of transformation procedures and the establishment of cloning vectors and transposons in the systems described a major step has been taken to shed some light on the mechanism of pathogenicity. Table 2 provides a summary of the pathogenicity genes and their possible functions in the systems described. A very promising development is certainly the fact that projects have been funded recently to determine the sequence of the whole genome for *Clavibacter michiganensis* ssp. *michiganensis*, *C. m.* ssp. *sepedonicus*, and *Leifsonia xyli* ssp. *xyli*. Hopefully, the data provided by the genome projects and the improvement of genetic tools and methods will bring more attention to Gram-positive phytopathogenic bacteria and induce more scientists to focus their energy on these organisms.

Table 2. Pathogenicity genes and function

Species	Pathogenicity related genes	Function
Clavibacter michiganensis sbsp. *michiganensis*	*celA*	Endo-β-1,4-glucanase
	pat-1	Serine-protease (putative)
Clavibacter michiganensis ssp. *sepedonicus*	*celA*	Endo-β-1,4-glucanase
Leifsonia xyli	–	–
Streptomyces scabies, *S. acidiscabies*	*txtA*, *txtB*	Peptidesynthetases for the phytotoxin thaxtomin
	necA	?
Rhodococcus fascians	*att*-locus	?
		Enzymes for arginine and β-lactam biosynthesis
	hyp-locus	?
	fas-locus: *ipt*	Isopentenyl-transferase
	fasR	Regulatory protein of the AraC-family

References

Ahlemeyer J, Eichenlaub R (2001) Genetics of phytopathogenic bacteria. Progress in Botany 62. Springer, Berlin Heidelberg New York, pp 98–113

Baker KF (1950) Bacterial fasciation disease of ornamental plants in California. Plant Dis Rep 34:121–126

Beimen A, Bermpohl A, Meletzus D, Eichenlaub R, Barz W (1992) Accumulation of phenolic compounds in leaves of tomato plants after infection with *Clavibacter michiganense* subsp. *michiganense* strains differing in virulence. Z Naturforsch 47c:898–909

Benhamou N (1991) Cell surface interactions between tomato and *Clavibacter michiganense* subsp. *michiganense*: localization of some polysaccharides and hydroxyproline-rich glycoproteins in infected host leaf tissues. Physiol Mol Plant Pathol 38:15–38

Bermpohl A, Dreier J, Eichenlaub R (1996) Exopolysaccharides in the pathogenic interaction of *Clavibacter michiganensis* subsp. *michiganensis* with tomato plants. Microbiol Res 151:391–399

Brown NA (1927) Sweet pea fasciation, a form of crowngall. Phytopathology 17:29–30

Brumbley SM, Petrasovits LA, Birch RG, Taylor PWJ (2002) Transformation and transposon mutagenesis of *Leifsonia xyli* subsp. *xyli*, causal organism of ratoon stunting disease of sugarcane. Mol Plant Microbe Interact 15:262–268

Bukhalid RA, Loria R (1997) Cloning and expression of a gene from *Streptomyces scabies* encoding a putative pathogenicity factor. J Bacteriol 179:7776–7783

Bukhalid RA, Young Chung S, Loria R (1998) *nec1*, a gene conferring a necrogenic phenotype, is conserved in plant-pathogenic *Streptomyces* spp. and linked to a transposase pseudogene. Mol Plant Microbe Interact 11:960–967

Carlsson RR, Vidaver AK, Siebler JE (1982) Pathogenicity, bacteriocins and colony morphology of *Corynebacterium michiganense* ssp. *tesselarius* isolated from wheat seed. Phytopathology 72:992

Chater KF (1998) Taking a genetic scalpel to the *Streptomyces* colony. Microbiology 144:1465–1478

Chen TA, Su HJ, Raju BC, Huang WC (1977) A new spiroplasm isolated from Bermudagrass (*Cynodon dactylon* L. Pers.) Proc Am Phytopathol Soc 4:171

Clark MC, Lawrence CH (1986) Characterization of a plasmid in isolates of *Corynebacterium sepedonicum*. Can J Microbiol 32:617–622

Cornelis K, Ritsema T, Nijsse J, Holsters M, Goethals K, El Jaziri M (2001) The plant pathogen *Rhodococcus fascians* colonizes the exterior and interior of aerial parts of plants. Mol Plant Microbe Interact 14:599–608

Cosgrove DJ (1998) Cell wall loosening by expansins. Plant Physiol 118:333–339

Crespi M, Messens E, Caplan AB, van Montagu M, Desomer J (1992) Fasciation induction by the phytopathogen *Rhodococcus fascians* depends upon a linear plasmid encoding a cytokinin synthase gene. EMBO J 11:795–804

Crespi M, Vereecke D, Temmermann W, van Montagu M, Desomer J (1994) The *fas* operon of *Rhodococcus fascians* encodes new genes required for efficient fasciation of host plants. J Bacteriol 176:2492–2501

Davis MJ, Gillaspie AG Jr, Harris RW, Lawson RW (1980) Ratoon stunting disease of sugarcane: isolation of the causal bacterium. Science 210:1365–1367

Davis MJ, Gillaspie AG Jr, Vidaver AK, Harris RW (1984) *Clavibacter*: a new genus containing some phytopathogenic coryneform bacteria, including *Clavibacter xyli* subsp. *xyli* sp. nov., subsp., nov. and *Clavibacter xyli* subsp. *cynodontis* subsp. nov., pathogens that cause ratoon stunting disease of sugarcane and bermudagrass stunting disease. Int J Syst Bacteriol 34:107–117

Desomer J, Crespi M, van Montagu M (1991) Illegitimate integration of non-replicative vectors in the genome of *Rhodococcus fascians* upon electrotransformation as an insertional mutagenesis system. Mol Microbiol 5:2115–2124

Dimock M, Turner J, Lampel J (1993) Endophytic organisms for delivery of genetically engineered microbial pesticides in plants. In: Kim L (ed) Advanced engineered pesticides. Marcel Dekker, New York, pp 85–97

Dreier J, Bermpohl A, Eichenlaub R (1995) Southern-hybridization and PCR for specific detection of phytopathogenic *Clavibacter michiganensis* subsp. *michiganensis*. Phytopathology 85:462–468

Dreier J, Meletzus D, Eichenlaub R (1997) Characterization of the plasmid encoded virulence region *pat*-1 of the phytopathogenic *Clavibacter michiganensis* subsp. *michiganensis*. Mol Plant Microbe Interact 10:195–206

Evtushenko LI, Dorofeeva LV, Subbotin SA, Cole JR, Tiedje JM (2000) *Leifsonia poae* gen nov., sp nov., isolated from nematode galls on *Poa annua*, and reclassification of "*Corynebacterium aquaticum*" Leifson 1962 as *Leifsonia aquatica* (ex Leifson 1962) gen. nov., nom. Rev., comb. nov. and *Clavibacter xyli* Davis et al. 1984 with two subspecies as *Leifsonia xyli* (Davis et al. 1984) gen. nov., comb. nov. Int J Syst Evol Microbiol 50:371–380

Fahey JW, Dimock MB, Tomasino SF, Taylor JM, Carlson PS (1991) Genetically engineered endophytes as biocontrol agents: a case study from industry. In: Andrews JH, Hirano SS (eds) Microbial ecology of leaves. Springer, Berlin Heidelberg New York, pp 401–411

Fatmi M, Schaad NW (1988) Semiselective agar medium for isolation of *Clavibacter michiganensis* subsp. *michiganensis* from tomato seed. Phytopathology 781:121–126

Gartemann K-H, Eichenlaub R (2001) Isolation and characterization of IS*1409*, an insertion element of 4-chlorobenzoate-degrading *Arthrobacter* sp. strain TM1, and development of a system for transposon mutagenesis. J Bacteriol 183:3729–3736

Gitaitis RD, Beaver RW, Voloudakis AE (1991) Detection of *Clavibacter michiganenis* subsp. *michiganensis* in symptomless tomato transplants. Plant Dis 75:834–838

Goethals K, Vereecke D, Jaziri M, van Montagu M, Holsters M (2001) Leafy gall formation by *Rhodococcus fascians*. Annu Rev Phytopathol 39:27–52

Goodfellow M (1984) Reclassification of *Corynebacterium fascians* (Tilford) Dowson in the genus *Rhodococcus*, as *Rhodococcus fascians* comb. nov. Syst Appl Microbiol 5:225–229

Goodfellow M (1989) Suprageneric classification of Actinomycetes. In: Williams ST, Sharpe ME, Holt JG (eds) Bergey's manual of systematic bacteriology, vol 4. Williams and Wilkins, Baltimore, pp 2333–2339

Goyer C, Vachon J, Beaulieu C (1998) Pathogenicity of *Streptomyces scabies* mutants altered in thaxtomin A production. Phytopathology 88:442–445

Haapalainen M, Karp M, Metzler MC (1996) Isolation of strong promoters from *Clavibacter xyli* subsp. *cynodontis* using a promoter probe plasmid. Biochim Biophys Acta 1305:130–134

Haapalainen M, Nissinen R, Metzler MC (1997) Pathogenicity factors and molecular biology of the plant pathogenic genus *Clavibacter*. Recent Res Dev Microbiol 1:209–216

Haapalainen ML, Kobets N, Piruzian E, Metzler MC (1998) Integrative vector for stable transformation and expression of a beta-1,3-glucanase gene in *Clavibacter michiganensis* subsp. *cynodontis*. FEMS Microbiol Lett 162:1–7

Healy FG, King RR, Loria R (1997) Identification of thaxtomin A non-producing mutants of *Streptomyces scabies*. Phytopathology 87:41

Healy FG, Bukhalid RA, Loria R (1999) Characterization of an insertion sequence element associated with genetically diverse plant pathogenic *Streptomyces* spp. J Bacteriol 181:1562–1568

Healy FG, Wach M, Krasnoff SB, Gibson DM, Loria R (2000) The *txtAB* genes of the plant pathogen *Streptomyces acidiscabies* encode a peptide synthetase required for phytotoxin thaxtomin A production and pathogenicity. Mol Microbiol 38:794–804

Hughes CG (1974) The economic importance of ratoon stunting disease. Proc Int Soc Sugarcane Technol 15:213–217

Jahr H, Bahro R, Burger A, Ahlemeyer J, Eichenlaub R (1999) Interactions between *Clavibacter michiganensis* and its host plants. Environ Microbiol 1:113–118

Jahr H, Dreier J, Meletzus D, Bahro R, Eichenlaub R (2000) The endo-β-1,4-glucanase of *Clavibacter michiganensis* subsp. *michiganensis* is a pathogenicity determinant required for induction of bacterial wilt of tomato. Mol Plant Microbe Interact 13:703–714

Kaminék M (1992) Progress in cytokinin research. Trends Biotechnol 10:159–164

King RR, Lawrence CH, Clark MC (1991) Correlation of phytotoxin production with pathogenicity of *Streptomyces scabies* isolates from scab infected potato tubers. Am Potato J 68:675–680

King RR, Lawrence CH, Calhoun LA, Ristaino JB (1994) Isolation and characterization of thaxtomin-type phytotoxins associated with *Streptomyces ipomoeae.* J Agric Food Chem 42:1791–1794

Kirchner O, Gartemann K-H, Zellermann E-M, Eichenlaub R, Burger A (2001) A highly efficient transposon mutagenesis system for the tomato pathogen *Clavibacter michiganensis* subsp. *michiganensis.* Mol Plant Microbe Interact 14:1312–1318

Klämbt D, Thies G, Skoog F (1966) Isolation of cytokinins from *Corynebacterium fascians.* Proc Natl Acad Sci USA 56:52–59

Laine MJ, Nakhei H, Dreier J, Lehtilä K, Meletzus D, Eichenlaub R, Metzler MC (1996) Stable transformation of the gram-positive bacterium *Clavibacter michiganensis* subsp. *sepedonicus* with several cloning vectors. Appl Environ Microbiol 62:1500–1506

Laine MJ, Haapalainen M, Wahlroos T, Kankare K, Nissinen R, Kassuwi S, Metzler MC (2000) The cellulase encoded by the native plasmid of *Clavibacter michiganensis* ssp. *sepedonicus* plays a role in virulence and contains an expansin-like domain. Physiol Mol Plant Pathol 57:221–233

Lambert DH, Loria R (1989a) *Streptomyces acidiscabies* sp. nov. Int J Syst Bacteriol 39:393–396

Lambert DH, Loria R (1989b) *Streptomyces scabies* sp. nov. nom. rev. Int J Syst Bacteriol 39:387–392

Lampel JS, Canter Gl, Dimock MB, Kelly JL, Anderson JJ, Uratani BB, Foulkes JS, Turner JT (1994) Integrative cloning, expression, and stability of the CryIA(c) gene from *Bacillus thuringiensis* subsp. *kurstaki* in a recombinant strain of *Clavibacter xyli* subsp. *cynodontis.* Appl Environ Microbiol 60:51–508

Lawrence CH, Clark MC, King RR (1990) Induction of common scab symptoms in aseptically cultured potato tubers by the vivotoxin, thaxtomin. Phytopathology 80:606–608

Loria R, Bukhalid RA, Fry BA, King RR (1997) Plant pathogenicity in the genus *Streptomyces.* Plant Dis 81:836–846

Maes T, Vereecke D, Ritsema T, Cornelis K, Ngo Thi Thu H, van Montagu M, Holsters M, Goethals K (2001) The *att* locus of *Rhodococcus fascians* strain D188 is essential for full virulence on tobacco through the production of an autoregulatory compound. Mol Microbiol 42:13–28

Manzer F, Genereux H (1981) In: Hobson WJ (ed) Compendium of potato disease. American Phytpathology Society, St Paul, MN

McCulloch L (1925) *Aplanobacter insidiosum* n. sp., the cause of an alfalfa disease. Phytopathology 15:496–497

Meletzus D, Eichenlaub R (1991) Transformation of the phytopathogenic bacterium *Clavibacter michiganense* subsp. *michiganense* by electroporation and development of a cloning vector. J Bacteriol 173:184–190

Meletzus D, Bermpohl A, Dreier J, Eichenlaub R (1993) Evidence for plasmid encoded virulence factors in the phytopathogenic bacterium *Clavibacter michiganensis* subsp. *michiganensis* NCPPB382. J Bacteriol 175:2131–2136

Metzler MC, Zhang Y-P, Chen T-A (1992) Transformation of the gram-positive bacterium *Clavibacter xyli* subsp. *cynodontis* by electroporation with plasmids from the incP incompatibility group. J Bacteriol 174:4500–4503

Metzler MC, Laine MJ, DeBoer SH (1997) Status of molecular biological research on the plant pathogenic genus *Clavibacter.* FEMS Microbiol Lett 150:1–8

Miller HJ, Janse JD, KamermanW, Muller PJ (1980) Recent observations of leafy gall in Liliaceae and some other families. Neth J Plant Pathol 86:55–68

Miyajima K, Tanaka F, Takeuchi T, Kuninaga S (1998) *Streptomyces turgidiscabies* sp. nov. Int J Syst Bacteriol 48:495–502

Mogen BD, Oleson AE (1987) Homology of pCS1 plasmid sequences with chromosomal DNA in *Clavibacter michiganense* subsp. *sepedonicum*: evidence for the presence of a repeated sequence and plasmid integration. Appl Environ Microbiol 53:2476–2481

Mogen BD, Oleson AE, Sparks RB, Gudmestad NC, Secor GA (1988) Distribution and partial characterization of pCS1, a highly conserved plasmid present in *Clavibacter michiganense* subsp. *sepedonicum.* Phytopathology 78:1381–1386

Mogen BD, Oleson HR, Sparks RB, Gudmestad NC, Oleson AE (1990) Genetic variation in strains of *Clavibacter michiganense* subsp. *sepedonicum*: polymorphism in restriction fragments containing a highly repeated sequence. Phytopathology 80:90–96

Morris RO (1986) Genes specifying auxin and cytokinin biosynthesis in phytopathogens. Annu Rev Plant Physiol 37:509–538

Murai N, Skoog F, Doyle ME, Hanson RS (1980) Relationships between cytokinin production, presence of plasmids, and fasciation caused by strains of *Corynebacterium fascians.* Proc Natl Acad Sci USA 77:619–623

Nissinen R, Kassuwi S, Peltola R, Metzler MC (2001) In planta-complementation of *Clavibacter michiganensis* subsp. *sepedonicus* strains deficient in cellulase production or HR induction restores virulence. Eur J Plant Pathol 107:175–182

Person LH, Martin W (1940) Soil rot of sweet potatoes in Louisiana. Phytopathology 30:913–926

Rai PV, Strobel GA (1968a) Phytotoxic glycopeptides produced by *Corynebacterium michiganense.* II. Biological properties. Phytopathology 59:53–57

Rai PV, Strobel GA (1968b) Phytotoxic glycopeptides produced by *Corynebacterium michiganeense.* I. Methods of preparation, physical and chemical characterization. Phytopathology 59:47–52

Roberto F, Kosuge T (1987) In: Fox EW, Jacobs M (eds) Molecular biology of plant growth control. Alan R Liss, New York, pp 371–380

Sasaki J, Chijimatsu M, Suzuki K-I (1998) Taxonomic significance of 2,4-diaminobutyric acid isomers in the cell wall peptidoglycan of actinomycetes and reclassification of *Clavibacter toxicus* as *Rathayibacter toxicus* comb. nov. Int J Syst Bacteriol 48:403–410

Scarbrough E, Armstrong DJ, Skoog F, Frihart CR, Leonard NJ (1973) Isolation of *cis*-zeatin from *Corynebacterium fascians* cultures. Proc Natl Acad Sci USA 70:3825–3829

Schuster ML (1975) Leaf freckles and wilt of corn incited by *Corynebacterium nebraskense.* Res Bull 270 Agric Exp Stn University of Nebraska, Lincoln

Shirakawa T, Sasaki T, Ozaki K (1991) Ecology and control of tomato bacterial canker and detection of its pathogen. Jpn Agric Res Q (JARQ) 25:27–32

Stackebrandt E, Rainey FA, WardRainey NL (1997) Proposal for a new hierarchic classification system, *Actinobacteria* classis nov. Int J Syst Bacteriol 47:479–491

Stapp C (1961) Bacterial plant pathogens. Oxford University Press, Oxford

Strider DL (1969) Bacterial canker of tomato caused by *Corynebacterium michiganense*: a literature review and bibliography. NC Agric Exp Stn Tech Bull 193

Suzuki K, Suzuki M, Sasaki J, Park YH, Komagata K (1999) *Leifsonia* gen. nov., a genus for 2,4-diaminobutyric acid-containing *actinomycetes* to accommodate "*Corynebacterium aquaticum*" Leifson 1962 and *Clavibacter xyli* subsp. *cynodontis* Davis et al. 1984. J Gen Appl Microbiol 45:253–262

Taylor J, Stearman RS, Uratani BB (1993) Development of a native plasmid as a cloning vector in *Clavibacter xyli* subsp. *cynodontis.* Plasmid 29:241–244

Temmermann W, Vereecke D, Dreesen R, van Montagu M, Holsters M, Goethals K (2000) Leafy gall formation is controlled by *fasR*, an AraC-type regulatory gene in *Rhodococcus fascians.* J Bacteriol 182:5832–5840

Thimann KV, Sachs T (1966) The role of cytokinins in the "fasciation disease" caused by *Corynebacterium fascians.* Am J Bot 53:731–739

Thompson ET (1986) The toxicity of a number of different bacteriocides to *Clavibacter michiganense* subsp. *michiganense* Smith 1910 Jensen 1934 comb. nov. basonym *Corynebacterium michiganense* subsp. *michiganense* AL and to the tomato plant, *Lycopersicon esculentum.* J Appl Bacteriol 61:427–436

Tilford PE (1936) Fasciation of sweet peas caused by *Phytomonas fascians* n. sp. J Agric Res 53:383–394

Turner JT, Lampel JS, Stearman RS, Sundin GW, Gunyuzlu P, Anderson JJ (1991) Stability of the delta-endotoxin from *Bacillus thuringiensis* subsp. *kurstaki* in a recombinant strain of *Clavibacter xyli* subsp. *cynodontis.* Appl Environ Microbiol 57:3522–3528

Uratani BB, Alcorn SC, Tsang BH, Kelly JL (1995) Construction of secretion vectors and use of heterologous signal sequences for protein secretion in *Clavibacter michiganensis* subsp. *cynodontis.* Mol Plant Microbe Interact 8:892–898

Van Den Bulk RW, Löffler HJM, Dons JJM (1989) Effect of phytotoxic compounds produced by *Clavibacter michiganense* subsp. *michiganensis* on resistant and susceptible tomato plants. Neth J Plant Pathol 95:107–117

Vereecke D, Cornelis K, Temmermann W, Jaziri M, van Montagu M, Holsters M, Goethals K (2002) Chromosomal locus that affects pathogenicity of *Rhodococcus fascians.* J Bacteriol 184:1112–1120

Wallis FM (1977) Ultrastructural histopathology of tomato plants infected with *Corynebacterium michiganense.* Physiol Plant Pathol 11:333–342

Dr. Annette Burger
Prof. Dr. Rudolf Eichenlaub
Lehrstuhl Gentechnologie/Mikrobiologie
Universität Bielefeld
Universitätsstrasse 25
33501 Bielefeld, Germany

e-mail: Annette.Burger@biologie.uni-bielefeld.de
e-mail: eichenlaub@biologie.uni-bielefeld.de

Population Genetics: Aspects of Biodiversity

Rob O'Neill, Rod Snowdon, and Wolfgang Köhler

1 Introduction

Population genetics has a significant role to play in the analysis and description of biodiversity in plant communities and populations. This paper aims to give a general overview of population genetics methods that can be used to quantify biodiversity, and methods that can be used to determine genetic diversity within and among plant populations. In order to understand the basic idea of biodiversity measurements, it is important not only to be familiar with common methods employed in assessing genetic diversity, but also to have a general understanding of the mathematics used to translate experimental information into a form suitable for assessing biodiversity. Such aspects will be covered in general terms with reference to examples of their application to various fields of plant genetics in recent years.

The term biodiversity is often applied to describe quite different concepts, such as number of species or genetic diversity. Hence, before defining experimental methods used for biodiversity estimates, one must begin by defining what level of diversity is actually being studied. Because this definition can take various forms, for the purposes of this review we will present our own view on what constitutes biodiversity and its connection to genetic diversity. With respect to this definition, various methods used for the detection of genetic diversity will be described along with an explanation of selected measurements of genetic diversity, and examples of how this data can be used as a tool for conservation, breeding, evolution and ecology.

2 Genetic Diversity vs. Biodiversity

The simplests, and most common, explanation for biological diversity, or "biodiversity", is the "variety of life" (Huston 1994; Gaston 1996; Waldman and Shevah 2000). This variety manifests itself in numerous

Progress in Botany, Vol. 64

ways, most of which can be categorised as phylogenetic, morphological, or genetic variation (Owens and Bennett 2000).

Biodiversity is sometimes defined as a description of variation at all levels of biological perception, which, when the abundance of each variety is taken into account, ultimately incorporates the total variety of life on earth. However, a glance through some of the vast range of literature covering different aspects of biodiversity reveals a division of opinion as to whether or not biodiversity represents purely a 'snapshot' of the variety of biological entities at any given time, or if the concept also encompasses the processes and complexes formed as a result of the interaction between landscape and living organisms, and between the living organisms themselves (Gaston 1996). There can be no doubt that landscape and climate play a crucial role in determining the number of species present in a community or habitat, and their ability to interact intra- and interspecifically. From a population genetics point of view, however, biodiversity is generally broadly defined not as variety of species, but rather in terms of the genetic diversity present within individual species. The two concepts are of course extremely closely linked, since environment can have a major influence on the makeup of the gene pool within species living in any given habitat. For example, the industrial and technological revolutions of the last two centuries have had an enormous destructive effect on wetland and forest ecosystems (Waldman and Shevah 2000) with an accompanying reduction in species diversity in each case. Besides the extinction of populations and species, however, such processes can also lead to a reduction in the size of remaining populations, and the consequence is generally a corresponding reduction in the size of the gene pool.

Such examples demonstrate the close association that is often inferred between biodiversity and conservation. Certainly, it is true that the two can often be treated synonymously, as conservation is the main method used today to preserve biodiversity. However, theoretically, they are distinct concepts and habitat conservation alone cannot guarantee the conservation of biodiversity from a genetic diversity viewpoint. Measurement of genetic diversity and utilisation of diversity data for preservation of biodiversity demand detailed knowledge of population dynamics and the genetic forces acting on the population(s) under study.

Population genetics attempts to uncover the degree of variation existing in a population (in the form of different genotypes or phenotypes), and establish why this variation is present and what factors are affecting its dynamics within the population. The tools population geneticists use to detect such variation can be broadly divided into classical (Mendelian) and molecular genetic techniques.

Classical population genetics, or Mendelian genetics, involves the observation and assessment of morphological data, which normally take the form of characters giving rise to a distinct external phenotype that can be directly visualised by the researcher. While such techniques are still useful today, the number of morphological markers that can be scored for any given object is limited, and information that is useful for population genetics can only be obtained if sufficient morphological variation is present within the population. Depending on the population size it is unlikely that easily scorable morphological markers alone will be sufficient to enable distinction among all individuals, and for this reason molecular genetic markers have today been almost universally adopted as the major method for assessing genetic diversity among and between populations.

In the past decade, a wealth of molecular genetic techniques has been developed that allow assessment of considerable numbers of markers in a relatively short timeframe. Depending on the scale of the observations, with appropriate data analysis it is possible to apply molecular marker techniques for the quantification and comparison of genetic diversity within populations, between populations, between species, and between taxa.

In the following, we will concentrate on the first two possibilities, because from a population genetics perspective of biodiversity it is only of interest to analyse and compare individuals that exchange genes, or have the potential to do so. In other words, we are interested in the dynamics or relationships within single species.

3 Detection of Genetic Diversity

The major prerequisite for the measurement of genetic diversity is the identification within the population or populations under study of variation, or polymorphism, that has its basis in genetic variation. In other words, observations of heritable markers are used to establish the genetic similarities or differences between more or less related individuals. Logically, the degree to which the genetic relatedness of any two individuals can be determined is dependent on the number of polymorphic phenotype traits available: The more characteristics that can be scored, the better the distinction that can be achieved. Here we will consider the advantages that molecular genetic markers have over classical, morphological markers, and briefly outline the advantages and disadvantages of some of the major molecular marker techniques used today to gather information about genetic diversity in plant population genetics studies.

a) Morphological vs. Molecular Markers

Morphological traits have their basis in genetic alterations that, in the course of evolution, have led to visible differences in phenotype. Estimates from data analysis of the complete sequence of the model plant *Arabidopsis thaliana* generally presume the presence of only some 25,000 genes in total (Alonso-Blanco and Koorneef 2000), and it is widely agreed that even considerably more complex crop plants are unlikely to possess a significantly higher number of genes. When one considers that only an extremely small fraction of the genes which have been identified to date code for traits that are manifested in externally observable phenotypes (i.e. classical morphological markers), it becomes clear that the level of information that can be realistically obtained from morphological markers is limited. For targeted studies of allelic variation at specific genes, however, analysis of morphological or biochemical markers can still give useful information about genetic diversity (e.g. Gilliland et al. 2000) and correlations of morphological data to molecular marker information can be useful where marker studies are intended to obtain information on specific traits which may, for example, be of interest to plant breeders (e.g., Rebourg et al. 2001). However, use of morphological markers alone immediately excludes analysis of those portions of the genome containing non-coding sequences, which in higher plants can often account for more than 95% of the complete DNA sequence. The assumption that polymorphism in these sequence regions has no expression in the visible, morphological phenotype, is in another respect advantageous: Mutations in non-coding regions of the genome confer no, or little, selective advantage or disadvantage, and hence their inheritance can generally be assumed to conform to Mendelian laws. Moreover, the potential for new allelic forms is in no way influenced by the number of genes, thus the potential for polymorphism is directly proportional to the rate of natural mutation in the DNA sequence. Of course such variations can also occur in coding regions. When these result in amino acid substitutions in enzymatic proteins, the consequence can be observable polymorphism in the electrophoretic properties of the protein subunits. Such variations, known as allozyme electrophoretic markers, were the first major class of "molecular" markers to be widely used for large-scale analysis of genetic diversity within and among plant (and animal) populations.

b) Allozyme Markers

Allozyme (isoenzyme) markers were the method of choice for many plant population geneticists throughout the 1970s and 1980s (for a review, see Hamrick and Godt 1989), because in most cases they represented the cheapest and most effective tool for analyses of "neutral" variation among large numbers of individuals. Electrophoretic variation in proteins detectable by biochemical staining reactions can be directly related to allelic variation (all alleles present in heterozygotes can be directly observed and counted), and allele frequency data can subsequently be used to compare individuals and groups of individuals both within and among populations. However, nucleotide substitutions in the DNA sequence do not necessarily result in a change in protein amino acid sequence and therefore cannot be detected by allozyme electrophoresis. Moreover, as with classical morphological markers, the number of allozyme markers is somewhat limited and their utility is limited by their low resolution power in populations or species with only a narrow genetic base. Thus, with the rapid advances in molecular genetic techniques over the last two decades, allozyme techniques have become comparatively rare in the midst of DNA-based marker technologies with the potential to resolve genetic differences directly at the DNA sequence level. In terms of genetic diversity, the information content and quality is considerably higher with most DNA markers than with allozymes; however, due to their simplicity and low cost, genetic diversity studies using isoenzyme markers are still found relatively frequently in the published literature. Recent examples include studies of genetic structure and genetic diversity in populations of rice (Li and Rutger 2000), wheat (Moghaddam et al. 2000), *Viola* (Culley and Wolfe 2001), *Pinus* (Epperson and Cheung 2001), *Raphanus* (Huh and Ohnishi 2001), *Pterostylis* (Sharma et al. 2001) and a threatened Asteraceae (Neel and Ellstrand 2001), to name but a few.

c) DNA Markers

DNA markers reveal polymorphisms in a DNA sequence, or the presence or absence of a particular DNA sequence at a particular site in the genome (normally a restriction site or PCR primer binding site). Alterations in these can create or remove restriction sites or cause nonhomology between primer sequences and the target DNA, whereas sequence insertions or deletions between such sites can also lead to a length polymorphism in a given stretch of DNA. In most cases these polymorphisms manifest themselves in variation in the length of homologous length DNA fragments and can thus be visualised and quantified by electrophoretic separation of the fragments. In its simplest form, however, a DNA polymorphism may comprise no more than the substitution of a single nucleotide in a defined DNA segment.

DNA markers are normally not readily observable in the external plant phenotype, because only a very small proportion happen to occur directly within a gene. As described above, however, the fact that their presence in non-coding sequences generally has no selective advantage or disadvantage means that plant genomes, which in many cases contain a vast proportion of non-coding DNA, have the potential to contain an extremely large number of DNA polymorphisms. Hence, mutations in just a small percentage of the genome result in a wealth of potential molecular markers (Paterson et al. 1991). Today, plant population genetics to a large extent involves the detection of various forms of DNA polymorphisms in the form of DNA markers, along with appropriate data analysis to describe the relationships within and among populations on the basis of the available marker information.

Plants contain three different genomes, constituting the nuclear, chloroplast, and mitochondrial DNA, and each genome possesses specific properties that can influence population genetics studies. The nuclear genome is inherited either biparentally or uniparentally, depending on the mating system, a factor with important consequences for gene flow studies. Chloroplast DNA (cpDNA), on the other hand, is usually inherited uniparentally (maternally) and is highly conserved in comparison to the nuclear and mitochondrial genomes (Avise 1994), a property that endears it to phylogenetic studies at the taxonomic level. For example, Rossetto et al. (2001) used analysis of cpDNA to provide evidence for intergeneric relationships in Australian Vitaceae. Mitochondrial DNA (mtDNA) is subject to frequent structural rearrangements that make its use more limited to the population level (Karp et al. 1996). This was exemplified in a study on the endangered species *Penstemon haydenii* (Caha et al. 1998) where cpDNA analysis did not show any polymorphism among samples from nine populations using RFLPs. However, mtDNA analysis on the same populations allowed various haplotypes to be identified.

The first step in a population genetics study employing DNA markers is the choice of an appropriate marker system. Many factors must be considered, including the purposes for which the information is required and the way the data obtained may be influenced by the population structure or the mating system of the species in question. Of course, in many cases, cost and labour constraints play an equally important role in deciding which marker system is most appropriate. Here we will briefly describe and evaluate some DNA marker techniques which are commonly used today for studies of genetic diversity within and among populations.

Restriction Fragment Length Polymorphism

Restriction fragment length polymorphism (RFLP) analysis reveals sequence differences at DNA restriction sites. Polymorphisms are observed by hybridisation of a labelled probe to homologous sequences within DNA restriction fragments separated by agarose gel electrophoresis. In this way a set of probe-specific bands is produced from the complete set of restriction fragments arising from digestion of, for example, the complete genomic DNA. An RFLP marker is an allelic variant of a polymorphic restriction fragment, which can arise because of sequence insertions or deletions within restriction fragments, or from sequence alterations that add or remove restriction sites.

RFLPs are co-dominant markers that have been used extensively for genetic mapping of various crop plant species; however, several drawbacks limit their use for extensive population genetics studies. The necessity to clone restriction fragments makes development of RFLP probes labour-intensive and time-consuming, meaning that the number of markers available for a given species is generally relatively low. Moreover, each marker will generally reveal only a single locus in a diploid organism, not all of which will be polymorphic in the population under investigation. Another restriction is the requirement for relatively large quantities of DNA, which can be difficult to obtain, particularly from small, rare or older plants. Furthermore, in many cases, RFLP polymorphisms can only be effectively detected by radioactively labelled probes, hence special laboratory facilities are necessary and cost and safety factors must be considered. For these reasons, RFLP markers are generally not counted amongst the most suitable for studies of genetic diversity, particularly due to the limited number of polymorphic loci that can be studied in a single experiment. Exceptions arise occasionally; for example, Cornide et al. (2000) used RFLP analyses for assessment of sugarcane genetic resources and Metais et al. (2000) described the use of RFLP markers to study genetic diversity between commercial *Phaseolus* bean lines. Most emphasis has shifted to other methods, however, particularly since the development in the past two decades of various marker applications that take advantage of PCR technology. As a rule these alternatives have the potential to generate considerably higher quantities of useful marker data at much lower cost, without great safety considerations and from a significantly smaller quantity of tissue.

Randomly Amplified Polymorphic DNA

Randomly amplified polymorphic DNA (RAPD) markers (Williams et al. 1990) are PCR amplification products that are generated from genomic DNA between the homologous binding sites of short, oligonucleotide primers in a standard PCR reaction. For any given 10mer primer, homologous sites the required distance apart will occur randomly throughout a genome with a mean frequency of about one every million base pairs (Weir 1996). In most higher plant genomes, RAPD-PCR results in between 5 and 30 amplification products with lengths between 200 and 2000 kb, which can be separated by standard agarose gel electrophoresis and visualised by staining with ethidium bromide. No prior knowledge of the DNA sequence is necessary and a virtually unlimited quantity of ubiquitous oligonucleotide primers is available, from which the best-amplified and most polymorphic can be rapidly selected in screenings of population subsamples. In comparison to other molecular marker methods, the RAPD technique is comparatively cheap and the amount of information that can be gathered by a single experimenter in a short space of time is relatively high. The only specialised equipment necessary are a thermocycler and an apparatus for agarose gel electrophoresis, which today are present in most standard molecular genetics laboratories.

On the other hand, RAPD-PCR suffers from poor reliability and reproducibility, mainly due to the short length of the primers and the relatively non-stringent PCR conditions that are necessary to obtain suitable amplification products. Besides the general danger of contamination that exists for PCR reactions in general, because of the short primers the number of RAPD amplification products can alter drastically with even minor differences in various amplification parameters. Hence their reproducibility is often questioned, particularly where experiments are repeated in different laboratories. For studies of genetic diversity where it can be ensured that PCR conditions are identical for all samples investigated, however, RAPD markers are a price worthy option, requiring little DNA and enabling the generation of reasonably large quantities of genotype data. After its advent the RAPD technique was quickly adopted by plant population geneticists, and the considerable volume of published studies describing the use of RAPDs for elucidation of diversity and population genetics relationships demonstrate the popularity of the technique. Recent examples of the application of RAPD markers to estimate genetic diversity and genetic relatedness are as wide-ranging as wild rice (Qian et al. 2001), Brazilian wheat genotypes (De Freitas et al. 2000), wild potato populations in the USA (Del Rio et al. 2001), Chilean red clover (Campos-de-Quiroz and Ortega-Klose 2001), apple (Goulao et al. 2001), *Prunus* (Jordano and Godoy 2000), cultivated and wild tea plants (Lai et al. 2001), and an endangered Spanish *Rosmarinus* species (Martin and Bermejo 2000).

It should be remembered that RAPDs are dominant markers, because the phenotype classes simply comprise the presence or absence of an amplification product, and therefore homozygotes cannot be distinguished from heterozygotes at any given locus. This can lead to bias in allele frequency calculations, particularly for outcrossing species, hence statistical manipulation is necessary to compensate for this. Furthermore, a single RAPD band may in fact contain DNA from more than one locus with the same fragment length.

Simple Sequence Repeats (Microsatellites)

Simple sequence repeat (SSR) sequences, also known as microsatellites, are a class of variable number tandem repeats comprised of a short (1–6 bp) nucleotide motif repeated up to 100 times or more. SSRs, which exist in eukaryotic nuclear genomes and in the chloroplast genome of some plants (see Powell et al. 1995), are of great interest to population geneticists because of their high mutation rate, which has been variously estimated between 10^{-2} (Dallas 1992) and 10^{-6} (e.g. Udupa and Baum 2001), presumed to be due to frequent frameshift mutations as a result of the simple repeat nature of the sequence. Such a high rate of mutation results in highly polymorphic markers, allowing genetic discrimination of very closely related individuals with only a relatively low number of markers. The co-dominant nature of SSR markers allows the distinction between heterozygotes and homozygotes, and their dimeric phenotype makes the scoring of genotypes and calculation of allele frequencies extremely simple (homozygotes display a single band, heterozygotes two bands). Because of the ultimate restriction in the length of microsatellites (and therefore the number of possible alleles), they are not so informative for distantly related taxa (Goldstein et al. 1995); however, for studies of genetic variation within and among populations of the same species, SSR markers are an extremely valuable tool. Moreover, for genetic diversity assessment at the subpopulation level the high numbers of alleles per locus is a significant advantage, particularly in species with a high degree of inbreeding and in small or fractured populations.

SSR markers are generated by PCR amplification of specific microsatellite loci using primers developed from the sequences flanking the simple repeat. This means, however, that primers must first be developed for the species in question by isolating, cloning and sequencing individual SSR loci. Microsatellite primer development is thus a costly and time-consuming process. In cases where sequence information is already available for polymorphic SSR primers, on the other hand, the technique is then relatively cheap. In many cases, SSR polymorphisms can be satisfactorily distinguished in a high-resolution agarose gel; however, because microsatellite polymorphisms can also result from the addition or deletion of only a single copy of the repeat motif, in the case of dimeric or trimeric repeat motifs it is often advisable to separate amplification products by polyacrylamide gel electrophoresis.

A large number of studies have demonstrated the utility of SSR variation for investigation and analysis of population structure and genetic diversity in various plant species.

Diverse examples selected from the recent literature include evaluations of diversity in endemic and introduced soybean (Brown-Guedira et al. 2000, Narvel et al. 2000), variation in apple hybrids (Hokanson 2001), a study of clonal diversity and population substructure in the marine flowering plant *Zostera marina* (Reusch et al. 2000), a study of species diversity in barley (Saghai Maroof et al. 1994), and numerous investigations of population structure in the model plant *A. thaliana* (Todokoro et al. 1995, Innan et al. 1997, Loridon et al. 1998, Virk et al. 1999). Interestingly, it has been shown that appropriate statistical analysis of only five linked microsatellite loci can provide more information on coalescent dates than approximately 15 kb of DNA sequence (Wilson and Balding 1998).

Intersimple Sequence Repeats

Intersimple sequence repeat (ISSR) markers, first described by Gupta et al. (1994) and Zietkiewicz et al. (1994), are PCR products amplified by primers complementary to SSR repeat motifs with an extra 1–3 base pairs that anchor the primer at the 3′ or 5′ end. The amplification products therefore comprise stretches of DNA between random simple sequence repeats. In contrast to SSR markers, the ISSR technique requires no prior sequence information and hence the development of these markers is considerably less expensive and time-consuming than is the case for SSR primer development. ISSR markers therefore represent a potentially useful, cheap tool for genetic mapping, particularly because together with SSRs they are generally held to be relatively abundant in gene-rich genomic regions. Application of ISSR markers for plant genetics in general is described by Godwin et al. (1997). To date their use for genetic diversity studies has been only moderate in comparison with other DNA marker methods. Culley and Wolfe (2001) determined the population genetic structure of *Viola pubescens* by ISSR analysis, Fang and Rose (1999) described use of such markers to show phylogenetic relationships among citrus germplasm, and Leroy et al. (2000) studied genetic diversity in *Brassica oleracea* accessions. Salimath et al. (1995) assessed genetic diversity in finger millet (*Eleusine coracana*) and, in a direct comparison with RAPDs and RFLPs, reported favourably on the use of ISSRs for plant genome analysis.

Amplified Fragment Length Polymorphism

Amplified fragment length polymorphisms (AFLP) were first described by Zabeau and Vos (1993). The method is based on selective amplification of restriction fragments by annealing known adapter sequences to restriction fragment ends, which then act as highly specific primer binding sites for subsequent PCR amplification (see Vos et al. 1995). An excellent review of AFLP methodology and applications is provided by Mueller and Wolfenbarger (1999).

Selectivity in the AFLP amplification procedure is achieved by the use of primers including between one and three extra nucleotides in addition to the adapter sequence, so that only the subset of restriction fragments containing these bases directly adjacent to the restriction site is amplified. For most plant genomes the genomic DNA is double-digested using both a frequently cutting restriction enzyme (with a 4-bp restriction site) and also a rare cutter (with a 6-bp restriction site). Because a restriction site for a 4-cutter enzyme occurs on average every 256 bp in the genome, the resulting restriction fragments have a length suitable for PCR amplification (<1000 bp). Furthermore, on average one-sixteenth of the products can be expected to have one end digested with the 6-cutting enzyme. For plant genomes with a 1C DNA content between about 800 (e.g., diploid *Brassica* species) and 5000 Mbp (e.g. barley), this results in a total of between 3.1

and 20 million fragments, of which around 195,000–1.2 million have one end cut with the rare-cutting enzyme. Selective PCR with primers containing three extra, randomly selected nucleotides ("+3-primer") will amplify on average 1/4096th of these. In order to visualise only those amplification products from restriction fragments cut with the rare-cutting enzyme, and not the considerably more numerous fragments cut at both ends with the 4-cutter, the primer specific to the adapter sequence of the rare-cutter restriction site is labelled with either a fluorescent or radioactive marker. This can be expected to give an average of around 50 (*Brassica*) to 300 (barley) labelled amplification products for any given combination of +3-primers. The size and number of the resulting AFLP products make them ideally suited to size fractionation and visualisation as bands by polyacrylamide gel electrophoresis. For extremely large (e.g., conifers) or considerably smaller (e.g., fungi) genomes it is possible to adjust the final quantity of bands either by varying the number of selective nucleotides in the AFLP primers or by using restriction enzymes that cut more or less frequently. The number of polymorphic AFLP markers can also be dependent on the specific restriction enzymes used. For example, Breyne et al. (1999) revealed that in *Arabidopsis thaliana* ecotypes SacI/MseI digestion gave rise to fewer polymorphic markers than EcoRI/MseI digestion, reflecting the GC content of the genome.

Depending on the structure of the species or population under investigation, up to around 80% of all AFLP bands can show length polymorphisms (seeMiyashita et al. 1999; Krauss 2000; Sawkins et al. 2001), hence an extremely large number of polymorphic markers can be obtained in a single experiment. Moreover, the use of long (approx. 20 bp), restriction-site-specific PCR primers makes AFLP amplification products highly reproducible, both between experiments in the same laboratory and also between different laboratories. A comparative study by Jones et al. (1997) in eight different European laboratories revealed only one single scoring difference out of a total of 172 markers when all laboratories analysed the same samples with the same primer combinations. The amount of genomic DNA required for AFLP analysis is very small, and because a single restriction digest can be used for hundreds of amplifications with different primer combinations, the number of markers that can be produced is virtually unlimited. The requirement for DNA restriction and ligation enzymes makes AFLP analyses relatively expensive, particularly for extensive population studies. Also, special equipment or facilities are required for handling and analysis of fluorescent or radioactively labelled amplification products. Despite these drawbacks, however, the quantity and quality of genotype information that can be obtained using this marker system with a relatively low experimental input is extremely high in comparison with other methods. Although AFLP markers, like RAPD markers, are dominantly inherited, the sheer volume of reproducible, polymorphic marker data that can be achieved by AFLP analysis makes this technique highly applicable not only for genetic mapping studies, but also for in-depth studies of genetic diversity at the population and subpopulation level. In recent years, an enormous wealth of information has been obtained by AFLP analysis to describe genetic diversity in a vast number of plant species ranging from the model plant *Arabidopsis thaliana* (Breyne et al. 1999; Miyashita et al. 1999; Erschadi et al. 2000), to major agricultural plants including rice (e.g. Fuentes et al. 1999), wheat (Manifesto et al. 2001), barley (Pakniyat et al. 1997), *Lolium* (Cresswell et al. 2001), sunflower (Hongtrakul et al. 1997), *Beta* (Hansen et al. 1999), and tobacco (Ren and Timko 2001).

Single Nucleotide Polymorphisms

Single nucleotide polymorphisms (SNPs) are the most abundant form of DNA polymorphism in most organisms, and have been estimated to occur on average once per 250–1000 bp, accounting for approximately 90% of DNA sequence variants in the human genome (Brookes 1999; Gupta et al. 2001). Various methods used for the detection and scoring of SNPs are reviewed in detail by Gupta et al. (2001) and Landegren et al. (1998), respectively. Most SNP protocols involve target sequence PCR amplification, which is time-consuming and costly but has great potential for improvement through automation and with the application of the miniature array (DNA-chip) technique. A common problem inherent with SNPs at present is that pseudogenes, which are highly similar in sequence to the target gene, may also be processed. Brookes (1999) estimated that this could result in false scoring of up to 20% of SNP assays.

To date the major employment of SNPs has been in the detection of genes playing a role in human diseases. However, their use in genomic studies of higher plants has accelerated in recent years, particularly since detailed sequence and candidate gene information has been made available through the complete sequencing of the *A. thaliana* genome. SNPs can be expected to play a prominent role in diversity studies in plants in future, particularly where candidate genes for disease resistance or other important agronomic traits are concerned. One of the exciting prospects that SNPs have to play in population genetics is that, due to their extent throughout the genome, meiotic recombination events are less likely to occur between them in comparison to genes or other larger sequences of DNA. Thus, several SNPs in a sequence of DNA may retain complete linkage disequilibrium for a considerable period of time.

DNA Sequencing

The complete sequence of the DNA is the only way to reveal all polymorphisms embedded in any given genomic region, and thus DNA sequencing provides the ultimate tool for describing the genetic diversity at the DNA level. The resolution of sequencing for studies of genetic diversity was evidenced by Hardtke et al. (1996), who used cpDNA and genomic DNA sequences to estimate genetic similarity among *Arabidopsis* ecotypes. Separate calculations based on the nuclear and the non-nuclear sequences revealed the existence of small clusters of very closely related ecotypes separated from each other by extensive sequence divergence.

Despite the enormously detailed information that can be gathered, sequencing cannot normally be widely applied for genetic diversity studies on a population level. Even with automation, screening large numbers of individuals is time-consuming and expensive. Sequencing of targeted gene regions is limited by primer availability and, as for SNPs, there is a danger of pseudogenes arising when conserved primers are used.

To summarise, the choice of marker system and strategy should reflect the focus of the investigation. For example, hypervariable microsatellite loci are very appropriate for studies between individuals in a population, but become less informative as one increases the taxonomic level of comparison, or the time scale. Likewise, RFLP analysis of cpDNA can show little variation between taxa and over a long time period (Caha et al. 1998). Thus, the sensitivity of the marker system should be weighed against the variability (mutation rate, crossing over rate) of the targeted DNA region. Furthermore, an important consideration is that multiple marker systems are likely to give a much broader coverage of the genome, and thus more accurate results, than a single marker system alone (Davierwala et al. 2000).

Powell et al. (1996) described the concept of the "Marker Index" as a measure of the statistical power of a molecular marker method combined with the amount of information gained per analysis. Despite the fact that AFLPs may produce a lower proportion of polymorphic bands than SSR markers, for example, they give considerably more polymorphic bands per assay and thus have the highest Marker Index under this system of measurement. On the other hand, microsatellites have always revealed the highest levels of polymorphism when compared with other markers (Pejic et al. 1998) and thus may be more suitable to elucidate allelic differences in highly inbred populations or species. In order not to overlook these and other important factors, consultation of appropriate statistical literature (e.g., Weir 1996) is strongly recommended during the planning of molecular marker studies of population genetics diversity.

4 Measures of Genetic Diversity

The expression used to convey genetic diversity for any given study depends on the parameters being evaluated and the individuals or the populations being measured or compared. Different conditions apply, for example, for measurements of the genetic diversity within a given population or between populations, and for measurements based on morphological or molecular data. Obviously, morphological data can also divulge genetic diversity, but epistasis can alter or obscure the underlying genetic variation.

a) Fixation Indices

For within-population diversity studies, the simplest measures of genetic diversity are calculated from the number and frequency of alleles in the population and the proportion of heterozygotes (or heterozygous loci). If a population is in Hardy-Weinberg equilibrium then genotype frequencies are proportional to the allele frequencies in a binomial relationship. However, most populations exhibit a substructure of some form, an observation that led Wright (1943) to the development of "F-statistics". These attempt to quantify the loss of heterozygosity in a population due to the non-random mating of individuals within the population and the consequent inbreeding effects.

For within-population analysis, Wright's inbreeding coefficient F is calculated as:

$$F = (H_0 - H_I)/H_0$$

where H_0 is the expected heterozygosity and H_I the observed heterozygosity in a given population. From a biological point of view, F represents the reduction in heterozygosity in a population due to inbreeding, in comparison to a randomly mating population with the same allele frequencies. Other influences affecting this value include gene flow, selection, mutation rate and genetic drift (Petit et al. 2001), demonstrating the dependence of the inbreeding coefficient on evolutionary factors.

Wright suggested the idea of a hierarchical population structure with three levels: subpopulations ($_S$), regional populations ($_I$), and the total population ($_T$). Comparing any two of these population levels shows the decrease in heterozygosity due to non-random mating in one population relative to another, more inclusive, level of population. The fixation index F_{ST} is calculated as follows:

$$F_{ST} = (H_T - H_{ST})/H_T$$

where H_T is the average heterozygosity for the total population and H_S is the heterozygosity averaged over all subpopulations. As the allele frequencies deviate, the difference between H_T and H_S increases and therefore serves as a measure of genetic distance between populations. Measures can also be obtained for F_{IT} and F_{IS} by using the appropriate average heterozygosity values. In much the same way quantitative traits can also be used as a measure of genetic diversity. In this case quantitative information is recorded and compared instead of allele frequencies, and the corresponding fixation index is represented as Q_{ST} (Merilä and Crnokrak 2001). High values of any of these measurements are an indication for subpopulations with little interpopulation gene flow, i.e. high genetic variation between populations, whereas lower values indicate that the allelic variation in the total population is more or less completely reflected in the subpopulations, a sign of more or less unrestricted movement of genes between subpopulations.

It is of particular importance for conservationists to realise that endangered species are often under strict selection conditions and therefore could show little variation in quantitative traits that might be under selection, but a higher variation in selectively neutral molecular markers (Petit et al. 2001). In some highly inbred plant populations, however, Q_{ST} indices have been found to be approximately equal to F_{ST} indices (e.g. Bonnin et al. 1997; Hardy et al. 2000).

b) Discrete and Continuous Data

Information obtained from morphological observations and molecular genetic techniques can take the form of either discrete (qualitative) or continuous (quantitative) data. Discrete data of DNA marker information can be classified into two forms depending on whether the method used delivers band-sharing or sequence data. Shared bands are scored simply according to their presence or absence and can be conveniently recorded in a one-zero matrix for statistical analysis, whereas single-locus techniques result in information that may be variously scored as nucleotides (SNP markers, sequence information), as DNA lengths (microsatellites) or, in the case of allozymes, according to net electrophoretic mobility. The complete data from a population can be summarised on the basis of allele frequencies (co-dominant markers) or band frequencies (dominant markers) for all loci investigated, whereby each individual is classified by a numerical genotype describing the presence or absence of all alleles present in the population. This individual genotype information serves as the basis for statistical comparisons between individuals within and among populations, and ultimately for the estimation of genetic diversity based on the available molecular marker data. The statistical tests applied depend not only on the type of marker system and the extent of the data, but also on what is appropriate for the structure of the population being examined.

Both sequence and shared band data can also be used to estimate genetic distances. The genetic distance between two samples (individuals or populations) is a quantitative estimate of how divergent they are genetically (Avise 1994). The objective of creating phylogenies is to determine the most common ancestor for groups of samples: the more distant the common ancestor, the more genetically diverse the samples.

The genetic structure of plant populations is affected strongly by their phylogenetic histories (Schaal and Leverich 2001). Thus it is important to have accurate theoretical models that can use contemporary data obtained with molecular markers to infer historical events such as founder effects and bottlenecks as well as elucidating the roles of migration, genetic drift and other demographic trends.

c) Calculation of Genetic Distances

There is some resistance by systematists to the idea of genetic distance measures. This stems from the fact that under any distance measure two entities can appear either highly divergent or as "sister taxa", without sharing any particular feature which would otherwise lead to that conclusion (Goldstein et al. 2000). Despite this, measurements and representations of genetic distance have become the most common way, and are indeed perhaps the most appropriate way available at present, to describe the level of genetic diversity present within a species or population based on morphological or molecular marker data. There are several methods available for phylogenetic reconstruction, at both the intra- and interspecific level. It is beyond the scope of this review to give exact details on the statistical analyses behind the various genetic distance estimates commonly used for diversity studies; however, an evaluation of the different statistical techniques with emphasis on plant populations, paying particular attention to the various statistical software available for their implementation, can be found in Labate (2000). In all cases the type of method that can be used for phylogenetic analysis depends on the type of data available.

Genetic data recorded as presence or absence of bands can be transformed into a matrix of pairwise distances and analysed by subjecting the distance matrix to clustering algorithms such as UPGMA (unweighted pair group method using arithmetic averages) or neighbour-joining. Sequence data, however, can be analysed using either parsimony-based methods or maximum likelihood methods. The results of such analyses are usually shown as dendrograms. It is important to remember that in representing genetic relationships using (rooted) trees presumes a phylogeny in the given populations. The position of each individual or subpopulation is dependent on all other individuals or subpopulations that are included in the analysis; in other words, the spatial position of the individual or subpopulation is always dependent on the genotypes included in the analysis. Adding new data may result in a completely different tree (Nei 1996). On the other hand, in an unrooted tree or a principle component analysis, each individual or subpopulation is assigned a specific value in a two- or three-dimensional space displaying the distance between the groups without underlying a phylogenetic relationship (Weir 1996).

Maximum likelihood methods consider each individual nucleotide or amino acid in sequence data independently and calculate a tree representing the highest probability of the observed data set being derived from it under a certain model, while with parsimony-based methods the tree that best describes the smallest amount of evolutionary changes (in the form of nucleotides or amino acids) is used to explain difference observed between population samples (Karp et al. 1996).

5 Role of Molecular Markers in Biodiversity Studies

The development of molecular marker techniques in recent decades has allowed access to the wealth of information contained in the DNA sequences of all organisms and now allows plant breeders, ecologists, geneticists and conservationists to increase the power of their studies on all botanical forms. Despite the proliferation of marker techniques and the huge advances in statistical evaluation brought about by the accessibility of modern computers and the development of specialised, user-friendly software, some questions still remain concerning the relevance of molecular marker information to the overall fitness of individuals or populations.

Undoubtedly, the application of molecular marker techniques has enabled biologists to examine the genetic structure of individuals and populations to a far more advanced level than was possible a few decades ago. However, there is still a large gap in our knowledge of how genetic information transforms into measures of individual or population fitness. Some marker systems, such as AFLPs, microsatellites and SNPs, may reveal large amounts of statistical variation within and between populations; however, this may not necessarily reflect relevant biological variation, especially when the markers in question are selectively neutral. Conversely, other methods with lower genetic resolution may reveal little variation despite biologically significant differences within the population. Choice of polymorphic markers within a particular marker system may also be relevant. Virk et al. (2000) analysed rice with AFLPs to test whether or not mapped markers were more useful for assessing genetic diversity and concluded that it is appropriate to use unmapped markers provided that the marker type has been shown to have a wide distribution over the genome.

In general, however, molecular markers are the most convenient method for detailed investigations of genetic relationships within and among populations and, in most cases, markers allow good estimates of the true genetic situation. To name just one example, Bonnin et al. (1997) found a positive correlation between variation in quantitative traits and the amount of polymorphism detected by molecular markers in natural populations of the selfing plant species *Medicago truncatula*. Due to the power of marker techniques compared to conventional Mendelian traits, inferences as to the amount of non-random mating within populations can be elucidated (Pryor et al. 2001). Indeed, knowing the frequencies of alleles within populations and subpopulations enables a much more accurate picture of how populations are interacting – for example, the geographical distances of mating, the occurrence of bottlenecks or estimates of gene flow – even if the markers selected are neutral and have little biological meaning for fitness.

In plant breeding programs, marker-assisted selection (MAS) is beginning to take hold as a cost-viable method of screening, especially when traditional phenotypic selection is difficult. Molecular markers can shorten the breeding process by rapid detection of desirable characteristics, facilitating the introgression of these characters from wild species into domesticated strains. However, the technology is still expensive enough that it cannot yet be applied to the demands of large-scale breeders and effective markers are still only available for a small number of important traits in major crop species. On the other hand, application of molecular markers for genetic diversity studies in crop plants has the potential to identify distantly related germplasm of potentially high value for the construction of heterotic groups for hybrid breeding. As an example, Mayes et al. (2000) used RFLP to assess genetic structure for an oil palm breeding programme. Such approaches are particularly important today in crops such as oilseed rape, where the genetic basis is extremely narrow. Moreover, genetic diversity of crop species is essential for yield stability and improvement. If cultivars possess a low level of genetic diversity, they are then more susceptible to losses due to diseases or pests (Jordan et al. 1998). Barrett et al. (1998) compared AFLP and pedigree-based genetic diversity assessment on wheat cultivars and found that AFLP was better able to identify cultivars, making the marker data a very useful breeding tool for parent plant selection.

A major source of genetic diversity, not only for plant breeding but also for the conservation of plant germplasm in general, are genebank collections. In general, the aim of germplasm conservation is to maximise the genetic variation stored in genebanks (Del Rio et al. 2001), for which molecular marker techniques represent an important tool. The dimensions of in situ plant collections are always limited by cost and space constraints, particularly where seed material needs to be regularly regenerated to retain germination ability, and hence it is impossible for genebanks to retain all possible forms of a given species. With the help of marker analyses, however, it is possible to estimate the genetic diversity present in a particular species and to select reference genotypes that represent the broadest possible range of variation present within a manageable number of genotypes. The cassava core collection, for example, which had been based on isozyme analysis, place of origin and morphology data and which was selected to represent the potential genetic diversity of the crop, was studied using microsatellites and AFLPs by Chavarriaga-Aguirre et al. (1999), who demonstrated that the traditional selection and markers had been highly effective in selecting unique genotypes for the core.

Genetic variability data can also give information about wild populations of endangered plant species, enabling decisions to be made regarding their conservation. In devising conservation strategies the breeding system of the species must be taken into account. For example, selfing species tend to have a higher differentiation between populations and a lower level of heterozygosity compared to outcrossing species (Zawko et al. 2001).

For ecological and evolutionary studies, molecular markers can often assist in the exact description of populations when morphological characters do not suffice. For example, Blattner and Badani Méndez (2001) used RAPD markers as evidence against a second centre of barley domestication in Morocco. Conservation efforts on the endangered desert plant *Agave victoriae-reginae* using allozyme data resulted in the discovery of three distinct groups of populations (Martinez-Palacios et al. 1999), and it was possible to make inferences about the effectiveness of animal pollinators given the F_{ST}. Similarly, Williams and Waser (1999) used allozyme data to infer gene flow in *Delphinium* and to estimate pollination range and isolation distances, and Gaudeul et al. (2000) used AFLP to assess genetic diversity in the endangered alpine plant *Eryngium alpinum*. In another application, RFLPs and RAPDs have also been used in the assessment of weed species for the purposes of biological weed control (Nissen et al. 1995). Hernandez et al. (2001) recently discovered a retrotransposon-like element in olive that was informative for species identification. In general, it can be said that inferences about evolution and dispersion of plant species are often best obtained by marker analysis.

6 Summary

The huge variety of botanical species for which studies involving analysis of genetic diversity by various molecular marker technologies have been published in recent years demonstrates the enormous potential that these techniques have in the description of plant species from a population genetics perspective. With appropriate choice of the marker system and careful data handling and interpretation, it is possible to use molecular marker data to make inferences regarding gene flow and population structure which can assist in decision-making with regard not only to plant breeding, but also for conservation purposes. Since the widespread development of modern accessible molecular marker methods it has become possible for population geneticists to obtain an unprecedented detailed insight into the relationships within and among plant populations from the DNA sequence level upwards and to use this data to make inferences from evolutionary and ecological perspectives. With further developments in the high-throughput automation of existing molecular marker techniques, and with further new markers and rapid advances in sequencing technologies, the amount of genetic data available for any given plant species will almost certainly increase even more in the near future. The challenge for population genetics in coming years will therefore reside to a large extent in the processing and evaluation of this data with respect to its relevance from a biological perspective. Just how much useful information can be gained from the DNA of a given species remains to be seen, but it is necessary already today to consider the relevance of such information for diversity studies and the ways in which new genetic information can be applied to solve problems facing botanists in the future.

References

Alonso-Blanco C, Koorneef M (2000) Naturally occurring variation in *Arabidopsis*: an underexploited resource for plant genetics. Trends Plant Sci 5:22–26

Avise JC (1994) Molecular markers, natural history and evolution. Chapman and Hall, London

Barrett BA, Kidwell KK, Fox PN (1998) Comparison of AFLP and pedigree-based genetic diversity assessment methods using wheat cultivars from the Pacific Northwest. Crop Sci 38:1271–1278

Blattner FR, Badani Méndez AG (2001) RAPD data do not support a second centre of barley domestication in Morocco. Genet Resour Crop Evol 48:13–19

Bonnin I, Prosperi JM, Olivieri I (1997) Comparison of quantitative genetic parameters between two natural populations of a selfing plant species, *Medicago truncatula* Gaertn. Theor Appl Genet 94:641–651

Breyne P, Rombaut D, Van Gysel A, Van Montagu M, Gerats T (1999) AFLP analysis of genetic diversity within and between *Arabidopsis thaliana* ecotypes. Mol Gen Genet 261:627–634

Brookes AJ (1999) The essence of SNPs. Gene 234:177–186

Brown-Guedira GL, Thompson JA, Nelson RL, Warburton ML (2000) Evaluation of genetic diversity of soybean introductions and North American ancestors using RAPD and SSR markers. Crop Sci 40:815–823

Caha CA, Lee DJ, Stubbendieck J (1998) Organellar genetic diversity in *Penstemon haydenii* (Scrophulariaceae): an endangered plant species. Am J Bot 85:1704–1709

Campos-de-Quiroz H, Ortega-Klose F (2001) Genetic variability among elite red clover (*Trifolium pratense* L.) parents used in Chile as revealed by RAPD markers. Euphytica 122:61–67

Chavarriaga-Aguirre P, Maya MM, Tohme J, Duque MC, Iglesias C, Bonierbale MW, Kresovich S et al. (1999) Using microsatellites, isozymes and AFLPs to evaluate genetic diversity and redundancy in the cassava core collection and to assess the usefulness of the DNA-based markers to maintain germplasm collections. Mol Breed 5:263–273

Cornide MT, Coto O, Calvo D, Canales E, Esquivel FD, Oramas GP (2000) Molecular markers for the identification and assisted management of genetic resources for sugarcane breeding. Plant Varieties Seeds 13:113–123

Cresswell A, Sackville Hamilton NR, Roy AK, Viegas BMF (2001) Use of amplified fragment length polymorphism markers to assess genetic diversity of *Lolium* species from Portugal. Mol Ecol 10:229–241

Culley TM, Wolfe AD (2001) Population genetic structure of the cleistogamous plant species *Viola pubescens* Aiton (Violaceae), as indicated by allozyme and ISSR molecular markers. Heredity 86:545–556

Dallas JF (1992) Estimation of microsatellite mutation rates in recombinant inbred strains of mouse. Mamm Genome 3:452–456

Davierwala AP, Chowdari KV, Kumar S, Reddy APK, Ranjekar PK, Gupta VS (2000) Use of three different marker systems to estimate genetic diversity of Indian elite rice varieties. Genetica 108:269–284

De Freitas LB, Jerusalinsky L, Bonatto SL, Salzano FM (2000) Extreme homogeneity among Brazilian wheat genotypes determined by RAPD markers. Pesqui Agropecu Bras 35:2255–2260

Del Rio AH, Bamberg JB, Huaman Z, Salas A, Vega SE (2001) Association of ecogeographical variables and RAPD marker variation in wild potato populations of the USA. Crop Sci 41:870–878

Epperson BK, Chung MG (2001) Spatial genetic structure of allozyme polymorphisms within populations of *Pinus strobus* (Pinaceae). Am J Bot 88:1006–1010

Erschadi S, Haberer G, Schöniger M, Torres-Ruiz RA (2000) Estimating genetic diversity of *Arabidopsis thaliana* ecotypes with amplified fragment length polymorphisms (AFLP). Theor Appl Genet 100:633–640

Fang DQ, Roose ML (1999) Inheritance of intersimple sequence repeat markers in citrus. J Hered 90:247–249

Fuentes JL, Escobar F, Alvarez A, Gallego G, Duque MC, Ferrer M, Deus JE, Tohme JM (1999) Analyses of genetic diversity in Cuban rice varieties using isozyme, RAPD and AFLP markers. Euphytica 109:107–115

Gaston KJ (1996) Biodiversity. Blackwell Science, Cambridge

Gaudeul M, Taberlet P, Till-Bottraud I (2000) Genetic diversity in an endangered alpine plant, *Eryngium alpinum* L. (Apiaceae), inferred from amplified fragment length polymorphism markers. Mol Ecol 9:1625–1637

Gilliland TJ, Coll R, Calsyn E, De Loose M, van Eijk MJT, Roldan-Ruiz I (2000) Estimating genetic conformity between related ryegrass (*Lolium*) varieties. 1. Morphology and biochemical characterisation. Mol Breed 6:569–580

Godwin ID, Aitken EAB, Smith LW (1997) Application of inter simple sequence repeat (ISSR) markers to plant genetics. Electrophoresis 18:1524–1528

Goldstein DB, Linares AR, Cavalli-Sforza LL, Feldman MW (1995) An evaluation of genetic distances for use with microsatellite loci. Genetics 139:463–471

Goldstein PZ, DeSalle R, Amato G, Vogler AP (2000) Conservation genetics at the species boundary. Conserv Biol 14:120–131

Goulao L, Cabrita L, Oliveira CM, Leitao JM (2001) Comparing RAPD and AFLP (TM) analysis in discrimination and estimation of genetic similarities among apple (*Malus domestica* Borkh.) cultivars - RAPD and AFLP analysis of apples. Euphytica 119:259–270

Gupta M, Chyi Y-S, Romero-Severson J, Owen JL (1994) Amplification of DNA markers from evolutionarily diverse genomes using single primers of simple-sequence repeats. Theor Appl Genet 89:998–1006

Gupta PK, Roy JK, Prasad M (2001) Single nucleotide polymorphisms: a new paradigm for molecular marker technology and DNA polymorphism detection with emphasis on their use in plants. Curr Sci 80:524–535

Hamrick JL, Godt MJ (1989) Allozyme diversity in plants. In: Brown AHD, Clegg MT, Kahler AL, Weir BS (eds) Plant population genetics, breeding, and genetic resources. Sinauer Association, Sunderland, MA, pp 43–63

Hansen M, Kraft T, Christiansson M, Nilsson N-O (1999) Evaluation of AFLP in *Beta*. Theor Appl Genet 98:845–852

Hardtke CS, Muller J, Berleth T (1996) Genetic similarity among *Arabidopsis thaliana* ecotypes estimated by DNA sequence comparison. Plant Mol Biol 32:915–922

Hardy OJ, Vanderhoeven S, Meerts P, Vekemans X (2000) Spatial autocorrelation of allozyme and quantitative markers within a natural population of knapweeds. J Evol Biol 13:656–667

Hernandez P, de la Rosa R, Rallo L, Martin A, Dorado G (2001) First evidence of a retrotransposon-like element in olive (*Olea europaea*): implications in plant variety identification by SCAR-marker development. Theor Appl Genet 102:1082–1087

Hokanson SC, Lamboy WF, Szewc-McFadden AK, McFerson JR (2001) Microsatellite (SSR) variation in a collection of *Malus* (apple) species and hybrids. Euphytica 188:281–294

Hongtrakul V, Huestis GM, Knapp SJ (1997) Amplified fragment length polymorphisms as a tool for DNA fingerprinting sunflower germplasm: genetic diversity among oilseed inbred lines. Theor Appl Genet 95:400–407

Huh MK, Ohnishi O (2001) Allozyme diversity and population structure of Japanese and Korean populations of wild radish, *Raphanus sativus* var. hortensis f. raphanistroides (Brassicaceae). Genes Genet Syst 76:15–23

Huston MA (1994) Biological diversity. Cambridge University Press, Cambridge

Innan H, Terauchi R, Miyashita NT (1997) Microsatellite polymorphism in natural populations of the wild plant *Arabidopsis thaliana*. Genetics 146:1441–1452

Jones CJ, Edwards KJ, Castaglione S, Winfield MO, Sala F, van de Wiel C, Bredemeijer G et al. (1997) Reproducibility testing of RAPD, AFLP and SSR markers in plants by a network of European laboratories. Mol Breed 3:381–390

Jordan DR, Tao YZ, Godwin ID, Henzell RG, Cooper M, McIntyre CL (1998) Loss of genetic diversity associated with selection for resistance to sorghum midge in Australian sorghum. Euphytica 102:1–7

Jordano P, Godoy JA (2000) RAPD variation and population genetic structure in *Prunus mahaleb* (Rosaceae), an animal-dispersed tree. Mol Ecol 9:1293–1305

Karp A, Seberg O, Buiatti M (1996) Molecular techniques in the assessment of botanical diversity. Ann Bot 78:143–149

Krauss SL (2000) Accurate gene diversity estimates from amplified fragment length polymorphism (AFLP) markers. Mol Ecol 9:1241–1245

Labate JA (2000) Software for population genetic analyses of molecular marker data. Crop Sci 40:1521–1528

Lai JA, Yang WC, Hsiao JY (2001) An assessment of genetic relationships in cultivated tea clones and native wild tea in Taiwan using RAPD and ISSR markers. Bot Bull Acad Sin 42:93–100

Landegren U, Nilsson M, Kwok PY (1998) Reading bits of genetic information: methods for single-nucleotide polymorphism analysis. Genome Res 8:769–776

Leroy XJ, Leon K, Branchard M (2000) Characterisation of *Brassica oleracea* L. by microsatellite primers Plant Syst Evol 225:235–240

Li Z, Rutger JN (2000) Geographic distribution and multilocus organization of isozyme variation of rice (*Oryza sativa* L.). Theor Appl Genet 101:379–387

Loridon K, Cournoyer B, Goubley C, Depeiges A, Picard G (1998) Length polymorphism and allele structure of trinucleotide microsatellites in natural accessions of *Arabidopsis thaliana*. Theor Appl Genet 97:591–604

Manifesto MM, Schlatter AR, Hopp HE, Suarez EY, Dubcovsky J (2001) Quantitative evaluation of genetic diversity in wheat germplasm using molecular markers. Crop Sci 41:682–690

Martin JP, Bermejo JEH (2000) Genetic variation in the endemic and endangered *Rosmarinus tomentosus* Huber-Morath & Maire (Labiatae) using RAPD markers. Heredity 85:434–443

Martinez-Palacios A, Eguiarte LE, Furnier GR (1999) Genetic diversity of the endangered endemic *Agave victoriae-reginae* (Agavaceae) in the Chihuahuan Desert. Am J Bot 86:1093–1098

Mayes S, Jack PL, Corley RHV (2000) The use of molecular markers to investigate the genetic structure of an oil palm breeding programme. Heredity 85:288–293

Merilä J, Crnokrak P (2001) Comparison of genetic differentiation at marker loci and quantitative traits. J Evol Biol 14:892–903

Metais I, Aubry C, Hamon B, Jalouzot R, Peltier D (2000) Description and analysis of genetic diversity between commercial bean lines (*Phaseolus vulgaris* L.). Theor Appl Genet 101:1207–1214

Miyashita NT, Kawabe A, Innan H (1999) DNA variation in the wild plant *Arabidopsis thaliana* revealed by amplified fragment length polymorphism analysis. Genetics 152:1723–1731

Moghaddam M, Ehdaie B, Waines JG (2000) Genetic diversity in populations of wild diploid wheat *Triticum urartu* Tum. ex. Gandil. revealed by isozyme markers. Genet Resour Crop Evol 47:323–334

Mueller UG, Wolfenbarger LL (1999) AFLP genotyping and fingerprinting. Trends Ecol Evol 14:389–394

Narvel JM, Fehr WR, Chu WC, Grant D, Shoemaker RC (2000) Simple sequence repeat diversity among soybean plant introductions and elite genotypes. Crop Sci 40:1452–1458

Neel MC, Ellstrand NC (2001) Patterns of allozyme diversity in the threatened plant *Erigeron parishii* (Asteraceae). Am J Bot 88:810–818

Nei M (1996) Phylogenetic analysis in molecular evolutionary genetics. Annu Rev Genet 30:371–403

Nissen SJ, Masters RA, Lee DJ, Rowe ML (1995) DNA-based marker systems to determine genetic diversity of weedy species and their application to biocontrol. Weed Sci 43:504–513

Owens IPF, Bennett PM (2000) Quantifying biodiversity: a phenotypic perspective. Conserv Biol 14:1014–1022

Pakniyat H, Powell W, Baird E, Handley LL, Robinson D, Scrimgeour CM, Nevo E, Hackett CA, Caligari PDS, Forster BP (1997) AFLP variation in wild barley (*Hordeum spontaneum* C. Koch) with reference to salt tolerance and associated ecogeography. Genome 40:332–341

Paterson AH, Tanksley SD, Sorrells ME (1991) DNA markers in plant improvement. Adv Agron 46:39–90

Pejic I, Ajmone-Marsan P, Morgante M, Kozumplick V, Castiglioni P, Taramino G, Motto M (1998) Comparative analysis of genetic similarity among maize inbred lines detected by RFLPs, RAPDs, SSRs, and AFLPs. Theor Appl Genet 97:1248–1255

Petit C, Fréville H, Mignot A, Colas B, Riba M, Imbert E, Hurtrez-Boussés S et al. (2001) Gene flow and local adaptation in two endemic plant species. Biol Conserv 100:21–34

Powell WG, Morgante M, Doyle JJ, McNicol JW, Tingey SV, Rafalski AJ (1995) Polymorphic simple sequence repeat regions in chloroplast genomes: applications to the population genetics of pines. Proc Natl Acad Sci 92:7759–7763

Powell W, Morgante M, Andre C, Hanafey M, Vogel J, Tingey S, Rafalski A (1996) The utility of RFLP, RAPD, AFLP and SSRP (microsatellite) markers for germplasm analysis. Mol Breed 2:225–238

Pryor KV, Young JE, Rumsey FJ, Edwards KJ, Bruford MW, Rogers HJ (2001) Diversity, genetic structure and evidence of outcrossing in British populations of the rock fern *Adiantum capillus-veneris* using microsatellites. Mol Ecol 10:1881–1894

Qian W, Ge S, Hong DY (2001) Genetic variation within and among populations of a wild rice *Oryza granulata* from China detected by RAPD and ISSR markers. Theor Appl Genet 102:440–449

Rebourg C, Gouesnard B, Charcosset A (2001) Large scale molecular analysis of traditional European maize populations. Relationships with morphological variation. Heredity 86:574–587

Ren N, Timko MP (2001) AFLP analysis of genetic polymorphism and evolutionary relationships among cultivated and wild *Nicotiana* species. Genome 44:559–571

Reusch TBH, Stam WT, Olsen JL (2000) A microsatellite-based estimation of clonal diversity and population subdivision in *Zostera marina*, a marine flowering plant. Mol Ecol 9:127–140

Rossetto M, Jackes BR, Scott KD, Henry RJ (2001) Intergeneric relationships in the Australian Vitaceae: new evidence from cpDNA analysis. Genet Resour Crop Evol 48:307–314

Saghai Maroof MA, Biyashev RM, Yang GP, Zhang Q, Allard RW (1994) Extraordinarily polymorphic microsatellite DNA in barley: species diversity, chromosomal locations, and population dynamics. Proc Natl Acad Sci USA 91:5466–5470

Salimath SS, Deoliveira AC, Godwin ID, Bennetzen JL (1995) Assessment of genome origins and genetic diversity in the genus *Eleusine* with DNA markers. Genome 38:757–763

Sawkins MC, Maass BL, Pengelly BC, Newbury HJ, Ford-Lloyd BV, Maxted N, Smith R (2001) Geographical patterns of genetic variation in two species of *Stylosanthes* Sw. using amplified fragment length polymorphism. Mol Ecol 10:1947–1958

Schaal BA, Leverich WJ (2001) Plant population biology and systematics. Taxon 50:679–695

Sharma IK, Jones DL, Young AG, French CJ (2001) Genetic diversity and phylogenetic relatedness among six endemic *Pterostylis* species (Orchidaceae; series Grandiflorae) of Western Australia, as revealed by allozyme polymorphisms. Biochem Syst Ecol 29:697–710

Todokoro S, Terauchi R, Kawano S (1995) Microsatellite polymorphisms in natural populations of *Arabidopsis thaliana* in Japan. Jpn J Genet 70:543–554

Udupa SM, Baum M (2001) High mutation rate and mutational bias at $(TAA)_n$ microsatellite loci in chickpea (*Cicer arietinum* L.). Mol Genet Genomics 265:1097–1103

Virk PS, Pooni HS, Syed NH, Kearsey MJ (1999) Fast and reliable genotype validation using microsatellite markers in *Arabidopsis thaliana*. Theor Appl Genet 98:462–464

Virk PS, Newbury HJ, Jackson MT, Ford-Lloyd BV (2000) Are mapped markers more useful for assessing genetic diversity? Theor Appl Genet 100:607–613

Vos P, Hogers R, Bleeker M, Reijans M, van de Lee T, Hornes M, Frijters A et al. (1995) AFLP: a new technique for DNA fingerprinting. Nucleic Acids Res 23:4407–4414

Waldman M, Shevah Y (2000) Biological diversity - an overview. Water Air Soil Pollut 123:299–310

Weir BS (1996) Genetic data analysis II. Sinauer Associates, Sunderland, MA

Williams CF, Waser NM (1999) Spatial genetic structure of *Delphinium nuttallianum* populations: inferences about gene flow. Heredity 83:541–550

Williams JGK, Kubelic AR, Livak KJ, Rafalsky JA, Tingey SV (1990) DNA polymorphisms amplified by arbitrary primers are useful as genetic markers. Nucleic Acids Res 18:6531–6535

Wilson IJ, Balding DJ (1998) Genealogical inference from microsatellite data. Genetics 150:499–510

Wright S (1943) Isolation by distance. Genetics 28:114–138

Zabeau M, Vos P (1993) Selective restriction fragment amplification: a general method for DNA fingerprinting. European Patent Application, Publication #0534858-A1, Office europeèn des brevets, Paris

Zawko G, Krauss SL, Dixon KW, Sivasithamparam K (2001) Conservation genetics of the rare and endangered *Leucopogon obtectus* (Ericaceae). Mol Ecol 10:2389–2396

Zietkiewicz E, Rafalski A, Labuda D (1994) Genome fingerprinting by simple sequence repeat (SSR)-anchored polymerase chain reaction amplification. Genomics 20:176–183

Rob O'Neill
Wolfgang Köhler
Biometrie und Populationsgenetik
Justus-Liebig-Universität Gießen
Interdisziplinäres Forschungszentrum
Heinrich-Buff-Ring 26–32
35392 Gießen, Germany

Rod Snowdon
Pflanzenzüchtung
Justus-Liebig-Universität Gießen
Interdisziplinäres Forschungszentrum
Heinrich-Buff-Ring 26–32
35392 Gießen, Germany

e-mail: wolfgang.koehler@agrar.uni-giessen.de

Strategies of Breeding for Durable Disease Resistance in Cereals

Wolfgang Friedt, Kay Werner, Bettina Pellio, Claudia Weiskorn, Marco Krämer, and Frank Ordon

1 Introduction

Grain crops, like rice (*Oryza sativa*), maize (*Zea mays*), wheat (*Triticum aestivum*), barley (*Hordeum vulgare*), and – to a lesser extent – sorghum (*Sorghum bicolor*), oats (*Avena sativa*), and rye (*Secale cereale*) are of major importance for animal feeding and human nutrition (cf. FAO 2001). The grain yield and yield stability of these and other crops like potato and soybean are essential for ensuring human nutrition on a worldwide basis. However, in summary, yield losses of these crops due to biotic stresses are impressive. For example, Oerke et al. (1994) estimated that about 70% of the whole potential production would be lost if pests went uncontrolled and the Natural Resources Institute (1992) estimated a loss of 30–40% corresponding to about US$ 300 billion/year (cf. Nelson 2001). The most cost-effective and environmentally friendly approach of avoiding such yield losses is the exploitation and use of genetic diversity of crop plants regarding reaction to major pathogens, i.e., breeding for pest and disease resistance. With respect to the mode of inheritance, race-specific "qualitative" types of resistance (monogenic resistances) have to be distinguished from non-race-specific, quantitative resistance (oligo- or polygenic resistances). Introgression of these resistances from unadapted germplasms, related or even more distant species, respectively, is achieved by sexual recombination, i.e., crossing of parental lines followed by phenotypic selection in the segregating offspring. In this case the success of breeding depends entirely on extensive field or glasshouse tests for resistance to the respective pathogen(s). However, with respect to monogenic resistances, closely linked markers are available in cereals already (e.g., Ordon et al. 1998; Graner et al. 2000) and many quantitative trait loci (QTL) for disease resistance have been identified, e.g., in barley (cf. Hayes et al. 2002). Respective molecular markers facilitate a more efficient selection for mono- and oligogenic resistance(s) and corresponding improvement of yield stability. Besides this, the rising efficiency in the transformation of cereals and the increasing number of isolated resistance genes, as well as the identification of genes involved in resistance pathways, will offer new possi-

Progress in Botany, Vol. 64

bilities in resistance breeding in the future. Furthermore, pathogen resistance in cereals may be enhanced by the use of induced resistance (IR) that has proven its efficiency already in dicots. The present review will bring different aspects of breeding for durable resistance into focus, i.e., an efficient use of monogenic resistances by pyramiding, the manipulation of quantitative disease resistance by QTL analyses, the application of induced resistance, and strategies for enhancing resistance by genetic engineering, respectively.

2 Resistance Gene Accumulation – Gene Pyramiding

Breeding for resistance based on monogenic resistance is a two-step approach consisting of the identification of a resistance donor within the gene pool followed by the transfer of corresponding genes into basic materials, breeding lines and novel cultivars (Graner et al. 2000). Therefore, on the one hand extensive screening programs are required as reported, e.g., for the barley yellow mosaic virus disease complex (Proeseler et al. 1989; Ordon et al. 1993). On the other, systems and strategies ensuring reliable resistance gene detection and selection procedures are needed. Besides phenotypic selection the availability of closely linked genetic markers like restriction fragment length polymorphisms (RFLPs), random amplified polymorphic DNAs (RAPDs), amplified fragment length polymorphisms (AFLPs) and microsatellites (simple sequence repeats, SSRs) permit fast and reliable detection of respective resistance genes enabling marker assisted selection (MAS) procedures (Mohan et al. 1997). Meanwhile, numerous resistance genes and resistance loci, respectively, have been localized and mapped using molecular markers for various crops and pathosystems (Ordon et al. 1998). For example, Graner et al. (2000) have provided a summary of barley genes that confer resistance against different fungal and viral pathogens (and which have been tagged by molecular markers).

With regard to the potential risk of a resistance breakdown, the durability of disease resistance is of special importance concerning the economy of plant breeding. Disease resistance durability, which is defined as the adequacy of the resistance throughout the useful lifetime expected for a variety (Leach et al. 2001), has to be evaluated concerning its effectiveness in a cultivar under conditions of widespread cultivation in an environment favoring the respective disease (Wang et al. 1994). Monogenic (qualitative) resistance (Flor 1971), i.e., resistance conferred by a single major gene (Leach et al. 2001), is assumed to be non-durable in most cases, with exceptions like *mlo* resistance, due to the generally high mutation rate of plant pathogens (Kiyosawa 1982; Kloppers and Pretorius 1997; Takken and Joosten 2000), thus leading to the selection of new strains of the pathogen that are able to overcome the effects of single resistance genes. Additionally, the phenomenon of a disease resistance breakdown is increased by the large-scale and long-term cultivation of varieties carrying single resistance genes enabling the pathogen to overcome the resistance (Singh et al. 2001). Therefore, ongoing attempts in screening for new sources of resistance (Huang et al. 1997a), followed by the detection, tagging and incorporation in adapted cultivars, are needed. In this respect the gene pool of cereals suitable for the introgression of resistance genes has been considerably enlarged by cell and tissue culture techniques, e.g., wide crosses via embryo rescue (Hammer 1997). On the other hand, strategies for prolonging the useful lifetime of these genes in cultivars in combination with creating broader resistance spectra have to be developed.

In this respect, several methods for disease control by resistance breeding have been discussed (Kiyosawa 1982), one of them focusing on the combination of resistance genes. This task, which is often termed 'pyramiding', is defined as the accumulation of host resistance genes in a single line or cultivar (Nelson 1978) and has been taken into consideration repeatedly during the last decades (Schaefer et al. 1963; Nelson 1978; Pedersen 1988). The coeval mutation of a pathogen for virulence to two or more genes is supposed to be more unlikely than for single genes and therefore the effectiveness of combined genes is presumed to be more durable compared with their sole occurrence in different genotypes (Schaefer et al. 1963). Besides the prolongation of resistance durability, the strategy of gene pyramiding is also applied with the intention of increasing the level and spectrum of disease resistance (Huang et al. 1997a; Kinane and Jones 2000) leading to a broad-spectrum resistance (Liu et al. 2000). This item is of special interest due to the fact that resistance genes often differ in their specificity against the various pathogen strains so that their combination often results in a broader spectrum of resistance than that conferred by each gene individually (Kloppers and Pretorius 1997; Bender et al. 2000; Singh et al. 2001). Attempts for pyramiding resistance genes have meanwhile been carried out for several crops and pathosystems, e.g., in barley (Brown et al. 1996; Raman et al. 1999; Saeki et al. 1999; Pellio et al. 2000; Werner et al. 2000), wheat (Dweikat et al. 1997; Kloppers and Pretorius 1997; Liu et al. 2000), and rice (Yoshimura et al. 1995; Huang et al. 1997a; Hittalmani et al. 2000; Singh et al. 2001). In most cases two- and three-gene combinations with regard to homozygosity at the respective resistance loci have been performed resulting in a wider spectrum of the respective disease resistance.

For example, in barley (*Hordeum vulgare* L.), pyramiding attempts have been carried out for different scald resistance genes resulting in improved resistances to the fungal pathogen *Rhynchosporium secalis* in comparison to lines carrying a single resistance gene: While Brown et al. (1996) performed the pair-wise combination of genes for resistance to scald by the use of linked isozyme markers, Raman et al. (1999) detected AFLP markers with linkage to a scald resistance gene suitable for pyramiding attempts.

Concerning the barley yellow mosaic virus complex, which can only be overcome by breeding of resistant cultivars, attempts have been made to combine the resistance genes *rym4*, *rym5*, *rym9*, and *rym11*, differing in their reaction against the several virus strains (Ordon et al. 1999), in one breeding line by the use of PCR-based markers (Pellio et al. 2000; Werner et al. 2000). For this purpose strategies based on both segregating progeny (F_2) and DH-lines have been used (Fig. 1). Doubled-haploid populations have the advantage that homozygous recessive genotypes are much more frequent than in F_2 populations and dominant markers are as informative as co-dominant ones due to the lack of heterozygous genotypes (Werner et al. 2000). Saeki et al. (1999) reported on pyramiding of genes *rym3* and *rym5* in Japan by using isozyme markers and phenotypic selection procedures.

With regard to stripe rust of barley (*Puccinia striiformis* f. sp. *hordei*), Castro et al. (2000) created pyramids of resistance QTLs on chromosomes 4H, 1H, and 5H with the objective of determining the relationship between QTL number and level of resistance. Pyramiding was carried out by the use of microsatellite markers and reduction of disease severity was observed in genotypes with positive alleles at two or more loci. Kloppers and Pretorius (1997) observed similar results when they investigated effects of combinations amongst the genes *Lr13*, *Lr34*, and *Lr37* conferring resistance against leaf rust (*Puccinia recondita* f. sp. *tritici*) in wheat. The two-gene combinations, which have been generated by phenotypic selection only, showed higher levels of resistance than observed in lines carrying the individual genes based on the different specificity of these genes against various pathotypes (Kloppers and Pretorius 1997). Dweikat et al. (1997) identified RAPD markers linked to 11 genes conferring resistance to the Hessian fly [*Mayetiola destructor* (Say)] – a damaging insect in wheat – and tested their usefulness for gene pyramiding in the pathosystem wheat/Hessian fly; an approach which would not be applicable without markers due to difficulties in detecting multiple genes by phenotypic testing (Dweikat et al. 1997). Successful pyramiding of different genes for powdery mildew (*Blumeria graminis*) resistance in wheat (*Triticum aestivum* L. em. Thell) by the use of molecular markers was illustrated by Liu et al. (2000), who combined resistance genes *Pm2*, *Pm4a*, and *Pm21* in a two-by-two manner.

In rice, pyramiding attempts focus on the interactions rice–bacterial blight (*Xanthomonas oryzae*) and rice–rice blast (*Pyricularia oryza*). In both cases, combinations of resistance genes have been carried out by the use of RFLP- and PCR-based markers (Yoshimura et al. 1995; Huang et al. 1997a; Hittalmani et al. 2000; Singh et al. 2001).

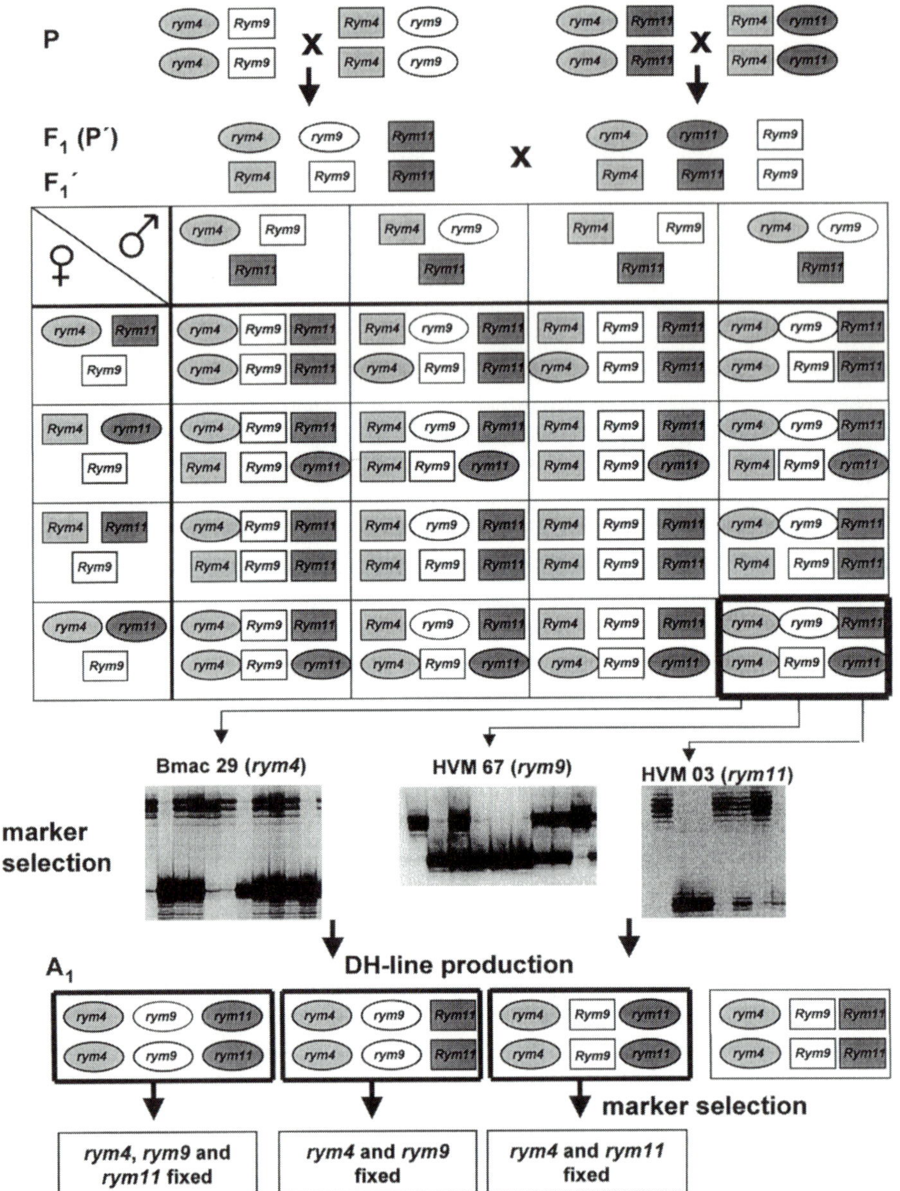

Fig. 1. Gene pyramiding strategy in barley to combine the resistance genes *rym4*, *rym9*, and *rym11* differing in their reaction against different virus strains of the barley yellow mosaic virus complex (Ordon et al. 1999) in one breeding line by the use of PCR-based markers. (Pellio et al. 2000; Werner et al. 2000)

Comparing the different reported pyramiding attempts it is obvious that the availability of molecular markers closely linked to the respective resistance gene(s) is an essential prerequisite for an efficient combination of resistance genes (Brown et al. 1996; Sanchez et al. 2000; Werner et al. 2000). By phenotypic selection only it appears to be very demanding, difficult or even impossible to reliably combine and detect multiple resistance genes in one genotype (Brown et al. 1996; Dweikat et al. 1997), due to dominance and epistatic effects and/or simple masking of the effect of one gene by another (Huang et al. 1997; Hittalmani et al. 2000; Singh et al. 2001).

The task of introgression and detection of resistance gene combinations into modern cultivars has already been carried out successfully, as can be seen regarding the rice cultivar 'Jefferson' possessing a gene pyramid concerning rice blast resistance (McClung et al. 1997) and concerning bacterial blight resistance based on the combination of three genes into an Indian rice cultivar (Singh et al. 2001). Similar success has been reported for the wheat–powdery mildew pathosystem, where several two- and three-gene combinations have been detected by phenotypic testing in wheat cultivars grown in China (Xia et al. 1995; Huang et al. 1997b). In addition, resistance gene combinations have been successfully constructed in a pair-wise manner in an elite wheat cultivar based on molecular marker assisted selection (Liu et al. 2000).

3 Polygenic Resistance – QTL Mapping

In contrast to major-gene resistance it is the consensus, and to some extent supported by historical evidence, that quantitative resistance is more durable per se. Therefore, quantitative resistance may become more important in the future, especially as in some crop species the gene pool is already largely depleted of effective major resistance genes. In the past, studies on quantitatively inherited resistance have mainly focused on epidemiological studies and biometrics, but today efficient tools, i.e., molecular marker techniques and sophisticated software packages, are available that facilitate the (partial) dissection of quantitative resistance into Mendelian loci followed by the determination of their effect(s) and interactions (Michelmore 1995). An overview of QTL mapping and quantitative disease resistance is given by Young (1996, 1999). With respect to QTL mapping, the development of locus-specific PCR-based markers, especially SSRs, e.g., in barley (Ramsay et al. 2000, Macaulay et al. 2001) or wheat (Roeder et al. 1998), and the availability of the AFLP technique (Vos et al. 1995) facilitating an accelerated construction of genetic maps in comparison to RFLPs (Graner et al. 1991), has enhanced QTL detection in recent years. Consequently, the number of disease resistance QTLs has risen considerably (cf. Ordon et al. 1998, Hayes et al.

2002). An overview of QTLs mapped in cereals for disease resistance until the end of 1997 is given by Ordon et al. (1998).

Since that time, for example, in barley at least three additional QTLs conferring resistance to leaf rust at the seedling stage (*Puccinia hordei)* explaining 55% of the phenotypic variance and five QTLs for resistance in the adult stage explaining 60% of the phenotypic variance with two QTLs in common have been identified (Qi et al. 1998; see Table 1). In subsequent studies it turned out that only one QTL had a substantial effect at both developmental stages and it was shown that some of the QTLs exhibit isolate-specific effects (Qi et al. 1999). In addition, one QTL for resistance to leaf rust has been identified in the position of the *Rph16* gene (Kicherer et al. 2000). With regard to stripe rust (*Puccinia striiformis* f. sp. *hordei*), different QTLs have been identified by Toojinda et al. (1998, 2000). Besides these highly important pathogens, QTLs have been detected for resistance to bacterial leaf streak caused by *Xanthomonas campestris* pv. *hordei* (El Attari et al. 1998) and to *Pyricularia oryzae* (Sato et al. 2001). In addition to these QTLs for fungal diseases, QTLs for tolerance against BYDV have recently been identified (Toojinda et al. 2000; Scheurer et al. 2001). For BYDV-PAV and BYDV-MAV seven QTLs have been detected and a substantial QTL × environment interaction was observed (Toojinda et al. 2000). Scheurer et al. (2001) used a German isolate of BYDV-PAV and detected two QTLs explaining about 46% of the phenotypic variance on chromosomes 2HL and 3HL in the region of the *Ryd2* gene. It is interesting to note that some QTLs are mapped in chromosomal regions of barley where previously major resistance genes had been identified. This has, e.g., been shown for powdery mildew (Backes et al. 1996), yellow and leaf rust (Thomas et al. 1995; Kicherer 2000) and net blotch (Richter et al. 1998) resistances, contributing to the theory that QTLs may be allelic (defeated) versions of "qualitative" resistance genes with an intermediate phenotype (for details, cf. Young 1996).

In comparison to barley the genome of wheat (*Triticum aestivum*) is much more complex. However, recently, QTLs for powdery mildew (Keller et al. 1999; Chantret et al. 2000), leaf rust (Messmer et al. 2000), and *Pyrenophora tritici repentis* (Effertz et al. 2001) have been mapped. Due to the production of mycotoxins, *Fusarium* head blight (FHB) is of special importance in wheat and barley today and is subject to many QTL analyses. A QTL associated with the GliD1b allele on chromosome 1D of wheat has been detected (Ittu et al. 2000). A major QTL for FHB resistance has also been detected on chromosome 3BS (Waldron et al. 1999; Anderson et al. 2001). Beyond that, QTLs for FHB resistance have been detected in barley, too (Zhu et al. 1999; Van Sanford 2001), and it turned out that QTLs in wheat and barley are located in a syntenic region of chromosome 3 (homologous group 3; Kolb et al. 2001).

Table 1. QTLs for resistance to various diseases of barley, their chromosomal location, the material used for mapping and corresponding reference(s) (modified according to NABGP http: //www.css.orst.edu/barley/nabgmp/nabgmp.htm)

Disease [Pathogen]	Barley chromosome[a, b]	Material	Reference(s)
Leaf rust [*Puccinia hordei*]	1, 2, 2, 4, 4, 6, 7, 7	Vada/L94	Qi et al. (1998, 1999)
Leaf rust [*Puccinia hordei*]	2, 6, 7	Harrington/TR306	Spaner et al. (1998)
Leaf rust [*Puccinia hordei*]	2, 2, 2, 4	Krona/HOR1063	Kicherer et al. (2000)
Leaf rust [*Puccinia hordei*]	3, 7	Blenheim/E224	Thomas et al. (1995)
Stripe rust [*Puccinia striiformis* f. sp. *hordei*]	1, 5, 7	Blenheim/E224	Thomas et al. (1995)
Stripe rust [*Puccinia striiformis* f. sp. *hordei*]	2, 3, 5, 6	Shyri/Galena	Toojinda et al. (2000)
Stripe rust [*Puccinia striiformis* f. sp. *hordei*]	4, 7	Cali-sib/Bowman	Chen et al. (1994)
Stripe rust (seedling) [*Puccinia striiformis* f. sp. *hordei*]	4, 6	Cali-sib/Bowman	Hayes et al. (1996)
Stem rust [*Puccinia graminis* f. sp.*tritici*]	1, 4	Harrington/TR306	Spaner et al. (1998)
Fusarium HB [*Fusarium graminearum*]	1, 2, 3, 3, 4, 5,	Gobern./CMB	Zhu et al. (1999)
Fusarium HB [*Fusarium graminearum*]	1, 1, 1, 2, 2, 2, 2, 3, 4, 5, 7	Chevron/M69	de la Pena et al. (1999)
Net blotch [*Pyrenophora teres* f. *teres*]	1, 2, 3, 3, 4, 6, 6, 7	Steptoe/Morex	Steffenson et al. (1996)
Net blotch [*Pyrenophora teres* f. *teres*]	1, 3, 4, 6, 7	Harrington/TR306	Spaner et al. (1998)
Net blotch [*Drechslera teres* f. *teres*]	1, 2, 2, 3, 3, 3, 4, 4, 6, 6, 6, 6	Arena/Hor 9088	Richter et al (1998)
Spot blotch [*Cochliobolus sativus*]	1, 5	Steptoe/Morex	Steffenson et al. (1996)

Table 1 (continued)

Disease [Pathogen]	Barley chromosome[a, b]	Material	Reference(s)
Scald [*Rhynchosporium secalis*]	1, 2, 3, 6, 6	Igri/Danilo	Backes et al. (1995)
Scald [*Rhynchosporium secalis*]	3, 3	Blenheim/E224	Thomas et al. (1995)
Scald [*Rhynchosporium secalis*]	3, 4, 6	Harrington/TR306	Spaner et al. (1998)
Leaf stripe [*Pyrenophora graminea*]	1, 2, 4	Proctor/Nudinka	Pecchioni et al. (1996)
Rice blast (seedling) [*Pyricularia oryzae* Cav.]	3, 4, 4, 7	Harrington/TR306	Sato et al. (2001)
Powdery mildew [*Blumeria graminis* f. sp. *hordei*]	1, 6	Igri/Danilo	Backes et al. (1995, 1996)
Powdery mildew [*Blumeria graminis* f. sp. *hordei*]	1, 7	Proctor/Nudinka	Heun et al. (1992)
Powdery mildew [*Blumeria graminis* f. sp. *hordei*]	4, 7	Harrington/TR306	Spaner et al. (1998)
Powdery mildew [*Blumeria graminis* f. sp. *hordei*]	5	Blenheim/E224	Thomas et al. (1995)
Kernel discoloration [*Bipolaris sorokiniana*, *Alternaria* spp., *F. graminearum*]	1, 1, 2, 2, 3, 3, 4, 4, 6, 6,	Chevron/M69	de la Pena et al. (1999)
Bacterial leaf streak [*Xanthomonas campestris* pv. *hordei*]	3, 3, 7	Steptoe/Morex	El Attari et al. (1998)
Barley yellow dwarf virus (BYDV)	1, 3, 4, 5	Shyri/Galena	Toojinda et al. (2000)
Barley yellow dwarf virus (BYDV-PAV)	1, 2, 4	Post/Nixe	Scheurer et al. (2001)
Barley yellow dwarf virus (BYDV-PAV)	2, 3	Post/Vixen	Scheurer et al. (2001)
Barley yellow mosaic virus (BaYMV)	1, 3, 4	Ko A/Mokusekko	Miyazaki et al. (2001)
Aphids [*R. maidis*, *R. padi*, *Schizaphis graminum*, *Sitobion akebiae*]	1, 5	Harrington/TR306	Moharramipour et al. (1997)

[a]Chromosome1=7H, 2=2H, 3=3H, 4=4H, 5=1H, 6=6H, 7=5H.
[b]Multiple chromosome nomination corresponds to the number of QTLs.

The few examples presented elucidate that the number of identified QTLs has risen considerably due to more efficient marker techniques and it may be assumed that further progress will be gained by using the vast amount of information provided by genomics and bioinformatics (cf. Young 1999). For example, in rice, it has already been demonstrated that ESTs showing high homology to disease resistance genes or defense response (DR) genes to some extent cluster in genomic regions harboring QTLs for disease resistance, i.e., against rice blast, bacterial blight and sheath blight (Wang et al. 2001). Similar results were obtained in wheat using DR genes concerning QTLs, e.g., for leaf rust and stem rust (Faris et al. 1999). Therefore, the large number of ESTs identified recently (Graner et al. 2000) will be a useful tool for QTL mapping in cereals in the future.

Provided respective QTLs have been mapped accurately and have proven their efficiency in repeated trials over years and locations, and in different populations, they can be treated like Mendelian factors and combined by marker-assisted selection (MAS) in applied breeding programs. However, in comparison with MAS for single genes, which is already applied to some extent, e.g., in barley breeding (Schiemann and Backes 2000), the application of QTLs in practical breeding is rather limited. However, the potential of marker-assisted introgression of QTLs has been demonstrated, e.g., by transferring QTLs for stripe rust into a genetically unrelated background (Toojinda et al. 1998). Beyond this, respective positive QTLs for resistance to *Puccinia striiformis* f. sp. *hordei* derived from different lines have been successfully combined (pyramiding, Castro et al. 2000). It may be assumed, therefore, that in the future reliable and easy-to-handle marker systems, accurate mapping procedures, the application of 'advanced backcross QTL analysis' (Bernacchi et al. 1998), and the use of information provided by genomics (e.g., ESTs) will facilitate a more efficient use of detection and application of QTLs for resistance in cereal breeding, thereby leading to longer lasting, i.e., more durable, disease resistance.

4 Induced Resistance

In addition to genetically fixed resistance, the use of induced resistance (IR) can be considered as an environmentally friendly approach to plant protection. Induced resistance is described as a localized pathogen infection after preceding treatment with microbial components and products or with structurally unrelated organic and inorganic compounds (Kuç 2001). Such locally or systemically induced resistance to infectious diseases was first reported by Chester (1933) and its potential for disease control was already described by Kuç et al. (1959), Loebenstein (1963), Ross (1966) and Ryals et al. (1994). In contrast to the gene for gene resistance (Flor 1971) as a pathogen-specific resistance, IR results in a broad-spectrum, long-lasting resistance and is reported to be most effective against fungi, less effective against bacteria, and least effective against systemic viruses (Kuç 2001). Induced resistance can on the one hand be initiated by biotic factors like necrotizing pathogens, non-pathogens or root colonizing bacteria, and on the other by various chemicals (Fought and Kuç 1996; Benhamou and Belanger 1998).

This type of resistance has been observed in a broad spectrum of plants belonging to all classes of higher plants (reviewed in Sticher et al. 1997). Most investigations on IR have been conducted in dicotyledonous species and only little is known about IR in cereals, although Parker and Coleman (1997) described that structural features of resistance gene products in monocots and dicots are the same. However, the mechanism(s) leading to IR are especially in cereals not well understood. Nevertheless, IR has been described to be effective, e.g., in wheat (*Triticum aestivum*) against *Blumeria graminis* f. sp. *tritici* and *Puccinia recondita*, in rice against *Magnaporthe grisea* (Smith and Métraux 1991; Schweizer et al. 1999) and against *Xanthomonas campestris* pv. *oryzae* (Jin et al. 1997), in barley against *Blumeria graminis* f. sp. *hordei* (Hwang and Heitefuss 1982; Sarhan et al. 1991), against *Puccinia striiformis* (Reiss 1986) and against *Drechslera teres* (Jørgensen et al. 1998). In contrast to *Arabidopsis thaliana*, root-colonizing bacteria leading to an enhanced level of resistance have not been reported for monocots to date (van Loon et al. 1998). In cereals, the best-studied resistance inducer seems to be acibenzolar-S-methyl (ASM; Oostendorp et al. 2001). Chemically induced resistance (cIR) was shown in wheat against *Puccina recondita*, *Septoria nodorum* and *Blumeria graminis* (Görlach et al. 1996), in maize against *Peronosclerospora sorghi* (Morris et al. 1998), in rice against *Magnaporthe grisea* and *Xanthomonas oryza* (Shimura et al. 1981), and in barley against *Blumeria graminis* (Kogel et al. 1995), whereas no significant effects in barley against barley yellow dwarf virus (BYDV-PAV) were observed (Weiskorn et al. 2001). The level of protection gained by IR is to some extent varied (Manhandhar et al. 1999; Ton et al. 1999), but is considered to be affected by the genotype, the environment, and the inducing agent, respectively.

Genotypic variation in inducible resistance has been described in several plants, e.g., carnation (van Peer et al. 1991), cucumber (Liu et al. 1995), radish (Leeman et al. 1995), soybean (Underwood et al. 2000), poplar (Havill and Raffa 1999) and *Arabidopsis* (Ton et al. 2001). In cucumber (Liu et al. 1995) and wheat (Stadnik and Buchenauer 1999), it was shown that IR is mainly expressed in plants with a low level of genetically determined resistance and only to a lesser extent in resistant cultivars. In wheat, the penetration efficiency of *Blumeria graminis* was markedly decreased in the susceptible cv. Monopol after ASM treatment, but not in cv. Zentos which has a high level of quantitative resistance. An enhancement of phenylalanine ammonium lyase (PAL) activity was found to be involved in both, quantitative resistance and in ASM-induced resistance. Accordingly, wheat mutants with a lack of PAL activity were not able to express IR after ASM treatment (Stadnik and Buchenauer 2000), whereas, in rice, activity of PAL was not enhanced after induction of systemic resistance (Smith and Métraux 1991), indicating that there are at least two pathways in monocots leading to IR. Similarly, Loebler and Lee (1998) suggested that there are two signaling pathways in barley interfering with extracellular and intracellular jasmonate, respectively. However, in cereals, defense signaling pathways are only partially understood and even less is known about defense mechanisms (Piffanelli et al. 1999).

Between constitutive and inducible resistance either positive, or negative, or no trend in relationship was found for many plant pathogen and plant herbivore systems. In radish (Leeman et al. 1995), susceptible and resistant genotypes were able to establish IR on the same level. Moreover, in the soybean–Mexican bean beetle system no significant correlation was found between levels of induced and constitutive resistance measured in the same genotype (Underwood et al. 2000). Such genotypes with different genetics regarding inducibility can eventually be useful sources for breeding cultivars with a better IR response.

Recently, in barley, many gene products differentially expressed after treatment with chemical inducers have been described, most of them with an unknown function called barley chemically induced (Bci) genes (Besser et al. 2000). Two genes were found to be homologous to those previously found to be involved in induced resistance, i.e., the lipoxygenase gene *Lox2:Hv:1* (Vorös et al. 1998; Hause et al. 1999) and various thionin genes (Kogel et al. 1995). Some of these Bci genes were also found to be homologous to induced genes in wheat (Wci-genes), indicating that Bci genes might be useful cIR markers in cereals (Besser et al. 2000). In wheat, the Wci genes were activated by chemical inducers like ASM, 2,6-dichloroisonicotinic acid (INA), and salicylic acid (SA), but no PR1-like genes (being common markers for IR in dicots) were induced (Görlach et al. 1996; Molina et al. 1999). In rice, chemical induction resulted in the expression of a gene encoding a nucleotide-binding site and a leucine-rich repeat (Sakamoto et al. 1999), which are known to be involved in race-specific resistance reactions. The availability of mutants altered in the induction and expression of IR for the model plant *Arabidopsis thaliana* offers the opportunity for detailed genetic analyses on the mode of action and genetic determination of IR. For example, Ton et al. (2001) investigated to what extent the expression of IR depends on the level of basal resistance using *Pseudomonas fluorescens* as the inducing agent and *P. syringae* pv. *tomato* (*Pst*) as the challenging pathogen. Genetic analysis of crosses between inducible and non-inducible ecotypes of *Arabidopsis* revealed that IR inducibility and basal resistance against *Pst* were inherited as monogenic dominant traits that are genetically linked. Both types of resistance seem to be affected by the same locus mapped on chromosome III. These results suggest that in *Arabidopsis* the genes involved in quantitative resistance are involved in IR, too. To the best of our knowledge, there are currently no corresponding results concerning cereals.

In comparison to dicots, knowledge of IR on cereals is rather limited and no information on the mode of inheritance of the response to an IR treatment is available as being a prerequisite for genetically optimizing this type of pathogen resistance. Besides this, IR has proven its efficiency in cereals mainly in laboratory tests and examples of the efficient practical use of induced resistance as a method of plant disease control are few. However, with the rise in genomic information in cereals and knowledge on signaling pathways followed by the genetic optimization of the IR response, this approach may become an alternative strategy against a wide range of pathogens in the future.

5 Genetic Engineering: Transformation of Cereals

In principal, genetic engineering is a very efficient tool for creating novel genetic variation with respect to disease resistance, avoiding the need for sexual introgression of resistance genes. Genetic engineering or modification of crops can most simply be defined as the transfer of genetic material from a different species (e.g., plant, microbe or animal) or of synthesized genes into a target plant. The main reason for using this method in crop improvement is that genetic engineering enables the application of several genetic manipulations that are not possible via conventional plant breeding methods. Typical examples of already characterized resistance genes are presented in Table 2. For the production of transgenic plants, basically three major components are required: (1) availability of cells or tissue that are competent for plant regeneration, (2) a system to deliver DNA into these regenerable cells, and (3) a method for selecting or identifying transgenic cells, tissue, organs, or plants at a satisfactory frequency (cf. Birch 1997).

a) *Agrobacterium*-Mediated Transformation

The first successful genetic engineering of plants was achieved in 1983 using *Agrobacterium tumefaciens* as a vector for DNA delivery into tobacco (De Block et al. 1984; Horsch et al. 1984) and has been extended by optimization of different technologies to over 120 species in at least 35 families (Birch 1997). *Agrobacterium*, often referred to as nature's ge-

Table 2. Classes of characterized resistance genes regarding their protein structure (mod. Hubert et al. 2001)

Class	Gene	Host	Pathogen	Protein structure[a]	Reference
1	L6	Flax	*Melampsora lini*	TIR-NBS-LRR	Lawrence et al. (1995)
	M	Flax	*Melampsora lini*	TIR-NBS-LRR	Anderson et al. (1997)
	N	Tobacco	Tobacco Mosaic Virus	TIR-NBS-LRR	Whitham et al. (1996)
	RPP5	*Arabidopsis*	*Peronospora parasitica*	TIR-NBS-LRR	Parker et al. (1997)
	I2	Tomato	*Fusarium oxysporium*	NBS-LRR	Simons et al. (1998)
	Mla	Barley	*Blumeria graminis*	NBS-LRR	Zhou et al. (2001)
	RPM1	*Arabidopsis*	*Pseudomonas syringae*	NBS-LRR	Grant et al. (1996)
	RPS2	*Arabidopsis*	*Pseudomonas syringae*	NBS-LRR	Bent et al. (1994); Mindrinos et al. (1994)
	Xa1	Rice	*Xanthomonas oryzae*	NBS-LRR	Yoshimura et al. (1998)
2	*Cf-2/5*	Tomato	*Cladosporium fulvum*	LRR-TM	Dixon et al. (1998)
	Cf-4/9	Tomato	*Cladosporium fulvum*	LRR-TM	Jones et al. (1994); Thomas et al. (1997); Takken et al. (1999)
	$HS1^{pro-1}$	Beet	*Heterodera schachtii*	LRR-TM	Cai et al. (1997)
3	*Pto*	Tomato	*Pseudomonas syringae*	Protein kinase	Martin et al. (1993)
4	*Xa-21*	Rice	*Xanthomonas oryzae*	LRR-TM-kinase	Song et al. (1995)
5	*mlo*	Barley	*Blumeria graminis*	Membrane protein	Büschges et al. (1997)
6	*Hm1*	Maize	*Cochliobolus carbonum*	Toxin reductase	Johal and Briggs (1992)

[a]NBS, nucleotide binding site; LRR, leucine-rich repeat; TIR, domain with homology to the *Toll* gene of *Drosophila* and the *Interleukin-1* receptor of mammals; TM, transmembrane domain.

netic engineer, naturally causes crown gall disease by transferring genes from a tumor-inducing (Ti) plasmid into the plant genome. The mechanism of DNA transfer is encoded by a series of virulence (*vir*) genes. The transferred DNA (T-DNA) is a defined part of the Ti plasmid that is flanked by 25-bp directly repeated border sequences and does not encode any genes important for the transfer process itself. Genes from the Ti plasmid that are integrated in the plant chromosome are expressed at high levels in the plant (Zambryski et al. 1989). This method offers a number of advantages including the defined integration of the foreign gene copies into the host genome, and a preferred fitting into transcriptionally active regions of the chromosomes (Koncz et al. 1989). Broad-leafed plants such as tobacco and tomato were easiest to transform and show no difficulties in infection, because the bacteria can colonize and form tumors on its natural host range of dicotyledonous plants (De Cleene and De Ley 1976).

Agrobacterium-mediated gene transfer has been applied less successful in monocotyledonous plants and led Potrykus (1990) to conclude that this approach had very little potential for these species, although at that time a few studies had already been published providing evidence for the production of transgenic cereal tissues (Raineri et al. 1990; Gould et al. 1991). In general, monocot cells and tissues are relatively recalcitrant to in vitro regeneration and some of the promoters that exhibit high activity in dicot transformation systems show a low level of activity in monocot cells and tissues. Therefore, the establishment of genotype-specific regeneration systems and transformation protocols has been seen as an important task in cereal genetic engineering (e.g., Jähne et al. 1994; Machii et al. 1998). In recent years, significant progress has been achieved in this respect and a reliable transformation protocol for cereals via *Agrobacterium* was first reported in the 1990s for different rice varieties (Chan et al. 1992; Hiei et al. 1994), then maize (Ishida et al. 1996), wheat (Cheng et al. 1997), barley (Tingay et al. 1997) and sorghum (Zhao et al. 2000).

When compared to the 35S promoter from cauliflower mosaic virus, which has proven its excellence in dicot transformation systems, various other promoters have been tested because of its low activity in monocot plants. The development of the promoters from the rice actin gene *Act1* (McElroy et al. 1990) and the maize ubiquitin gene *Ubi1* (Christensen et al. 1992) have provided higher frequencies of recovery of transformed callus in rice (Li et al. 1997) and barley (Tingay et al. 1997), respectively. In this context, rice is currently regarded as a model monocotyledonous plant to study gene expression and to introduce agronomically useful genes (for review, see Tyagi et al. 1999).

There is no indication that the mechanism of DNA transfer for cereals acts in a different way to that observed for dicotyledonous plants. Bidney et al. (1992) found that the transformation was enhanced by wounding tissues prior to the application of *Agrobacterium*, and, following this method, Tingay et al. (1997) injured immature barley embryos by bombarding the scutellum surface with gold microprojectiles resulting in a transformation efficiency of 4.2%, giving transgene numbers ranging from single copy insertion to at least 10 copies. Likewise, Hansen and Chilton (1996) developed a hybrid method of biolistics and *Agrobacterium*-mediated transformation, termed 'agrolistics'. In their system, DNA to be delivered into maize cells by particle bombardment is flanked by T-DNA border sequences and co-transferred with genes encoding the polypeptides VirD1 and VirD2, to stimulate the *Agrobacterium*-mediated T-DNA integration into the genome. Increased public attention regarding the utilization of antibiotic and herbicide resistance genes as selectable markers (SM) in plant transformation has accentuated the requirement for selectable marker-free transgenics.

A number of techniques are available today to remove these genes from the genome (e.g., Dale and Ow 1991), and Matthews et al. (2001) established a method for marker gene elimination from transgenic barley using a standard binary vector with a plasmid carrying two T-DNAs, which were located adjacent to each other with no intervening region. Since the development of the binary vector system in the early 1980s (Hoekema et al. 1983), many other vector systems have been constructed (e.g., Komari et al. 1996; Toki 1997). Komari et al. (1996) designed in rice some unique plasmids that carry two well-separated T-DNA segments, one carrying the SM gene and the other the gene of interest. This protocol was found to be very efficient, as the co-transformation with the two T-DNAs was greater than 47%. The integration and segregation of T-DNAs were verified by molecular analysis and allow the generation of marker-free transformants. Using this method in barley, Jacobsen et al. (2000) reported that co-transformation of the two genes occurred in 66% of cases leading to the recovery of 16% selectable marker-free transformants or 1% of co-cultivated immature embryos.

b) Particle Bombardment and Other Transformation Methods

Genetic engineering of plants using direct DNA delivery methods was achieved soon after the first success with *Agrobacterium tumefaciens*. Toriyama et al. (1988) and Zhang and Wu (1988) were the first to report recovery of transgenic rice plants by using direct DNA transfer into protoplasts by the polyethylene glycol (PEG) method that was established by Paszkowski et al. (1984). Later, transgenic barley plants derived from embryogenic cell suspension cultures were obtained using this method (Funatsuki et al. 1995). The use of electroporation to introduce DNA into the protoplasts was reported in maize (Rhodes et al. 1988), and Shimamoto et al. (1989) were the first to report recovery of fertile transgenic rice plants. Although the mechanisms of both systems are not fully understood, and despite limitations like somaclonal variations or genotype dependence of protoplast isolation, these methods have been used because they are technically simple and inexpensive (cf. Repellin et al. 2001). Besides *Agrobacterium*-mediated transformation of plant genomes, the particle bombardment ('biolistics') gene delivery into plant tissues and callus has proven to be the most successful transformation method for the genetic improvement of several cereal crops soon after its development (Sanford et al. 1987). This physical method of introducing DNA into cells has overcome the difficulties resulting from problems in regeneration of cereal plantlets from protoplasts. Up to now, fertile transgenic plants have been successfully obtained in rice (Christou et al. 1991), barley (Wan and Lemaux 1994), wheat (Becker et al. 1994), maize (Brettschneider et al. 1997), rye (Castillo et al. 1994), and oats (Gless et al. 1998). Inactivation of transgene expression seems to be especially problematic in cereal crops, where the most frequently used transformation methods typically result in complex, multicopy integration patterns and it has been observed that more than 50% of T_1 plants exhibit transgene silencing (for review, see Iyer et al. 2000).

c) Genetically Engineered Resistance to Pathogens

Resistance gene products may serve as receptors for pathogen *Avr* factors or indirect recognition through a co-receptor (Staskawicz et al. 1995). Efforts to clone those genes involved in disease resistance were achieved with several procedures and provide the feasibility to utilize them as tailored broad-spectrum resistance to diseases in transgenic plants (Bent 1996). As shown in Table 2, the proteins encoded by most of the characterized resistance genes can be sorted into different classes (for review, see Hulbert et al. 2001). Potential approaches to use these genes for engineering disease resistance are discussed by Rommens and Kishore (2000). For example, Song et al. (1995) have cloned the rice gene *Xa21* conferring resistance against the pathogen *Xanthomonas oryzeae* by a map-based cloning approach. This is the first resistance gene to be cloned and transferred via particle bombardment in any cereal crop plant. In transgenic plants it was found to be effective against multiple *Xanthomonas* isolates (Wang et al. 1996), and it was possible to transform the elite Indica rice cultivar 'IR 72' with *Xa21* through particle bombardment resulting in two T_1 plants with high levels of resistance (Tu et al. 1998). However, in these transformants, gene silencing or inactivation was observed which is considered to be a complex phenomenon in plants. In addition to *Xa21*, other resistance genes like the *Mlo*-locus (Büschges et al. 1997), the *Rar*1-locus (Lahaye et al. 1998) and the *Mla*-locus (Wei et al. 1999, Haltermann et al. 2001; Zhou et al. 2001) have been isolated, facilitating new breeding strategies. For example, the different alleles of the *Mla*-locus, which cannot be combined by recombination, may be combined in one breeding line (pyramiding) or one cultivar may be transformed with different alleles and those near isogenic lines (NILs) will be grown in a mixture having a strong impact on population dynamics of the pathogen (Schulze-Lefert, pers. comm.).

Moreover, the coat proteins of many plant viruses have been transformed into a wide range of plant species to obtain viral protection. Barley yellow dwarf virus-PAV (BYDV-PAV) is one of the most widespread viruses of cereals worldwide and can cause production losses of about 15% in barley, 17% in wheat and 25% in oats (Lister and Ranieri 1995). Attempts have been made to introduce synthetic resistance genes into barley and oats using coat protein sequences of BYDV-PAV (Wan and Lemaux 1994; McGrath et al. 1997) to obtain resistant plants. However, this resistance proved to be unstable. The discovery that virus resistance and gene silencing in plants can be induced by simultaneous expression of sense and antisense RNA (Waterhouse et al. 1998) was employed in an attempt to generate barley with protection against BYDV-PAV (Wang et al. 2000). According to this study, the barley cv. 'Golden Promise' was transformed with a transgene encoding a hairpin (hp) RNA derived from BYDV-PAV polymerase sequences and it was found that one-third of these transformed plants had extreme resistance against this virus disease. These results indicate that immunity would be robust in the field and may be useful in minimizing losses in cereal production worldwide. The development of pest-resistant crops was initiated by the discovery that the bacterium *Bacillus thuringiensis* (Bt) produces endotoxins highly toxic to insects, but not harmful to mammals. In this respect, transgenic Bt-resistant maize plants were the first cultivars tolerant to European corn borer that were tested under field conditions (Armstrong et al. 1995). Besides the identification and cloning of potentially useful disease resistance genes of cereal crops, the development of transgenic plants with a new or broadened resistance spectrum entails additional challenges. There are limitations in the acceptance and commercialization of genetically engineered crops these days, especially in Europe. Nevertheless, transgenic plants with enhanced disease resistance will in the near future be an essential prerequisite for the unraveling and understanding of plant–microbe interactions and plant pathogen resistance. In the long run, transgenic crop plants could become a very valuable component of future disease management programs, in both agriculture and horticulture.

6 Summary and Future Prospects

Substantial pathogen-derived yield losses are frequently observed in grain crops. Therefore, efficient breeding for resistance is essential with respect to human and animal nutrition and ecologically friendly food production. Resistance breeding itself is a complex task due to the rapid adaptation of pathogens to resistance genes, on the one hand, and to the problems of reliable selection for quantitative disease resistance, on the other. In this respect, molecular markers are considered to be very useful tools. They are already available for many resistance genes in cereals, some derived even from the secondary gene pool. Besides this, easy-to-handle PCR-based markers have been developed for QTLs against important diseases. Apart from marker-assisted selection (MAS) procedures, respective markers facilitate pyramiding of single genes or QTLs for resistance(s), thereby creating more effective or longer lasting disease resistance. In particular, quantitative types of disease resistance are more accessible today by molecular markers and will be even more approachable in the future due to the rise in information gathered on cereal genomes. For example, in the future, resistant lines may be primarily selected by their expression profile(s) on microarrays (Young 1999). Deeper knowledge of plant resistance genes and QTLs for disease resistance will facilitate the development of efficient strategies for resistance breeding. In the future these strategies may be complemented by the use of induced resistance, although only little information on monocots is available in comparison to dicots up to now. Last, but not least, genetic engineering is considered to be an important tool for resistance breeding in the future, opening up new ways of a more systematic and directed improvement of grain crops. Apart from using pathogen-derived resistance or toxic substances, isolated plant resistance genes offer the opportunity of creating new alleles in vitro, or combining known alleles, thereby improving resistance.

References

Anderson JA, Stack RW, Liu S, Waldron BL, Fjeld AD, Coyne C, Moreno-Sevilla B, Fetch JM, Song QJ, Cregan PB, Frohberg CJ (2001) DNA markers for *Fusarium* head blight resistance QTLs in two wheat populations. Theor Appl Genet 102:1164–1168

Anderson PA, Lawrence GJ, Morrish BC, Ayliffe MA, Finnegan EJ, Ellis JG (1997) Inter-activation of the flax rust resistance gene M associated with loss of a repeated unit within the leucin-rich repeat coding region. Plant Cell 9:641–651

Armstrong CL, Parker GB, Pershing JC, Brown SM, Sanders PR, Duncan DR, Stone T, Dean DA, DeBoer DL, Hart J, Howe AR, Morrish FM, Pajeau ME, Petersen WL, Reich BJ, Rodriguez R, Santino CG, Sato SJ, Schuller W, Sims SR, Stehling S, Tarochione LJ, Fromm ME (1995) Field evaluation of European corn borer control in progeny of 173 transgenic corn events expressing an insecticidal protein from *Bacillus thuringiensis*. Crop Sci 35:550–557

Backes G, Graner A, Foroughi-Wehr B, Fischbeck G, Wenzel G, Jahoor A (1995) Localization of quantitative trait loci (QTL) for agronomic important characters by the use of a RFLP map in barley (*Hordeum vulgare* L.). Theor Appl Genet 90:294–302

Backes G, Schwarz G, Wenzel G, Jahoor A (1996) Comparison between QTL analysis on powdery mildew resistance in barley based on detached primary leaves and on field data. Plant Breed 115:419–421

Becker D, Brettschneider R, Lörz H (1994) Fertile transgenic wheat from microprojectile bombardment of scutellar tissue. Plant J 5:299–307

Bender CM, Pretorius ZA, Kloppers FJ, Spies JJ (2000) Histopathology of leaf rust infection and development in wheat genotypes containing *Lr12* and *Lr13*. J Phytopathol 148:65–76

Benhamou N, Belanger R (1998) Benzothiadiazole-mediated induced resistance to *Fusarium oxysporum* f. sp. *Radicislycopersici* in tomato. Plant Physiol 118:1203–1212

Bent AF (1996) Plant disease resistance genes: function meets structure. Plant Cell 8:1757–1771

Bent AF, Kunkel BN, Dahlbeck D, Brown KL, Schmidt R, Giraudat J, Leung J, Staskwicz BJ (1994) *RPS2* of *Arabidopsis thaliana*: a leucin-rich repeat class of plant disease resistance genes. Science 265:1856–1860

Bernacchi D, Beck-Bunn T, Emmatty D, Eshed Y, Inai S, Lopez J, Petiard V, Sayama H, Uhlig J, Zamir D, Tanksley S (1998) Advanced backcross QTL analysis of tomato. II. Evaluation of near isogenic lines carrying single-donor introgressions for desirable wild QTL-alleles derived from *Lycopersicon hirsutum* and *L. pimpineelifolium*. Theor Appl Genet 97:170–180

Besser K, Jarosch B, Langen G, Kogel KH (2000) Expression analysis of genes induced in barley after chemical activation reveals distinct disease resistance pathways. Mol Plant Pathol 1:277–286

Bidney D, Scelonge C, Martich J, Burrus M, Sims L, Huffman G (1992) Microprojectile bombardment of plant tissues increases transformation frequency by *Agrobacterium tumefaciens*. Plant Mol Biol 18:301–313

Birch RG (1997) Plant transformation: problems and strategies for practical application. Annu Rev Plant Phys 48:297–326

Brettschneider R, Becker D, Lörz H (1997) Efficient transformation of scutellar tissue of immature maize embryos. Theor Appl Genet 94:737–748

Brown AHD, Garvin DF, Burdon JJ, Abott DC, Read BJ (1996) The effect of combining scald resistance genes on disease levels, yield and quality traits in barley. Theor Appl Genet 93:361–366

Büschges R, Hollrichter K, Panstruga R, Simons G, Wolter M, Frijiters A, van Daelen R, van der Lee T, Diergaarde P, Groendijk J, Töpsch S, Vos P, Salamini F, Schulze Lefert P (1997) The barley *Mlo* gene: a novel control element of plant pathogen resistance. Cell 88:695–705

Cai D, Kleine M, Kifle S, Harloff HJ, Sandal NN, Marcker KA, Klein-Lankhorst RM, Salentijn EMJ, Lange W, Stiekema WJ, Wyss U, Grundler FMW, Jung C (1997) Positional cloning of a gene for nematode resistance in sugar beet. Science 275:832–834

Castillo AM, Vasil V, Vasil IK (1994) Rapid production of fertile transgenic plants of rye (*Secale cereale* L.) Bio/Technology 12:1366–1371

Castro A, Corey A, Filichkin T, Hayes P, Sandoval-Islas JS, Vivar H (2000) Stripe rust resistance QTL pyramids in barley. Proc 8th Int Barley Genet Symp, 22–27 Oct, Adelaide, Australia. Contrib Papers, vol II, pp 86–88

Chan M-T, Lee T-M, Chang H-H (1992) Transformation of indica rice (*Oryza sativa* L.) mediated by *Agrobacterium tumefaciens*. Plant Cell Physiol 33:577–583

Chantret N, Sourdille P, Röder M, Tavaud M, Bernard M, Dousinault G (2000) Location and mapping of the powdery mildew resistance gene *MlRE* and detection of a resistance QTL by bulked segregant analysis (BSA) with microsatellites in wheat. Theor Appl Genet 100:1217–1224

Chen F, Prehn D, Hayes PM, Mulrooney D, Corey A, Vivar H (1994) Mapping genes for resistance to barley stripe rust (*Puccinia striiformis* f. sp. *hordei*). Theor Appl Genet 88:215–219

Cheng M, Fry JE, Pang S, Zhou H, Hironaka CM, Duncan DR, Conner TW, Wan Y (1997) Genetic transformation of wheat mediated by *Agrobacterium tumefaciens*. Plant Physiol 115:971–980

Chester K (1933) The problem of acquired physiological immunity in plants. Q Rev Biol 8:275–324

Christensen AH, Sharrock RA, Quail PH (1992) Maize polyubiquitin genes: structure, thermal perturbation of expression and transcript splicing, and promoter activity following transfer to protoplasts by electroporation. Plant Mol Biol 18:675–689

Christou P, Ford TL, Kofron M (1991) Production of transgenic rice (*Oryza sativa* L.) plants from agronomically important *indica* and *japonica* varieties via electric discharge particle acceleration of exogenous DNA into immature zygotic embryos. Bio/Technology 9:957–962

Dale EC, Ow DW (1991) Gene transfer with subsequent removal of the selection gene from the host genome. Proc Natl Acad Sci USA 88:10558–10562

De Block M, Herrera-Estella L, Van Montagu M, Schell J, Zambryski P (1984) Expression of foreign genes in regenerated plants and their progeny. EMBO J 3:1181–1189

De Cleene M, De Ley J (1976) The host range of crown gall. Bot Rev 42:389–466

De la Pena RC, Smith KP, Capettini F, Muelbauer GJ, Gallo-Meagher M, Dill-Macky R, Somers DA, Rasmuson DC (1999) Quantitative trait loci associated with resistance to *Fusarium* head blight and kernel discoloration in barley. Theor Appl Genet 99:561–569

Dixon MS, Hatzixanthis K, Jones DA, Harrison K, Jones JD (1998) The tomato *Cf-5* disease resistance gene and six homologs show pronounced allelic variation in leucinrich repeat copy number. Plant Cell 10:1915–1925

Dweikat I, Ohm H, Patterson F, Cambron S (1997) Identification of RAPD markers for 11 Hessian fly resistance genes in wheat. Theor Appl Genet 94:419–423

Effertz RJ, Anderson JA, Francl LJ (2001) Restriction fragment length polymorphism mapping of resistance to two races of *Pyrenophora tritici-repentis* in adult and seedling wheat. Phytopathology 91:572–578

El Attari H, Hayes PM, Rebai A, Barrault G, Dechamp-Guillaume G, Sarrafi A (1998) Potential of doubled-haploid lines and localization of quantitative trait loci (QTL) for partial resistance to bacterial leaf streak (*Xanthomonas campestris* pv. *hordei*) in barley. Theor Appl Genet 96:95–100

FAO (2001) FAOSTAT Agriculture Data. Available: http: //apps.fao.org/

Faris JD, Li WL, Liu DJ, Chen PD, Gill BS (1999) Candidate gene analysis of quantitative disease resistance in wheat. Theor Appl Genet 98:219–225

Flor H (1971) Current status of the gene-for-gene concept. Annu Rev Phytopathol 9:275–296

Fought L, Kuç J (1996) Lack of specificity in plant extracts and chemicals as inducers of systemic resistance in cucumber plants to anthracnose. J Phytopathol 144:1–6

Funatsuki H, Kuroda M, Lazzeri PA, Müller E, Lörz H, Kishinami I (1995) Fertile transgenic barley generated by direct transfer to protoplasts. Theor Appl Genet 91:707–712

Gless C, Lörz H, Jähne-Gärtner A (1998) Transgenic oat plants obtained at high efficiency by microprojectile bombardment of leaf base segments. J Plant Physiol 152:151–157

Görlach J, Volrath S, Knauf-Beiter G, Hengy G, Beckhove U, Kogel KH, Oostendorp M, Staub T, Ward E, Kessmann H, Ryals J (1996) Benzothiadiazole, a novel class of inducers of systemic acquired resistance, activates gene expression and disease resistance in wheat. Plant Cell 8:629–643

Gould J, Devery M, Hasegawa O, Ulian EC, Peterson G, Smith RH (1991) Transformation of *Zea mays* L. using *Agrobacterium tumefaciens* and the shoot apex. Plant Physiol 95:426–434

Graner A, Jahoor A, Schondelmaier J, Siedler H, Pillen K, Fischbeck G, Wenzel G, Herrman RG (1991) Construction of an RFLP map in barley. Theor Appl Genet 83:250–256

Graner A, Michalek W, Streng S (2000) Molecular mapping of genes conferring resistance to viral and fungal pathogens. Proc 8th Int Barley Genet Symp, 22–27 Oct, Adelaide, Australia. Inv Papers, vol I, pp 45–52

Grant MR, Godiard L, Straube E, Ashfield T, Lewald J, Sattler A, Innes RW, Dangl JL (1996) Structure of the *Arabidopsis RPM1* gene enabling dual specificity disease resistance. Science 269:843–846

Haltermann D, Zhou F, Wie F, Wise RP, Schulze-Lefert P (2001) The Mla6 coiled-coil, NBS-LRR protein confers AvrMla6-dependent resistance specificity to *Blumeria graminis* f. sp. *hordei* in barley and wheat. Plant J 25:335–348

Hammer K (1997) Genpools - Struktur, Verfügbarkeit und Bearbeitung für die Züchtung. Schr Genet Ressour 8:4–14

Hansen G, Chilton M-D (1996) "Agrolistic" transformation of plant cells: integration of T-strands generated *in planta*. Proc Natl Acad Sci USA 93:14978–14983

Hause B, Vorös K, Kogel KH, Beßer K, Wasternack C (1999) A jasmonate-responsive lipoxygenase of barley leaves is induced by plant activators but not by pathogens. J Plant Physiol 154:459–462

Havill NP, Raffa KF (1999) Effects of elicitation treatment and genotypic variation on induced resistance in *Populus*: impacts on gypsy moth (Leptidoptera: Lymantriidae) development and feeding behaviour. Oecologia 120:295–303

Hayes PM, Prehn D, Vivar H, Blake T, Comeau A, Henry I, Johnston M, Jones B, Steffenson B (1996) Multiple disease resistance loci and their relationship to agronomic and quality loci in a spring barley population. J.QTL. http: //probe.nalusda. gov: 8000/otherdocs/jqtl/index.html

Hayes PM, Castro A, Mequez-Cedillo L, Corey A, Henson C, Jones B, Kling J, Mather D, Matus I, Rossi C, Sato K (2002) A summary of published barley QTL reports. http: //www.css.orst.edu/barley/nabgmp/qtlsum.htm

Heun M (1992) Mapping quantitative powdery mildew resistance of barley using a restriction fragment length polymorphisms map. Genome 32:1019–1025

Hiei Y, Ohta S, Komari T, Kumashiro (1994) Efficient transformation of rice (*Oryza sativa* L.) mediated by *Agrobacterium* and sequence analysis of the boundaries of the T-DNA. Plant J 6:271–282

Hittalmani S, Parco A, Mew TV, Zeigler RS, Huang N (2000) Fine mapping and DNA marker-assisted pyramiding of three major genes for blast resistance in rice. Theor Appl Genet 100:1121–1128

Hoekema A, Hirsch PR, Hooykaas PJJ, Schilperoot RA (1983) A binary plant vector strategy based on separation of *vir*- and T-regions of the *Agrobacterium tumefaciens* Ti-plasmid. Nature 303:179–180

Horsch RB, Fraley RT, Rogers SG, Sanders PR, Lloyd A, Hoffmann N (1984) Inheritance of functional foreign genes in plants. Science 223:496–498

Huang B, Angeles ER, Domingo J, Magpantay G, Singh S, Zhang G, Kumaravadivel N, Benett J, Khush GS (1997a) Pyramiding of bacterial blight resistance genes in rice: marker-assisted selection using RFLP and PCR. Theor Appl Genet 95:313–320

Huang XQ, Hsam SLK, Zeller FJ (1997b) Identification of powdery mildew resistance genes in common wheat (*Triticum aestivum* L. em Thell.). IX. Cultivars, land races and breeding lines grown in China. Plant Breed 116:233–238

Hulbert SH, Webb CA, Smith SM, Sun Q (2001) Resistance gene complexes: evolution and utilization. Annu Rev Phytopathol 39:285–312

Hwang BK, Heitefuss R (1982) Induced resistance of spring barley to *Erysiphe graminis* f. sp. *hordei*. Phytopathol Z 103:41–47

Ishida Y, Saito H, Ohta S, Hiei Y, Komari T, Kumashiro T (1996) High efficiency transformation of maize (*Zea mays* L.) mediated by *Agrobacterium tumefaciens*. Nat Biotechnol 14:745–750

Ittu M, Saulescu NN, Hagima I, Ittu G, Mustatea P (2000) Association of fusarium head blight resistance with gliadin loci in a winter wheat cross. Crop Sci 40:62–67

Iyer LM, Kumpatla SP, Chandrasekharan MB, Hall TC (2000) Transgene silencing in monocots. Plant Mol Biol 43:323–346

Jaehne A, Becker D, Brettschneider R, Lörz H (1994) Regeneration of transgenic, microspore-derived, fertile barley. Theor Appl Genet 89:525–533

Jacobsen J, Matthews P, Abbott D, Wang M, Waterhouse P (2000) Improvement of barley quality using genetic engineering. Proc 8th Int Barley Genet Symp, 22–27 Oct, Adelaide, Australia. Inv Pap, vol I, pp 121–123

Jin QL, Liu NZ, Qiu JL, Li DB, Wang J (1997) A truncated fragment of hairpin$_{pss}$ induces systemic resistance to *Xanthomonas campestris* pv. *oryzae* in rice. Physiol Mol Plant P 51:243–257

Johal GS, Briggs SP (1992) Reductase activity encoded by the *HM1* disease resistance gene in maize. Science 258:985–987

Jones DA, Thomas CM, Hammond-Kosack KE, Balint-Kurti PJ, Jones JDG (1994) Isolation of the tomato *Cf-9* gene for resistance to *Cladosporium fulvum* by transposon tagging. Science 266:789–793

Jørgensen JH, Lübeck PS, Thordal-Christensen H, de Neergård E, Smedegård-Petersen V (1998) Mechanisms of induced resistance in barley against *Drechslera teres*. Phytopathology 88:698–707

Keller M, Keller B, Schachermayer G, Winzeler M, Schmid JE, Stamp P, Messmer MM (1999) Quantitative trait loci for resistance against powdery mildew in a segregating wheat x spelt population. Theor Appl Genet 98:903–912

Kicherer S, Backes G, Walther U, Jahoor A (2000) Localising QTLs for leaf rust resistance and agronomic traits in barley (*Hordeum vulgare* L.). Theor Appl Genet 100:881–888

Kinane JT, Jones PW (2000) Components of partial resistance to powdery mildew in wheat mutants. Eur J Plant Pathol 106:607–616

Kiyosawa S (1982) Genetics and epidemiological modeling of breakdown of plant disease resistance. Annu Rev Phytopathol 20:93–117

Kloppers FJ, Pretorius ZA (1997) Effects of combinations amongst genes *Lr13*, *Lr34* and *Lr37* on components of resistance in wheat to leaf rust. Plant Pathol 46:737–750

Kogel KH, Ortel B, Jarosch B, Atzorn R, Schiffer R, Wasternack C (1995) Resistance in barley against powdery mildew (*Erysiphe graminis* f. sp. *hordei*) is not associated with enhanced levels of endogenous jasmonates. Eur J Plant Pathol 101:319–332

Kolb FL, Bai GH, Mühlbauer GJ, Anderson JA, Smith KP, Fedak G (2001) Host resistance genes for fusarium head blight: mapping and manipulation with molecular markers. Crop Sci 41:611–619

Komari T, Hiei Y, Saito Y, Murai N, Kumashiro T (1996) Vectors carrying two separate T-DNAs for co-transformation of higher plants mediated by *Agrobacterium tumefaciens* and segregation of transformants free from selection markers. Plant J 10:165–174

Koncz C, Martini N, Meyerhofer R, Koncz-Kalman Z, Körber H, Redei GP, Schell J (1989) High-frequency T-DNA-mediated gene tagging in plants. Proc Natl Acad Sci USA 86:8467–8471

Kuç J (2001) Concepts and direction of induced systemic resistance in plants and its application. Eur J Plant Pathol 107:7–12

Kuç J, Barnes E, Daftsios A, Williams E (1959) The effect of amino acids on susceptibility of apple varieties to scab. Phytopathology 49:313-315

Lahaye T, Shirasu K, Schulze Lefert P (1998) Chromosome landing at the barley *Rar1* locus. Mol Gen Genet 260:92–101

Lawrence GJ, Finnegan EJ, Ayliffe MA, Ellis JG (1995) *L6* gene in flax rust resistance is related to the *Arabidopsis* bacterial resistance gene *RPS2* and the tobacco viral resistance gene *N*. Plant Cell 7:1195–1206

Leach JE, Cruz CMV, Bai J, Leung H (2001) Pathogen fitness penalty as a predictor of durability of disease resistance genes. Annu Rev Phytopathol 39:187–224

Leemann M, Van Pelt JA, Den Ouden FM, Heinsbroek M, Bakker M, Schippers B (1995) Biocontrol of fusarium wilt of radish by lipopolysaccharides of *Pseudomonas fluorescens*. Phytopathology 85:1021–1027

Li Z, Upadhyaya NM, Meena S, Gibbs AJ, Waterhouse PM (1997) Comparison of promoters and selectable marker genes for the use in indica rice transformation. Mol Breed 3:1–14

Lister RM, Ranieri R (1995) Distribution and economic importance of barley yellow dwarf. In: D'Arcy CJ, Burnett PA (eds) Barley yellow dwarf: 40 years of progress, APS Press, St Paul, Minnesota, pp 29–53

Liu J, Liu D, Tao W, Li W, Wang Chen P, Cheng S, Gao D (2000) Molecular marker-facilitated pyramiding of different genes for powdery mildew resistance in wheat. Plant Breed 119:21–24

Liu L, Kloepper JW, Tuzun S (1995) Induction of systemic resistance in cucumber by plant growth-promoting rhizobacteria: duration of protection and effect of protection and root colonization. Phytopathology 85:1064–1068

Loebenstein G (1963) Further evidence on systemic resistance induced by localized necrotic virus infections in plants. Phytopathology 53:306–308

Löbler M, Lee J (1998) Jasmonate signalling in barley. Trends Plant Sci 3:8–9

Macaulay M, Ramsay L, Powell W, Waugh R (2001) A representative, highly informative 'genotyping set' of barley SSRs. Theor Appl Genet 102:801–809

Machii H, Mizuno H, Hirabayashi T, Li H, Hagio T (1998) Screening of wheat genotypes for high callus induction and regeneration capability from anther and immature embryo cultures. Plant Cell Tissue Org 53:67–74

Manhandhar HK, Mathur SB, Smedegård-Petersen V, Thordal-Christensen H (1999) Accumulation of transcripts for pathogenesis-related proteins and peroxidase in rice plants triggered by *Pyricularia oryza*, *Bipolaris sorokinana* and UV light. Physiol Mol Plant P 55:289–295

Martin GB, Brommenschenkel SH, Chunwongse J, Frary A, Ganal MW, Spivey R, Wu T, Earle ED (1993) Map-based cloning of a protein kinase gene conferring disease resistance in tomato. Science 262:1432–1436

Matthews PR, Wang M-B, Waterhouse PM, Thornton S, Fieg SJ, Gubler F, Jacobsen JV (2001) Marker gene elimination from transgenic barley, using co-transformation with adjacent 'twin T-DNAs' on a standard *Agrobacterium* transformation vector. Mol Breed 7:195–202

McClung AM, Marchetti MA, Webb BD, Bollich CN (1997) Registration of 'Jefferson' rice. Crop Sci 37:629–630

McElroy D, Zhang W, Cao C, Wu R (1990) Isolation of an efficient *actin* promoter for use in rice transformation. Plant Cell 2:163–171

McGrath PF, Vincent JR, Lei C-H, Lister RM, Torbert K, Pawlowski W, Wan Y, Lemaux PG, Rines H, Somers DA (1997) Coat protein mediated resistance to isolates of barley yellow dwarf virus in oats and barley. Eur J Plant Pathol 103:695–710

Messmer MM, Seyfarth R, Keller M, Schachermayer G, Winzeler M, Zanetti S, Feuillet C, Keller B (2000) Genetic analysis of durable leaf rust resistance in winter wheat. Theor Appl Genet 100:419–431

Michelmore RW (1995) Molecular approaches to manipulation of disease resistance genes. Annu Rev Phytopathol 33:393–428

Mindrinos M, Katagiri F, Yu GL, Ausubel FM (1994) The A. thaliana disease resistance gene RPS2 encodes a protein containing a nucleotide-binding site and leucin-rich repeats. Plant Cell 78:1089–1099

Miyazaki C, Osanai E, Saeki K, Ito K, Konishi T, Sato K, Saito A (2001) Mapping of quantitative trait loci conferring resistance to barley yellow mosaic virus in a Chinese barley landrace Mokusekko 3. Breed Sci 51:171–177

Mohan M, Nair S, Bhagwat A, Krishna TG, Yano M, Bhatia CR, Sasaki T (1997) Genome mapping, molecular markers and marker-assisted selection in crop plants. Mol Breed 3:87–103

Moharramiphour S, Tsumuki H, Sato K, Yoshida H (1997) Mapping resistance to cereal aphids in barley. Theor Appl Genet 94:592–596

Molina A, Görlach J, Volrath S, Ryals J (1999) Wheat genes encoding two types of PR-1 proteins are pathogen inducible but do not respond to activators of systemic acquired resistance. Mol Plant Microbe Interact 12:53–58

Morris SW, Vernooij B, Titatarn S, Starrett M, Thomas S, Wiltse CC, Frederiksen RA, Bhandhufalck A, Hulbert S, Uknes S (1998) Induced resistance response in maize. Mol Plant Microbe Interact 11:643–658

Natural Resources Institute (1992) A synopsis of integrated pest management in developing countries in the tropics. Natural Resources Institute, Chatham

Nelson R (2001) Biotic stresses in crops. In: Nösberger J, Geiger HH, Struik PC (eds) Crop science: progress and prospects. CABI, Wallingford

Nelson RR (1978) Genetics of horizontal resistance to plant diseases. Annu Rev Phytopathol 16:359–78

Oerke EC, Dehne HW, Schönbeck F, Weber A (1994) Crop production and crop protection: estimated losses in major food and cash crops, Elsevier, Amsterdam

Oostendorp M, Kunz W, Dietrich B, Staub T (2001) Induced disease resistance in plants by chemicals. Eur J Plant Pathol 107:19–28

Ordon F, Götz R, Friedt W (1993) Genetic stocks to barley yellow mosaic viruses (BaMMV, BaYMV, BaYMV-2) in Germany. Barley Genet Newslett 22:46–49

Ordon F, Wenzel W, Friedt W II (1998) Recombination: molecular markers for resistance genes in major grain crops. Progress in Botany 59. Springer, Berlin Heidelberg New York, pp 49–79

Ordon F, Schiemann A, Pellio B, Dauck V, Bauer E, Streng S, Friedt W, Graner A (1999) Application of molecular markers in breeding for resistance to the barley yellow mosaic virus complex. J Plant Dis Prot 106:256–264

Parker JE, Coleman MJ (1997) Molecular intimacy between proteins specifying plant-pathogen recognition. Trends Biochem Sci 22:291–296

Parker JE, Coleman MJ, Szabò V, Frost LN, Schmidt R, van der Biezen EA, Moores T, Dean C, Daniels MJ, Jones JDG (1997) The *Arabidopsis* downy mildew resistance gene *Rpp5* shares similarity to the toll and interleucin-1 receptor with *N* and *L6*. Plant Cell 9:879–894

Paszkowski J, Shillito RD, Saul M, Mandak V, Hohn T, Hohn B, Patrykus I (1984) Direct gene transfer to plants. EMBO J 3:2717–2722

Pecchioni N, Faccioli P, Toubia-Rahme H, Vale G, Terzi V (1996) Quantitative resistance to barley leaf stripe (*Pyrenophora graminea*) is dominated by one major locus. Theor Appl Genet 93:97–101

Pedersen WL (1988) Pyramiding major genes for resistance to maintain residual effects. Annu Rev Phytopathol 26:369–378

Pellio B, Werner K, Friedt W, Graner A, Ordon F (2000) Resistance to the barley yellow mosaic virus complex – from Mendelian genetics towards map based cloning. Czech J Genet Plant Breed 36:84–87

Piffanelli P, Devoto A, Schulze-Lefert P (1999) Defence signalling pathways in cereals. Curr Opin Plant Biol 2:295–300

Potrykus I (1990) Gene transfer to cereals: an assessment. Bio/Technology 8:535–542

Proeseler G, Hartleb H, Kophanke D, Lehmann CO (1989) Resistenzeigenschaften im Gersten- und Weizensortiment Gatersleben. 27. Prüfung von Gersten auf ihr Verhalten gegenüber dem Milden Gerstenmosaik-Virus (barley mild mosaic virus), Gerstengelbmosaik-Virus (barley yellow mosaic virus), *Drechslera teres* (Sacc.) Shoem. und *Puccinia hordei* Otth. Kulturpflanze 37:145–154

Qi X, Niks E, Stam P, Lindhout P (1998) Identification of QTLs for partial resistance to leaf rust (*Puccinia hordei*) in barley. Theor Appl Genet 96:1205–1215

Qi X, Jiang G, Chen W, Niks RE, Stam P, Lindhout P (1999) Isolate-specific QTLs for partial resistance to *Puccinia hordei* in barley. Theor Appl Genet 99:877–844

Raineri DM, Bottino P, Gordon MP, Nester EW (1990) *Agrobacterium*-mediated transformation of rice (*Oryza sativa* L.). Bio/Technology 8:33–37

Raman H, Read BJ, Brown AHD, Abbott DC (1999) Molecular markers and pyramiding of multiple genes for resistance to scald in barley. Proc 9th Aust Barley Tech Symp, Melbourne

Ramsay L, Macaulay M, degli Ivanissevich S, MacLean K, Cardle L, Fuller J, Edwards KJ, Tuvesson S, Morgante M, Massari A, Maestri E, Marmiroli N, Sjakste T, Ganal M, Powell W, Waugh R (2000) A simple sequence repeat-based linkage map of barley. Genetics 156:1997–2005

Reiss E (1986) Senkung des Gelbrostbefalls anfälliger Gerstenpflanzen nach Applikation von Infiltraten aus gelbrostinfizierten, resistenten Gerstenblättern. Arch Phytopathol Pfl 22:271

Repellin A, Båga M, Jauhar PP, Chibbar RN (2001) Genetic enrichment of cereal crops via alien gene transfer: new challenges. Plant Cell Tissue Org 64:159–183

Rhodes CA, Pierce DA, Mettler LJ, Mascarenhas D, Detmer JJ (1988) Genetically transformed maize plants from protoplasts. Science 240:204–207

Richter K, Schondelmaier J, Jung C (1998) Mapping of quantitative loci affecting *Drechslera teres* resistance in barley with molecular markers. Theor Appl Genet 97:1225–1234

Roeder MS, Korzun V, Wendehake K, Plaschke J, Tixier MH, Leroy P, Ganal MW (1998) A microsatellite map of wheat. Genetics 149:2007–2023

Rommens CM, Kishore GM (2000) Exploiting the full potential of disease resistance genes for agricultural use. Curr Opin Biotechnol 11:120–125

Ross A (1966) Systemic effects of local lesion formation. In: Beemster A, Dykstra S (eds) Virus of plants. North Holland, Amsterdam, pp 127–150

Ryals J, Uknes S, Ward E (1994) Systemic acquired resistance. Plant Physiol 104:1109–1112

Saeki K, Miyazaki C, Hirota N, Saito A, Ito K, Konishi T (1999) RFLP mapping of BaYMV resistance gene *rym3* in barley (*Hordeum vulgare*). Theor Appl Genet 99:727–732

Sakamoto K, Tada Y, Yokozeki Y, Akagi H, Hayashi N, Fujimura T, Ichikawa N (1999) Chemical induction of disease resistance in rice is correlated with the expression of a gene encoding a nucleotide binding site and a leucine-rich repeat. Plant Mol Biol 40:847–855

Sanchez AC, Brar DS, Huang N, Li Z, Khush GS (2000) Sequence tagged site marker-assisted selection for three bacterial blight resistance genes in rice. Crop Sci 40:792–797

Sanford JC, Klein TM, Wolf ED, Allen N (1987) Delivery of substances into cells and tissues using a particle bombardment process. Particu Sci Technol 5:27–37

Sarhan ART, Király Z, Sziráki I, Smedegård-Petersen V (1991) Increased levels of cytokinins in barley leaves having the systemic acquired resistance to *Biploaris sorokiniana* (Sacc.) Shoemaker. J Phytopathol 113:101–108

Sato K, Inukai T, Hayes PM (2001) QTL analysis of resistance to the rice blast pathogen in barley (*Hordeum vulgare*). Theor Appl Genet 102:916–920

Schaefer JF, Caldwell RM, Patterson FL, Compton LE (1963) Wheat leaf rust combinations. Phytopathology 53:569–5673

Scheurer KS, Friedt W, Huth W, Waugh R, Ordon F (2001) QTL analysis of tolerance to a German strain of BYDV-PAV in barley (*Hordeum vulgare* L.). Theor Appl Genet 103:1074–1083

Schiemann A, Backes G (2000) The use of molecular markers in practical barley breeding. Proc 8th Int Barley Genet Symp, 22–27 Oct, Adelaide, Australia. Contrib Papers, vol III, pp 42–44

Schweizer P, Schlagenhauf E, Schaffrath U, Dudler R (1999) Different patterns of host genes are induced in rice by *Pseudomonas syringae*, a biological inducer of resistance, and the chemical inducer benzothiadiazole (BTH). Eur J Plant Pathol 105:659–665

Shimamoto K, Terada R, Izawa T, Fujimoto H (1989) Fertile transgenic rice plants regenerated from transformed protoplasts. Nature 338:274–276

Shimura M, Iwata M, Tashiro N, Sekizawa Y, Suzuki Y, Mase S, Watanabe T (1981) Anticonidial germination factors induced in the presence of probenazole and properties of four active substances. Agric Biol Chem 45:1431–1435

Simons G, Groenendijk J, Wijbrandi J, Rejans M, Groenen J, Diergaarde P, van der Lee T, Bleeker M, Onstenk J, de Both M, Haring M, Mes J, Cornelissen B, Zabeau M, Vos P (1998) Dissection of the Fusarium *I2* gene cluster in tomato reveals six homologs and one active gene copy. Plant Cell 10:1055–1068

Singh S, Sidhu JS, Huang N, Vikal Y, Li Z, Brar DS, Dhaliwal HS, Khush GS (2001) Pyramiding three bacterial blight resistance genes (*xa5*, *xa13* and *Xa21*) using marker-assisted selection into indica rice cultivar PR106. Theor Appl Genet 102:1011–1015

Smith JA, Métraux JP (1991) *Pseudomonas syringae* pv. *syringae* induces systemic resistance to *Pyricola oryza* in rice. Physiol Mol Plant P 37:20–27

Song W-Y, Wang G-L, Chen L-L, Kim H-S, Pi L-Y, Holsten T, Gardner J, Wang B, Zhai W-X, Zhu L-H, Fauquet C, Ronald P (1995) A receptor kinase-like protein encoded by the rice disease resistance gene *Xa21*. Science 270:1804–1806

Spaner D, Shugar LP, Choo TM, Falak I, Briggs KG, Legge WG, Falk DE, Ullrich SE, Tinker NA, Steffenson BJ, Mather DE (1998) Mapping of disease resistance loci in barley based on visual assessment of naturally occurring symptoms. Crop Sci 38:843–850

Stadnik MJ, Buchenauer H (1999) Accumulation of autofluorogenic compounds at the penetration site of *Blumeria graminis* f. sp. *tritici* is associated with both benzothiadiazole-induced and quantitative resistance of wheat. J Phytopathol 147:615–622

Stadnik MJ, Buchenauer H (2000) Inhibition of phenylalanine ammonia-lyase suppresses the resistance induced by benzothiadiazole in wheat to *Blumeria graminis* f. sp. *tritici*. Physiol Mol Plant P 57:25–34

Staskawicz BJ, Ausubel FM, Baker B, Ellis JG, Jones JDG (1995) Molecular genetics of plant disease resistance. Science 268:661–667

Steffenson BJ, Hayes PM, Kleinhofs A (1996) Genetics of seedling and adult plant resistance to net blotch (*Pyrenophora teres* f. *teres*) and spot blotch (*Cochliobolus sativus*) in barley. Theor Appl Genet 92:552–558

Sticher L, Mauch-Mani B, Métraux JP (1997) Systemic acquired resistance. Annu Rev Plant Pathol 35:235–270

Takken FLW, Joosten MHAJ (2000) Plant resistance genes: their structure, function and evolution. Eur J Plant Pathol 106:699–713

Takken FL, Thomas CM, Joosten MH, Goldstein C, Westerink N, Hille J, Nijkamp HJJ, De Wit PJGM, Jones JDG (1999) A second gene at the tomato *Cf-4* locus confers resistance to *Cladosporium fulvum* through recognition of a novel avirulence determinant. Plant J 20:279–288

Thomas WTB, Powell W, Waugh R, Chalmers KJ, Barua UM, Jack P, Lea V, Forster BP, Swanston JS, Ellis RP, Hanson PR, Lance RCM (1995) Detection of quantitative trait loci for agronomic, yield, grain and disease characters in spring barley (*Hordeum vulgare* L.). Theor Appl Genet 91:1037–1047

Thomas CM, Jones DA, Parniske M, Harrison K, Balint-Kurti PJ, Hatzixanthis K, Jones JDG (1997) Characterization of the tomato *Cf-4* gene for resistance to *Cladosporium fulvum* identifies sequences that determine recognition specificity in *Cf-4* and *Df-9*. Plant Cell 9:2209–2224

Tingay S, McElroy D, Kalla R, Fieg S, Wang M-B, Thornton S, Brettell R (1997) *Agrobacterium tumefaciens*-mediated barley transformation. Plant J 11:1369–1376

Toki S (1997) Rapid and efficient *Agrobacterium*-mediated transformation in rice. Plant Mol Biol 15:16–21

Ton J, Pieterse CMJ, Boon JJ (1999) Changes in chemical composition related to fungal infection and induced resistance to carnation and radish investigated by pyrolysis mass spectrometry. Physiol Mol Plant P 55:297–311

Ton J, Davison S, Van Loon LC, Pieterse CMJ (2001) Hereditability of rhizobacteria-mediated induced systemic resistance and basal resistance in Arabidopsis. Eur J Plant Pathol 107:63–68

Toojinda T, Baird E, Booth A, Broers L, Hayes P, Powell P, Thomas W, Vivar H, Young G (1998) Introgression of quantitative trait loci (QTLs) determining stripe rust resistance in barley: an example of marker-assisted line development. Theor Appl Genet 96:123–131

Toojinda T, Broers LH, Chen XM, Hayes PM, Kleinhofs A, Korte J, Kudrna D, Leung H, Line RF, Powell W, Ramsay L, Vivar H, Waugh R (2000) Mapping quantitative and qualitative disease resistance genes in a doubled haploid population of barley (*Hordeum vulgare*). Theor Appl Genet 101:580–589

Toriyama K, Arimoto Y, Uchimiya H, Hinata K (1988) Transgenic rice plants after direct gene transfer into protoplasts. Bio/Technology 6:1072–1074

Tu J, Ona I, Zhang Q, Mew TW, Kush GS, Datta SK (1998) Transgenic rice variety IR72 with *Xa21* is resistant to bacterial blight. Theor Appl Genet 97:31–36

Tyagi AK, Mohanty A, Chaudhury A, Maheswari SC (1999) Transgenic rice: a valuable monocot system for crop improvement and gene research. Crit Rev Biotech 19:41–79

Underwood N, Morris W, Gross K, Lockwood JR (2000) Induced resistance to Mexican bean beetles in soybean: variation among genotypes and lack of correlation with constitutive resistance. Oecologia 122:83–89

Van Loon LC, Bakker PAHM, Pieterse CMJ (1998) Systemic resistance induced by rhizosphere bacteria. Annu Rev Phytopathol 36:453–483

Van Peer R, Niemann GJ, Schippers B (1991) Induced resistance and phytoalexin accumulation in biological control of fusarium wilt of carnation by *Pseudomonas* sp. strain WCS417r. Phytopathology 81:728–734

Van Sanford D, Anderson J, Campbell K, Costa J, Cregan P, Griffey C, Hayes P, Ward R (2001) Discovery and deployment of molecular markers linked to fusarium head blight resistance: an integrated system for wheat and barley. Crop Sci 41:638–644

Vorös K, Feussner I, Kühn H, Lee J, Graner A, Löbler M, Parthier B, Wasternack C (1998) Characterization of a methyl-jasmonate-inducible lipoxygenase from barley (*Hordeum vulgare* cv Salome) leaves. Eur J Biochem 251:36–44

Vos P, Hogers R, Bleeker M, Reijans M, Van de Le T, Hornes M, Frijters A, Pot J, Peleman J, Kuiper M, Zabeau M (1995) AFLP: a new technique for DNA fingerprinting. Nucleic Acids Res 23:4407–4414

Waldron BL, Moreno-Sevilla B, Anderson JA, Stack RW, Frohbergy RC (1999) RFLP mapping of QTL for fusarium head blight resistance in wheat. Crop Sci 39:805–811

Wan Y, Lemaux PG (1994) Generation of large numbers of independently transformed fertile barley plants. Plant Physiol 104:37–48

Wang GL, Mackill DJ, Bonman JM, McCouch SR, Champoux MC, Nelson RJ (1994) RFLP mapping of genes conferring complete and partial resistance to blast in a durably resistant rice cultivar. Genetics 136:1421–1434

Wang G-L, Song W-L, Ruan D-L, Sideris S, Ronald PC (1996) The cloned gene *Xa21* confers resistance to multiple *Xanthomonas oryzae* pv. *oryzae* isolates in transgenic plants. Mol Plant Microbe Interact 9:850–855

Wang M-B, Abbott DC, Waterhouse PM (2000) A single copy of a virus-derived transgene encoding hairpin RNA gives immunity to barley yellow dwarf virus. Mol Plant Pathol 1:347–356

Wang Z, Taramino G, Yang D, Liu G, Tingey SV, Miao GH, Wang GL (2001) Rice ESTs with disease-resistance gene- or defense-response gene-like sequences mapped to regions containing major resistance genes or QTLs. Mol Gen Genet 265:302–310

Waterhouse PM, Graham MW, Wang M-B (1998) Virus resistance and gene silencing in plants can be induced by simultaneous expression of sense and antisense RNA. Proc Natl Acad Sci USA 95:13959–13964

Wei F, Gobelman-Werner K, Morroll SM, Kurth J, Mao L, Wing R, Leister D, Schulze-Lefert P, Wise RP (1999): The *Mla* (powdery mildew) resistance cluster is associated with three NBS-LRR gene families and suppressed recombination within a 240-kb DNA interval on chromosome 5S (1HS) of barley. Genetics 153:1929–1948

Weiskorn C, Krämer M, Huth W, Friedt W, Ordon F (2002) Investigations on the efficiency of systemic acquired resistance (SAR) against a German isolate of BYDV-PAV in barley (*Hordeum vulgare* L.). J Plant Dis Prot (in press)

Werner K, Friedt W, Ordon F (2000) Strategies for pyramiding resistance genes against the barley yellow mosaic virus complex based on molecular markers and DH-lines. Proc 8th Int Barley Genet Symp, Adelaide, 22–27 Oct, Australia. Contrib Papers, vol II, pp 200–202

Whitham S, McCormick S, Baker B (1996) The *N* gene of tobacco confers resistance to tobacco mosaic virus in transgenic tomato. Proc Natl Acad Sci USA 93:8776–8781

Xia XC, Hsam SLK, Stephan U, Yang TM, Zeller FJ (1995) Identification of powdery mildew resistance genes in common wheat (*Triticum aestivum* L.). VI. Wheat cultivars grown in China. Plant Breed 114:174–175

Yoshimura S, Yoshimura A, Iwata N, McCouch SR, Abenes ML, Baraoidan MR, Mew TW, Nelson RJ (1995) Tagging and combining bacterial blight resistance genes in rice using RAPD and RFLP markers. Mol Breed 1:375–378

Yoshimura S, Yamanouchi U, Katayose Y, Toki S, Wang ZX, Kono I, Kurata N, Yano M, Iwata N, Sasaki T (1998) Expression of *Xa1*, a bacterial blight-resistance gene in rice, is induced by bacterial inoculation. Proc Natl Acad Sci USA 95:1663–1668

Young ND (1996) QTL mapping and quantitative disease resistance in plants. Annu Rev Phytopathol 34:479–501

Young ND (1999) A cautiously optimistic vision for marker-assisted breeding. Mol Breed 5:505–510

Zambryski P, Tempe J, Schell J (1989) Transfer and function of T-DNA genes from Agrobacterium Ti and Ri plasmids in plants. Cell 56:193–201

Zhang W, Wu R (1988) Efficient regeneration of transgenic rice plants from rice protoplasts and correctly regulated expression of the foreign gene in the plants. Theor Appl Genet 6:835–840

Zhao ZY, Cai TS, Tagliani L, Miller M, Wang N, Pang H, Rudert M, Schroeder S, Hondred D, Seltzer J, Pierce D (2000) *Agrobacterium*-mediated sorghum transformation. Plant Mol Biol 44:789–798

Zhou F, Kurth J, Wie F, Elliott C, Valè G, Yahiaoui N, Keller B, Sommerville S, Wise R, Schulze-Lefert P (2001) Cell-autonomous expression of barley *Mla1* confers race specific resistance to the powdery mildew fungus via Rar1-independent signaling pathway. Plant Cell 13:337–350

Zhu H, Gilchrist L, Hayes P, Kleinhofs A, Kudrna D, Liu Z, Prom L, Steffenson B, Toojinda T, Vivar H (1999) Does function follow form? Principal QTLs for *Fusarium* head blight (FHB) resistance are coincident with QTLs for inflorescence traits and plant height in a doubled-haploid population of barley. Theor Appl Genet 99:1221–1232

Prof. Dr. Dr. h.c. Wolfgang Friedt
Dipl.-Ing. agr. Kay Werner
Dipl.-Ing. agr. Bettina Pellio
Dipl.-Hort. Claudia Weiskorn
Dipl.-Ing. agr. Marco Krämer
PD Dr. Frank Ordon
Institut für Pflanzenbau und Pflanzenzüchtung I
Lehrstuhl für Pflanzenzüchtung, IFZ
Heinrich-Buff-Ring 26–32
35392 Giessen, Germany

e-mail: wolfgang.friedt@agrar.uni-giessen.de

Physiology

Coordination of V-ATPase and V-PPase at the Vacuolar Membrane of Plant Cells*

Martina Drobny, Elke Fischer-Schliebs, Ulrich Lüttge, and Rafael Ratajczak

1 Introduction

The vacuole is one of the most conspicuous multifunctional compartments within a plant cell. In most mature plant cells the central vacuole occupies as much as 90% of the cell volume, or even more in plants with crassulacean acid metabolism (CAM). The vacuole is involved in the control of cell volume and cell turgor, the regulation of cytoplasmic ion concentrations, and pH, sequestration of toxic ions and detoxification of xenobiotics, and transient storage of metabolites (Sze et al. 1992; Blumwald and Gelli 1997; Martinoia and Ratajczak 1997; Wink 1997). Molecules stored in the vacuole include inorganic ions such as sodium, potassium, calcium, magnesium, heavy metals, chloride, and nitrate; organic molecules such as carbohydrates, organic acids, amino acids, polyamines, peptides; and secondary plant products. Regulation of content and volume of plant vacuoles depends on the coordinated action of transporters and channels located in the vacuolar membrane, i.e. the tonoplast (Maeshima 2001). Since transport in most cases has to be driven against a concentration gradient, energisation of the tonoplast is required. This task is achieved by two key transport proteins of the tonoplast: the vacuolar H^+-translocating adenosine triphosphatase (V-ATPase, EC 3.6.1.34) and the vacuolar H^+-translocating pyrophosphatase (V-PPase, EC 3.6.1.1). At the expense of energy released by hydrolysing phosphate bonds of ATP or inorganic pyrophosphate, PP_i, the two H^+ pumps generate an electrochemical H^+ gradient across the tonoplast that provides the driving force for secondary-active solute transport. V-ATPase and V-PPase play a pivotal role in solute transport into and out of the vacuolar compartment and are crucial to processes such as cell enlargement and plant growth, signal transduction, protoplasmic ion and pH homeostasis, and regulation of metabolic pathways (Sze et al. 1992; Rea and Poole 1993; Lüttge et al. 1995a; Ratajczak and Wilkins 2000).

* In memory of Rafael Ratajczak whose research was on V-ATPase and V-PPase. He died on 18 May, 2002.

Progress in Botany, Vol. 64

The central importance of V-ATPase and V-PPase for cell compartmentation and vacuolar function requires a strict control of their activities. Moreover, transport activities of the plasma membrane and the central vacuolar membrane must be coordinated in response to environmental cues and hormonal signals. For the plasma membrane H^+-translocating adenosine triphosphatase (P-ATPase, EC 3.6.3.6), during the last decades advances have been made in understanding enzyme regulation (Palmgren 2001). Despite the importance of V-ATPase and V-PPase in plant cell physiology, relatively little is known about the regulation of their activity in vivo. Various mechanisms have been suggested to be involved in the regulation process of enzyme activities, but how these mechanisms are interrelated to allow a precise control of enzyme action is presently not well understood (Sze et al. 1999; Maeshima 2000; Ratajczak 2000; Ratajczak and Wilkins 2000; Dietz et al. 2001). Our knowledge of coordination of the two H^+-translocating enzymes arises from observations of reactions to environmental and biotic factors that provoke stress and during plant development. Physiological responses of enzyme activity, protein amount, and in some cases specific transcript amounts of V-ATPase subunits or V-PPase, have been recorded. However, these observations have often been conflicting. Hence, our present view of if and how V-ATPase and V-PPase cooperate in a plant cell is still obscure. In this review we will give a summary of the actual state of research on V-ATPase and V-PPase and their cooperation in plant cells.

2 V-ATPase

a) Holoenzyme Structure and Subunit Composition

The V-ATPase is a multi-heteromeric enzyme complex that is structurally and evolutionarily related to the F-type ATP synthases (F-ATPase, EC 3.6.3.14) of chloroplasts, mitochondria, and bacteria (Nelson and Taiz 1989; Gogarten et al. 1992; Weber and Senior 1997). In analogy to the F-ATPase, the V-ATPase holoenzyme consists of two domains, a membrane peripheral domain (V_1) and an integral membrane domain (V_O). Electron microscope analysis of the negatively stained V-ATPase has revealed a 'head and stalk' structure for the V_1 domain of the enzyme (Klink and Lüttge 1991; Lee Taiz and Taiz 1991; Dschida and Bowman 1992; Getz and Klein 1995), with diameters ranging between 9.0 and 11.8 nm for the V-ATPase head and 2.0 and 3.1 nm for the V-ATPase stalk, respectively (Fig. 1; see Lüttge and Ratajczak 1997). The stalk connects the head with the membrane integral V_O domain that can be visualised by freeze-fracture electron microscopy as spherical structures of 6.5–9.1 nm in diameter (Klink et al. 1990; Klink and Lüttge 1992; Mariaux et al. 1994).

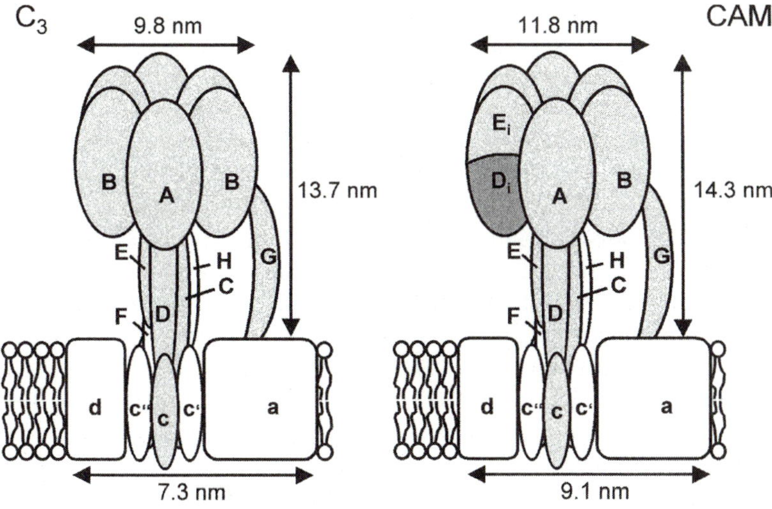

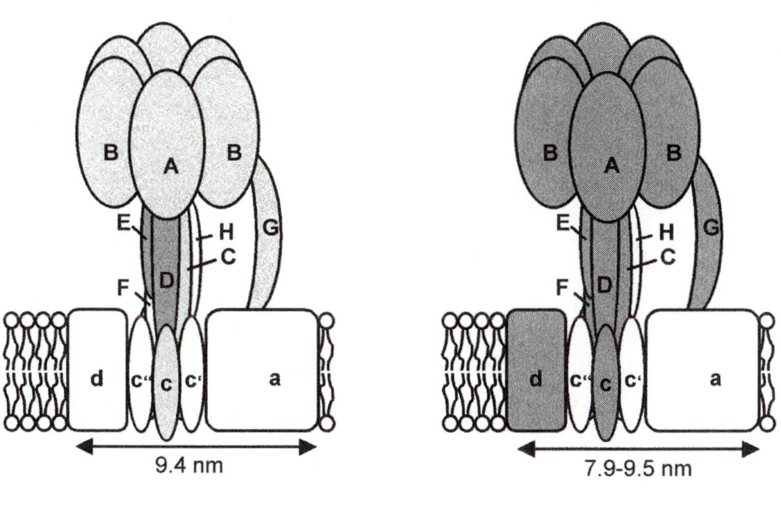

Fig. 1. Structural models of the plant V-ATPase of *Mesembryanthemum crystallinum* L. performing C_3-photosynthesis or crassulacean acid metabolism (CAM) induced by salt stress, *Kalanchoë daigremontiana* Hamet et Perrier de la Bâthie and *Nicotiana tabacum* L. The V-ATPase models are based on the subunit composition of the yeast enzyme (Stevens and Forgac 1997). V-ATPase subunits identified by amino acid sequencing and/or MALDI analysis are given in *dark grey*, V-ATPase subunits identified by immunological detection are given in *light grey*. Subunits not identified to date in plants shown are given in *white*. The dimensions of the V_1 domain and the intramembranous V_O domain refer to data obtained by freeze-fracture studies (Mariaux et al. 1994; Rockel et al. 1994; Lüttge et al. 2001) and analyses of negatively stained V-ATPase molecules (Lüttge and Ratajczak 1997)

While earlier electron microscope studies gave evidence for only a single central stalk connecting the V-ATPase head to V_O, in later studies the existence of one or more peripheral stalks for the V-ATPase (Boekema et al. 1997, 1998, 1999; Wilkens et al. 1999; Ubbink-Kok et al. 2000), as well as for the F-ATPase (Böttcher et al. 1998; Wilkens and Capaldi 1998), were reported for different organisms. For the plant V-ATPase, three-dimensional projection maps derived from electron microscopic images of the complete V-ATPase holoenzyme of *Kalanchoë daigremontiana* were documented for the first time only recently (Domgall et al. 2002). The three-dimensional information revealed a complex stalk architecture with a central stalk surrounded by three peripheral stalks of different sizes and shapes.

In V-ATPase research the enzyme of *Saccharomyces cerevisiae* has been extensively used in the past as a suitable model system for identification of genes encoding V-ATPase subunits and their role in enzyme functions by the use of deletion mutants with a phenotype, which can only be complemented by an acidic growth medium (Nelson and Nelson 1990; Nelson and Harvey 1999). In *S. cerevisiae*, so far 13 genes have been identified encoding for V-ATPase subunits (Table 1; for review, see Forgac 2000). These subunits are suggested to compose the V-ATPase holoenzyme complex with a subunit stoichiometry of $\underline{A}_3\underline{B}_3\underline{C}_1\underline{D}_1\underline{E}_1\underline{F}_x\underline{G}_y\underline{H}_z$ residing within the V_1 domain and $\underline{a}_1\underline{d}_1\underline{c}''_1\ (\underline{c},\ \underline{c}')_6$ residing within the V_O domain. Not all genes homologous to the yeast V-ATPase genes have been found in the plant genome so far. Beside genes encoding for subunits A, B, E, and c which were cloned much earlier for several plant species (Lüttge and Ratajczak 1997), recently four additional plant genes encoding for V-ATPase subunits have been identified: C (Tavakoli et al. 1999), D (Kluge et al. 1999), G (Rouquié et al. 1998), and F (Mayer et al. 1999). After complete sequencing of the *Arabidopsis thaliana* genome in the year 2000, eleven of the V-ATPase subunits of the yeast enzyme can now also be documented for the plant V-ATPase by complete sequence information: A, B, C, D, E, F, G, H, a, d, and c (see Ratajczak and Wilkins 2000; Maeshima 2001, and SwissProt protein database information). However, the existence of genes in the plant genome does not necessarily prove the incorporation of the respective subunits into the V-ATPase holoenzyme. On the protein level, in purified V-ATPase fractions of *Avena sativa* (Ward and Sze 1992), *Hordeum vulgare* (DuPont and Morrissey 1992), *Pyrus communis* (Hosaka et al. 1994), and *Nicotiana tabacum* (Drobny et al. 2002), up to ten distinct polypeptides have been found. The identification of individual polypeptides as V-ATPase subunits in different plant species was based either on apparent molecular mass comparison of electrophoretically separated polypeptides with subunits identified in other organisms, immunological evidence, or sequence comparison (Table 1). Figure 1 shows, representative for the plant enzyme, the V-ATPases of *Mesembryanthemum crystallinum* with

Table 1. Subunit composition of the V-ATPase and function of subunits in plants, mammalians, and yeast cells. For information about mammalian (coated vesicle) and yeast V-ATPase subunits, see Forgac (2000)

Subunit	Calculated molecular mass (kDa) Plant	Coated vesicles	Yeast	Yeast gene	Subunit function
V_1					
A	63–72	73	69	*VMA1*	ATP-binding, catalytic
B	52–60	58	57	*VMA2*	ATP-binding, non-catalytic
C	37–52	40	42	*VMA5*	Activity, assembly
D	29–42	34	32	*VMA8*	Activity, assembly, coupling
E	27–40	33	27	*VMA4*	Activity, assembly
F	13–14	14	14	*VMA7*	Activity, assembly
G	12–13	15	13	*VMA10*	Activity, assembly
H	51–54	50	54	*VMA13*	Activity (not assembly), regulation
V_O					
a	95–115	100	95	*VPH1/STV1*	Proton translocation, assembly, targeting
d	32–36	38	36	*VMA6*	Activity, assembly
c	16–20	17	17	*VMA3*	Proton translocation, DCCD-binding site
c′	–	17	17	*VMA11*	Proton translocation, DCCD-binding site
c″	?[a]	19	23	*VMA16*	Proton translocation

[a]The question mark refers to a subunit of the V-ATPase that has not been unequivocally identified as a subunit of the plant enzyme.

C_3 photosynthesis or CAM, *K. daigremontiana*, and *N. tabacum*. Currently, the V-ATPase of *N. tabacum* is the best-studied plant V-ATPase on the protein level (Drobny et al. 2002). Subunits A, B, C, D, E, F, G, c, and d were unequivocally identified by amino acid sequencing and matrix-assisted laser desorption/ionisation mass spectrometry (MALDI-MS). Only four of the yeast V-ATPase subunits, i.e., a, c′, c″, and H, have not been documented on the protein level as constituents of the V-ATPase of *N. tabacum* (Fig. 1).

One fascinating aspect of V-ATPase research has been the quest for precise locations of V-ATPase subunits in the holoenzyme with the aim of disclosing structure–function relations. Different experimental strategies were applied to obtain information on the assignment of V-ATPase subunits to the V_1 or V_O domain and their site of incorporation within the holoenzyme complex, i.e. covalent cross-linking of V-ATPase subunits (Adachi et al. 1990; Tomashek et al. 1997; Xu et al. 1999), dissociation experiments using cold inactivation or treatment with chaotropic reagents (Lai et al. 1988; Ward and Sze 1992; Lüttge et al. 1995a; Supekova et al. 1995), co-immunoprecipitation (Puopolo et al. 1992; Landolt-Marticorena et al. 2000), and differential expression of V-ATPase subunits in yeast mutant strains (Tomashek et al. 1997; Leng et al. 1998). The most intensively studied V-ATPases with respect to subunit arrangement are the V-ATPases of *S. cerevisiae* and clathrin-coated vesicles of *Bos taurus* (for review, see Stevens and Forgac 1997). Thus, they have served as a basis for modelling the plant V-ATPase of which only little is known in this respect, especially for the stalk region of the V_1 domain (Ratajczak and Wilkins 2000).

For the plant V-ATPase holoenzyme a molecular mass of about 400–650 kDa has been determined by size-exclusion chromatography (Mandala and Taiz 1986, Ward and Sze 1992). Subunits A, B, and *c* have been identified in the V-ATPase complex of all plant species studied so far and regarded as essentially required for enzyme function (Lüttge and Ratajczak 1997). The catalytic subunit A (63–72 kDa) and the regulatory subunit B (52–60 kDa) form the head of the V_1 domain with three copies each of the two subunits. Hexameric structures have been shown for V-ATPase heads of *M. crystallinum* (Kramer et al. 1995) and *Manduca sexta* (Radermacher et al. 1999) by rotational image analysis of negatively stained V-ATPases and are believed to represent three subunits A and three subunits B alternatively arranged as documented before for α-subunits and β-subunits of the F-ATPase (Abrahams et al. 1994).

In analogy to the γ-subunit of the F-ATPase, because of its high amount of α-helices, subunit D (29–42 kDa) is assumed to build the central stalk of the enzyme and to be involved in coupling of ATP-hydrolysis and H^+ transport (Nelson et al. 1995). Also, subunits C, E, F, G, and H seem to be stalk subunits of the V-ATPase (Forgac 2000). While the exact localisation and function of the subunits C (37–52 kDa), E (27–40 kDa), F (13–14 kDa), and H (51–54 kDa) are still unknown, subunit G is thought to be part of the peripheral V-ATPase stalk because of significant sequence similarity to the F-ATPase b-subunit (Supekova et al. 1996). Although lacking an apparent transmembrane domain (Hunt and Bowman 1997), subunit G seems to be in tight contact with the membrane peripheral V_O domain, as shown by V_1-V_O dissociation experiments with the V-ATPase of *A. sativa* and *M. crystallinum* (Ward and Sze 1992; Emig 1999). Subunit H seems to have multiple binding capabilities, as evidenced by the crystal structure of the yeast subunit (Sagermann et al. 2001) interacting probably with V_1 and V_O subunits (Xu et al. 1999; Landolt-Marticorena et al. 2000). Functionally, in *S. cerevisiae*, subunit H is thought to inhibit V-ATPase activity of deassembled cytosolic V_1 domains of the enzyme (Parra et al. 2000), while it supports enzyme activity in the holoenzyme (Ho et al. 1993). In plants, subunit H has not yet been unequivocally identified to be part of the functional V-ATPase, although polypeptides with apparent molecular masses of about 50 kDa have been detected in purified fractions of different plant species (Parry et al. 1989; Fischer-Schliebs et al. 1997a; Kawamura et al. 2000).

Subunit c (16–20 kDa) is the major subunit of the membrane integral V_O domain. Since V-ATPase subunit c has about twice the molecular mass of the F-ATPase subunit c, it has been suggested that it emerged by gene duplication and fusion from the F-ATPase proteolipid (Mandel et al. 1988). Subunit c is a highly hydrophobic protein containing four membrane-spanning α-helical domains, and it is suggested that it is directly involved in H^+ transport. In *S. cerevisiae*, a glutamate residue in position 137 of subunit c, which is located in the membrane-spanning domain IV, has been shown to be essential for H^+-transport activity. Replacement of glutamate-137 by other amino acids (with the exception of aspartate) abolished V-ATPase activity (Noumi et al. 1991). Glutamate-137 is suggested to be the binding site of *N,N'*-dicyclohexylcarbodiimide (DCCD) which inhibits V-ATPase activity (Lai et al. 1991). DCCD-binding studies revealed the existence of at least six copies of subunit c per V-ATPase holoenzyme of *H. vulgare* (Kaestner et al. 1988). In the yeast V-ATPase the V_O domain additionally contains at least one copy each of the proteolipid subunits c' (17 kDa) and c'' (23 kDa), both of which are homologues to subunit c and participate in H^+ translocation (Hirata et al. 1997). The proteolipid c' has not yet been cloned from plants, while there is some evidence for the existence of genes encoding subunit c'' in *A. thaliana* (Stevens and Forgac 1997). A database search with the sequence of *S. cerevisiae* subunit c'' (Hirata et al. 1997) led to a partial sequence information (180 amino acids, SwissProt T08586) encoding a "probable H^+ translocating ATPase 18 K chain" of *A. thaliana* exhibiting 57% sequence similarity to the yeast subunit c''.

Subunit a (100 kDa) is the only V-ATPase subunit that exists in the yeast genome in the form of two isoforms, *Vph1* (Manolson et al. 1992) and *Stv1* (Manolson et al. 1994). Although subunit a does not exhibit sequence similarity to any F-ATPase subunit, it seems to be a functional analogue to the F-ATPase subunit a (Leng et al. 1996). Subunit a is assumed to be located in the peripheral stalk of the yeast V-ATPase holoenzyme (Landolt-Marticorena et al. 1999). According to mutational analysis in yeast, it was suggested that subunit a participates in V-ATPase assembly and H^+ transport (Leng et al. 1996, 1998). In plants, a cDNA encoding a homologue to the Vph1p/Stv1p subunit of yeast has been cloned in *Citrus limon* (Aviezer-Hagai et al. 2000) and *A. thaliana* (PIRonly T06068). A polypeptide of 95–115 kDa was detected in fractions of purified V-ATPases of several plant species (Parry et al. 1989; DuPont and Morrissey 1992; Fischer-Schliebs et al. 1997a). It is still unclear if this polypeptide actually corresponds to subunit a since it was not found in purified, active V-ATPases isolated from other plants (Ward and Sze 1992; Warren et al. 1992). However, a 100-kDa polypeptide associated with a fraction of free V_O complexes in *A. sativa* indicated a primary role of subunit a in the assembly of the V-ATPase holoenzyme complex rather than in the enzyme activity (Li and Sze 1999).

Regarding subunit d (32–36 kDa), little is known in plants. Polypeptides with apparent molecular masses of 32–36 kDa have been detected in several plant species (Parry at al. 1989; Ward and Sze 1992; Hosaka et al. 1994) that might be analogues of the yeast Vma6p. In *Beta vulgaris*, a 44-kDa polypeptide was labelled with an anti-Vma6p polyclonal antibody indicating that in plants subunit d might have a higher molecular mass than expected from the yeast enzyme (Bauerle et al. 1998). Also, in partially purified V-ATPases of *Vigna radiata* and *Pisum sativum*, polypeptides with molecular masses of 44 and 45 kDa, respectively, have been detected and may represent subunit d (Kawamura et al. 2000). In *N. tabacum*, subunit d was unequivocally identified by MALDI-MS exhibiting a molecular mass of 40 kDa (Drobny et al. 2002). Despite the absence of putative transmembrane helices (Wang et al. 1988), subunit d seems to be tightly associated with V_O (Zhang et al. 1992).

To date, some additional polypeptides have been exclusively found in mammalian V-ATPases, i.e., AC45 (Supek et al. 1994; Xu et al. 1999), M9.2 (Ludwig et al. 1998), and M9.7 (Merzendorfer et al. 1999). Their occurrence in the plant V-ATPase still has to be clarified. With the exception of AC45, these proteins are homologous to the assembly factor Vma21p of the yeast V-ATPase, that is localised to the endoplasmic reticulum (ER) but not to the mature V-ATPase complex of vacuolar membranes (Graham et al. 1998; Graham and Stevens 1999).

b) Enzyme Activity

The V-ATPase is characterised by two distinct catalytic activities, i.e., ATP hydrolysis and H^+ translocation. The pH optimum of both activities is in the range 6.5–8.0 (Brauer et al. 1992, and references cited therein). Divalent metal–nucleotide complexes serve as substrates of the V-ATPase with a specificity of ATP>>GTP>NTP and $Mg^{2+} \geq Mn^{2+} >> Ca^{2+}$, Co^{2+}. Apparent K_m values of plant V-ATPases for $MgATP^{2+}$ range between 0.2 and 0.8 mM (see Lüttge and Ratajczak 1997, and references cited therein). Detailed characterisation of substrate kinetics of the partially purified and reconstituted V-ATPase of *K. daigremontiana* showed two distinct K_m values of 0.77 mM and 2 μM, respectively, indicating either two or more catalytic centres or cooperativity between nucleotide binding sites (Warren et al. 1992). Screening several plant species for specific ATP-hydrolysis activities of the V-ATPase revealed a range as wide as 24–226 μmol ATP $mg_{V\text{-}ATPase}^{-1}$ h^{-1} depending on plant species and tissue (Fischer-Schliebs et al. 1997a). Substrate turnover rates of the

V-ATPase of 30–68 ATP s^{-1} have been reported (Warren et al. 1992; Lüttge and Ratajczak 1997).

Anion sensitivity is a characteristic feature of V-ATPases (Wang and Sze 1985). Enzyme activity is stimulated by millimolar concentrations of chloride and inhibited by millimolar concentrations of nitrate (Bennett and Spanswick 1983; Sze et al. 1992). The sensitivity of the V-ATPase to nitrate has been used frequently in the past for identification of the V-ATPase. Patch clamp studies with whole vacuoles indicate that nitrate interacts with the cytoplasmic side of the V-ATPase since nitrate applied to the lumenal side does not inhibit the V-ATPase activity (Hedrich et al. 1989). For the V-ATPase of *Neurospora crassa*, it was suggested that inhibition of activity by nitrate is due to oxidation of essential cysteine residues in the V_1 domain (Dschida and Bowman 1995, see below).

The macrolide antibiotics Bafilomycin A_1 and Concanamycin A are two inhibitors of the V-ATPase that are, compared to nitrate, more specific and effective with only few exceptions (Müller et al. 1996; Ratajczak et al. 1998). Bafilomycin A_1 and Concanamycin A inhibit the V-ATPase already at concentrations as low as 10^{-9} and 10^{-10} M, respectively (Bowman et al. 1988; Dröse et al. 1993). The target of Bafilomycin A_1 inhibition is the V_O domain as supported by inhibitor studies with reconstituted V_O domains (Crider et al. 1994; Zhang et al. 1994). It has been hypothesised that the insensitivity of the V-ATPase to Bafilomycin A_1 in *C. limon* fruit tonoplasts or leaves of *H. vulgare* in the stage of senescence indicate differences in V_O structure (Ratajczak and Wilkins 2000). Other inhibitors of the V-ATPase include mainly protein-modifying agents. These have been reviewed elsewhere (Lüttge and Ratajczak 1997).

c) Regulation of the V-ATPase

The regulation of the V-ATPase is definitely a process of high complexity. It requires the control of expression of multiple genes encoding various subunits, assembly and targeting of the V-ATPase holoenzyme to functionally diverse subcellular compartments, and modulation of activity in response to developmental, physiological and environmental stimuli. A cartoon summarising mechanisms suggested to be involved in regulation of V-ATPase is shown in Fig. 2. Aspects of gene expression and the role of isoforms of V-ATPase subunits will be discussed below. Enzyme assembly and targeting have recently been reviewed in detail (Ratajczak 2000; Ratajczak and Wilkins 2000).

One of several post-translational mechanisms posed for V-ATPase regulation is the controlled assembly and disassembly of the V-ATPase

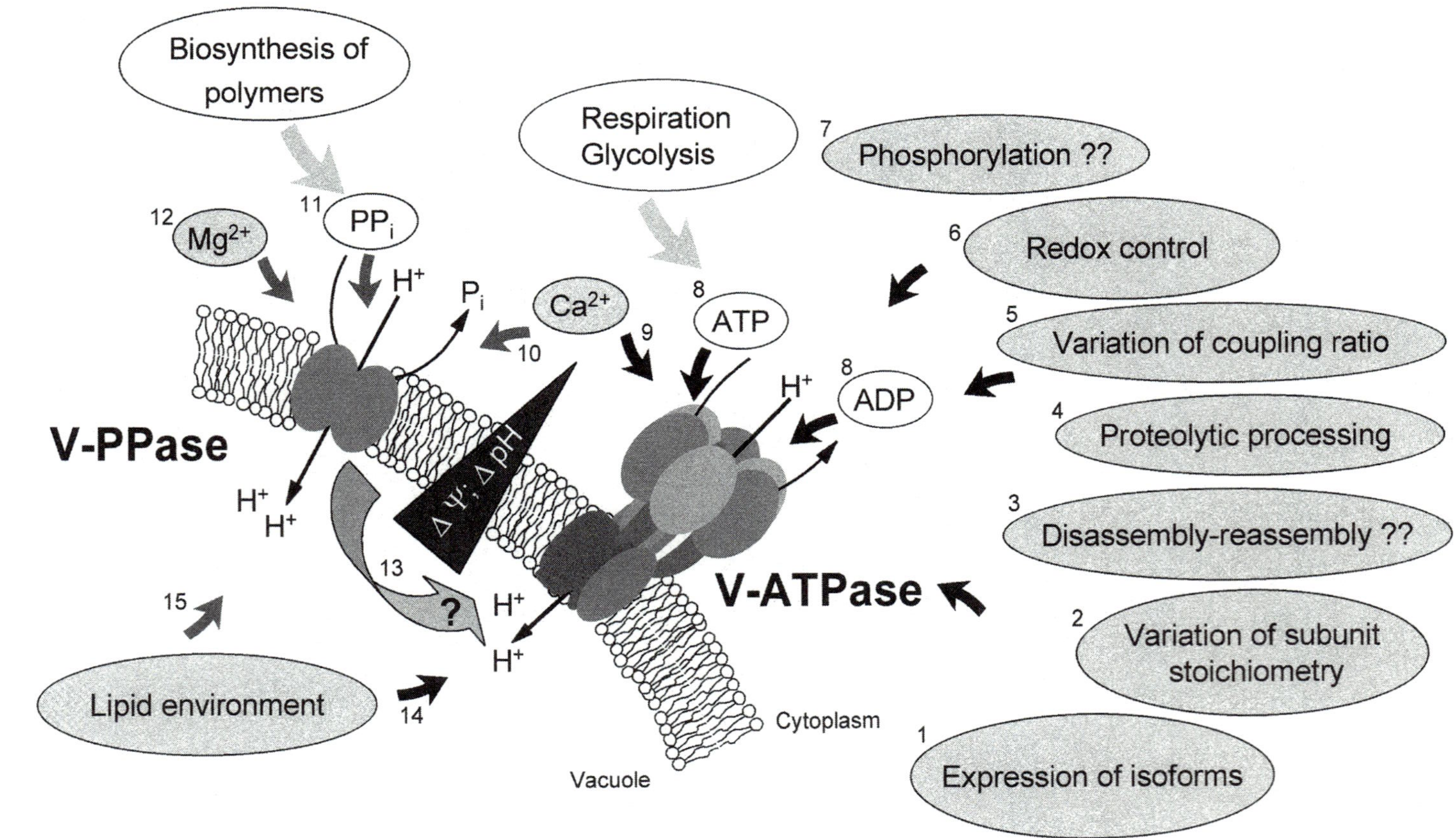

Biosynthesis of polymers
Respiration Glycolysis
7 Phosphorylation ??
6 Redox control
5 Variation of coupling ratio
4 Proteolytic processing
3 Disassembly-reassembly ??
2 Variation of subunit stoichiometry
1 Expression of isoforms
12 Mg2+
11 PPi
H+
Pi
10
Ca2+
9
8 ATP
8 ADP
H+
Δψ; ΔpH
13 ?
14
H+
H+
H+
H+
15
Lipid environment
V-PPase
V-ATPase
Cytoplasm
Vacuole

holoenzyme (Parra and Kane 1993; Kane 1995; Graf et al. 1996). Depriving yeast cells of glucose caused rapid dissociation of the V_1 domain from the V-ATPase. Upon re-addition of glucose the V_O domain readily associated again to form an active V-ATPase holoenzyme. Re-assembly does not require de novo protein synthesis, indicating that V_1–V_O assembly to form an active V-ATPase holoenzyme utilises an existing pool of V_1 domains. In plants, there is no solid evidence so far for a regulation of the V-ATPase by reversible dissociation of the enzyme complex. However, low-temperature treatment of chilling-sensitive *V. radiata* seedlings leads to the loss of peripheral subunits from the membrane and inactivation of the V-ATPase (Matsuura-Endo et al. 1992). Inactivation of the V-ATPase is reversible and was restored independently of protein synthesis after the temperature was returned to 26 °C (Yoshida 1991).

Disulphide bond formation between cysteine-254 and cysteine-532 of the subunit A, located in close proximity to the catalytic site of the V-ATPase, reversibly inhibits V-ATPase activity in *B. taurus* (Feng and Forgac 1992, 1994). Moreover, a significant fraction of the V-ATPase, i.e. more than 50%, in native clathrin-coated vesicles seems to exist in this reversibly inactivated, disulphide-bonded state in vivo (Feng and Forgac 1992). There is also evidence in plants for regulation of V-ATPase activity by the oxidation/reduction of disulphide bonds. Blue-light treatment in the presence of flavins of *Zea mays* coleoptiles leads to a reversible inhibition of V-ATPase activity (Krauss et al. 1987). Inactivation of the V-ATPase has been reported to occur by SH-group modifying reagents, such as oxidised glutathione, thioredoxin, and during isolation of membrane vesicles in the absence of dithiothreitol (DTT; Hager and Biber 1984; Dietz et al. 1998; Tavakoli et al. 2001). Protection of the *Z. mays* V-ATPase from inhibition by SH-group modifying reagents by MgATP (Hager and Lanz 1989) and reversion of inactivation of the *V. radiata* V-ATPase by the adenine analogue 7-chloro-4-nitrobenzo-2-oxa-1,3-

Fig. 2. Mechanisms and potential elements of the complex cellular regulation network of V-ATPase and V-PPase activities in plant cells. *Numbers beside arrows* denote references to experimental evidence or discussion of respective topics: *1* Tsiantis et al. (1996); Lehr et al. (1999); Fischer-Schliebs et al. (2000); Kawamura et al. (2000); *2* Lüttge and Ratajczak (1997); *3* Yoshida (1991); Matsuura-Endo et al. (1992); *4* Krisch et al. (2000); Ratajczak (2000); *5* Moriyama and Nelson (1988); Brauer et al. (1993); Davies et al. (1994); Müller et al. (1997); Fischer-Schliebs et al. (2000); *6* Hager and Biber (1984); Krauss et al. (1987); Hager and Lanz (1989); Dietz et al. (1998); Tavakoli et al. (2001); *7* Zocchi (1985); Scherer et al. (1988); Martiny-Baron et al. (1992); *8* Brauer and Tu (1994); Dietz et al. (1998); *9* Pfeifer (1995); Bressan et al. (1998); *10* Maeshima (1991); Rea et al. (1992); Darely et al. (1998); *11* Leigh et al. (1992); Zhen et al. (1997); *12* Maeshima (1991); Leigh et al. (1992); *13* Marquardt-Jarczyk and Lüttge (1990b); Fischer-Schliebs et al. (1997b); *14* Yamanishi and Kasamo (1992, 1993, 1994, 2001); Kasamo et al. (2000); *15* Rea and Poole (1986); Maeshima and Yoshida (1989)

diazole (NBD-Cl) by addition of DTT (Yamanishi and Kasamo 1992) confirmed the close location of essential cysteine residues near the catalytic site of subunit A. ATP hydrolysis and ATP-dependent H^+-transport activity of microsomal membranes of *H. vulgare* were inhibited by the oxidant H_2O_2. Reduction of disulphides by reduced glutathione reactivated the enzyme (Tavakoli et al. 2001). Regulation of the V-ATPase was related to disulphide bond formation not only in subunit A, but also in subunit E in *H. vulgare*, since the activity loss of the V-ATPase was accompanied by shifts in electrophoretic mobility of both subunits and was reversed upon reductive reactivation (Tavakoli et al. 2001). Oxidative inactivation of the V-ATPase was found to be strongly dependent on the pH value of the medium (Dietz et al. 2001). Whereas H_2O_2 had no effect on H^+-transport activity at pH 6.5, it severely inhibited the V-ATPase at pH 8.0. Thus downregulation of the V-ATPase by oxidative inactivation may occur in vivo only when the cytoplasm is not acidified, which is essential for the enzyme to function in pH homeostasis.

Evidence for specific proteolytic processing of the V-ATPase as a mode of enzyme regulation comes from experiments with *M. crystallinum*. During salt stress-induced transition from C_3 photosynthesis to CAM, two new polypeptides appeared, denoted E_i (28 kDa) and D_i (32 kDa), that are closely associated with the V-ATPase holoenzyme (Bremberger et al. 1988; Bremberger and Lüttge 1992). The appearance of subunits D_i and E_i was correlated with altered dimensions of the V_1 domain (Lüttge and Ratajczak 1997; see Fig. 1). By protein sequencing, subunit D_i was shown to be a fragment of subunit B and occurred in the V-ATPase in vivo as well (Zhigang et al. 1996). Subunit B fragmentation in vitro has been shown to be mediated by a specific protease or reactive oxygen species (Krisch et al. 2000). Incorporation of subunit D_i into the enzyme complex was not accompanied by a significant change of specific ATP hydrolysis activity; however, it increased the stability of the V-ATPase towards detergents and high salt concentrations and thus may play a role in regulation of the V-ATPase in response to CAM induction under salinity (Lüttge et al. 1995a; Ratajczak 2000; see Sect. 5).

Availability of substrate and changes in the cytosolic ATP/ADP ratio have been suggested to control V-ATPase activity in vivo (Dietz et al. 1998). As mentioned above, the apparent K_m values of plant V-ATPases for $MgATP^{2-}$ range between 0.2 and 0.8 mM and it has been suggested that possibly two or more catalytic centres or cooperativity between nucleotide binding sites exist (Warren et al. 1992). Evidence that ADP interacted at the catalytic site for ATP and thus acts as a competitive inhibitor of the V-ATPase came from experiments examining the interaction of ADP or the reactive adenine analogue NBD-Cl and ATP (Rausch et al. 1985; Randall and Sze 1987; Yamanishi and Kasamo 1992). Noncompetitive inhibition of the V-ATPase by ADP via binding of ADP to a site distinct from the catalytic site for ATP was suggested by Brauer and

Tu (1994). The existence of a distinct ADP-binding site on the V-ATPase leading to inhibition at minute ADP concentrations may have interesting significance for the kinetic regulation of the V-ATPase. One role for the ADP inhibitory site may be to reduce the rate of ATP-dependent H^+ transport below that which would have occurred by only a reduction in substrate availability. This could lead to a complete shutdown of the V-ATPase under low ATP and high ADP levels in the cell.

There is some evidence that the V-ATPase may be regulated via phosphorylation/dephosphorylation of the enzyme. Reports on effects of phosphorylation on V-ATPase enzyme action are, however, contradictory. Whereas in *Z. mays* microsomes Ca^{2+}-dependent phosphorylation of membrane proteins correlated with a decrease in V-ATPase activity (Zocchi 1985), in *Cucurbita pepo* tonoplast vesicles a concomitant stimulation of H^+-transport activity and Ca^{2+}-independent lysophospholipid-activated phosphorylation of subunit B of the V-ATPase was observed (Scherer et al. 1988; Martiny-Baron et al. 1992).

From the viewpoint of calcium as an ubiquitous signal in plants, the idea of Ca^{2+} and Ca^{2+} signalling as an element of V-ATPase regulation is intriguing. Calcium and calmodulin have been suggested to participate in regulating V-ATPase activity (Pfeifer 1995). In analogy to yeast, it has recently been proposed that Ca^{2+} signalling may play a role in V-ATPase expression under salinity stress (Bressan et al. 1998).

One powerful mechanism of rapidly controlling V-ATPase activity is through modulation of the coupling efficiency of H^+ transport and ATP hydrolysis. Based on the similar architecture of F-ATPases and V-ATPases and high homology of several subunits of the enzymes, one believes that the V-ATPase may function through a rotational mechanism similar to the F-ATPase (Grabe et al. 2000). According to this model, ATP hydrolysis at the catalytic subunit A leads to a rotation of the central stalk formed by subunits D and E. This rotation is transduced to the proteolipid ring consisting of subunits c which would rotate relative to subunit a located in the membrane. By rotational movements of the proteolipid ring, protons will be channelled through the membrane via binding at the cytosolic side and release to the vacuolar lumen from a pivotal glutamate residue of subunit c (for details, see Junge et al. 1997; Pänke and Rumber 1999). In F-ATPases, co-rotation of other subunits is prevented by a single peripheral stalk connecting F_1 and F_O. According to recent results showing three peripheral stalks for the V-ATPase of *K. daigremontiana*, the plant enzyme might be constructed in a more robust way (Domgall et al. 2002). In V-ATPases three peripheral stalks are connected via subunit d to the membrane integrated part of subunit a forming a stable static connection between V_1 and V_O. Variation of coupling ratios could be one mechanism of regulating the V-ATPase in vivo. In early studies, for thermodynamic reasons, the coupling ratio ($H^+_{transported}/ATP_{hydrolysed}$) of the plant V-ATPase was considered to be 2

(Lüttge et al. 1981; Smith et al. 1982; Bennett and Spanswick 1984). Patch clamp studies later reported variable coupling ratios of the V-ATPase ranging between 1.75 and 3.28 $H^{+}_{transported}/ATP_{hydrolysed}$ depending on the pH gradient between cytosol and vacuolar lumen (Davies et al. 1994). Brauer et al. (1993) proposed an indirect coupling mechanism based on their observation that H^+ transport and ATP hydrolysis activity of the *Z. mays* V-ATPase are inhibited by nitrate to different extents. Moriyama and Nelson (1988) suggested an intrinsic coupling or 'slip' of the V-ATPase. Varying coupling ratios might be due to changes of the 'slip' or uncoupling rates of the V-ATPase holoenzyme, as reported by Müller et al. (1997). In *C. limon* they observed a 2.0- to 2.4-fold intrinsic uncoupling of the V-ATPase in the epicotyl compared with the fruit. Varying V-ATPase coupling ratios could originate from the incorporation of a variable number of proteolipids in the proteolipid ring, as was suggested to occur in the V-ATPase of *M. crystallinum* during the shift of plants from C_3 photosynthesis to CAM (for review, see Lüttge and Ratajczak 1997; see also Sect. 5). Alternatively, incorporation of distinct isoforms of subunit c in the proteolipid ring might also lead to variation of V-ATPase coupling ratios, as was suggested to occur in *N. tabacum* under high and low nitrate nutrition, respectively (Fischer-Schliebs et al. 2000).

The V_O domain of the V-ATPase is embedded in the lipid bilayer and the lipid environment has been shown to be crucial for enzyme activity (Yamanishi and Kasamo 1992, 1993, 1994). Removal of lipids leads to inactivation of the purified V-ATPase but its activity can be restored by exogenous addition of phospholipids (Yamanishi and Kasamo 1992). Lipids may interact hydrophobically with the V_O domain of the enzyme and may support conformational changes occurring during enzyme function (see above). Activation of the V-ATPase was studied using various classes of phospholipids. Glycolipids isolated from tonoplast fractions of chill-treated cells inhibited H^+-transport activity of the V-ATPase reconstituted into phospholipid liposomes (Kasamo et al. 2000) counteracting the activation of the enzyme by tonoplast phospholipids (Yamanishi and Kasamo 2001). Glycolipid inhibition of the V-ATPase was suggested to result from a decrease in membrane fluidity of the surface region of the lipid bilayer since the activity of the reconstituted enzyme correlated to fluidity of this particular bilayer region.

d) Role of Subunit Isoforms

In contrast to yeast V-ATPase subunits which are encoded by only one gene each, with the exception of subunit a (Manolson et al. 1994), in the mammalian (Merzendorfer et al. 2000; Schoonderwoert and Martens 2001), as well as in the plant genome (Sze et al. 1992; Lüttge and Ratajczak 1997; Ratajczak 2000), a number of V-ATPase subunit genes have been identified as building multigene families encoding isoforms of the respective subunit. Up to now, isoforms of subunits A, B, D, E, G, c, and d have been reported for various plant species (Rouquié et al. 1998; Kawamura et al. 2000; Salanoubat et al. 2000; for reviews, see Lüttge and Ratajczak 1997; Ratajczak 2000).

Regulation of V-ATPase activity by differential expression of subunit isoforms in different plant tissues or in response to environmental factors is an intriguing possibility that is supported by increasing experimental evidence (Ratajczak 2000; Ratajczak and Wilkins 2000). Expression studies indicate that regulation of genes encoding V-ATPase subunit isoforms at the transcriptional level play a role in determining the subunit composition of the V-ATPase holoenzyme (Ratajczak and Wilkins 2000). Early evidence of a tissue-specific and developmentally regulated expression of subunit c came from studies with *Gossypium hirsutum* (Hasenfratz et al. 1995) and *A. thaliana* (Perera et al. 1995; Sze et al. 1999). Tissue specificity of subunit E and subunit D isoforms of the V-ATPase and the existence of isotype enzymes that exhibited differences in kinetic properties were also reported for *P. sativum* and *V. radiata* (Kawamura et al. 2000). Recently, two and three isoforms, respectively, of subunit E have been identified in *K. daigremontiana* and *N. tabacum* by MALDI-MS analysis (Ratajczak et al. 2002; Drobny et al. 2002). However, completely different isoforms of subunit E were found in leaf tissue implying that either more than one V-ATPase isotype with respect to subunit E is residing in the vacuolar membrane in leaf tissue or that isotypes are situated in different intracellular membranes or the plasma membrane within the same cell. Whereas in *P. sativum* the occurrence of different isoforms of subunit E in the holoenzyme has an impact on kinetic parameters of ATP hydrolysis activity, for *N. tabacum* this still needs to be analysed. Immunological screening of the V-ATPase of various plant species for structural differences revealed distinct classes of subunit c, that were regarded to represent subunit c isoforms. The existence of isotype enzymes correlated with differences in enzyme activity (Fischer-Schliebs et al. 1997a).

Incorporation of individual subunits into the holoenzyme complex is regarded as a possible response of the V-ATPase to environmental factors (Lüttge and Ratajczak 1997). Beside isoforms with *house-keeping* functions, the occurrence of specific stress-related isoforms has been suggested (Ratajczak 2000) that may be expressed under stress as high salinity (Lehr et al. 1999) or variation in plant nutrition (Fischer-Schliebs et al. 2000). In *N. tabacum*, immunological differences of subunit c in response to variation in nitrate nutrition have been observed possibly due to the expression of different isoforms of subunit c (Fischer-Schliebs et al. 2000). In the halophyte *M. crystallinum* on exposure of plants to salinity, a salt-inducible transcript of subunit c in roots has been suggested to exist beside an isoform that is preferentially expressed in leaves (Tsiantis et al. 1996). In *M. crystallinum*, immunologically distinguishable forms of subunit c in V-ATPases of different intracellular membranes have also been observed (Emig 1999).

3 V-PPase

a) Molecular Characteristics

In addition to the V-ATPase, the tonoplast contains a second primary-active H^+ pump, the V-PPase (Zhen et al. 1997; Maeshima 2000; Ratajczak and Wilkins 2000). The V-PPase is one of several membrane-associated PPases identified to date. Others include the H^+-translocating PP_i synthase in the energy coupling membranes of the photosynthetic purple non-sulphur bacterium *Rhodospirillum rubrum* (Baltscheffsky et al. 1998), and the PPase activities described for plant submitochondrial particles (Vianello et al. 1991) and thylakoid membranes (Jiang et al. 1997). Among these, the plant V-PPase is, besides the bacterial H^+-PP_i synthase, the most thoroughly investigated enzyme. The V-PPase is ubiquitous in higher plants and has been found in green algae such as *Nitella* (Shimmen and MacRobbie 1987) and *Chara corallina* (Nakanishi et al. 1999), and marine algae like *Acetabularia acetabulum* (Ikeda et al. 1991, 1999), as well as in fungi like *Dictyostelium discoideum* (MacDonald and Weeks 1988). While the V-PPase had formerly been thought to be restricted to plants and subject to pronounced sequence conservation, it is clear now that this is not the case. Membrane-bound PPases with similarity to the plant enzyme have also been found in the parasitic protozoans *Trypanosoma cruzi* (Scott et al. 1998) and *Leishmania donovari* (Rodrigues et al. 1999) as well as in the thermopilic archaeon *Pyrobaculum aerophilum* (Drozdowicz et al. 1999). Furthermore, a pyrophosphatase activity could be measured at the cytoplasmic membrane of the aerobic bacterium *Syntrophus gentianae* (Schöcke and Schink 1998).

The V-PPase was originally cloned from *A. thaliana* (Sarafian et al. 1992). Up to now, cDNAs for the V-PPase have been cloned from ten plant species (Nakanishi and Maeshima 1998; Baltscheffsky et al. 1999; Nakanishi et al. 1999; Suzuki et al. 1999; for review, see Zhen et al. 1997). The amino acid sequences of the V-PPase are highly conserved among different plant species with more than 85% sequence similarity and the least conserved region in the amino-terminal part of the protein. On the other hand, no striking homology exists between membrane-bound V-PPase and soluble PPases and H^+-ATPases, respectively (Rea and Poole 1993; Baltscheffsky et al. 1999). However, structural and functional relations between the plant V-PPase and the membrane-integral H^+-PP_i

synthase of *R. rubrum* indicate a common phylogenetic origin (Rea and Poole 1993; Zhen et al. 1997; Baltscheffsky et al. 1998).

The enzyme consists of a single intrinsic membrane protein with 760–775 amino acids and a molecular mass of 79–80 kDa. An apparent molecular mass of 64–72 kDa was determined in membrane fractions isolated from various plant species after sodium dodecyl sulphate (SDS) polyacrylamide gel electrophoretic fractionation (for reviews, see Leigh et al. 1994; Zhen et al. 1997). Hydrophilicity plots and topographic modelling of amino acid sequences of V-PPases revealed a highly hydrophobic membrane protein with 13 transmembrane domains and the N- and C-termini located in the vacuolar lumen (Sarafian et al. 1992; Zhen et al. 1997; Nakanishi et al. 2001). In contrast to the multimeric V-ATPase, the functional form of the V-PPase seems to be a homodimer of the single polypeptide, as indicated by gel filtration and chemical cross-linking experiments (Sato et al. 1991; Tzeng et al. 1996). From heterologously expressed (Kim et al. 1994) and reconstituted V-PPases (Britten et al. 1992; Sato et al. 1994), it is evident that the monomer alone is also sufficient for PP_i hydrolysis and H^+ translocation (Zhen et al. 1997).

The V-PPase constitutes up to 5–10% of the total tonoplast protein (Maeshima and Yoshida 1989; Ratajczak et al. 1995), and thus is three to four times less abundant at the tonoplast compared with the V-ATPase on the basis of protein amount. However, since the molecular mass of the V-PPase dimer is approximately four times lower than the molecular mass of the V-ATPase holoenzyme, almost equal numbers of active enzymes are present at the tonoplast. Thus, it is not surprising that the V-PPase is able to generate an electrochemical gradient across the tonoplast that is similar to or larger than the gradient created by the V-ATPase (Pope and Leigh 1987; Maeshima and Yoshida 1989; Johannes and Felle 1990; Terrier et al. 1998). The steady-state pH gradient, which the V-PPase can generate across the tonoplast from the neutral cytoplasmic pH against the vacuolar interior, has been shown to be approximately 3 pH units (Davies et al. 1993). This indicates that both the V-ATPase and V-PPase play a major role in vacuolar functions. Both can be visualised by freeze-fracture electron microscopy. By partial purification and selective reconstitution of the V-ATPase and V-PPase into liposomes, it was deduced from freeze-fracture replicas that in *K. daigremontiana* the membrane intrinsic particles of the V-PPase are smaller than those of the V-ATPase with diameters of 6.7–7.2 and 9.1 nm, respectively (Mariaux et al. 1994).

b) Physiological Role of the V-PPase

It is still unclear why there are two primary active H^+ pumps at the tonoplast of plant cells, although much work has been done to discover specific metabolic functions of the two enzymes since the question was first raised (Rea and Sanders 1987). Various functions have been suggested for the V-PPase (Ratajczak and Fischer-Schliebs 1997; Zhen et al. 1997; Maeshima 2000; Ratajczak and Wilkins 2000).

One function, of course, is the energisation of the tonoplast for secondary-active transport events, alternatively or in concert with the V-ATPase by H^+ transport driven by PP_i hydrolysis. Hence, the V-PPase could energise secondary-active transport processes under conditions of limited cytoplasmic ATP concentration (Carystinos et al. 1995; Darley et al. 1995). The metabolism of PP_i has been recognised as a central and distinctive feature of plants with PP_i being a potential key metabolite in plant systems (Stitt 1998). Besides the V-PPase, two other enzymes in the cytosol utilise PP_i as an energy source, the UDP-glucose pyrophosphorylase and the fructose-6-phosphate:phosphotransferase. All three enzyme reactions are redundant in that they can be replaced by ATP-dependent reactions. Thus, the alternative use of ATP or PP_i may be a general strategy in plant energy metabolism leading to a higher flexibility (Stitt 1998). Pyrophosphate is produced as a by-product in the activation or polymerisation steps of a wide range of biosynthetic pathways in plant cells, e.g. synthesis of DNA, RNA, aminoacyl-tRNA (protein synthesis), ADP-glucose (starch synthesis), UDP-glucose (cellulose synthesis), fatty acyl-CoA (β-oxidation of fatty acid), and in the synthesis of many secondary plant compounds. Since PP_i-dependent reactions operate close to the thermodynamic equilibrium, the use of PP_i by the V-PPase might be of extraordinary importance in cells containing high amounts of PP_i in the cytoplasm since the enzyme is one of only few enzymes hydrolysing cytoplasmic PP_i to P_i in plant cells (Weiner et al. 1987; Geigenberger and Stitt 1993; Geigenberger et al. 1993). Thus, the V-PPase could be the dominating H^+ pump with respect to the V-ATPase in young and growing plant tissues, where biosynthetic pathways are a predominant feature of cell metabolism (Maeshima 2000). Conversely, the V-PPase may also synthesise PP_i using the trans-tonoplast H^+ gradient generated by the V-ATPase (Façanha and de Meis 1998). In this context, the V-PPase could be involved in the regulation of metabolic pathways depending on the cytoplasmic PP_i concentration.

A further possible function of the V-PPase was suggested to be primary-active transport of K^+ ions from the cytoplasm into the vacuole. Evidence for direct K^+ transport via the V-PPase came from patch clamp studies on vacuoles of *B. vulgaris* (Davies et al. 1992) and *Chenopodium rubrum* (Obermeyer et al. 1996). However, the electrophysiological data have not been verified by biochemical investigations so far using tonoplast vesicles or purified V-PPase reconstituted into liposomes (Sato et al. 1994; Ros et al. 1995). After recognising the profound influence of the buffer system on the K^+ stimulation of PP_i hydrolysis activity of the V-PPase used in the respective experiments, a critical re-examination of the V-PPase-catalysed K^+ transport has been proposed (Gordon-Weeks et al. 1997).

Finally, there is evidence that the V-PPase might be involved in activation of the V-ATPase in respect to CAM performance in *Kalanchoë* species (Marquardt-Jarczyk and Lüttge 1990a; Fischer-Schliebs et al. 1997b). The initial ATP-dependent H^+-transport activity of the V-ATPase was stimulated when membrane vesicles of *K. daigremontiana* and *K. blossfeldiana* were pre-energised by the V-PPase. Conversely, pre-energisation of membrane vesicles by the V-ATPase did not affect V-PPase activity. Co-immunoprecipitation of V-ATPase and V-PPase using antibodies directed against subunit A of the V-ATPase holoenzyme indicated that the two enzymes may be closely associated in the membrane (Fischer-Schliebs et al. 1997b).

Recently, in *A. thaliana*, a new type of a H^+-translocating pyrophosphatase, AVP2, has been discovered. AVP2 shares only 36% identity with AVP1 and other plant V-PPases and their isoforms. In contrast to AVP1 and other plant V-PPases, AVP2 is insensitive to K^+ but shows high sensitivity to inhibition by Ca^{2+} (Drozdowicz et al. 2000). The structural and functional characteristics of AVP2 indicate that plants may not only contain one, but two distinct subclasses of V-PPases: a K^+-activated AVP1-like (type I) V-PPase and a K^+-insensitive Ca^{2+}-inhibited AVP2-like (type II) V-PPase (Drozdowicz et al. 2000). Hypothesising a possible subcellular localisation of AVP2 in mitochondrial or plastid membranes, it has been intriguingly speculated that AVP2 may function in these organelles in PP_i synthesis rather than in PP_i hydrolysis and thus contribute to augmentation of the cytosolic PP_i concentration (Drozdowicz et al. 2000).

c) Enzyme Activity and Regulation of the V-PPase

Specific activities of the V-PPase of vacuolar membranes vary between plant species, tissues, and assay conditions (for review, see Maeshima 2000). For purified V-PPase, specific activities between 180 and 510 $\mu mol\ mg_{protein}\ h^{-1}$ have been reported (Becker et al. 1995; Maeshima 2000). The pH optimum of H^+-transport activity and PP_i hydrolysis of the V-PPase is relatively broad: between 6.7 and 8.5 (Marquardt and Lüttge 1987; Brauer et al. 1992). A key characteristic of the V-PPase is a near obligatory requirement for millimolar concentrations of cytosolic K^+ for activity (White et al. 1990; Davies et al. 1991). The K_m value for K^+ stimulation of PP_i hydrolysis and H^+-transport activity has been reported to be 1.27 mM (Gordon-Weeks et al. 1997) and, hence, with cytosolic K^+ concentrations ranging between 50 and 100 mM, K^+ is unlikely to regulate V-PPase activity in vivo. For optimal enzyme activity, the V-PPase requires PP_i and Mg^{2+} as well as K^+. The responses of the enzyme to changes in total Mg^{2+} and PP_i concentrations are complex (for reviews, see Leigh et al. 1994; Zhen et al. 1997). For V-PPase activity, Mg^{2+} is essential to form an Mg-PP_i complex and to keep an active conformation of the enzyme. The true substrate of the V-PPase is the Mg_2PP_i complex (Leigh et al. 1992). Activation of the enzyme by free Mg^{2+} has been attributed to cooperative binding of Mg^{2+}, whereas cytosolic Mg^{2+} has been suggested to be a possible element of V-PPase regulation in plant cells (White et al. 1990). The apparent K_m values for Mg^{2+} have been reported to be 42 μM (Maeshima 1991) or 20–23 μM (Gordon-Weeks 1996). For the V-PPase of *V. radiata*, two different Mg^{2+}-binding sites have been detected: a low-affinity (K_m=0.25–0.46 mM) and a high-affinity (K_m=23–31 μM) binding site (Baykov et al. 1993). Leigh et al. (1992) applied a more complex reaction kinetic model to data obtained with the V-PPase of *Avena sativa* in order to elucidate the roles of free Mg^{2+} and various $MgPP_i$ complexes. They concluded that at cytosolic concentrations of free Mg^{2+} and PP_i of approximately 0.4 mM (Yazaki et al. 1988) and 0.2 to 0.3 mM (Weiner et al. 1987), respectively, the Mg_2PP_i complex, beside being the true substrate of the V-PPase, is also a non-competitive inhibitor of the enzyme and that the V-PPase is activated by Mg^{2+} by non-cooperative binding and competitively inhibited by free PP_i (Leigh et al. 1992; see Fig. 2 for potential factors involved in regulation of V-PPase).

Ca^{2+} has also been suggested to be involved in regulating the V-PPase activity. Calcium through the formation of $CaPP_i$ was reported to be a potent inhibitor of the V-PPase with a K_i of 17 μM (Maeshima 1991). However, Rea and coworkers (1992) proposed that in vivo modulation of cytosolic-free Ca^{2+} might constitute a specific mechanism for regulation of the V-PPase involving Ca^{2+} binding to a discrete regulatory site. Regulation of the V-PPase by Ca^{2+} has been discussed for *Vicia faba* guard cells with respect to dark-light correlated changes in cytosolic free Ca^{2+} and accumulation of sucrose as a stomatal osmoticum through control of cytosolic PP_i (Darely et al. 1998).

The H^+/PP_i coupling ratio of the V-PPase operating in the direction of PP_i hydrolysis has been determined to be 1 (Johannes and Felle 1990; Schmidt and Briskin 1993). However, with respect to enzyme function in vivo, the question of the H^+/PP_i coupling ratio of the V-PPase has drawn some attention in the past and has been discussed by Rea and Sanders (1987) and Leigh et al. (1994). Coupling ratios of 2 or higher would allow the V-PPase to operate in the synthetic direction and to catalyse PP_i synthesis from P_i even at trans-tonoplast ΔpH values of as low as 2 units. Experiments with native membrane vesicles of *Z. mays* indicated that the V-PPase can indeed synthesise PP_i using the trans-tonoplast H^+ gradient generated by the V-ATPase (Façanha and de Meis 1998).

The V-PPase is a highly hydrophobic integral membrane protein. Although the influence of lipids on the V-PPase activity has not been studied as thoroughly as for the V-ATPase (see above), the lipid environment has been shown to be crucial for enzyme activity (Rea and Poole 1986; Maeshima and Yoshida 1989).

The V-PPase is inhibited by potassium fluoride (Marquardt-Jarczyk and Lüttge 1990a), an inhibitor of inorganic pyrophosphatases in general. Diphosphonates are specific potent inhibitors of the V-PPase with aminomethylenediphosphonate as the most effective one of this group of molecules (Zhen et al. 1997).

d) Role of V-PPase Isoforms

As a consequence of isolation of cDNAs encoding the V-PPase from plants other than *Arabidopsis*, the existence of multiple genes encoding the V-PPase protein was recognised. Analyses of multiple cDNA isolates from *B. vulgaris* have revealed two classes, designated *BVP1* and *BVP2* (Kim et al. 1994). In *N. tabacum* three different classes of cDNA, named *TVP5*, *TVP9*, and *TVP31*, were found (Lerchl et al. 1995b). In *Oryza sativa*, two cDNA clones have been identified, i.e., *OVP1* and *OVP2* (Sakakibara et al. 1996). The nucleotide sequence of different clones is highly homologous within the coding region, i.e. in *N. tabacum* the nucleotide sequence of the coding regions of *TVP5*, *TVP9*, and *TVP31* are 79–89% identical, but differ strongly in untranslated regions. Therefore, these different genes may be differentially and individually regulated.

It has been hypothesised that multiple cDNA clones of the V-PPase encode isoforms of the enzyme. In *N. tabacum* a decrease in steady-state transcript levels is found for all three isoforms, proceeding from sink to source leaves but with up to a nine-fold more pronounced decrease in *TVP9*-specific transcript than in *TVP5* and *TVP31* (Lerchl et al. 1995b). Similar levels of the *TVP5*-specific transcripts are found in midribs, stems and roots, whereas *TVP9*- and *TVP31*-specific transcripts are much more abundant in roots (Lerchl et al. 1995b). In sepals and petals, specific transcript levels of *TVP5* and *TVP31* are higher than of *TVP9*. In *O. sativa*, *OVP2* is expressed in higher levels in calli than roots and shoots compared with *OVP1* (Sakakibara et al. 1996). In *B. vulgaris*, analyses of light- and dark-grown seedlings, leaves, roots, and storage roots of adult plants failed to reveal consistent differences between steady-state levels of expression of *BVP1* and *BVP2* (Kim et al. 1994).

Nothing is known so far about the significance of V-PPase isoforms or their possible specific location on membranes beside the tonoplast (see Sect. 4). Differences in specific transcript levels have been reported. However, cell specificity in transcription, V-PPase protein accumulation, and the enzymatic activity of each isoform remains to be determined to unveil the importance of distinct V-PPase isoforms in different developmental stages of plants and in their response to changes in environmental conditions.

4 Cellular Localisation of V-ATPase and V-PPase

V-ATPases are not exclusively localised in vacuolar membranes. Immunochemical and biochemical studies have revealed that V-ATPases are sited also in the ER, clathrin-coated vesicles, and the Golgi apparatus, the plasma membrane and other membranes of the secretory pathway (Depta et al. 1991; Herman et al. 1994; Oberbeck et al. 1994; Robinson et al. 1996; Ratajczak et al. 1999). The function of the V-ATPase at membranes beside the tonoplast is still a matter of debate, and the wide distribution of the V-ATPase within the cell has frequently been discussed in respect to assembly, targeting and turnover of the V-ATPase holoenzyme (Ratajczak 2000; Ratajczak and Wilkins 2000). To date, the only example for the presence of V-ATPase polypeptides at the plasma membrane of a plant cell is the pea cotyledon (Robinson et al. 1996). It is not clear whether the plasma membrane located V-ATPase is transitory in nature, en route to the vacuole in analogy to the integral membrane proteins of the lysosome (Braulke 1996), or if this is a consequence of the fusion of Golgi-derived vesicles during cell plate formation.

Since the first membrane-bound pyrophosphatase activity was found in 1975 in homogenates from higher plants (Karlson 1975) and later localised in and isolated from plant vacuolar membranes (Walker and Leigh 1981; Rea and Poole 1986; Maeshima and Yoshida 1989), the V-PPase has been regarded, besides the V-ATPase, as a marker for the tonoplast of plant cells. As for the V-ATPase, immunochemical and biochemical studies have, however, documented the V-PPase in the plasma membrane and in subcellular membranes besides the tonoplast, i.e. in *Z. mays* (Oberbeck et al. 1994), *Ricinus communis* (Williams et al. 1992; Long et al. 1995), *P. sativum* (Robinson et al. 1996), *Chlamydomonas reinhardtii* (Robinson et al. 1998), *Brassica oleracea* (Ratajczak et al. 1999), and *Agrobacterium tumefaciens* induced stem tumors in *R. communis* (Langhans et al. 2001). In some plants the immunocytochemically documented presence of the V-PPase at the plasma membrane was supported by the presence of PP_i hydrolysis activity and immunological detection in plasma membrane fractions (Robinson et al. 1996; Ratajczak et al. 1999). In one case (Ratajczak et al. 1999) it was shown that the plasma membrane located V-PPase, although active in PP_i hydroysis, was incapable of creating and maintaining a H^+ gradient. This again evoked a discussion on the function of the V-PPase at the plasma membrane (Robinson 1996; Davies et al. 1997) and revived the idea that for thermodynamic reasons plasma membrane sited V-PPase may operate in the synthetic mode when elevation of the cytosolic PP_i level is required (Davies et al. 1997; Stitt 1998). This could be particularly important in companion cells of sieve elements (Langhans et al. 2001) where PP_i is needed for sucrose mobilisation (Lerchl et al. 1995a; Stitt 1998).

Intriguingly, the presence of V-ATPase and V-PPase in the plasma membrane seems to vary in different plant species. While in pea cotyledon cells (Robinson et al. 1996) both enzymes have been shown to be present at the plasma membrane, in cells of cauliflower inflorescence (Ratajczak et al. 1999), suspension-cultured tobacco BY-2 cells (Sikora et al. 1998), and *C. reinhardtii* (Robinson et al. 1998), only V-PPase protein has been detected. The presence of the V-PPase in the plasma membrane may simply be a consequence of membrane trafficking, as has been speculated for the V-ATPase. Alternatively, the two enzymes, especially the V-PPase, may have distinct functions in the plasma membrane.

5 Responses of V-ATPase and V-PPase to Environmental Factors

The V-ATPase has been recognised as an enzyme both serving stress responses and undergoing stress-related modifications. Due to these stress-related changes, the V-ATPase was also named 'eco-enzyme' (Lüttge et al. 1995b). While the V-ATPase is directly involved in ecophysiological adaptations at the molecular level, the role of the V-PPase under stress conditions remains enigmatic (Maeshima 2000; Ratajczak and Wilkins 2000). The V-PPase has rather been conceived as an enzyme of high significance in growth and development of plants. However, also for the V-PPase, there are cases supporting regulation in response to environmental stimuli.

The best-studied environmental factor influencing V-ATPase expression and activity is salinity. Salt stress-induced up-regulation of the V-ATPase seems to be a major element in salt tolerance in halophytes (Lüttge and Ratajczak 1997; Ratajczak 2000; Ratajczak and Wilkins 2000). For all plants studied so far at the transcriptional level an increase in transcript amount of subunits A, B, and c in response to salinity has been reported (Ratajczak 2000, and references cited therein). Coordination of the expression of individual subunits of the V-ATPase in response to salinity was documented for *B. vulgaris* roots (Kirsch et al. 1996; Lehr et al. 1999) and *Daucus carota* suspension cells (Löw and Rausch 1996; Rausch et al. 1996), where subunits A and c increased upon salt stress. In leaves of *M. crystallinum*, a coordinated increase in transcript levels of subunits A, B, E, F, and c occurred under long-term salt treatment (Rockel et al. 1998; Golldack and Dietz 2001). Short-term exposure of plants to high salinity for only a few hours led to an increase in subunit c message only, while levels of subunits A, B, and E remained almost constant (Dietz and Arbinger 1996; Löw et al. 1996; Tsiantis et al. 1996).

In contrast to the transcriptional level, on the protein level the response of the V-ATPase to salinity is diverse (Ratajczak 2000). Different patterns of responses have been found. In roots of *Triticum aestivum*, the amount of V-ATPase holoenzyme protein decreased upon salt treatment (Wang et al. 2000). In *N. tabacum* cell-suspension cultures, subunit *A* protein decreased (Reuveni et al. 1990). In *H. vulgare* plants, the abundance of subunits A, B, and E did not significantly change under salinity stress (Hurkman et al. 1988; Dietz et al. 1995; Mariaux et al. 1997). The V-ATPase protein amount of cultured cells of *D. carota* (Löw and Rausch 1996) and roots of *Helianthus annuus* (Ballesteros et al. 1996) showed no response of individual V-ATPase subunits to moderate salt treatment. Whereas in *M. crystallinum* suspension cells the protein amount of subunits A, B, E, and d was not affected by salinity (Vera-Estrella et al. 1999), in leaves of *M. crystallinum* the V-ATPase protein amount increased during the salt stress-induced shift of C_3 photosynthesis to CAM. This increase in protein amount was correlated with an increase in subunits A, B, and c, with the most pronounced change for the proteolipid subunit c (Rockel et al. 1998). The response to salinity stress in most plant species analysed was found to be accompanied by an increase in V-ATPase activity (for references, see Ratajczak 2000; Ratajczak and Wilkins 2000). In some cases changes in ATP hydrolysis activity did not correlate with changes in H^+-transport activity indicating a variation of the coupling ratio of the enzyme. This may be due to regulation of the V-ATPase via a 'slip' mechanism, variation of the numbers of subunit c incorporated into the holoenzyme, or the incorporation of subunit c isoforms as mentioned above (Sect. 2.c; Ballesteros et al. 1996; Löw and Rausch 1996; Mariaux et al. 1997).

Several studies have shown that the plant V-ATPase may also undergo structural changes in response to environmental factors or changes in requirements for tonoplast energisation. In *M. crystallinum*, the salt-stress-induced transition to CAM is accompanied by alteration of subunit composition of the holoenzyme (Ratajczak et al. 1994; Lüttge et al. 1995a) with the two new subunits E_i and D_i described in Section 2.c above. An increase in the diameter of the V_O domain was visualised by electron microscopic analysis of intramembrane particles on freeze-fracture replicas of tonoplast vesicles of *M. crystallinum* and *K. blossfeldiana* during the C_3/CAM transition (Rockel et al. 1994; Mariaux et al. 1997). This enlargement of the V_O domain was proposed to be due to incorporation of a higher number of copies of subunit c per V-ATPase holoenzyme complex in the CAM state, which was supported by a higher staining intensity of subunit c in silver-stained gels. Incorporation of an additional new polypeptide into the V-ATPase holoenzyme complex in response to salinity has also been reported for the partially purified V-

ATPase of *Citrus sinensis* (35 kDa; Bañuls et al. 1995). Interestingly, in the V-ATPase of *C. limon* fruit cells, a tissue-specific expression of several polypeptides in the molecular mass range of 30 kDa has been detected which were absent in the V-ATPase of *C. limon* hypocotyl tissue (Müller et al. 1997).

Besides salinity, some other environmental factors have been reported to influence V-ATPase expression and activity. Low-temperature treatment at 0 °C of chilling-sensitive *V. radiata* seedlings for 24 h leads to a decrease in H^+-transport activity (Matsuura-Endo et al. 1992). At 0 °C, inactivation of the V-ATPase was reversible within 24 h but irreversible by treating seedlings at 0 °C for 48 h (Yoshida 1991; Matsuura-Endo et al. 1992). Blue-light treatment of *Z. mays* coleoptiles leads to reversible inhibition of V-ATPase activity possibly via oxidation of essential SH-groups of the enzyme (Krauss et al. 1987, see above). In *Z. mays* seedlings, light-induced inhibition of cell extension growth was correlated with a decrease in steady-state transcript levels of subunit $\underline{A}$ and $\underline{c}$ (Viereck et al. 1996). The H^+-transport activity of the V-ATPase of *H. vulgare* was inhibited by the toxic metal salt $AlCl_3$ (Kasai et al. 1993a).

Information on the response of the V-PPase to salinity stress is conflicting. A decrease in V-PPase activity was reported for *V. radiata* root tissue upon salt treatment (Nakamura et al. 1992). In seedlings of *Vigna unguiculata*, salt stress inactivated the V-PPase whereas the V-ATPase expression was induced (Otoch et al. 2001). The V-PPase activity of *Suaeda maritima* plants (Leach et al. 1990) and cultured cells of *N. tabacum* (Reuveni et al. 1990) was not affected by salinity. In *Suaeda salsa*, V-ATPase protein amount and activity increased upon salt stress whereas the V-PPase remained constant at a low activity level (Wang et al. 2001). On the other hand, in NaCl-adapted cells of *D. carota* and *Acer pseudoplatanus*, V-PPase activity was induced (Colombo and Cerana 1993; Zingarelli et al. 1994). In *H. annuus*, V-PPase hydrolytic activity and H^+-transport activity were activated to different degrees suggesting that for the V-PPase also there may be modulation of the coupling ratio under certain metabolic conditions as a mechanism of enzyme regulation in analogy to the V-ATPase (Ballesteros et al. 1996).

In *M. crystallinum*, in contrast to the V-ATPase, the V-PPase activity and protein amount decrease when plants are exposed to salt stress leading to a shift from C_3 photosynthesis to CAM (Bremberger et al. 1988; Rockel et al. 1994). CAM expression in plants due to plant development and aging in the absence of salinity leads to the same response (Rockel et al. 1994). Thus, the decline in V-PPase activity and protein amount seemed to be correlated to CAM expression and not to salinity per se. However, experiments with *K. blossfeldiana*, where CAM induction is elicited by shortening of the photoperiod, revealed that results obtained with *M. crystallinum* cannot be generalised. In *K. blossfeldiana*, CAM induction under short-day treatment was accompanied by an increase in V-PPase activity and protein amount (Fischer-Schliebs et al. 1998). This indicates that down-regulation of the V-PPase is not a general feature of CAM. As stated earlier, the V-PPase may serve in concert with the V-ATPase in energising the tonoplast by hydrolysing cytosolic PP_i. For thermodynamic reasons, due to the high electrochemical H^+ gradient created at the tonoplast during the night with vacuolar pH values approaching 3.0 and lower, it is unlikely that the V-PPase contributes much to organic acid accumulation in the late dark periods during the CAM cycle.

The V-PPase seems to be of particular importance under stress conditions such as anoxia or chilling which induce an increase in V-PPase transcript and protein levels (Carystinos et al. 1995; Darley et al. 1995). Carystinos and coworkers showed that in *O. sativa* seedlings under anoxia the transcript amount of the V-PPase increased and decreased again when seedlings were returned to air. V-PPase activity and protein amount in isolated tonoplast vesicles increased with time of anoxia or chilling at 10 °C. Incubation of *V. radiata* seedlings at 4 °C led to a two-fold higher V-PPase activity whereas the V-ATPase activity remained constant (Darley et al. 1995).

Studies on *Secale cereale* plants indicate that growth under conditions of mineral nutrition deficiency lead to an increase in V-PPase activity (Kasai et al. 1998). Increased V-PPase activity was not correlated to an increase in protein amount indicating that an activation of the enzyme occurred under nutrient deficiency.

6 Regulation by Phytohormones

There is increasing evidence that, as elements of signal transduction networks, phytohormones are involved in regulation of V-ATPase and V-PPase activity (Ratajczak and Wilkins 2000). The in vivo treatment of roots of *H. vulgare* with abscisic acid (ABA) caused an increase in protein amount and activity of the V-PPase, whereas the V-ATPase activity was increased without any change in the protein amount. Treatment of seedlings with cytokinin decreased V-PPase activity without a significant effect on the V-ATPase (Kasai et al. 1993b). In a gibberellin-deficient tomato mutant, transcript amounts of subunit $\underline{c}$ and proteins of subunit $\underline{A}$ and $\underline{B}$ increased in response to exogenous gibberellin (GA) application (Cooley et al. 1999). The results suggest that the V-ATPase plays a role in GA-regulated germination of tomato seeds. Salt stress caused an increase in subunit $\underline{A}$ mRNA in both tobacco cells (Narasimhan et al. 1991) and tomato leaves (Binzel 1995; Binzel and Dunlap 1995). While ABA could at least partially mimic this effect in tobacco (Narasimhan et al.1991), it did not appear to be involved in the response to salt in tomato (Binzel and Dunlap 1995). Löw et al. (1996) found that mRNA abundance of subunits A and c were differentially affected by tissue age and salt stress, while the abundance of subunit B mRNA was unaffected by the same factors. Swanson and Jones (1996) demonstrated that GA induces vacuolar acidification in barley aleurone cells, but did not detect significant differences in V-ATPase protein content among control, GA-treated, and ABA-treated aleurone cells. They suggested that other mechanisms, such as cytosolic pH or redox state, might regulate the activity of the V-ATPase. While expression of a V-ATPase in tomato seeds clearly is dependent on at least a minimal level of GA, additional post-transcriptional mechanisms seem to be involved in regulating V-ATPase activity.

7 Responses of V-ATPase and V-PPase in Growth and Development

The V-PPase has been recognised to play a pivotal role in young, growing tissues in which many biosynthetic pathways lead to the production of large amounts of PP_i (Maeshima 2000), and where accumulation of K^+ ions is necessary as a part of the osmotic driving force for cell growth (Davies et al. 1992). From this perspective, one would expect a dominant V-PPase activity compared to the V-ATPase in meristems, sink tissues, or metabolically highly active tissues in general. However, the patterns of V-PPase and V-ATPase protein amount and enzyme activity found are diverse and reflect a more complex relationship of V-PPase and V-ATPase in vacuolar functions. In hypocotyls of *Cucurbita moschata* and *V. radiata* (Maeshima 1990; Suzuki and Kasamo 1993; Nakanishi and Maeshima 1998), as well as during the early stages of fruit development in *P. communis* (Shiratake et al. 1997), a higher V-PPase activity compared to V-ATPase activity was indeed detected. During the prolonged growth of hypocotyls a decrease in V-PPase activity was observed while V-ATPase activity remained unchanged (Nakanishi and Maeshima 1998), decreased (Maeshima 1990), or increased (Suzuki and Kasamo 1993). The level of V-PPase transcript amount was low in tissues of young pear fruits and increased with maturation (Suzuki et al. 1999). However, in tonoplast vesicles isolated from pear fruits at different stages of development, V-PPase protein amount and activity remained constant (Shiratake et al. 1997). By using a GUS-reporter assay system, a tissue-specific regulation of V-PPase expression in developing pollen of *A. thaliana* was reported recently implying a role for the V-PPase in pollen maturation (Mitsuda et al. 2001). Studies with *M. crystallinum* leaves showed a decrease in V-PPase activity and protein amount while activity and protein amount of the V-ATPase increased during plant aging and C_3/CAM transition (Bremberger and Lüttge 1992). In metabolically highly active *R. communis* stem tumors induced by infection with *A. tumefaciens*, V-ATPase and V-PPase protein amount and activity remained constant at a high level, while in non-infected stem tissue protein amount and activity of both enzymes decreased with aging (Fischer-Schliebs et al. 1998). CAM induction in young leaves of *K. blossfeldiana* by a shortening of the photoperiod led to an increase in protein amount and activity of both V-ATPase and V-PPase, while, under long-day conditions, protein amount and activity of both enzymes declined (Fischer-Schliebs et al. 1998). In rapidly elongating cotton fibres, the V-ATPase protein amount accumulates to maximum levels 15 days after anthesis and declines at the onset of secondary wall synthesis (Smart et al. 1998). Interestingly, the V-PPase is constitutively expressed at a high level with its activity approaching a maximum at 20 days after anthesis, clearly indicating that the role of the V-PPase differs from that of the V-ATPase. The differential increase in V-PPase activity relative to V-PPase transcript and protein amount during fibre expansion indicates post-translational regulation of the enzyme.

8 Coordination of V-ATPase and V-PPase

A major focus of current studies is to understand how higher plants regulate the specific tissue or cell-type demands of H^+-electrochemical gradients at the plasma membrane and the tonoplast (Sze et al. 1999; Ratajczak and Wilkins 2000; Dietz et al. 2001; Palmgren 2001). Since Rea and Sanders posed the question in 1987 as to why two H^+-translocating enzymes exist at the vacuolar membrane, the physiological role of the V-PPase has been a matter of debate. Why do plant cells need two enzymes performing the same task, both establishing and maintaining an electrochemical gradient by transport of protons into the vacuolar lumen? Is it the requirement for two differentially regulated enzymes to fulfil the same task because of its extraordinary basic importance? Are there distinct functions for the V-PPase due to the special metabolism of plants leading to the formation or use of inorganic pyrophosphate? Many experiments have been designed with the aim of finding an explanation for the coexistence of V-ATPase and V-PPase at the vacuolar membrane. Particular attention has been directed to levels of enzymatic activity, enzyme protein, and transcript amount of V-ATPase and V-PPase during plant development, cell differentiation and in response to changes in environmental stimuli. However, the picture that has evolved is still fuzzy. The V-ATPase has been recognised as an enzyme both serving stress responses and undergoing stress-related modifications (Lüttge et al. 1995b). The physiological role of the V-PPase and its function in adaptation and survival of plants in stress situations are still unclear (Ratajczak and Wilkins 2000). Different patterns of behaviour of the V-ATPase and the V-PPase during plant development or in response to environmental stimuli have been found, and it is still not understood how the V-ATPase and the V-PPase complement each other. It is currently accepted that the V-ATPase plays the predominant role in the maintenance of the transmembrane electrochemical H^+ gradient, whereas the V-PPase seems to serve as an ancillary backup system for the pumping of protons (Rea and Poole 1993). However, stresses such as chilling and anoxia lead to induction of gene expression of the V-PPase (Carystinos et al. 1995; Darley et al. 1995). The cellular PP_i content unlike that of ATP remains stable during marked changes in respiratory state (Weiner et al. 1987, Dancer and ap Rees 1989). Hence, the V-PPase may exert a key role in survival of plants under conditions of limited ATP supply. Moreover, the V-PPase seems to be of particular importance in young, growing tissues with a high PP_i production during the biosynthesis of macromolecules (Maeshima 2000). The role of PP_i as an energy source alternative to ATP has gained some attention in recent years (Stitt 1998). A general view is that PP_i hydrolysis provides a driving force for many biochemical reactions, such as starch synthesis in chloroplasts, where an active pyrophosphatase hydrolyses PP_i produced in the reaction catalysed by ADP-glucose pyrophosphorylase. In the cytosol, degradation of sucrose by sucrose synthase leads to the production of UDP-glucose, and UDP-glucose pyrophosphorylase will need to operate in the oppposite, i.e. PP_i-consuming, direction. This route of sucrose mobilisation will occur in tissues that are importing sucrose for cell wall synthesis or starch synthesis and requires a steady production of PP_i. In this respect, the idea that the V-PPase may operate in the direction of PP_i synthesis at the expense of the electrochemical H^+ gradient at the tonoplast generated by the V-ATPase is intriguing and subsequently leads to the possibility that the cytosolic pH, or transport across the tonoplast, might interact with the control of cytosolic PP_i levels or carbohydrate metabolism (Stitt 1987). Evidence for PP_i synthesis by reversal of the V-PPase in vitro was derived from experiments with tonoplast vesicles of *Z. mays* coleoptiles and seeds (Façanha and de Meis 1998). The reversibility of the V-PPase in vivo has been claimed to occur in cells with extreme acidic vacuoles such as juice cells of an acid variety of *C. limon* during the process of maturation of citrus fruits (Marsh et al. 2001). The citrus V-ATPase has been found to possess unique kinetic and biochemical characteristics that enable the enzyme to establish a large ΔpH gradient at the vacuolar membrane (Müller et al. 1996, 1997, 1999). Moreover, the tonoplast of juice cells seems to be less permeable to protons when compared to tonoplast membranes of epicotyl tissue allowing the maintenance of a steeper steady-state ΔpH in fruit tissue. Under these circumstances, for thermodynamic reasons, the V-ATPase can still operate in the hydrolytic mode transporting protons into the vacuolar lumen, whereas the V-PPase will favour synthesis of PP_i (Rea and Poole 1987).

A better understanding of the V-PPase function will certainly open new avenues in disclosing the mode of cooperation of the V-ATPase and V-PPase in vacuolar functions in diverse metabolic conditions of different stages of plant development or differentiation. Moreover, the significance of V-PPase isoforms or their possible specific location on membranes beside the tonoplast needs to be unveiled. Cell specificity in transcription, V-PPase protein accumulation, and the enzymatic activity of individual isoforms remains to be determined. Antisense suppression of one of the V-PPase isoforms in *N. tabacum* has not yet resulted in phenotype alterations (Lerchl 1995). Investigations with tobacco have been made difficult by the existence of three V-PPase isoforms in *N. tabacum* that might complement each other in their function. Generation of *V-PPase* antisense plants of *A. thaliana* do not offer a satisfying alternative any longer, since a second gene encoding a new type of a H^+-translocating pyrophosphatase, AVP2, has been discovered (Drozdowicz et al. 2000). For the V-ATPase, the determination of the number of genes encoding individual subunits, how genes are regulated, and how differential expression of genes may influence subunit composition, subcellular distribution, assembly of the holoenzyme, and the activity of the enzyme are still open questions. Several factors and mechanisms on the transcriptional and translational levels have been identified as potential regulatory elements in the control of V-ATPase, including Ca^{2+}, ATP/ADP ratio, reduction/oxidation of essential SH-groups, phosphorylation, variation of coupling ratios, expression of isoforms, variation of subunit stoichiometry, proteolytic processing of V-ATPase subunits, disassembly/reassembly of the V_O and V_1 domains, and the immediate lipid environment of the enzyme. For the V-PPase, only Ca^{2+}, Mg^{2+}, and the availability of the substrate PP_i, have been suggested to control enzyme activity in vivo. Certainly, the lipid environment will also have a pronounced influence (Fig. 2). The most attention so far has been directed to Ca^{2+} as an element of V-PPase regulation (Rea and Poole 1993). Cytosolic Ca^{2+} is indeed an attractive but speculative candidate for controlling the coordination of the V-ATPase and V-PPase, since the V-PPase is subject to pronounced inhibition by micromolar Ca^{2+} concentration (Maeshima 1991; Rea et al. 1992; Darely et al. 1998), whereas V-ATPase activity is rather insensitive to Ca^{2+} (Tu et al. 1989; Rea et al. 1992). However, response mechanisms to signals such as vacuolar lumenal pH and membrane voltage still need to be identified for both the V-ATPase and V-PPase. Experiments with isolated tonoplast vesicles indicate that the V-PPase may have a higher sensitivity to small inside positive membrane potentials than the V-ATPase (Marquardt-Jarczyk and Lüttge 1990b).

To date, we are still ignorant about whether and how one regulating factor might be linked to the other to fulfill the complex task of controlling and coordinating V-ATPase and V-PPase in vacuolar function. The discovery of signalling cascades or networks that regulate the V-ATPase and V-PPase in response to hormonal and environmental cues will be the challenge of future work.

References

Abrahams JP, Leslie AGW, Lutter R, Walker JE (1994) Structure at 2.8 Å resolution of F_1-ATPase from bovine heart mitochondria. Nature 370:621–628

Adachi I, Puopolo K, Marquez-Sterling N, Arai H, Forgac M (1990) Dissociation, cross-linking and glycosylation of the coated vesicle proton pump. J Biol Chem 265:967–973

Aviezer-Hagai K, Nelson H, Nelson N (2000) Cloning and expression of cDNA encoding plant V-ATPase subunits in the corresponding yeast null mutants. Biochim Biophys Acta 1459:489–498

Ballesteros E, Donaire JP, Belver A (1996) Effects of salt stress on H^+-ATPase and H^+-PPase activities of tonoplast-enriched vesicles isolated from sunflower roots. Physiol Plant 97:259–268

Baltscheffsky M, Nadanaciva S, Schultz A (1998) A pyrophosphate synthase gene: molecular cloning and sequencing of the cDNA encoding the inorganic pyrophosphate synthase from *Rhodospirillum rubrum*. Biochim Biophys Acta 1364:301–306

Baltscheffsky M, Schultz A, Baltscheffsky H (1999) H^+-PPases: a tightly membrane-bound family. FEBS Lett 457:527–533

Bañuls J, Ratajczak R, Lüttge U (1995) NaCl stress enhances proteolytic turnover of the tonoplast H^+-ATPase of *Citrus sinensis*: appearance of a 35 kDa fragment of subunit A still exhibiting ATP-hydrolysis activity? Plant Cell Environ 18:1341–1344

Bauerle C, Magembe C, Briskin DP (1998) Characterization of a red beet protein homologous to the essential 36-kilodalton subunit of the yeast V-type ATPase. Plant Physiol 117:859–867

Baykov A, Bakuleva NP, Rea P (1993) Steady-state kinetics of substrate hydrolysis of vacuolar H^+-pyrophosphatase. A simple three-state model. Eur J Biochem 217:755–762

Becker A, Canut H, Lüttge U, Maeshima M, Marigo G, Ratajczak R (1995) Comparison of the behaviour of the tonoplast inorganic H^+-pyrophosphatase from cells of *Catharanthus roseus* and leaves of *Mesembryanthemum crystallinum* performing C_3-photosynthesis and the obligate CAM-plant *Kalanchoë daigremontiana*. J Plant Physiol 146:88–94

Bennett AB, Spanswick RM (1983) Optical measurements of ΔpH and $\Delta\psi$ in corn root membrane vesicles. Kinetic analysis of Cl^- effects on a proton-translocating ATPase. J Membr Biol 71:95–107

Bennett AB, Spanswick R (1984) H^+-ATPase activity from storage tissue of *Beta vulgaris*. II. H^+/ATP stoichiometry of an anion sensitive H^+-ATPase. Plant Physiol 74:545–548

Binzel ML (1995) NaCl-induced accumulation of tonoplast and plasma membrane H^+-ATPase message in tomato. Physiol Plant 94:722–728

Binzel ML, Dunlap JR (1995) Abscisic acid does not mediate NaCl-induced accumulation of 70 kDa subunit tonoplast H^+-ATPase message in tomato. Planta 179:563–568

Blumwald E, Gelli A (1997) Secondary inorganic ion transport at the tonoplast. In: Leigh RA, Sanders D (eds) The plant vacuole. Advances in botanical research, vol 25. Academic Press, San Diego, pp 401–417

Boekema EJ, Ubbink-Kok T, Lolkema JS, Brisson A, Konings WN (1997) Visualization of a peripheral stalk in V-type ATPase: evidence for the stator structure essential to rotational catalysis. Proc Natl Acad Sci USA 94:14291–14293

Boekema EJ, Ubbink-Kok T, Lolkema JS, Brisson A, Konings WN (1998) Structure of V-type ATPase from *Clostridium fervidus* by electron microscopy. Photosyn Res 57:267–273

Boekema EJ, van Breemen JFL, Brisson A, Ubbink-Kok T, Konings WN, Lolkema JS (1999) Connecting stalks in V-type ATPase. Nature 401:37–38

Böttcher B, Schwarz L, Gräber P (1998) Direct indication for the existence of a double stalk in CF_1F_O. J Mol Biol 281:757–762

Bowman EJ, Siebers A, Altendorf K (1988) Bafilomycin: a class of inhibitors of membrane ATPases from microorganisms, animal cells, and plant cells. Proc Natl Acad Sci USA 85:7972–7976

Brauer D, Tu SI (1994) Effects of ATP analogs on the proton pumping by the vacuolar H^+-ATPase from maize roots. Physiol Plant 91:442–448

Brauer D, Conner DN, Tu SI (1992) Effects of pH on proton transport by vacuolar pumps from maize roots. Physiol Plant 86:63–70

Brauer D, Tu SI, Hsu AF, Patterson D (1993) Evidence for an indirect coupling mechanism for the nitrate-sensitive proton pump from corn root tonoplast. Physiol Plant 89:588–591

Braulke T (1996) Origin of lysosomal proteins. In: Lloyd JB, Maison RW (eds) Subcellular biochemistry, vol 27. Biology of the lysosome. Plenum Press, New York, pp 15–49

Bremberger C, Lüttge U (1992) Dynamics of tonoplast proton pumps and other tonoplast proteins of *Mesembryanthemum crystallinum* L. during the induction of crassulacean acid metabolism. Planta 188:575–580

Bremberger C, Haschke H-P, Lüttge U (1988) Separation and purification of the tonoplast ATPase and pyrophosphatase from plants with constitutive and inducible crassulacean acid metabolism. Planta 175:465–470

Bressan RA, Hasegawa PM, Prado JM (1998) Plants use calcium to resolve salt stress. Trans Plant Sci 3:411–412

Britten CJ, Zhen R-G, Kim EJ, Rea PA (1992) Reconstitution of transport function of the vacuolar H^+-translocating inorganic pyrophosphatase. J Biol Chem 267:21850–21855

Carystinos GD, MacDonald HR, Monroy AF, Dhindsa RS, Poole RJ (1995) Vacuolar H^+-translocating pyrophosphatase is induced by anoxia or chilling in seedlings of rice. Plant Physiol 108:641–649

Colombo R, Cerana R (1993) Enhanced activity of tonoplast pyrophosphatase in NaCl-grown cells of *Daucus carota*. J Plant Physiol 142:226–507

Cooley MB, Yang H, Dahal P, Mella RA, Downie AB, Haigh AM, Bradford KJ (1999) Vacuolar H^+-ATPase is expressed in response to gibberellin during tomato seed germination. Plant Physiol 121:1339–1347

Crider BP, Xie XS, Stone DK (1994) Bafilomycin inhibits proton flow through the H^+-channel of vacuolar proton pumps. J Biol Chem 269:17379–17381

Dancer J, ap Rees T (1989) The effects of 2,4-dinitrophenol and anoxia on the inorganic pyrophosphate content of spadix of *Arum maculatum* and root apices of *Pisum sativum*. Planta 178:421–424

Darely CP, Skiera LA, Northrop FD, Sanders D, Davies JM (1998) Tonoplast inorganic pyrophosphatase in *Vicia faba* guard cells. Planta 206:272–277

Darley CP, Davies JM, Sanders D (1995) Chill-induced changes in the activity and abundance of the vacuolar proton-pumping pyrophosphatase from mung bean hypocotyls. Plant Physiol 109:659–665

Davies JM, Rea PA, Sanders D (1991) Vacuolar proton-pumping pyrophosphatase in *Beta vulgaris* shows vectorial activation by potassium. FEBS Lett 278:666–668

Davies JM, Poole RJ, Rea PA, Sanders D (1992) Potassium transport into plant vacuoles energised directly by a proton-pumping inorganic pyrophosphatase. Proc Natl Acad Sci USA 89:11701–11705

Davies JM, Poole RJ, Sanders D (1993) The computed free energy change of hydrolysis of inorganic pyrophosphate and ATP: apparent significance for inorganic-pyrophosphate-driven reactions of intermediate metabolism. Biochim Biophys Acta 1141:29–36

Davies JM, Hunt I, Sanders D (1994) Vacuolar H^+-pumping ATPase variable transport coupling ratio controlled by pH. Proc Natl Acad Sci USA 91:8547–8551

Davies JM, Darley CP, Sanders D (1997) Energetics of the plasma membrane pyrophosphatase. Trends Plant Sci 2:9–10

Depta H, Holstein SEH, Robinson DG, Lützelschwab M, Michalke W (1991) Membrane markers in highly purified clathrin-coated vesicles from *Cucurbita* hypocotyls. Planta 183:434–442

Dietz K-J, Arbinger B (1996) cDNA sequence and expression of subunit E of the vacuolar H^+-ATPase in the inducible crassulacean acid metabolism plant *Mesembryanthemum crystallinum*. Biochim Biophys Acta 1281:134–138

Dietz K-J, Rudloff S, Ageorges A, Eckerskorn C, Fischer K, Arbinger B (1995) Subunit E of the vacuolar H^+-ATPase of *Hordeum vulgare* L.: cDNA cloning, expression and immunological analysis. Plant J 8:521–529

Dietz K-J, Heber H, Mimura T (1998) Modulation of the vacuolar H^+-ATPase by adenylates as basis for the transient CO_2-dependent acidification of the leaf vacuole upon illumination. Biochim Biophys Acta 1373:87–92

Dietz KJ, Tavakoli N, Klugc C, Mimura T, Sharma SS, Harris GC, Chardonnes AN, Gollack D (2001) Significance of the V-type ATPase for the adaptation to stressful growth conditions and its regulation on the molecular and biochemical level. J Exp Bot 52:1969–1980

Domgall I, Venzke D, Lüttge U, Ratajczak R, Böttcher B (2002) Three-dimensional model of a plant V-ATPase based on electron microscopy. J Biol Chem 227:13115–13121

Drobny M, Schnölzer M, Fiedler S, Lüttge U, Fischer-Schliebs E, Christian A-L, Ratajczak R (2002) Phenotypic subunit composition of the tobacco (*Nicotiana tabacum* L.) vacuolar type H^+-translocating ATPase. Biochim Biophys Acta (in press)

Dröse S, Bindseil KU, Bowman EJ, Siebers A, Zeeck A, Altendorf K (1993) Inhibitory effects of modified Bafilomycins and Concanamycins on P- and V-type adenosinetriphosphatases. Biochemistry 32:3902–3906

Drozdowicz YM, Lu Y-P, Patel V, Fitz-Gibbon S, Miller JH, Rea PA (1999) A thermostable vacuolar-type membrane pyrophosphatase from the archaeon *Pyrobaculum aerophilum*: implications for the origins of pyrophosphate-energised pumps. FEBS Lett 460:505–512

Drozdowicz YM, Kissinger JC, Rea PA (2000) AVP2, a sequence-divergent, K^+-insensitive H^+-translocating inorganic pyrophosphatase from *Arabidopsis*. Plant Physiol 123:353–362

Dschida WJ, Bowman BJ (1992) Structure of the vacuolar ATPase from *Neurospora crassa* as determined by electron microscopy. J Biol Chem 267:18783–18789

Dschida WJ, Bowman BJ (1995) The vacuolar ATPase: sulfite stabilization and the mechanism of nitrate inactivation. J Biol Chem 270:1557–1563

DuPont FM, Morrissey PJ (1992) Subunit composition and Ca-ATPase activity of the vacuolar ATPase from barley roots. Arch Biochem Biophys 294:341–346

Emig I (1999) Intrazelluläre Verteilung der V-ATPase in der C_3-CAM intermediären Pflanze *Mesembryanthemum crystallinum*. PhD Thesis, Darmstadt University of Technology, Darmstadt, Germany

Façanha AR, de Meis L (1998) Reversibility of H^+-ATPase and H^+-pyrophosphatase in tonoplast vesicles from maize coleoptiles and seeds. Plant Physiol 116:1487–1495

Feng Y, Forgac M (1992) A novel mechanism for regulation of vacuolar acidification. J Biol Chem 269:13224–13230

Feng Y, Forgac M (1994) Inhibition of vacuolar H^+-ATPase by disulfide bond formation between cysteine 254 and cysteine 532 in subunit *A*. J Biol Chem 267:19769–19772

Fischer-Schliebs E, Ball E, Berndt E, Besemfelder-Butz E, Binzel ML, Drobny M, Mühlenhoff D, Müller ML, Rakowski K, Ratajczak R (1997a) Differential immunological cross-reactions with antisera against the V-ATPase of *Kalanchoë daigremontiana* reveals structural differences of V-ATPase subunits of different plant species. Biol Chem 378:1131–1139

Fischer-Schliebs E, Mariaux J-B, Lüttge U (1997b) Stimulation of H^+-transport activity of vacuolar H^+-ATPase by activation of H^+-PPase in *Kalanchoë blossfeldiana*. Biol Plant 39:169–177

Fischer-Schliebs E, Ratajczak R, Weber P, Tavakoli N, Ullrich CI, Lüttge U (1998) Concordant time-dependent patterns of activities and enzyme protein amounts of V-PPase and V-ATPase in induced (flowering and CAM or tumour) and non-induced plant tissues. Bot Acta 111:130–136

Fischer-Schliebs E, Drobny M, Ball E, Ratajczak R, Lüttge U (2000) Variation in nitrate nutrition leads to changes in performance of the V-ATPase and immunological differences of proteolipid subunit c in tobacco (*Nicotiana tabacum* L.) leaves. Aust J Plant Physiol 27:639–648

Forgac M (2000) Structure, mechanism and regulation of the clathrin-coated vesicle and yeast vacuolar H^+-ATPases. J Exp Biol 203:71–80

Geigenberger P, Stitt M (1993) Sucrose synthase catalyses a readily reversible reaction in developing potato tubers and other plant tissues. Planta 189:329–339

Geigenberger P, Langenberger S, Wilke I, Heineke D, Heldt HW, Stitt M (1993) Sucrose is metabolised by sucrose synthase and glycolysis within the phloem complex of *Ricinus communis* L. seedlings. Planta 190:446–453

Getz HP, Klein M (1995) The vacuolar ATPase of red beet storage tissue: electron microscopic demonstration of the "head-and-stalk" structure. Bot Acta 108:14–23

Gogarten JP, Starke T, Kibak H, Fishmann J, Taiz L (1992) Evolution and isoforms of V-ATPase subunits. J Exp Biol 172:137–147

Golldack D, Dietz K-J (2001) Salt-induced expression of the vacuolar H^+-ATPase in the common ice plant is developmentally controlled and tissue specific. Plant Physiol 125:1643–1654

Gordon-Weeks R, Steele SH, Leigh RA (1996) The role of magnesium, pyrophosphate and their complexes as substrates and activators of the vacuolar H^+-pumping inorganic pyrophosphatase. Studies using ligand protection from covalent inhibitors. Plant Physiol 111:195–202

Gordon-Weeks R, Koren'kov VD, Steele SH, Leigh RH (1997) Tris is a competitive inhibitor of K^+ activation of the vacuolar H^+-pumping pyrophosphatase. Plant Physiol 114:901–905

Grabe M, Wang H, Oster G (2000) The mechanochemistry of V-ATPase proton pumps. Biophys J 78:2798–2813

Graf R, Harvey WR, Wieczorek H (1996) Purification and properties of a cytosolic V_1-ATPase. J Biol Chem 217:20908–20913

Graham LA, Stevens TH (1999) Assembly of the yeast vacuolar proton-translocating ATPase. J Bioenerg Biomembr 31:39–47

Graham LA, Hill KJ, Stevens TH (1998) Assembly of the yeast vacuolar H^+-ATPase occurs in the endoplasmic reticulum and requires a Vma12p/Vma22p assembly complex. J Cell Biol 142:39–49

Hager A, Biber W (1984) Functional and regulatory properties of H^+ pumps at the tonoplast and plasma membranes of *Zea mays* coleoptiles. Z Naturforsch 39:927–937

Hager A, Lanz C (1989) Essential sulfhydryl groups in the catalytic center of the tonoplast H^+-ATPase from coleoptiles of *Zea mays* L. as demonstrated by biotin-streptavidin-peroxidase system. Planta 180:116–122

Hasenfratz M-P, Tsou C-L, Wilkins TA (1995) Expression of two related vacuolar H^+-ATPase 16-kilodalton proteolipid genes is differentially regulated in a tissue-specific manner. Plant Physiol 108:1395–1404

Hedrich R, Kurkdjian A, Guern J, Flügge UI (1989) Comparative studies on the electrical properties of the H^+-translocating ATPase and pyrophosphatase of the vacuolar-lysosomal compartment. EMBO J 8:2835–2841

Herman EM, Li X, Su RT, Larsen P, Hsu HT, Sze H (1994) Vacuolar-type H^+-ATPases are associated with the endoplasmic reticulum and pro-vacuoles of root tip cells. Plant Physiol 106:1313–1324

Hirata R, Graham LA, Takatsuki A, Stevens TH, Anraku Y (1997) *VMA11* and *VMA16* encode second and third proteolipid subunits of the *Saccharomyces cerevisiae* vacuolar membrane H^+-ATPase. J Biol Chem 272:4795–4803

Ho MN, Hirata R, Umemoto N, Ohya Y, Takatsuki A, Stevens TH, Anraku Y (1993) *VMA13* encodes a 54-kDa vacuolar H^+-ATPase subunit required for activity but not assembly of the enzyme complex in *Saccharomyces cerevisiae.* J Biol Chem 268:18286–18292

Hosaka M, Kanayama Y, Shiratake K, Yamaki S (1994) Tonoplast H^+-ATPase of mature pear fruit. Phytochemistry 36:565–567

Hunt IE, Bowman BJ (1997) The intriguing evolution of the "*b*" and "*G*" subunits in F-type and V-type ATPases: isolation of the *vma*-10 gene from *Neurospora crassa.* J Bioenerg Biomembr 29:533–540

Hurkman WJ, Tanaka CK, DuPont FM (1988) The effects of salt stress on polypeptides in membrane fractions from barley roots. Plant Physiol 88:1263–1273

Ikeda M, Satoh S, Maeshima M, Mukohata Y, Moritani C (1991) A vacuolar ATPase and pyrophosphatase in *Acetabularia acetabulum.* Biochim Biophys Acta 1070:77–82

Ikeda M, Rahman MH, Oritani C, Umami K, Tanimura Y, Akagi R, Tanaka Y, Maeshima M, Watanabe Y (1999) A vacuolar H^+-pyrophosphatase in *Acetabularia acetabulum*: molecular cloning and comparison with higher plants and a bacterium. J Exp Bot 50:139–140

Jiang SS, Fan LL, Yang SJ, Kuo SY, Pan RL (1997) Purification and characterization of thylakoid membrane-bound inorganic pyrophosphatase from *Spinacia oleracea* L. Arch Biochem Biophys 346:105–112

Johannes E, Felle H (1990) Proton gradient across the tonoplast of *Riccia fluitans* as a result of the joint action of two electroenzymes. Plant Physiol 93:412–417

Junge W, Lill H, Engelbrecht S (1997) ATP synthase: an electrochemical transducer with rotatory mechanics. Trends Plant Sci 22:420–423

Kaestner KH, Randall SK, Sze H (1988) N,N'-Dicyclohexylcarbodiimide-binding proteolipid of the vacuolar H^+-ATPase of barley roots. J Biol Chem 263:1388–1393

Kane PM (1995) Disassembly and reassembly of the yeast vacuolar H^+-ATPase in vivo. J Biol Chem 270:17025–17032

Karlson J (1975) Membrane-bound potassium and magnesium ion stimulated inorganic pyrophosphatase from roots and cotyledons of sugar beet (*Beta vulgaris* L.). Biochim Biophys Acta 399:356–363

Kasai M, Sasaki M, Tanakamura S, Yamamoto Y, Matsumoto H (1993a) Possible involvement of abscisic acid in increases in activities of two vacuolar H^+-pumps in barley roots under aluminium stress. Plant Cell Physiol 34:1335–1338

Kasai M, Yamamoto Y, Maeshima M, Matsumoto H (1993b) Effects of in vivo treatment with abscisic acid and/or cytokinin on activities of vacuolar H^+-ATPase pumps of tonoplast enriched membrane vesicles prepared from barley roots. Plant Cell Physiol 34:1107–1115

Kasai M, Nakamura T, Kudo N, Sato H, Maeshima M, Sawada S (1998) The activity of the root vacuolar H^+-pyrophosphatase in rye plants grown under conditions deficient in mineral nutrients. Plant Cell Physiol 39:890–894

Kasamo K, Yamaguchi M, Nakamura Y (2000) Mechanism of the chilling-induced decrease in proton pumping across the tonoplast of rice cells. Plant Cell Physiol 41:840–849

Kawamura Y, Arakawa K, Maeshima M, Yoshida S (2000) Tissue specificity of *E* subunit isoforms of plant vacuolar H^+-ATPase and existence of isotype enzymes. J Biol Chem 275:6515–6522

Kim Y, Kim EJ, Rea PA (1994) Isolation and characterization of cDNAs encoding the vacuolar H^+-pyrophosphatase of *Beta vulgaris*. Plant Physiol 106:375–382

Kirsch M, Zhigang A, Viereck R, Löw R, Rausch T (1996) Salt stress induces an increased expression of V-type H^+-ATPase in mature sugar-beet leaves. Plant Mol Biol 32:543–547

Klink R, Lüttge U (1991) Electron-microscopic demonstration of a "head and stalk" structure of the leaf vacuolar ATPase in *Mesembryanthemum crystallinum* L. Bot Acta 104:122–131

Klink R, Lüttge U (1992) Quantification of visible structural changes of the V_OV_1-ATPase in the leaf-tonoplast of *Mesembryanthemum crystallinum* by freeze-fracture replicas prepared during the C_3-photosynthesis to CAM transition. Bot Acta 105:414–420

Klink R, Haschke H-P, Kramer D, Lüttge U (1990) Membrane particles, proteins and ATPase activity of tonoplast vesicles of *Mesembryanthemum crystallinum* in the C-3 and CAM state. Bot Acta 103:24–31

Kluge C, Golldack D, Dietz K-J (1999) Subunit *D* of the vacuolar H^+-ATPase of *Arabidopsis thaliana*. Biochim Biophys Acta 1419:105–110

Kramer D, Mangold B, Hille A, Emig I, Hess A, Ratajczak R, Lüttge U (1995) The head structure of a higher plant V-type H^+-ATPase is not always a hexamer but also a pentamer. J Exp Bot 46:1633–1636

Krauss W, Schiebel G, Eberl D, Hager A (1987) Blue light induced reversible inactivation of the tonoplast-type H^+-ATPase from corn coleoptiles in the presence of flavins. Photochem Photobiol 45:873–844

Krisch R, Rakowski K, Ratajczak R (2000) Processing of V-ATPase subunit *B* of *Mesembryanthemum crystallinum* L. is mediated in vitro by a protease and/or reactive oxygen species. Biol Chem 381:583–592

Lai S, Randall SK, Sze H (1988) Peripheral and integral subunits of the tonoplast H^+-ATPase from oat roots. J Biol Chem 263:16731–16737

Lai S, Watson JC, Hansen J, Sze H (1991) Molecular cloning and sequencing of cDNAs encoding the proteolipid subunit of the vacuolar H^+-ATPase from a higher plant. J Biol Chem 266:16078–16084

Landolt-Marticorena C, Kahr WH, Zawarinski P, Correa J, Manolson MF (1999) Substrate- and inhibitor-induced conformational changes in the yeast V-ATPase provide evidence for communication between the catalytic and proton-translocating sectors. J Biol Chem 274:26057–26064

Landolt-Marticorena C, Williams KM, Correa J, Chen W, Manolson MF (2000) Evidence that the NH_2 terminus of VpH1p, an integral subunit of the V_O sector of the yeast V-ATPase, interacts directly with the Vma13p subunits of the V_1 sector. J Biol Chem 275:15449–15457

Langhans M, Ratajczak R, Lützelschwab M, Michalke W, Wächter R, Fischer-Schliebs E, Ullrich C (2001) Immunolocalization of plasma-membrane H^+-ATPase and tonoplast-type pyrophosphatase in the plasma membrane of the sieve element-companion cell complex in the stem of *Ricinus communis* L. Planta 213:11–19

Leach RP, Rogers WJ, Wheeler KP, Flowers TJ, Yeo AR (1990) Molecular markers for ion compartmentation in cells of higher plants. I. Isolation of vacuoles of high purity. J Exp Bot 41:1079–1087

Lee Taiz S, Taiz L (1991) Ultrastructural comparison of the vacuolar and mitochondrial H^+-ATPase of *Daucus carota*. Bot Acta 104:117–121

Lehr A, Kirsch M, Viereck R, Schiemann J, Rausch T (1999) cDNA and genomic cloning of sugar beet V-type H^+-ATPase subunit *A* and *c* isoforms: evidence for coordinate expression during plant development and coordinate induction in response to high salinity. Plant Mol Biol 39:463–475

Leigh RA, Pope AJ, Jennings IR, Sanders D (1992) Kinetics of the vacuolar H^+-pyrophosphatase. The roles of magnesium, pyrophosphate, and their complexes as substrates, activators, and inhibitors. Plant Physiol 100:1698–1705

Leigh RA, Gordon-Weeks R, Steele SH, Koren'kov VD (1994) The H^+-pumping inorganic pyrophosphatase of the vacuolar membrane of higher plants. In: Blatt MR, Leigh RA, Sanders D (eds) Membrane transport in plants and fungi: molecular mechanism and control. Symposia of the Society of Experimental Biology, vol XLVIII. The Company of Biologists Limited, Cambridge, pp 61–75

Leng X-H, Manolson MF, Liu Q, Forgac M (1996) Site-directed mutagenesis of the 100-kDa subunit (Vph1p) of the yeast vacuolar (H^+)-ATPase. J Biol Chem 271:22487–22493

Leng X-H, Manolson MF, Forgac M (1998) Function of the COOH-terminal domain of Vph1p in activity and assembly of the yeast V-ATPase. J Biol Chem 273:6716–6723

Lerchl J (1995) Molekulare Analysen zum Saccharose- und Pyrophosphatstoffwechsel in transgenen Pflanzen. PhD Thesis, FU Berlin, Berlin, Germany

Lerchl J, Geigenberger P, Stitt M, Sonnewald U (1995a) Impaired photoassimilate partitioning caused by phloem specific removal of pyrophosphate can be complemented by a phloem specific cytosolic yeast-derived invertase in transgenic plants. Plant Cell 7:259–270

Lerchl J, König S, Zrenner R, Sonnewald U (1995b) Molecular cloning, characterization and expression analysis of isoforms encoding tonoplast bound proton translocating inorganic pyrophosphatase in tobacco. Plant Mol Biol 29:833–840

Li XH, Sze H (1999) A 100 kDa polypeptide associates with the V_O membrane sector but not with the active oat vacuolar H^+-ATPase, suggesting a role in assembly. Plant J 17:19–30

Löw R, Rausch T (1996) In suspension-cultured *Daucus carota* cells salt stress stimulates H^+-transport but not ATP hydrolysis of the V-ATPase. J Exp Bot 47:1725–1732

Löw R, Rockel B, Kirsch M, Ratajczak R, Hörtensteiner S, Martinoia E, Lüttge U, Rausch T (1996) Early salt stress effects on the differential expression of vacuolar H^+-ATPase genes in roots and leaves of *Mesembryanthemum crystallinum*. Plant Physiol 110:259–265

Long AR, Williams LE, Nelson SJ, Hall JL (1995) Localization of membrane pyrophosphatase activity in *Ricinus communis* seedlings. J Plant Physiol 146:629–638

Ludwig J, Krescher S, Brandt U, Pfeiffer K, Getlawi F, Apps DK, Schägger H (1998) Identification and characterization of a novel 9.2-kDa membrane sector-associated protein of vacuolar proton-ATPase from chromaffin granules. J Biol Chem 273:10939–10947

Lüttge U, Ratajczak R (1997) The physiology, biochemistry and molecular biology of the plant vacuolar ATPase. In: Leigh RA, Sanders D (eds) The plant vacuole. Advances in botanical research, vol 25. Academic Press, San Diego, pp 253–296

Lüttge U, Smith JAC, Marigo G, Osmond CB (1981) Energetics of malate accumulation in the vacuoles of *Kalanchoë tubiflora* cells. FEBS Lett 126:81–84

Lüttge U, Fischer-Schliebs E, Ratajczak R, Kramer D, Berndt E, Kluge M (1995a) Functioning of the tonoplast in vacuolar C-storage and remobilization in crassulacean acid metabolism. J Exp Bot 46:1377–1388

Lüttge U, Ratajczak R, Rausch T, Rockel B (1995b) Stress responses of tonoplast proteins: an example for molecular ecophysiology and the search for eco-enzymes. Acta Bot Neerl 44:343–362

Lüttge U, Fischer-Schliebs E, Ratajczak R (2001) The H^+-pumping V-ATPase of higher plants: a versatile "eco-enzyme" in response to environmental stress. Cell Biol Mol Lett 6:356–361

MacDonald JI, Weeks G (1988) Evidence for a membrane-bound pyrophosphatase in *Dictyostelium discoideum*. FEBS Lett 238:9–12

Maeshima M (1990) Development of vacuolar membranes during elongation of cells in mung bean hypocotyls. Plant Cell Physiol 31:311–317

Maeshima M (1991) H^+-translocating inorganic pyrophosphatase of plant vacuoles. Inhibition by Ca^{2+}, stabilization by Mg^{2+} and immunological comparison with other inorganic pyrophosphatases. Eur J Biochem 196:11–17

Maeshima M (2000) Vacuolar H^+-pyrophosphatase. Biochim Biophys Acta 1465:37–51

Maeshima M (2001) Tonoplast transporters: organization and function. Annu Rev Plant Physiol Plant Mol Biol 52:469–499

Maeshima M, Yoshida S (1989) Purification and properties of a vacuolar membrane proton-translocating inorganic pyrophosphatase from mung bean. J Biol Chem 264:20068–20073

Mandala S, Taiz L (1986) Characterization of the subunit structure of the maize tonoplast ATPase. J Biol Chem 261:12850–12855

Mandel M, Moriyama Y, Hulmes JD, Pan YC, Nelson H, Nelson N (1988) cDNA sequence encoding the 16-kDa proteolipid of chromaffin granules implies gene duplication in the evolution of H^+-ATPases. Proc Natl Acad Sci USA 85:5521–5524

Manolson MF, Proteau D, Preston RA, Stenbit A, Roberts BT, Hoyt MA, Preuss D, Mulholland J, Botstein D, Jones EW (1992) The VPH1 gene encodes a 95-kDa integral membrane polypeptide required for in vivo assembly and activity of the yeast vacuolar H^+-ATPase. J Biol Chem 267:14292–14303

Manolson MF, Wu B, Proteau D, Taillon BE, Roberts BT, Hoyt MA, Jones EW (1994) STV1 gene encodes functional homologue of 95-kDa yeast vacuolar H^+-ATPase subunit Vph1p. J Biol Chem 269:14064–14074

Mariaux J-B, Becker A, Kemna I, Ratajczak R, Fischer-Schliebs E, Kramer D, Lüttge U, Marigo G (1994) Visualization by freeze-fracture electron microscopy of intramembraneous particles corresponding to the tonoplast H^+-pyrophosphatase and H^+-ATPase of *Kalanchoë daigremontiana* Hamet et Perrier de la Bâthie. Bot Acta 107:321–327

Mariaux J-B, Fischer-Schliebs E, Lüttge U, Ratajczak R (1997) Dynamics of activity and structure of the tonoplast vacuolar-type H^+-ATPase in plants with differing CAM expression and in a C_3 plant under salt stress. Protoplasma 196:181–189

Marquardt G, Lüttge U (1987) Proton translocating enzymes at the tonoplast of leaf cells of the CAM plant *Kalanchoë daigremontiana*. II. The pyrophosphatase. J Plant Physiol 129:269–286

Marquardt-Jarczyk G, Lüttge U (1990a) PP_iase-activated ATP-dependent H^+-transport at the tonoplast of mesophyll cells of the CAM plant *Kalanchoë daigremontiana*. Bot Acta 103:203–213

Marquardt-Jarczyk G, Lüttge U (1990b) Different valinomycin effects on tonoplast ATPase and PP_iase of the CAM plant *Kalanchoë daigremontiana* measured by formation of H^+ gradients and membrane potentials. CR Acad Sci Paris 310:311–316

Marsh K, González, Echeverría E (2001) Partial characterzation of H^+-translocating inorganic pyrophosphatase from 3 citrus varieties differing in vacuolar pH. Physiol Plant 111:519–526

Martinoia E, Ratajczak R (1997) Transport of organic molecules across the tonoplast. In: Leigh RA, Sanders D (eds) The plant vacuole. Advances in botanical research, vol 25. Academic Press, San Diego, pp 365–400

Martiny-Baron G, Manolson MF, Poole RJ, Hecker D, Scherer GFE (1992) Proton transport and phosphorylation of tonoplast polypeptides from zucchini are stimulated by the phospholipid platelet-activating factor. Plant Physiol 99:1635–1641

Matsuura-Endo C, Maeshima M, Yoshida S (1992) Mechanism of the decline in vacuolar H^+-ATPase activity in mung bean hypocotyls during chilling. Plant Physiol 100:718–722

Mayer KFX, Schueller C, Wambutt R, Murphy G, Volckaert G, Pohl T, Düsterhöft A, Stiekema W, Entian K-D, Terryn N, Harris B, Ansorge W, Brandt P, Grivell LA, Rieger

M, Weichselgartner M, de Simone V, Obermaier B, Mache R, Müller M, Kreis M, Delseny M, Puigdomenech P, Watson M, Schmidtheini T, Reichert B, Portetelle D, Perez-Alonso M, Boutry M, Bancroft I, Vos P, Hoheisel J, Zimmermann W, Wedler H, Ridley P, Langham S-A, McCullagh B, Bilham L, Robben J, Van der Schueren J, Grymonprez B, Chuang Y-J, Vandenbussche F, Braeken M, Weltjens I, Voet M, Bastiaens I, Aert R, Defoor E, Weitzenegger T, Bothe G, Ramsperger U, Hilbert H, Braun M, Holzer E, Brandt A, Peters S, van Staveren M, Dirkse W, Mooijman P, Klein Lankhorst R, Rose M, Hauf J, Koetter P, Berneiser S, Hempel S, Feldpausch M, Lamberth S, Van den Daele H, De Keyser A, Buysshaert C, Gielen J, Villarroel R, De Clercq R, Van Montagu M, Rogers J, Cronin A, Quail M, Bray-Allen S, Clark L, Doggett J, Hall S, Kay M, Lennard N, McLay K, Mayes R, Pettett A, Rajandream M-A, Lyne M, Benes V, Rechmann S, Borkova D, Blöcker H, Scharfe M, Grimm M, Löhnert T-H, Dose S, de Haan M, Maarse AC, Schäfer M, Müller-Auer S, Gabel C, Fuchs M, Fartmann B, Granderath K, Dauner D, Herzl A, Neumann S, Argiriou A, Vitale D, Liguori R, Piravandi E, Massenet O, Quigley F, Clabauld G, Mündlein A, Felber R, Schnabl S, Hiller R, Schmidt W, Lecharny A, Aubourg S, Chefdor F, Cooke R, Berger C, Monfort A, Casacuberta E, Gibbons T, Weber N, Vandenbol M, Bargues M, Terol J, Torres A, Perez-Perez A, Purnelle B, Bent E, Johnson S, Tacon D, Jesse T, Heijnen L, Schwarz S, Scholler P, Heber S, Francs P, Bielke C, Frishman D, Haase D, Lemcke K, Mewes H-W, Stocker S, Zaccaria P, Bevan M, Wilson RK, de la Bastide M, Habermann K, Parnell L, Dedhia N, Gnoj L, Schutz K, Huang E, Spiegel L, Sekhon M, Murray J, Sheet P, Cordes M, Abu-Threideh J, Stoneking T, Kalicki J, Graves T, Harmon G, Edwards J, Latreille P, Courtney L, Cloud J, Abbott A, Scott K, Johnson D, Minx P, Bentley D, Fulton B, Miller N, Greco T, Kemp K, Kramer J, Fulton L, Mardis E, Dante M, Pepin K, Hillier L, Nelson J, Spieth J, Ryan E, Andrews S, Geisel C, Layman D, Du H, Ali J, Berghoff A, Jones K, Drone K, Cotton M, Joshu C, Antonoiu B, Zidanic M, Strong C, Sun H, Lamar B, Yordan C, Ma P, Zhong J, Preston R, Vil D, Shekher M, Matero A, Shah R, Swaby IK, O'Shaughnessy A, Rodriguez M, Hoffman J, Till S, Granat S, Shohdy N, Hasegawa A, Hameed A, Lodhi M, Johnson A, Chen E, Marra M, Martienssen R, McCombie WR (1999) Sequence and analysis of chromosome 4 of the plant *Arabidopsis thaliana*. Nature 402:769–777

Merzendorfer H, Huss M, Schmid R, Harvey WR, Wieczorek H (1999) A novel insect V-ATPase subunit M9.7 is glycosylated extensively. J Biol Chem 274:17372–17378

Merzendorfer H, Reineke S, Zhao X-F, Jacobmeier B, Harvey WR, Wieczorek H (2000) The multigene family of the tobacco hornworm V-ATPase: novel subunits *a*, *C*, *D*, *H* and putative isoforms. Biochim Biophys Acta 1467:369–379

Mitsuda N, Takeyasu K, Sato MH (2001) Pollen-specific regulation of vacuolar H^+-PPase expression by multiple cis-acting elements. Plant Mol Biol 46:185–192

Moriyama Y, Nelson N (1988) The vacuolar H^+-ATPase, a proton pump controlled by a slip. In: Stein WD (ed) The ion pumps: structure, function and regulation. Alan R Liss, New York, pp 387–394

Müller ML, Irkens-Kiesecker U, Rubinstein B, Taiz L (1996) On the mechanism of hyperacidification in lemon. Comparison of the vacuolar H^+-ATPase activities of fruits and epicotyls. J Biol Chem 271:1916–1924

Müller ML, Irkens-Kiesecker U, Kramer D, Taiz L (1997) Purification and reconstitution of the vacuolar H^+-ATPases from lemon fruits and epicotyls. J Biol Chem 272:12762–12770

Müller ML, Jensen M, Taiz L (1999) The vacuolar H^+-ATPase of lemon fruits is regulated by variable H^+/ATP coupling and slip. J Biol Chem 274:10706–10716

Nakamura Y, Kasamo K, Shimosato N, Sakata M, Ohta E (1992) Stimulation of the extrusion of protons and H^+-ATPase activities with the decline in pyrophosphatase activity of the tonoplast in intact mung bean roots under high-NaCl stress and its relation to external levels of Ca^{2+} ions. Plant Cell Physiol 32:139–149

Nakanishi Y, Maeshima M (1998) Molecular cloning of vacuolar H^+-pyrophosphatase and its developmental expression in growing hypocotyls of mung bean. Plant Physiol 116:589–597

Nakanishi Y, Matsuda N, Aizawa K, Kashiyama T, Yamamoto K, Mimura T, Ikeda M, Maeshima M (1999) Molecular cloning of the cDNA of vacuolar H^+-pyrophosphatase from *Chara corallina*. Biochim Biophys Acta 1418:245–250

Nakanishi Y, Saijo T, Wada Y, Maeshima M (2001) Mutagenic analysis of functional residues in putative substrate-binding site and acidic domains of vacuolar H^+-pyrophosphatase. J Biol Chem 276:7654–7660

Narasimhan ML, Binzel ML, Perez-Prat E, Chen Z, Nelson DE, Singh NK, Bressan RA, Hasegawa PM (1991) NaCl regulation of tonoplast ATPase 70-kilodalton subunit mRNA in tobacco cells. Plant Physiol 97:562–568

Nelson N, Harvey WR (1999) Vacuolar and plasma membrane proton-adenosinetriphosphatases. Physiol Rev 79:361–385

Nelson H, Nelson N (1990) Disruption of genes encoding subunits of yeast vacuolar H^+-ATPase causes conditional lethality. Proc Natl Acad Sci USA 87:3503–3507

Nelson N, Taiz L (1989) The evolution of H^+-ATPases. Trends Biochem Sci 14:113–116

Nelson H, Mandiyan S, Nelson N (1995) A bovine cDNA and a yeast gene (*VMA8*) encoding the subunit *D* of the vacuolar H^+-ATPase. Proc Natl Acad Sci USA 92:497–501

Noumi T, Beltran C, Nelson H, Nelson N (1991) Mutational analysis of the yeast vacuolar (H^+)-ATPase. Proc Natl Acad Sci USA 88:1938–1942

Oberbeck K, Drucker M, Robinson DG (1994) V-ATPase and pyrophosphatase in endomembranes of maize roots. J Exp Bot 45:235–244

Obermeyer G, Sommer A, Bentrup F-W (1996) Potassium and voltage dependence of the inorganic pyrophosphatase of intact vacuoles from *Chenopodium rubrum*. Biochim Biophys Acta 1284:203–212

Otoch MLO, Sobeira ACM, de Aragão MEF, Orellano EG, Lima MGS, de Melo DF (2001) Salt modulation of vacuolar H^+-ATPase and H^+-pyrophosphatase activities in *Vigna unguiculata*. J Plant Physiol 158:545–551

Pänke O, Rumber B (1999) Kinetic modeling of rotary CF_OF_1-ATP synthase: storage of elastic energy during energy transduction. Biochim Biophys Acta 1412:118–128

Palmgren MG (2001) Plant plasma membrane H^+-ATPases: powerhouses for nutrient uptake. Annu Rev Plant Physiol Plant Mol Biol 52:817–845

Parra KJ, Kane PM (1993) Reversible association between the V_1 and V_O domains of the yeast vacuolar H^+-ATPase is an unconventional glucose-induced effect. Mol Cell Biol 18:17064–7074

Parra KJ, Keenan KL, Kane PM (2000) The *H* subunit (Vma13p) of the yeast V-ATPase inhibits the ATPase activity of cytosolic V_1 complexes. J Biol Chem 275:21761–21767

Parry RV, Turner JC, Rea PA (1989) High purity preparations of higher plant vacuolar H^+-ATPase reveal additional subunits. J Biol Chem 264:20025–20032

Perera IY, Li X, Sze H (1995) Several distinct genes encode nearly identical 16 kDa proteolipids of the vacuolar H^+-ATPase from *Arabidopsis thaliana*. Plant Mol Biol 29:227–244

Pfeifer W (1995) Effects of W-7, W-5, Verapamil and Diltiazem on vacuolar proton transport. Comparison of vacuolar H^+-ATPase and H^+-PPase from roots of *Zea mays*. Physiol Plant 94:284–290

Pope AJ, Leigh RA (1987) Some characteristics of anion transport at the tonoplast of oat roots, determined from the effects of anions on pyrophosphate-dependent proton transport. Planta 172:91–100

Puopolo K, Sczekan M, Magner R, Forgac M (1992) The 40-kDa subunit enhances but is not required for activity of the coated vesicle proton pump. J Biol Chem 267:5175–5176

Radermacher M, Ruiz T, Harvey WR, Wieczorek H, Grüber G (1999) Molecular architecture of *Manduca sexta* V_1 ATPase visualised by electron microscopy. FEBS Lett 453:383–386

Randall S, Sze H (1987) Probing the catalytic subunit of the tonoplast H^+-ATPase from oat roots: binding of 7-chloro-4-nitrobenzo-2-oxa-1,3-diazole to the 72-kilodalton polypeptide. J Biol Chem 262:7135–7141

Ratajczak R (2000) Structure, function and regulation of the plant vacuolar H^+-translocating ATPase. Biochim Biophys Acta 1465:17–36

Ratajczak R, Wilkins TA (2000) Energizing the tonoplast. In: Robinson DG, Rogers JC (eds) Annual Plant Review, Sheffield Academic Press, Sheffield, pp 133–173

Ratajczak R, Richter J, Lüttge U (1994) Adaptation of the tonoplast V-type H^+-ATPase of *Mesembryanthemum crystallinum* to salt stress, C_3-CAM transition and plant age. Plant Cell Environ 17:1101–1112

Ratajczak R, Hille A, Mariaux J-B, Lüttge U (1995) Quantitative stress response of the V_O -V_1-ATPase of higher plants detected by immuno-electron microscopy. Bot Acta 108:505–513

Ratajczak R, Fischer-Schliebs E (1997) Control of higher plant activities by electrochemical proton gradients set up by H^+ pumps at the plasma membrane and the tonoplast. In: Greppin H, Penel C, Simon P (eds) Travelling shot on plant development. Rochat-Baumann. Imprimerie Nationale, Genève, pp 183–199

Ratajczak R, Feussner I, Hause B, Böhm A, Parthier B, Wasternack C (1998) Alterations of V-type H^+-ATPase during methyljasmonate-induced senescence in barley (*Hordeum vulgare* cv. Salomé). J Plant Physiol 152:199–206

Ratajczak R, Hinz G, Robinson DG (1999) Localization of pyrophosphatase in membranes of cauliflower inflorescence cells. Planta 208:205–211

Ratajczak R, Pfeifer T, Drobny M, Schnölzer M, Lüttge U (2002) Molecular evidence for the occurrence of H^+-transporting V-ATPase subunit *D* and two different isoforms of subunit *E* in leaves of the obligate CAM species *Kalanchoë daigremontiana.* Biol Plant (in press)

Rausch T, Ziemann-Roth M, Hilgenberg W (1985) ADP is a competitive inhibitor of ATP-dependent H^+-transport in microsomal membranes of *Zea mays* L. coleoptiles. Plant Physiol 77:881–885

Rausch T, Kirsch M, Löw R, Lehr A, Viereck R, Zhigang A (1996) Salt stress responses of higher plants: the role of proton pumps and Na^+/H^+-antiporters. J Plant Physiol 148:425–433

Rea PA, Poole RJ (1985) Proton-translocating inorganic pyrophosphatase in red beet (*Beta vulgaris* L.) tonoplast vesicles. Plant Physiol 77:46–52

Rea PA, Poole RJ (1986) Chromatographic resolution of H^+-translocating pyrophosphatase from H^+-translocating ATPase of higher plant tonoplast. Plant Physiol 81:126–129

Rea PA, Sanders D (1987) Tonoplast energization: two H^+ pumps, one membrane. Physiol Plant 71:131–141

Rea PA, Poole RJ (1993) Vacuolar H^+-translocating pyrophosphatase. Annu Rev Plant Physiol Plant Mol Biol 44:157–180

Rea PA, Britten CJ, Jennings IR, Calvert CM, Skiera LA, Leigh RA, Sanders D (1992) Regulation of vacuolar H^+-pyrophosphatase by free calcium. A reaction kinetic analysis. Plant Physiol 100:1706–1715

Reuveni M, Bennett AB, Bressan RA, Hasegawa PM (1990) Enhanced H^+ transport capacity and ATP hydrolysis activity of the tonoplast H^+-ATPase after NaCl adaptation. Plant Physiol 94:524–530

Robinson DG (1996) Pyrophosphatase is not (only) a vacuolar marker. Trends Plant Sci 1:330

Robinson DG, Haschke H-P, Hinz G, Hoh B, Maeshima M, Marty F (1996) Immunological detection of tonoplast polypeptides in the plasma membrane of pea cotyledons. Planta 198:95–103

Robinson DG, Hoppenrath M, Oberbeck K, Luykx P, Ratajczak R (1998) Localization of pyrophosphatase and V-ATPase in *Chlamydomonas reinhardtii.* Bot Acta 111:108–122

Rockel B, Ratajczak R, Becker A, Lüttge U (1994) Changed densities and diameters of intra-membrane tonoplast particles of *Mesembryanthemum crystallinum* in correlation with NaCl-induced CAM. J Plant Physiol 143:318–324

Rockel B, Lüttge U, Ratajczak R (1998) Changes in message amount of V-ATPase subunits during salt-stress induced C_3-CAM transition in *Mesembryanthemum crystallinum.* Plant Physiol Biochem 36:567–573

Rodrigues CO, Scott DA, Docampo R (1999) Presence of a vacuolar H^+-pyrophosphatase in promastigotes of *Leishmania donovari* and its localisation to a different compartment from the vacuolar H^+-ATPase. Biochem J 340:759–766

Ros R, Romieu C, Gibrat R, Grignon C (1995) The plant inorganic pyrophosphatase does not transport K^+ in vacuole membrane vesicles multilabeled with fluorescent probes for H^+, K^+, and membrane potential. J Biol Chem 270:4368–4374

Rouquié D, Tournaire-Roux C, Szponarski W, Rossignol M, Doumas P (1998) Cloning of the V-ATPase subunit *G* in plant: functional expression and sub-cellular localization. FEBS Lett 437:287–292

Sagermann M, Stevens TH, Matthews BW (2001) Crystal structure of the regulatory subunit *H* of the V-type ATPase of *Saccharomyces cerevisiae.* Proc Natl Acad Sci USA 98:7134–7139

Salanoubat M, Lemcke K, Rieger M, Ansorge W, Unseld M, Fartmann B, Valle G, Blöcker H, Perez-Alonso M, Obermaier B, Delseny M, Boutry M, Grivell LA, Mache R, Puigdomènech P, De Simone V, Choisne N, Artiguenave F, Robert C, Brottier P, Wincker P, Cattolico L, Weissenbach J, Saurin W, Quétier F, Schäfer M, Müller-Auer S, Gabel C, Fuchs M, Benes V, Wurmbach E, Drzonek H, Erfle H, Jordan N, Bangert S, Wiedelmann R, Kranz H, Voss H, Holland R, Brandt P, Nyakatura G, Vezzi A, D'Angelo M, Pallavicini A, Toppo S, Simionati B, Conrad A, Hornischer K, Kauer G, Löhnert T-H, Nordsiek G, Reichelt J, Scharfe M, Schön O, Bargues M, Terol J, Climent J, Navarro P, Collado C, Perez-Perez A, Ottenwälder B, Duchemin D, Cooke R, Laudie M, Berger-Llauro C, Purnelle B, Masuy D, de Haan M, Maarse AC, Alcaraz J-P, Cottet A, Casacuberta E, Monfort A, Argiriou A, Flores M, Liguori R, Vitale D, Mannhaupt G, Haase D, Schoof H, Rudd S, Zaccaria P, Mewes H-W, Mayer KFX, Kaul S, Town CD, Koo HL, Tallon LJ, Jenkins J, Rooney T, Rizzo M, Walts A, Utterback T, Fujii CY, Shea TP, Creasy TH, Haas B, Maiti R, Wu D, Peterson J, Van Aken S, Pai G, Militscher J, Sellers P, Gill JE, Feldblyum TV, Preuss D, Lin X, Nierman WC, Salzberg SL, White O, Venter JC, Fraser CM, Kaneko T, Nakamura Y, Sato S, Kato T, Asamizu E, Sasamoto S, Kimura T, Idesawa K, Kawashima K, Kishida Y, Kiyokawa C, Kohara M, Matsumoto M, Matsuno A, Muraki A, Nakayama S, Nakazaki N, Shinpo S, Takeuchi C, Wada T, Watanabe A, Yamada M, Yasuda M, Tabata S (2000) Sequence and analysis of chromosome 3 of the plant *Arabidopsis thaliana.* Nature 408:820–822

Sakakibara Y, Kobayashi H, Kasamo K (1996) Isolation and characterization of cDNAs encoding vacuolar H^+-pyrophosphatase isoforms from rice (*Oryza sativa* L.). Plant Mol Biol 31:1029–1038

Sarafian V, Potier M, Poole RJ (1992) Radiation-inactivation analysis of vacuolar H^+-ATPase and H^+-pyrophosphatase from *Beta vulgaris* L. Biochem J 283:493–497

Sato MH, Maeshima M, Ohsumi Y, Yoshida M (1991) Dimeric structure of H^+-translocating pyrophosphatase from pumpkin vacuolar membranes. FEBS Lett 290:177–180

Sato MH, Kasahara M, Ishii N, Homareda H, Matsui H, Yoshida M (1994) Purified vacuolar inorganic pyrophosphatase consisting of a 75-kDa polypeptide can pump H^+ into reconstituted proteoliposomes. J Biol Chem 269:6725–6728

Scherer GFE, Martiny-Baron G, Stoffel B (1988) A new set of regulatory molecules in plants: a plant phospholipid similar to platelet-activating factor stimulates protein kinase and proton translocating ATPase in membrane vesicles. Planta 175:241–253

Schmidt AL, Briskin DP (1993) Energy transduction in tonoplast vesicles from red beet (*Beta vulgaris* L.) storage tissue: H^+/substrate stoichiometries for the H^+-ATPase and H^+-PPase. Arch Biochem Biophys 301:165–173

Schöcke L, Schink B (1998) Membrane-bound proton-translocating pyrophosphatase of *Syntrophus gentianae*, a syntrophically benzoate-degrading fermenting bacterium. Eur J Biochem 256:589–594

Schoonderwoert VTG, Martens GJM (2001) Proton pumping in the secretory pathway. J Membr Biol 182:159–169

Scott DA, de Souza W, Benchimol M, Zhong L, Lu HG, Moreno SN, Docampo R (1998) Presence of a plant-like proton-pumping pyrophosphatase in acidocalcisomes of *Trypanosoma cruzi*. J Biol Chem 273:22151–22158

Shimmen T, MacRobbie EAC (1987) Demonstration of two proton translocating systems in tonoplast of permeabilized *Nitella* cells. Protoplasma 136:205–207

Shiratake K, Kanayama Y, Maeshima M, Yamaki S (1997) Changes in H^+-pumps and a tonoplast intrinsic protein of vacuolar membranes during the development of pear fruit. Plant Cell Physiol 38:1039–1045

Sikora A, Hillmer S, Robinson DJ (1998) Sucrose starvation causes a loss of immunologically detectable pyrophosphatase and V-ATPase in the tonoplast of suspension-cultured tobacco cells. J Plant Physiol 152:207–212

Smart LB, Vojdani F, Maeshima M, Wilkins TA (1998) Genes involved in osmoregulation during turgor-driven cell expansion of developing cotton fibers are differentially regulated. Plant Physiol 116:1539–1549

Smith JAC, Marigo G, Lüttge U, Ball E (1982) Adenine-nucleotide levels during crassulacean acid metabolism and the energetics of malate accumulation in *Kalanchoë tubiflora*. Plant Sci Lett 26:13–21

Stevens TH, Forgac M (1997) Structure, function and regulation of the vacuolar (H^+)-ATPase. Annu Rev Cell Dev Biol 13:779–808

Stitt M (1987) Fructose 2,6-bisphosphate and plant carbohydrate metabolism. Plant Physiol 84:201–204

Stitt M (1998) Pyrophosphate as an energy donor in the cytosol of plant cells: an enigmatic alternative to ATP. Bot Acta 111:167–175

Supek F, Supekova L, Mandiyan S, Pan YC, Nelson H, Nelson N (1994) A novel accessory subunit for vacuolar H^+-ATPase from chromaffin granules. J Biol Chem 269:24102–24106

Supekova L, Supek F, Nelson N (1995) The *Saccharomyces cerevisiae* vma10 is an intron-containing gene encoding a novel 13-kDa subunit of vacuolar H^+-ATPase. J Biol Cem 270:13726–13732

Supekova L, Sbia M, Supek F, Ma Y-M, Nelson N (1996) A novel subunit of vacuolar H^+-ATPase related to the *b* subunit of F-ATPases. J Exp Biol 199:1147–1156

Suzuki K, Kasamo K (1993) Effects of aging on the ATP- and pyrophosphate-dependent pumping of protons across the tonoplast isolated from pumpkin cotyledons. Plant Cell Physiol 34:613–619

Suzuki Y, Maeshima M, Yamaki S (1999) Molecular cloning of vacuolar H^+-pyrophosphatase and its expression during the development of pear fruit. Plant Cell Physiol 40:900–904

Swanson SJ, Jones RL (1996) Gibberellic acid induces vacuolar acidification in barley aleurone. Plant Cell 8:2211–2221

Sze H, Ward JM, Lai S (1992) Vacuolar H^+-translocating ATPases from plants: structure, function, and isoforms. J Bioenerg Biomembr 24:371–381

Sze H, Xuhang L, Palmgren MG (1999) Energization of plant cell membranes by H^+-pumping ATPases: regulation and biosynthesis. Plant Cell 11:677–689

Tavakoli N, Eckerskorn C, Golldack D, Dietz K-J (1999) Subunit *C* of the vacuolar H^+-ATPase of *Hordeum vulgare*. FEBS Lett 456:68–72

Tavakoli N, Kluge C, Golldack D, Mimura T, Dietz KJ (2001) Reversible redox control of plant vacuolar H^+-ATPase activity is related to disulfide bridge formation in subunit *E* as well as subunit *A*. Plant J 28:51–59

Terrier N, Deguilloux C, Sauvage F-X, Martinoia E, Romieu C (1998) Proton pumps and anion transport in grape berries (*Vitis vinifera* L.): the inorganic pyrophosphatase plays a predominant role in energization of the tonoplast. Plant Physiol Biochem 36:367–377

Tomashek JJ, Graham LA, Hutchins MU, Stevens TH, Klionsky DJ (1997) V_1-situated stalk subunits of the yeast vacuolar proton-translocating ATPase. J Biol Chem 272:26787–26793

Tsiantis MS, Bartholomew DM, Smith JAC (1996) Salt regulation of transcript levels for the *c* subunit of a vacuolar H^+-ATPase in the halophyte *Mesembryanthemum crystallinum*. Plant J 9:729–736

Tu S-I, Nungesser E, Brauer D (1989) Characterization of the effects of divalent cations on the coupled activities of the H^+-ATPase in tonoplast vesicles. Plant Physiol 90:1636–1643

Tzeng CM, Yang CY, Yang SJ, Jiang SS, Kuo SY, Hung SH, Ma JT, Pan RL (1996) Subunit structure of vacuolar proton-pyrophosphatase as determined by radiation inactivation. Biochem J 316:143–147

Ubbink-Kok T, Boekema EJ, van Breemen JFL, Brisson A, Konings WN, Lolkema JS (2000) Stator structure and subunit composition of the V_1/V_O Na^+-ATPase of the thermophilic bacterium *Caloramator fervidus*. J Mol Biol 296:311–321

Vera-Estrella R, Barkla BJ, Bohnert HJ, Pantoja O (1999) Salt stress in *Mesembryanthemum crystallinum* L. cell suspensions activates adaptive mechanisms similar to those observed in the whole plant. Planta 207:426–435

Vianello A, Zancani M, Bradiot E, Petrussa E, Macri F (1991) Proton pumping inorganic pyrophosphatase of pea stem submitochondrial particles. Biochim Biophys Acta 1060:299–302

Viereck R, Kirsch M, Löw R, Rausch T (1996) Down-regulation of plant V-type H^+-ATPase genes after light-induced inhibition of growth. FEBS Lett 384:285–288

Walker RR, Leigh RA (1981) Mg-dependent, cation-stimulated inorganic pyrophosphatase associated with vacuoles isolated from storage roots of red beets (*Beta vulgaris* L.). Planta 153:150–155

Wang B, Lüttge U, Ratajczak R (2001) Effects of salt treatment and osmotic stress on V-ATPase and V-PPase in leaves of the halophyte *Suaeda salsa*. J Exp Bot 52:2355–2365

Wang BS, Ratajczak R, Zhang JH (2000) Activity, amount and subunit composition of vacuolar-type H^+-ATPase and H^+-PPase in wheat roots under severe NaCl stress. J Plant Physiol 157:109–166

Wang SY, Moriyama Y, Mandel M, Hulmes JD, Pan YC, Danho W, Nelson H, Nelson N (1988) Cloning of cDNA encoding a 32-kDa protein. An accessory polypeptide of the H^+-ATPase from chromaffin granules. J Biol Chem 263:17638–17642

Wang Y, Sze H (1985) Similarities and differences between the tonoplast-type and the mitochondrial H^+-ATPases of oat roots. J Biol Chem 260:10434–10443

Ward JM, Sze H (1992) Subunit composition and organization of the vacuolar H^+-ATPase from oat roots. Plant Physiol 99:170–179

Warren M, Smith JAC, Apps DK (1992) Rapid purification and reconstitution of a plant vacuolar ATPase using Triton X-114 fractionation: subunit composition and substrate kinetics of the H^+-ATPase from the tonoplast of *Kalanchoë daigremontiana*. Biochim Biophys Acta 1106:117–125

Weber J, Senior AE (1997) Catalytic mechanism of F_1-ATPase. Biochim Biophys Acta 1319:19–58

Weiner H, Stitt M, Heldt HW (1987) Subcellular compartmentation of pyrophosphate and alkaline phosphatase in leaves. Biochim Biophys Acta 893:13–21

White PJ, Marshall J, Smith JA (1990) Substrate kinetics of the tonoplast H^+-translocating inorganic pyrophosphatase and its activation by free Mg^{2+}. Plant Physiol 93:1063–1070

Wilkens S, Capaldi RA (1998) Electron microscopic evidence of two stalks linking the F_1 and F_O parts of the *Escherichia coli* ATP synthase. Biochim Biophys Acta 1365:93–97

Wilkens S, Vasilyeva E, Forgac M (1999) Structure of the vacuolar ATPase by electron microscopy. J Biol Chem 274:31804–31810

Williams LE, Nelson SJ, Hall JL (1992) Characterization of solute transport in plasma-membrane vesicles isolated from cotelydons of *Ricinus communis* L. I. Adenosine triphosphatase and pyrophosphatase activities associated with plasma-membrane fraction isolated by phase partitioning. Planta 182:532–539

Wink M (1997) Compartmentation of secondary metabolites and xenobiotics in plant vacuoles. In: Leigh RA, Sanders D (eds) The plant vacuole. Advances in botanical research, vol 25. Academic Press, San Diego, pp 141–169

Xu T, Vasilyeva E, Forgac M (1999) Subunit interactions in the clathrin-coated vesicle vacuolar (H^+)-ATPase complex. J Biol Chem 274:28909–28915

Yamanishi M, Kasamo K (1992) Binding of 7-chloro-4-nitrobenzo-2-oxa-1,3-diazole to an essential cysteine residue(s) in the tonoplast H^+-ATPase from mung bean (*Vigna radiata* L.) hypocotyls. Plant Physiol 99:652–658

Yamanishi M, Kasamo K (1993) Modulation of activity of purified tonoplast H^+-ATPase from mung bean (*Vigna radiata* L.) hypocotyls by various phospholipids. Plant Cell Physiol 34:411–419

Yamanishi M, Kasamo K (1994) Effect of cerebroside and cholesterol on reconstitution of purified tonoplast H^+-ATPase from mung bean (*Vigna radiata* L.) hypocotyls in liposomes. Plant Cell Physiol 35:655–663

Yamanishi M, Kasamo K (2001) Modulation in activity of purified tonoplast H^+-ATPase by tonoplast glycolipids prepared from cultured rice (*Oryza sativa* L. var. Boro) cells. Plant Cell Physiol 42:516–523

Yazaki Y, Asukagawa N, Ishikawa Y, Ohta E, Sakata M (1988) Estimation of cytoplasmic free Mg^{2+} levels and phosporylation potentials in mung bean root tips by in vivo ^{31}P NMR spectroscopy. Plant Cell Physiol 29:919–924

Yoshida S (1991) Chilling-induced inactivation and its recovery of tonoplast H^+-ATPase in mung bean cell suspension cultures. Plant Physiol 95:456–460

Zhang J, Myers M, Forgac M (1992) Characterization of the V_O domain of the coated vesicle (H^+)-ATPase. J Biol Chem 267:9773–9778

Zhang JM, Feng Y, Forgac M (1994) Proton conduction and bafilomycin binding by the V_O domain of the coated vesicle V-ATPase. J Biol Chem 269:23518–23523

Zhen R-G, Kim EJ, Rea PA (1997) The molecular and biochemical basis of pyrophosphate-energized proton translocation at the vacuolar membrane. In: Leigh RA, Sanders D (eds) The plant vacuole. Advances in botanical research, vol 25. Academic Press, Oxford, pp 298–337

Zhigang A, Löw R, Rausch T, Lüttge U, Ratajczak R (1996) The 32 kDa tonoplast polypeptide D_i associated with the V-type H^+-ATPase of *Mesembryanthemum crystallinum* L. in the CAM state: a proteolytically processed subunit *B*? FEBS Lett 389:314–318
Zingarelli L, Anzani P, Lado P (1994) Enhanced K^+-stimulated pyrophosphatase activity in NaCl-adapted cells of *Acer pseudoplantanus*. Physiol Plant 91:510–516
Zocchi G (1985) Phosphorylation-dephosphorylation of membrane proteins controls the microsomal H^+-ATPase activity of corn roots. Plant Sci 40:153–159

Dr. Martina Drobny
Dr. Elke Fischer-Schliebs
Prof. Dr. Ulrich Lüttge
Darmstadt University of Technology
Institute of Botany
Schnittspahnstrasse 3–5
64587 Darmstadt, Germany

e-mail: fischer-schliebs@bio.tu-darmstadt.de

Ca^{2+} Mobilization from Internal Stores in Electrical Membrane Excitation in *Chara*

Gerhard Thiel, Michael Wacke, and Ilse Foissner

1 Introduction

Stimulus-induced changes in the concentration of free Ca^{2+} in the cytoplasm (Ca^{2+}_{cyt}) have been recognized as key events in numerous physiological processes in plants. Recent reviews report how Ca^{2+}_{cyt} links environmental stimuli as well as hormonal signals to cellular processes such as the control of ion channels, cytoplasmic streaming, secretion and gene expression (Sanders et al. 1999; Blatt 2000; Reddy 2001). With increasingly elaborated methods it has now become clear that the signaling cascades that involve changes in Ca^{2+}_{cyt} are rather more complex than initially thought. Ca^{2+}_{cyt} can be found to increase in response to distinct stimuli to elevated steady-state levels of different amplitude. In other cases, Ca^{2+}_{cyt} was found to oscillate with characteristic frequencies and amplitudes (e.g., McAinsh et al. 1995; Allen et al. 2000, 2001; Plieth 2001). However, it is not only the patterns of Ca^{2+}_{cyt} elevation that appear to be complex, the mechanisms underlying such stimulus-induced changes also appear to be heterogeneous. In this context, it was found that Ca^{2+} can arrive in the cytoplasm from the external medium via channels in the plasma membrane (White 2000; Plieth 2001), via stimulus-coupled release from diverse internal stores (Sanders et al. 1999; Blatt 2000; Reddy 2001), or even from both (Grabov and Blatt 1998).

As much as the complexity of Ca^{2+} signaling has forced us to review our simple ideas of signal transduction cascades in plant cells, new data on stimulus response coupling have also fostered new views on the mechanisms linking relevant signals to changes in cytoplasmic Ca^{2+}. One illustrative example for this is that the plasma membrane of plant cells harbors voltage-sensitive Ca^{2+} channels (White 2000). It is thought that these channels are activated – very much like in animal cells – by depolarizing voltages to catalyze Ca^{2+} influx into the cytoplasm. On the other hand, some recent experiments in guard cells have revealed that it is membrane hyperpolarization rather than membrane depolarization that results in an increase in Ca^{2+}_{cyt} due to an activation of plasma

Progress in Botany, Vol. 64

membrane channels (Grabov and Blatt 1998). The message to learn from this is that signaling cascades in plants my bear some surprises that are not anticipated from simple models.

The present review addresses the question of signal response coupling in a well-investigated plant model system, namely, the electrical excitation in the green alga *Chara*. Some recent investigations reveal a detailed insight into the relationships between electrical stimulation of the cells and second messenger mediated mobilization of Ca^{2+} from internal stores.

2 Ca^{2+}_{cyt} and the Action Potential

Some of the first experimental evidence for a role of Ca^{2+}_{cyt} in signal transduction in plants came from research on the giant green alga *Chara*. From a mass of indirect experiments it had long been anticipated that Ca^{2+}_{cyt} plays a key role in the process of membrane excitation (for review, see Beilby 1984). The hard experimental support for this hypothesis came from the first real Ca^{2+}_{cyt} measurements in plants using luminescent detection of the Ca^{2+}-sensitive protein aequorin. The pioneering work of Williamson and Ashley (1982) and Kikuyama and Tazawa (1983) demonstrated with this technique that Ca^{2+}_{cyt} rises transiently during an action potential (AP). From electrophysiological measurements it was known that the Cl^- channels, which carry the depolarizing current in these cells (Homann and Thiel 1994), are activated by elevated Ca^{2+}_{cyt} (Katsuhara and Tazawa 1992). So it is most reasonable to assume that an electrically triggered rise in Ca^{2+}_{cyt} is the primary event in membrane excitation because it activates the Cl^- channels and thus membrane depolarization.

3 Where Does the Ca^{2+} Come From?

The question about the origin of the Ca^{2+}, which rises in the cytoplasm during the AP, seemed to be answered easily. Already the first experiments by Williamson and Ashley (1982) showed that replacement of Ca^{2+} by Mg^{2+} in the extracellular medium resulted in a loss of excitability and abolished the transient rise in Ca^{2+}. Together with the fact that stimulation of the AP is voltage-dependent, it was concluded that the plasma membrane of these cells contain voltage-dependent Ca^{2+} channels and that their activation is the primary event in membrane excitation (Katsuhara and Tazawa 1992). Similar experiments have been conducted in numerous plant systems to tackle the question as to whether cytoplasmic Ca^{2+} is affected by an influx of Ca^{2+} via activation of plasma membrane channels.

Although straightforward and attractive as a hypothesis the idea of voltage-activated Ca^{2+} channels is not corroborated by a critical scrutiny of the experimental data. Thus, for example, it was found that the Ca^{2+} channel blocker La^{3+} abolished excitation in characean cells (e.g., Beilby 1984). However, the interpretation that this is due to the inhibition of a plasma membrane Ca^{2+} channel is hampered by the fact that this blocker becomes active – even at high concentrations – only after more than 10 min of pretreating cells (Beilby 1984). If the primary event of excitation was a Ca^{2+} influx via voltage-sensitive Ca^{2+} channels, the effect of the blocker should be immediate. The same line of criticism can be derived from experiments in which the blocking of excitation by removal of Ca^{2+} from the external medium was interpreted as a key observation in favor of a voltage-dependent Ca^{2+} channel. In line with this view were experiments in which the replacement of extracellular Ca^{2+} by Mg^{2+} resulted in a loss of excitability and transient rise in Ca^{2+} in *Chara* cells. Adding Ca^{2+} back to the medium resulted in a restoration of excitability (Williamson and Ashley 1982). However, it took about 10 min before the cells regained excitability after readdition of Ca^{2+}. This is much too long to view a Ca^{2+} channel as the primary event in excitation (Williamson and Ashley 1982). If a simple voltage-gated Ca^{2+} channel was responsible for excitation the process should be reactivated the very second Ca^{2+} is added back to the external medium. Taken together these data show that extracellular Ca^{2+} is essential for excitation; however, apparently, influx via plasma membrane channels is not the primary source for the elevation of Ca^{2+}_{cyt} during excitation.

Some early experimental data, which supported a role of Ca^{2+} release from internal stores, came from Kikuyama and Tazawa (1983). They reported Ca^{2+} measurements in tonoplast-free *Chara* cells with a biphasic Ca^{2+}_{cyt} transient during the AP. The suggestion that the initial transient reflects release of Ca^{2+} from internal stores was based on the finding that this initial transient was abolished after multiple APs, even with sufficient Ca^{2+} in the external medium.

Animal physiologists have developed an experimental method to discriminate between release of Ca^{2+} from internal stores and Ca^{2+} influx via plasma membrane Ca^{2+} channels as a source of changes in Ca^{2+}_{cyt} (Merrit et al. 1989). The basis of this approach is that Mn^{2+} generally behaves similarly to Ca^{2+} and passes Ca^{2+}-permeable channels. Furthermore, Mn^{2+} shows a high affinity reaction with the Ca^{2+}-reporting fluorescent dye fura-2 in that Mn^{2+} quenches the fluorescence of the dye. Hence, when Mn^{2+} is present in the external medium, activation of Ca^{2+} channels in the plasma membrane will cause an influx of Mn^{2+} into the cytoplasm and a quenching of the fura-2 fluorescence. In the case, however, that the rise in Ca^{2+} derives from release of internal stores, no such quenching should be observed.

The same technique was also used to address the origin of Ca^{2+}_{cyt} during membrane excitation in *Chara*. The experiments revealed that the rise in Ca^{2+} during an AP was not associated with a quenching of the fura-2 fluorescence when Mn^{2+} was present in the external medium (Plieth et al. 1998). The interpretation of these data, namely, that the Ca^{2+} originates from internal stores, was further substantiated by experiments in which *Chara* cells were incubated in solutions with mM concentrations of Mn^{2+}. This treatment resulted in a quenching of the fura-2 fluorescence, an observation which confirms the presence of Mn^{2+} and therefore probably also Ca^{2+}-permeable channels in the plasma membrane of *Chara* (Plieth et al. 1998). Most surprising, though, was the finding that after a long period of washing cells with Mn^{2+}-free solution, APs were then associated with quenching of the fura-2 fluorescence. These data were interpreted as evidence that the Mn^{2+}, which entered the cytoplasm during the incubation in high Mn^{2+} concentrations, was sequestered together with Ca^{2+} into cytoplasmic stores (Plieth et al. 1998). Upon electrical stimulation this stored Mn^{2+}/Ca^{2+} is mobilized. It is worth noting that the quenching of the fura-2 fluorescence recovered with a time constant of about 250 s after its appearance. The reason for this is not yet clear. However, it is tempting to speculate that this relaxation of the quenching reflects a removal of Mn^{2+} from the cytoplasm probably by sequestering Mn^{2+} into the cytoplasmic stores.

The interpretation of the data on fura-2 quenching as evidence for a Ca^{2+} release from internal stores during excitation has been criticized on the grounds that the quenching kinetics was slower than that of Ca^{2+} elevation (Kikuyama and Tazawa 1998). This criticism cannot be ignored but, since the rise in Ca^{2+}_{cyt} coincides with the decrease in fluorescence (Plieth et al 1998), it is still very likely that the quenching of the fura-2 fluorescence reports the release of Ca^{2+}/Mn^{2+} from internal stores. For a quantitative correlation of the Ca^{2+} release and fura-2 quenching kinetics it would be necessary to know more about the relative permeation of the relevant channels and the binding kinetics of fura-2 to Ca^{2+} and Mn^{2+}.

4 Ca^{2+} Is Mobilized in an All-or-None Fashion

It is a principle feature of APs that they have a sharp activation threshold. This also holds true for the AP in *Chara*. Recordings of membrane currents in *Chara* revealed that voltage pulses, which did or did not trigger an AP, only differed by a few mV (Beilby and Coster 1979). This is only to show how narrow this threshold is. A further feature of APs is that they exhibit a strict strength/duration relation (Lühring and Tazawa 1985). This means that excitation is triggered by the product of pulse strength and duration. In other words, long pulses with small amplitude are as effective as short pulses with high amplitude. In the past it was thought that the narrow threshold reflects the voltage dependency of the relevant ion channels in the plasma membrane (Beilby and Coster 1979). A recent investigation now sheds new light on this interpretation. By recording the cytoplasmic Ca^{2+} concentration with fura-2 in response to graded electrical stimulations, it was found that the elevation of Ca^{2+}_{cyt} also reveals the same sharp threshold (Wacke and Thiel 2001). While small pulses evoke absolutely no detectable change in Ca^{2+}_{cyt}, the entire Ca^{2+} response is observed after passing a narrow threshold. This means that the mobilization of Ca^{2+} from internal stores occurs in an all-or-none like fashion (Wacke and Thiel 2001). This observation has important consequences for the understanding of the excitation process. The steep dependency of the Ca^{2+}_{cyt} response on the stimulating pulse as well as the all-or-none type behavior of the Ca^{2+} response is no longer in accordance with the view that voltage-dependent channels are primarily responsible for excitation. Voltage-gated channels do not exhibit such a sharp activation threshold (Hille 1992). Hence this excludes Ca^{2+} influx via voltage-sensitive Ca^{2+} channels in the plasma membrane as the source of the bulk Ca^{2+}_{cyt} changes during excitation. Consequently, we now have to interpret the threshold previously assigned to the activation properties of plasma membrane channels as a threshold for Ca^{2+} mobilization.

5 A Second Messenger Must Be Involved in Linking Electrical Stimulation and Ca^{2+} Mobilization

The quantitative assessment of the relationship between electrical stimulation and Ca^{2+} mobilization demands that a second messenger must be formed in response to the electrical stimulus and that this second messenger is responsible for the release of Ca^{2+} from internal stores. The main argument in support of this hypothesis is that sub-threshold electrical pulses were unable to evoke any kind of Ca^{2+} elevation in the cytoplasm. However, if two sub-threshold pulses were applied in an appropriate sequence, cells responded at the last pulse with the full Ca^{2+} response (Wacke and Thiel 2001). Because these pulses can be separated by several seconds, once again this observation cannot be interpreted in the context of a voltage-stimulated influx of Ca^{2+}. First, Ca^{2+}_{cyt} was not observed to rise in the cytoplasm following a sub-threshold stimulation. Furthermore, the lifetime of Ca^{2+} in the cytoplasm must be expected to be shorter than seconds, because of the efficient buffer systems of the cytoplasm.

The most plausible interpretation for this "memory" is that an intermediate signaling molecule is formed in response to the electrical stimulus. This molecule has a lifetime of a few seconds. In the case of a sub-threshold pulse, the concentration of this molecule remains under the critical threshold for Ca^{2+} mobilization. As a consequence, Ca^{2+}_{cyt} remains unchanged. If then a second sub-threshold pulse is applied, the newly formed messenger molecule is added to the remains from the first stimulation. In this way two sub-threshold pulses can act in an additive manner to levels of the signaling molecule above a critical value and stimulate Ca^{2+} release from internal stores.

In particular, experiments that used multiple sub-threshold pulses have uncovered interesting kinetic features of the underlying system.

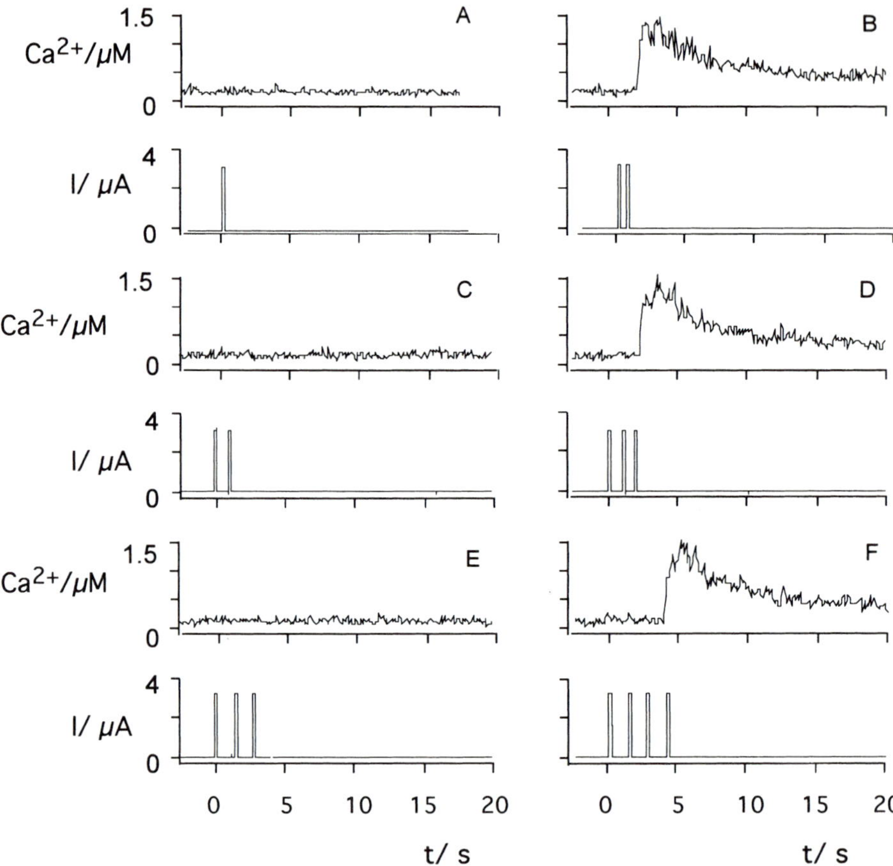

Fig. 1. Additive effect of sub-threshold pulses on stimulation of transient Ca^{2+} elevation. Ca^{2+}_{cyt} was measured in *Chara* internodal cells with the Ca^{2+}-sensitive dye fura-2-dextran. Plots illustrate measurements of Ca^{2+}_{cyt} in one internodal cell (*upper traces*) in response to variable pulse protocols (*lower traces*). A A single pulse is not effective, B a double pulse with 300-ms interval is effective in stimulating a Ca^{2+} response. C When the interval is increased to 700 ms, the pulses are no longer effective. D Addition of a third pulse after an interval of 700 ms again leads to stimulation of Ca^{2+} release at the last pulse. E If the pulse interval is increased to 1000 ms, the stimulation is no longer effective, F addition of a fourth pulse again makes the last pulse effective for Ca^{2+} mobilization

One finding was that sub-threshold pulses were only efficient in stimulating a rise in Ca^{2+} when the stimulations were not further apart than about 1 to 3 s (Wacke and Thiel 2001). The most simple explanation for this result was that the signaling molecule has a lifetime in the range of seconds. Thus a second pulse can only be efficient in triggering Ca^{2+} release if sufficient signaling molecules are left from the previous stimulation. This additive effect of sub-threshold pulses is illustrated in Fig. 1.

Surprising though was the observation that sub-threshold pulses also lost their efficiency in stimulating Ca^{2+} release if the two pulses were applied in too close proximity (Wacke and Thiel 2001). The best interpretation for this experimental result is that the production of the signaling molecule also can be rate-limiting.

6 Inositol-1,4,5-Trisphosphate ($InsP_3$) Is the Most Likely Candidate for the Second Messenger in Question

It was mentioned above that a second messenger with a lifetime in the order of some seconds must be involved as a link between the electrical stimulation and the mobilization of Ca^{2+} from internal stores. From work with animal cells we know a number of molecules that can act in response to relevant stimuli as second messenger for mobilization of Ca^{2+} from internal stores. Also in plant cells several of these molecules, including cyclic ADP-ribose, ryanodine and inositol-1,4,5-trisphosphate ($InsP_3$), were found to be active in mobilizing Ca^{2+} (Sanders et al. 1999; Blatt 2000; Reddy 2001).

The available experimental evidence suggests that $InsP_3$ is the second messenger in question for linking the electrical stimulus and Ca^{2+} mobilization in the AP in *Chara*. It was found that injection of $InsP_3$ into the cytoplasm of *Chara corallina* in some cases resulted in a stimulation of an action potential (Thiel et al. 1990). This does not proof that $InsP_3$ is also the physiological second messenger; nonetheless, it shows that the molecule has the potential to do so. More evidence for the role of $InsP_3$ in membrane excitation came from experiments with inhibitors of $InsP_3$ metabolism. The key enzyme for the production of $InsP_3$ is the phospholipase C (PLC). The activity of PLC can be inhibited by several drugs such as neomycin and U73122. To examine the effect of inhibition of $InsP_3$ on excitation, *Chara corallina* cells were challenged with both inhibitors (Biskup et al. 1999). By measuring the membrane currents during excitation it was found that both inhibitors resulted in a modification of the kinetics of the excitation current such that the peak of the current became progressively smaller and occurred later after start of the stimulation. Finally, cells completely lost their excitability. Ryanodin, on the other hand, had no effect on the membrane currents (Biskup et al. 1999).

The interpretation that an inhibition of $InsP_3$ synthesis results in a loss of Ca^{2+} mobilization can now be directly supported by measurements. The data in Fig. 2 show the transient rise in Ca^{2+} during an electrically stimulated action potential before and after addition of 100 μM neomycin to the bath medium. As predicted from the effect of the inhibitors on the excitation current, the transient Ca^{2+} elevation is abolished most likely as a consequence of the inhibition of $InsP_3$ synthesis. Nonetheless, from a critical view of the data, it must be noted that inhibition

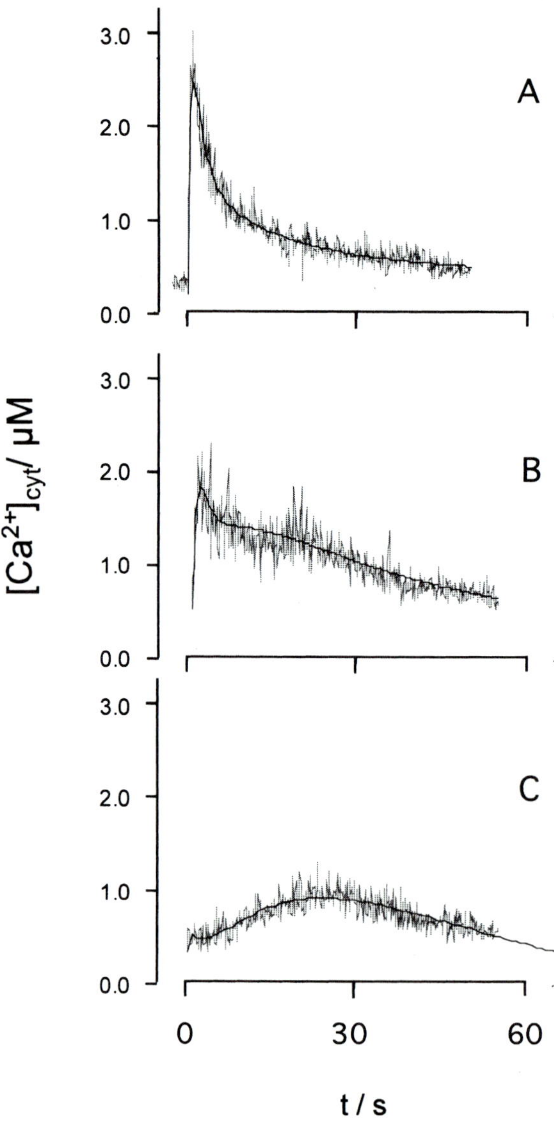

Fig. 2. Effect of neomycin on the electrically stimulated transient elevation of Ca^{2+}. Ca^{2+}_{cyt} was measured in *Chara* internodal cells with the Ca^{2+}-sensitive dye fura-2dextran. Transient changes in Ca^{2+} were elicited by 100 ms long pulses with an amplitude of 1 μA via extracellular electrodes. Ca^{2+} recordings before (A) and 20 s (B), 65 min (C) after addition of 100 μM neomycin to the bath medium

of $InsP_3$ synthesis only causes a gradual inhibition of Ca^{2+} mobilization. The Ca^{2+} response does not reveal an all-or-none behavior. This implies that $InsP_3$ is not the only factor which determines Ca^{2+} mobilization. In fact this notion is not all too surprising considering the known complex gating of the $InsP_3$-sensitive channel in animal cells (Tang et al. 1996).

Altogether, the experimental evidence highlights that the second messenger in question, that links the voltage stimulus and the mobilization of Ca^{2+} from internal stores, is $InsP_3$. This hypothesis would also fit the view that the respective second messenger has a lifetime in the order of some seconds. Measuring the lifetime of $InsP_3$ in animal cells for the molecule in the cytoplasm gave values in the order of 1.2–17 s (Wang et al. 1995; Fink et al. 1999). This is in the range of the lifetime expected for the second messenger involved in electrical excitation in *Chara*.

7 A Quantitative Model for $Ca^{2+}{}_{cyt}$ Changes in Electrical Excitation

Most of the details for spatiotemporal development of Ca^{2+} signals in animals have been extracted from kinetic models, which are able to successfully simulate the complex stimulus-induced changes in $Ca^{2+}{}_{cyt}$ (e.g., Goldbeter et al. 1990; Tang et al. 1996). Equivalent quantitative models for Ca^{2+} signaling in plants are so far rare (Bauer et al. 1998; Plieth 2001). Most of the ideas on oscillatory Ca^{2+} changes, for example, in guard cells and the hypothesis of Ca^{2+}-stimulated Ca^{2+} mobilization in plants in general are entirely descriptive (e.g., Ward and Schroeder 1994). It must therefore be stressed that a kinetic model, which was developed for the voltage-stimulated elevation of $Ca^{2+}{}_{cyt}$ in the AP in *Chara* (Wacke and Thiel 2001), presents a valuable tool to understand stimulus-induced Ca^{2+} signaling in a quantitative manner.

The experimentally observed complex dependency of Ca^{2+} mobilization on single or multiple electrical pulses in *Chara* could be explained by a mechanism in which the second messenger $InsP_3$ is produced transiently in a voltage-dependent manner (Fig. 3). This model was formulated as a straightforward kinetic model with three rate constants and the two pools of phosphatidyl-1,4,5,-bisphosphate (PIP_2) and $InsP_3$ connected in series and a voltage-dependent reaction k_2 determining the production of $InsP_3$. In this model, $InsP_3$ achieves mobilization of Ca^{2+} from internal stores in an all-or-none fashion once a threshold concentration of $InsP_3$ is exceeded. The mechanistic reason for the threshold is not yet understood, but it is likely that the complex activation of the Ca^{2+} channels in the internal stores by $InsP_3$ and Ca^{2+} (Tang et al. 1996) is responsible for this threshold-like behavior.

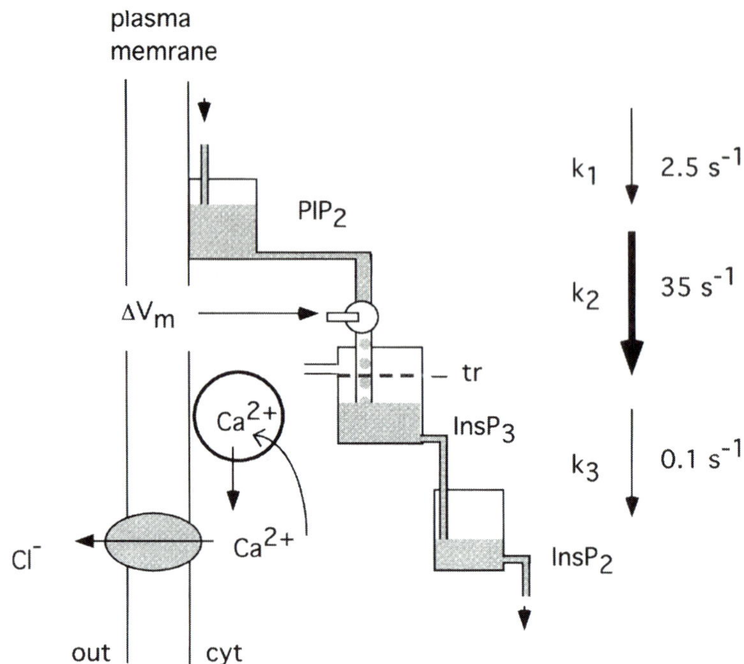

Fig. 3. Model for electrical membrane excitation involving the voltage-dependent production of the second messenger *$InsP_3$* and the subsequent mobilization of Ca^{2+} from internal stores. $InsP_3$ is produced with the rate constant k_2 from its precursor phosphatidyl 1,4,5-bisphosphate (*PIP_2*) and decomposed to inositol 1,4-bisphosphate (*$InsP_2$*) with the rate constant k_3. The rate constant k_2 is voltage dependent and when the $InsP_3$ concentration in the middle pool exceeds under electrical stimulation a critical threshold (*dashed line, tr*), Ca^{2+} is mobilized from internal stores. The elevation of $Ca^{2+}{}_{cyt}$ results in the activation of Ca^{2+}-sensitive Cl^- channels, and, thus, in membrane depolarization. $Ca^{2+}{}_{cyt}$ is removed from the cytoplasm via buffer systems probably by sequestering it back into the pools. The actual values for the rate constants, which are given in the figure, are estimated from fitting the model to experimental data

With this model the entire set of experimental data could be successfully simulated (Wacke and Thiel 2001). Thus, for example, we learn that the efficiency or failure of trains of sub-threshold electrical pulses depends on the relative velocities for the production and decomposition of $InsP_3$ as well as the availability of the precursor (i.e., PIP_2) of the second messenger $InsP_3$. For example, if two sub-threshold pulses are applied with too short intervals, stimulation fails to trigger Ca^{2+} mobilization. The model shows that stimulation fails because the pool of the $InsP_3$ precursor PIP_2 is drained so low after the first pulse that the second pulse is not able to produce sufficient new $InsP_3$ to exceed the threshold (Wacke and Thiel 2001). In the same way all other experimental data can be simulated. Furthermore, by using an indirect fitting procedure, it was possible to extract from the combination of model and experimental data the rate constants for the respective reactions. The estimated rate constants for the refilling of the PIP_2 pool, synthesis of $InsP_3$ and decomposition of $InsP_3$ to $InsP_2$ are shown in Fig. 3.

8 Identification of Ca^{2+} Stores

While it is quite clear that the $InsP_3$-gated channels in animals are located in the membrane of the endoplasmic reticulum (ER), the question as to the nature of the Ca^{2+} stores in plants has not yet been resolved. Naturally, the large vacuole of plant cells with its micromolar concentrations of Ca^{2+} has fostered the hypothesis that this compartment is the store of Ca^{2+} (Sanders et al. 1999; Reddy 2001). Patch clamp investigations on isolated vacuoles from higher plants have indeed suggested second messenger gated channels in the tonoplast of higher plants (Allen et al. 1995; Sanders et al 1995). However, there is also evidence from higher plants that other endo-compartments including the ER harbor stimulus-activated Ca^{2+} channels (Klüsener et al. 1995; Muir and Sanders 1997). In characean algae such intracellular Ca^{2+} stores other than the vacuole were detected in the context of mechano-sensitive stimulation of $Ca^{2+}{}_{cyt}$ (Tazawa et al. 1995). Fractionation techniques of the cytoplasm identified endo-compartments, probably the ER, but also the chloroplasts, as sources of Ca^{2+} release upon mechanic stimulation (Kikuyama and Tazawa 2001).

In the case of Ca^{2+} mobilization during the action potential in *Chara*, the vacuole can definitely be excluded as a functional Ca^{2+} store. The evidence against the vacuole as a Ca^{2+} store comes from experiments in which the Ca^{2+}-sensitive dye fura-2-dextran was injected into the cytoplasm and Mn^{2+} injected quantitatively into the vacuole of the same *Chara* cell. Upon triggering of an AP, Ca^{2+} was transiently elevated in the cytoplasm but without the fluorescence of the Ca^{2+}-sensitive dye being quenched (Plieth et al. 1998). Such a quenching would have been expected if Ca^{2+}/Mn^{2+} had been released from the vacuole.

Considering the morphology of *Chara*, in these cells the ER becomes the best candidate for a Ca^{2+} store involved in excitation. Electron and fluorescence microscopic examinations of characean algae uncover an extended network of cortical ER cisternae in close proximity to the plasma membrane (Fig. 4). The gap between the plasma membrane and the endoplasmic reticulum seems to be only a few nm (Fig. 4b). Examinations of the content of the cortical ER revealed that this compartment contains high concentrations of Ca^{2+} (Fig. 4c). This is not yet proof of the role of cortical ER as Ca^{2+} store. However, the presence of Ca^{2+} channels in the ER of plant cells (Klüsener et al. 1995), the high Ca^{2+} concentrations in this endo-compartment and the close proximity of cortical ER to the plasma membrane all support the view that this com-

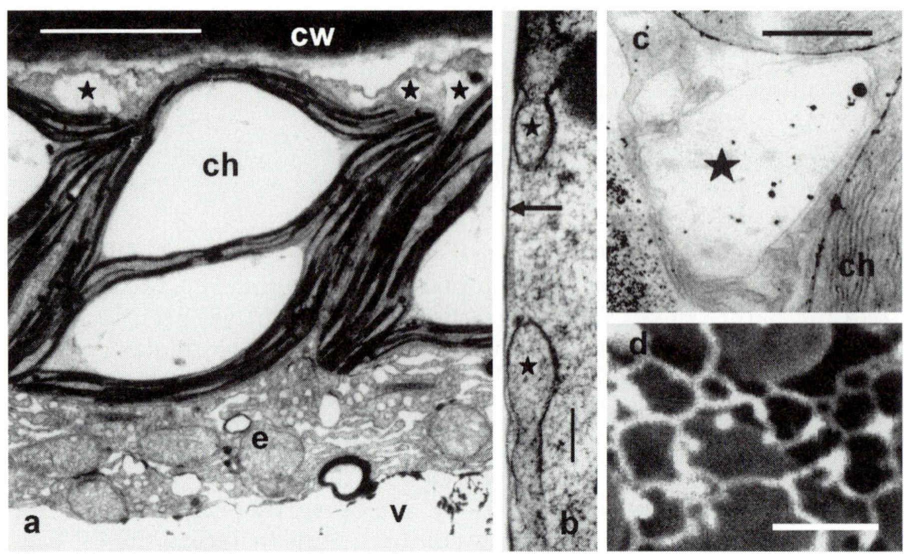

Fig. 4a–d. Electron (a–c) and fluorescence micrographs (d) of characean internodal cells. a Cross section through a conventionally fixed cell showing cell wall (*cw*), chloroplast layer (*ch*), endoplasm (*e*) and vacuole (*v*). *Asterisks* mark cortical ER cisternae (from Foissner 1989). b Detail of cortex in a high-pressure frozen and cryo-substituted cell. Note close contact of cortical ER cisternae (*asterisks*) with the plasma membrane (*arrow*). c Cortex of a cell fixed in the presence of potassium antimonate. Electron dense precipitates indicative of a high Ca^{2+} concentration are present in the cell wall (*cw*), along the chloroplast (*ch*) envelope and in an enlarged ER cisterna (*asterisk*). d Network of cortical ER cisternae visualized after staining with the fluorescent dye $DiOC_6$ (3,3-dihexyloxacarbocyanine). *Spherical structures* are mitochondria. *Bar* 2 μm (a), 150 nm (b), 500 nm (c), and 1.5 μm (d)

partment, or sub-compartments of the ER, are the functional Ca^{2+} stores in *Chara*.

9 Activity of Ca^{2+}-Sensitive Cl^- Channels as an Assay of Intracellular Ca^{2+} Release

An indirect, but powerful, way of monitoring changes in $Ca^{2+}{}_{cyt}$ near the plasma membrane with high spatiotemporal resolution relies on the recording of Ca^{2+}-sensitive channels by the patch clamp technique (Lipp and Niggli 1996). The advantage of these procedures is the ease of use, since Ca^{2+}-activated membrane currents can be measured relatively easily, e.g., in the cell attached configuration of the patch clamp technique. Another advantage of this approach is that the Ca^{2+} sensor – in this case the Ca^{2+}-sensitive channel – is harbored in the plasma membrane and thus "measures" only the Ca^{2+} concentration at the vicinity of the plasma membrane.

This indirect assessment of changes in the vicinity of the plasma membrane has been used extensively to uncover the elementary events of Ca^{2+} signaling in membrane excitation (Thiel et al. 1993; Homann and Thiel 1994). The principle of recording was the monitoring of those Cl^- channels which are activated during electrical membrane excitation. These channels are known to be activated by elevated concentrations of cytoplasmic Ca^{2+} (Okihara et al. 1991) and are, therefore, a good assay for addressing the question of near-membrane Ca^{2+} changes during excitation. Patch clamp recordings in the cell attached configuration revealed interesting insights into elementary Ca^{2+} events during excitation. It was first found that the respective Cl^- channels were activated in discrete "packages" (Thiel and Dityatev 1998). These packages revealed a distinct activation and inactivation kinetics, which was very similar to the changes in Cl^- conductance during the AP in the whole cell (Thiel et al 1997). This stresses that the transient Cl^- channel activity is the elementary unit of an AP or, in other words, the miniature form of an AP. Taking this transient activity of Cl^- channels as an indirect measure for the release of Ca^{2+} from internal stores, the following scenario can be deduced from the experimental results: The release of Ca^{2+} is favored by depolarizing voltages, but voltage alone is not sufficient for activation. This conclusion in the past already fostered the hypothesis that depolarizing voltages stimulate the production of an intermediate messenger molecule with a second long lifetime (Thiel et al. 1993; Thiel and Dityatev 1998) and that this messenger results in the release of Ca^{2+} from internal stores. Thus in this sense the measurements of Cl^- channel activity already predicted the direct measurements of $Ca^{2+}{}_{cyt}$ changes during excitation.

Recordings of the transient Cl^- channel activity revealed that within one patch something between a few channels to up to some hundred Cl^- channels of the same type were transiently activated (Homann and Thiel 1994). A detailed analysis of the variable transient Cl^- currents showed that the peak amplitudes of these currents could be grouped into distinct populations with approximately equidistant mean peak currents (Thiel and Dityatev 1998). Assuming that the activity of the respective Cl^- channels is determined by the release of Ca^{2+} from internal stores, this phenomenon suggests that the respective stores have a defined ("quantal") size. Upon stimulation, i.e., elevation of $InsP_3$, Ca^{2+} can be released from a single or from multiple stores. Individual stores appear to be located in close vicinity at a distance smaller than about 6–7 μm (Thiel and Dityatev 1998). Depending on the number of discharging stores, distinct numbers of Ca^{2+}-stimulated Cl^- channels activate.

10 Loose Ends and Hypothesis

The present hypothesis for an electrically stimulated mobilization of Ca^{2+} from internal stores that is mediated by the synthesis of the second messenger $InsP_3$ is able to explain a large body of experimental data. Nonetheless, there are still several questions to be answered and some experimental data are not directly explained by this model. One unresolved key question is how the synthesis of $InsP_3$ is related to the electrical stimulus. An important experimental observation, which is not yet covered by the model, is the fact that cells loose excitability after Ca^{2+} is removed from the external medium (Williamson and Ashley 1982). A plausible scenario, which could explain both the role of external Ca^{2+} and the link between membrane voltage and $InsP_3$ production, may be related to the properties of the phospholipase C. This enzyme is grouped in four families termed α–β–γ–δ. Members of the phospholipase C δ family, other than those of β and γ form, are not activated by GTP-binding proteins or phosphorylation, but by low concentrations of $Ca^{2+}{}_{cyt}$ (Pawelczyk and Matecki 1998). A phospholipase C was cloned from *Arabidopis* and potato that turned out to be sensitive to $Ca^{2+}{}_{cyt}$ (Hirayama et al. 1995; Kopka et al. 1998). Hence, it may be assumed that a gamma-type phospholipase C is present also in *Chara*. This enzyme may be activated by a very small influx of Ca^{2+} through voltage-sensitive plasma membrane channels. This influx of Ca^{2+} may cause a very local rise in $Ca^{2+}{}_{cyt}$ in the vicinity of the plasma membrane. Such a local rise in $Ca^{2+}{}_{cyt}$, which may not be picked up by the Ca^{2+}-recording technique, could then stimulate the production of $InsP_3$ and any further step illustrated above. The hypothesis would be consistent with the need of extracellular Ca^{2+} for triggering electrical excitation. It would further be in accordance with the electrophysiological data, which show a temporal uncoupling between membrane depolarization, activation of a Ca^{2+}-permeable channel with a very low conductance, and Cl^- channel activation (Thiel et al. 1993). The additional step of a Ca^{2+}-stimulated production of $InsP_3$ would be sufficient to explain this temporal uncoupling between voltage-dependent Ca^{2+} influx and global Ca^{2+} mobilization.

Acknowledgements. We thank Ulrike Homann (Darmstadt) and Christoph Plieth (Kiel) for critical reading of the manuscript.

References

Allen GJ, Muir SR, Sanders D (1995) Release of Ca^{2+} from individual plant vacuoles by both $InsP_3$ and cyclic ADP-ribose. Science 268:735–737

Allen GJ, Chu SP, Schumacher K, Shimazaki CT, Vafeados D, Kemper A, Hawke SD, Tallman G, Tsien RY, Harper JF, Chory J, Schroeder JI (2000) Alteration of stimulus-specific guard cell calcium oscillations and stomatal closing in Arabidopsis det3 mutant. Science 289:2338–2342

Allen GJ, Chu SP, Harrington CL, Schumacher K, Hoffmann T, Tang YY, Grill E, Schroeder JI (2001) A defined range of guard cell calcium oscillation parameters encodes stomatal movements. Nature 411:1053–1057

Bauer CS, Plieth C, Bethmann B, Popescu O, Hansen U-P, Simonis W, Schönknecht G (1998) Strontium-induced repetitive calcium spikes in a unicellular green alga. Plant Physiol 117:545–557

Beilby MJ (1984) Calcium in plant action potentials. Plant Cell Environ 7:415–421

Beilby MJ, Coster HGL II (1979) Two activation-inactivation transients in voltage clamp of the plasmalemma. Aust J Plant Physiol 6:323–335

Biskup B, Gradmann D, Thiel G (1999) Calcium release from $InsP_3$-sensitive stores initiates action potential in Chara. FEBS Lett 453:72–76

Blatt MR (2000) Ca^{2+} signalling and control of guard cell volume in stomatal movements. Curr Opin Plant Biol 3:196–204

Fink CC, Slepchenko B, Loew LM (1999) Determination of time-dependent inositol-1,4,5-trisphosphate concentrations during calcium release in smooth muscle cell. Biophys J 77:617–628

Foissner I (1989) Localization of calcium ions in wounded characean internodal cells. New Phytol 139:449–458

Goldbeter A, Dupont G, Berridg MJ (1990) Minimal model for signal-induced Ca^{2+} oscillations and for their frequency encoding through protein phosphorylation. Proc Natl Acad Sci USA 87:1461–1465

Grabov A, Blatt MR (1998) Membrane voltage initiates Ca^{2+} waves and potentiates Ca^{2+} increases with abscisic acid in stomatal guard cells. Proc Natl Acad Sci USA 95:4778–4783

Hille B (1992) Ionic channel of excitable membranes. Sinauer Associated, Sunderland, MA

Hirayama C, Ohto C, Mizoguchi T, Shinozaki K (1995) A gene encoding a phosphatidylinositol-specific phospholipase C is induced by dehydration and salt stress in *Arabidopsis thaliana*. Proc Natl Acad Sci USA 92:3903–3907

Homann U, Thiel G (1994) Cl^- and K^+ channel currents during the action potential in *Chara*. Simultaneous recording of membrane voltage and patch currents. J Membr Biol 141:297–309

Katsuhara M, Tazawa M (1992) Calcium-regulated channels and their bearing on physiological activities in characean cells. Philos Trans R Soc Lond B 338:19–29

Kikuyama M, Tazawa M (1983) Transient increase of intracellular Ca^{2+} during excitation of tonoplast-free *Chara cells*. Protoplasma 117:62–67

Kikuyama M, Tazawa M (1998) Temporal relationship between action potential and Ca^{2+} transient in characean cells. Plant Cell Physiol 39:1359–1366

Kikuyama M, Tazawa M (2001) Mechanosensitive Ca^{2+} release from intracellular stores in *Nitella flexilis*. Plant Cell Physiol 42:358–365

Klüsener B, Boheim G, Liss H, Engelberth J, Weiler EW (1995) Gadolinium-sensitive, voltage-dependent calcium release channels in the endoplasmic reticulum of a higher plant mechanoreceptor organ. EMBO J 14:2708–2714

Kopka J, Pical C, Gray JE, Müller-Röber B (1998) Temporal relationship between action potential and Ca^{2+} transient in characean cells. Plant Physiol 116:239–250

Lipp P, Niggli E (1996) A hierarchical concept of cellular and subcellular Ca^{2+}-signalling. Prog Biophys Mol Biol 65:265–296

Lühring H, Tazawa M (1985) Cytoplasmic Ca^{2+} has no effect on the excitability of *Chara* plasmalemma. Plant Cell Physiol 26:769–774

McAinsh MR, Webb AAR, Taylor JE, Hetherington AM (1995). Stimulus induced oscillations in guard cell cytosolic free calcium. Plant Cell J 9:297–304

Merrit JE, Jacob R, Hallam TJ (1989) Use of manganese to discriminate between calcium influx and mobilization from internal stores in stimulated human neurophils. J Biol Chem 264:377–382

Muir SR, Sanders D (1997) Inositol 1,4,5-trisphosphate-sensitive Ca^{2+} release across nonvacuolar membranes in cauliflower. Plant Physiol 114:1551–1521

Okihara K, Ohkawa T, Tsutsui I, Kasai M (1991) A Ca^{2+}- and voltage-dependent Cl^- sensitive anion channel in the *Chara* plasmalemma: a patch-clamp study. Plant Cell Physiol 32:593–602

Pawelczyk T, Matecki A (1998) Localization of phospholipase C delta3 in the cell and regulation of its activity by phospholipids and calcium. Eur J Biochem 257:169–177

Plieth C (2001) Plant calcium signaling and monitoring: pros and cons and recent experimental approaches. Protoplasma 218:1–23

Plieth C, Sattelmacher B, Hansen U-P, Thiel G (1998) The action potential in *Chara*: Ca^{2+} release from internal stores visualized by Mn^{2+} induced quenching of fura dextran. Plant J 13:167–175

Reddy ASN (2001) Calcium: silver bullet in signalling. Plant Sci 160:381–404

Sanders D, Muir SR, Allen GJ (1995) Ligand- and voltage-gated calcium release channels at the vacuolar membrane. Biochem Soc Trans 23:856–61

Sanders D, Brownlee C, Harper JF (1999) Communicating with calcium. Plant Cell 11:691–706

Tang Y, Stephenson JL, Othmer HG (1996) Simplification and analysis of models of calcium dynamics based on IP3-sensitive calcium channel kinetics. Biophys J 70:246–263

Tazawa M, Shimada K, Kikuyama M (1995) Cytoplasmic hydration triggers a transient increase in cytoplasmic Ca^{2+} concentration. Plant Cell Physiol 36:335–340

Thiel G, Dityatev A (1998) Transient activity of excitatory Cl^- channels in *Chara*: evidence for quantal release of a gating factor. J Membr Biol 163:183–191

Thiel G, MacRobbie EAC, Hanke DE (1990) Raising the intracellular level of inositol 1,4,5-trisphosphate changes plasma membrane ion transport in characean algae. EMBO J 9:1737–1741

Thiel G, Homann U, Gradmann D (1993) Microscopic elements of electric excitation in *Chara*, transient activity of Cl^- channels in the plasma membrane. J Membr Biol 134:53–66

Thiel G, Homann U, Plieth C (1997) Ion channel activity during the action potential in *Chara*: a new insight with new techniques. J Exp Bot 48: 609–622

Wacke M, Thiel G (2001) Electrically triggered all-or-none Ca^{2+}-liberation during action potential in the giant alga *Chara*. J Gen Physiol 118:11–21

Ward JM, Schroeder JI (1994) Calcium-activated K^+ channels and calcium-induced calcium release by slow vacuolar channels in guard cell vacuoles implicated in the control of stomatal closure. Plant Cell 6:669–683

Wang SS-H, Alousi AA, Thompson SH (1995) The life time of inositol 1,4,5-trisphosphate in single cells. J Gen Physiol 105:149–171

White PJ (2000) Calcium channels in higher plants. Biochim Biophys Acta 1465:171–89

Williamson RE, Ashley CC (1982) Free Ca^{2+} and cytoplasmic streaming in alga *Chara*. Nature 296:647–651

Gerhard Thiel
Michael Wacke
Institut für Botanik
Technische Universität Darmstadt
Schnittspahnstraße 3
64287 Darmstadt, Germany

e-mail: thiel@bio.tu-darmstadt.de

Ilse Foissner
Institut für Pflanzenphysiologie
Universität Salzburg
Hellbrunnerstraße 34
5020 Salzburg, Austria

Photosynthesis. Carbon Metabolism: Quantification and Manipulation

Grahame J. Kelly

"And, he gave it for his opinion, that whoever could make two ears of corn, or two blades of grass to grow upon a spot of ground where only one grew before, would deserve better of mankind, and do more essential service to his country, than the whole race of politicians put together."
Jonathan Swift in Gulliver's Travels, ca. 1727

This review is dedicated to Professor Martin Gibbs on the occasion of his 80th birthday

1 Introduction

Two philosophies have always guided research on photosynthetic carbon metabolism. The first is based on the desire to satisfy scientific curiosity by elucidating the chemical reactions and metabolic pathways involved, and the metabolic interconnections between chloroplasts, other parts of the cell, and the rest of the photosynthetic organism. An impressive overview in this regard has just been published (Cen et al. 2001). The second is the expectation that a thorough understanding of photosynthesis might in some way lead to benefits for humanity, the most obvious contender being faster growing crop plants that provide more food and fibre. As the years and decades have passed, the pendulum of emphasis has swung slowly but consistently away from the first and toward the second of these philosophies, and is perhaps now midway between the two. In this review, the literature relating to chemical reactions and metabolic pathways is given the most attention, but it will be noted that the latest research contains few new reactions and no new pathways. Is it possible that, finally, they have all been discovered? On the other hand, the literature relating to the second philosophy is given some attention because it deals with those topics that are linking photosynthetic carbon metabolism to recurring public news items of the type that have "GM foods" and "Kyoto Protocol" in their titles. The business of genetic manipulation of photosynthesis for increased productivity of food crops has been initiated (Sect. 5), but it is wise to point out from the

Progress in Botany, Vol. 64

start that the potential of this business is still largely unknown. The business of quantifying the biosphere's total photosynthesis, in both today's atmosphere and a future atmosphere containing extra CO_2, and of quantifying the sinks in which photosynthetically fixed carbon is sequestered on long-term scales, is also proceeding actively (Malhi and Grace 2000; Field 2001; Schlesinger and Lichter 2001), because these values are factored into the rules and calculations of the Kyoto Protocol (Schiermeier 2001). It is to be hoped, however, that readers of this review will find items that are fascinating not because of their relevance to these applied matters, but purely because they bring to light new tantalising facets of the wonder that is photosynthesis. The literature covered is predominantly that which has appeared since the last review in Volume 62 of Progress in Botany.

2 Carbon Metabolism Common to All Photosynthetic Cells

a) The Ribulose Bisphosphate (RuBP) Carboxylation System

Attention given to the Calvin cycle's CO_2-fixing enzyme ribulose-1,5-bisphosphate carboxylase/oxygenase (Rubisco) has been less than normal during the past 2 years, and an almost equal amount of attention has been given to the protein (Rubisco activase) that assists its activation (see below). Concerning Rubisco itself, Schlitter and Wildner (2000) have hypothesised that the enzyme's greater specificity for CO_2 (leading to photosynthesis) compared with that for O_2 (leading to photorespiration) is based on the rapidity with which the enzyme undergoes the conformational change that opens the niche in which the gaseous substrate binds. The content and activity of Rubisco in tissues are often difficult to estimate, so new approaches using tritium-labelled RuBP (Ferreira et al. 2000) and a trichloroacetic acid based procedure for determining the amount of enzyme lost during extraction (Rogers et al. 2001) may be welcomed by some researchers, including those giving the enzyme a major place in models directed toward quantifying the global carbon cycle (Bernacchi et al. 2001).

Algal and bacterial Rubiscos have become of special interest because their study has shown that the evolutionary diversity of the enzyme is much broader than indicated by studies of the plant enzyme. Some algae and photosynthetic bacteria contain an eight-large-subunit/eight-small-subunit Rubisco similar to the plant enzyme (termed the "green form I"), others contain a "red form I" in which the amino acid sequences of the two types of subunits are considerably different to those of the green form I, and some (but not all) dinoflagellates contain a "form II" Rubisco composed of eight large subunits only (Horken and Tabita 1999; Jenks and Gibbs 2000). Another difference, first observed in the 1980s, was that the two genes for the two types of subunit are both in the chloroplast in algae such as *Olisthodiscus luteus* (see Jenks and Gibbs 2000), in contrast to higher plants where the large subunit is encoded in the chloroplast but the small subunit is encoded in the nucleus. Whitney and Andrews (2001) recently attempted to create the *O. luteus* situation in a higher plant by relocating the Rubisco small subunit gene to the chloroplast in tobacco. They were moderately successful in that totally chloroplast-encoded Rubisco was obtained, but only at a level of about 1% of the total Rubisco; the tobacco synthesised the remaining 99% conventionally. A final note on algal Rubiscos relates to a field observation: in the dinoflagellate alga *Gonyaulax* sp., Rubisco moves into and out of pyrenoids in a circadian fashion (Nassoury et al. 2001), thereby contributing to this alga's earlier detected circadian rhythm in photosynthesis.

Recent research on the regulation of Rubisco activity in higher plants has centred almost exclusively on the protein Rubisco activase. The activase catalyses the early-morning dissociation of inhibitory sugar-Ps, such as RuBP (when previously bound to the inactive form of the enzyme) and, in some species, carboxyarabinitol-1-P, from Rubisco. These inhibitors maintain the enzyme in an inactive form during the night, and, as recently observed by Khan et al. (1999), may protect it against proteolytic breakdown. The activase presumably becomes effective only after sunrise because it undergoes a light-mediated activation very similar to that experienced by the Calvin cycle's fructose-1,6-bisphosphatase (Zhang et al. 2001). However, it seldom manages to fully activate all the Rubisco in leaves because its capacity is finite, and any synthesis of more of it would entail a trade-off of less Rubisco synthesis such that photosynthesis might decline rather than increase (Mott and Woodrow 2000). Clearly, all of a leaf's activase is needed to maintain photosynthesis, and this is dramatically demonstrated when the notable high-temperature sensitivity of the activase is examined. At temperatures around 40–45 °C (which are not unrealistic: it was 42 °C in Brisbane on Christmas Day last year!), the activase is thermally denatured (Salvucci et al. 2001) and also induced to associate with thylakoid membranes (Rokka et al. 2001). Consequent declines of about 50% in photosynthetic rates are observed (Crafts-Brandner and Law 2000; Crafts-Brandner and Salvucci 2000; Sharkey et al. 2001).

b) Other Enzymes of Chloroplast Carbon Metabolism

Recent research reports focused directly on the properties of the remaining ten enzymes of the Calvin cycle and other enzymes of chloroplast carbon metabolism have been few in number; one detailed study of spinach ribose-5-P isomerase has been provided by Jung et al. (2000). Two studies of the ubiquitous enzyme aldolase relate to stresses: the enzyme occurs as slightly different isozymes in chloroplasts, effectively indistinguishable except that certain isozymes become more predominant when the plant is stressed by salinity or high temperature (Michelis and Gepstein 2000; Yamada et al. 2000). A similar situation exists for chloroplast fructose-1,6-bisphosphatase (FBPase), in that the rice enzyme is salt-resistant when isolated from a salt-tolerant variety, but not when isolated from a salt-sensitive variety (Ghosh et al. 2001). It is well known that this chloroplast FBPase undergoes light-mediated reductive activation mediated by reduced thioredoxin *f*. The FBPase, initially inactive because a disulfide bridge severely disrupts the catalytic site (Chiadmi et al. 1999), transiently forms a mixed disulfide with the thioredoxin during the process in which the disulfide bridge is cleaved and reduced to two sulfhydryl groups (Balmer and Schürmann 2001). Thioredoxin is an incredible protein with many other roles, as pointed out in our 1998 review; a new addition is that it has a vital role in the complex interplay between stigma and pollen proteins that mediate self-incompatibility during sexual reproduction in many plant species (Cabrillac et al. 2001). Returning to leaves, another enzyme that undergoes light-mediated activation is P-ribulokinase, but it seems that this activation contributes to the regulation of photosynthesis only in low-light grown plants suddenly exposed to high light (Paul et al. 2000). Finally, three studies relating to the first two enzymes in the oxidative pentose-P pathway in chloroplasts have appeared. A new isoform of plastidic glucose-6-P dehydrogenase has been discovered by Wendt et al. (2000), while the delicate 6-P-gluconate dehydrogenase in chloroplasts has finally been purified from spinach by Krepinsky et al. (2001). A principal function of these two enzymes is to generate NADPH, the supply of which can be inadequate in situations such as low nitrogen availability (Robinson 2000).

c) The Three Respirations of Photosynthetic Cells: Mitochondrial Respiration, Chlororespiration, and Photorespiration

Photosynthetic cells are unique in possessing not one, but three respiratory activities. This makes their metabolism interesting, but also difficult to analyse. Haupt-Herting et al. (2001) have developed a new approach that may help to differentiate mitochondrial and photorespiratory CO_2 releases. During photosynthesis, mitochondrial respiration is reduced to a variable degree (Atkin et al. 2000), but it must continue, at least to some extent, because its effect on the cell's metabolite profile and redox status seems to be important for maintaining proper light-mediated regulation of enzymes and sucrose biosynthesis (Padmasree and Raghavendra 1999, 2001) and because the Krebs cycle is the immediate source of some of the carbon skeletons for amino acid biosynthesis following nitrate reduction (Cen et al. 2001). Not all of the CO_2 evolved from the Krebs cycle is necessarily lost from the plant: some may be recaptured by photosynthesis, especially in drought-stressed leaves (Haupt-Herting et al. 2001) and the glumes of cereal ears (Gebbing and Schnyder 2001).

Chlororespiration, like mitochondrial respiration, involves the oxidation of organic molecules and consequent generation of NAD(P)H which then feeds reducing equivalents into a membrane-located electron transport chain that delivers them finally to O_2. The major difference is that the membrane is the chloroplast's thylakoid membrane, rather than the mitochondrion's inner membrane (Fig. 1). Of course, being the thylakoid membrane, it is not difficult to envisage that some of the electron-transporting components might be the same as those used in photosynthesis. The best-established example is plastoquinone, and the central version of chlororespiration involves little more than two enzymes, the first of which catalyses the transfer of the reducing equivalents from NAD(P)H to this plastoquinone (see Carol and Kuntz 2001) while the second, a plastoquinol oxidase, catalyses a reaction between the resultant plastoquinol and O_2, regenerating the plastoquinone and releasing water (Cournac et al. 2000). A possible ancestor of the latter enzyme has been detected in a cyanobacterium by Büchel et al. (1998). Improved evidence for chlororespiration in leaf chloroplasts in darkness has become available (Feild et al. 1998), but of greater interest are three variants of the chlororespiratory process (Fig. 1). In the first, the organic molecule oxidised is not a carbohydrate (as in conventional respiration),

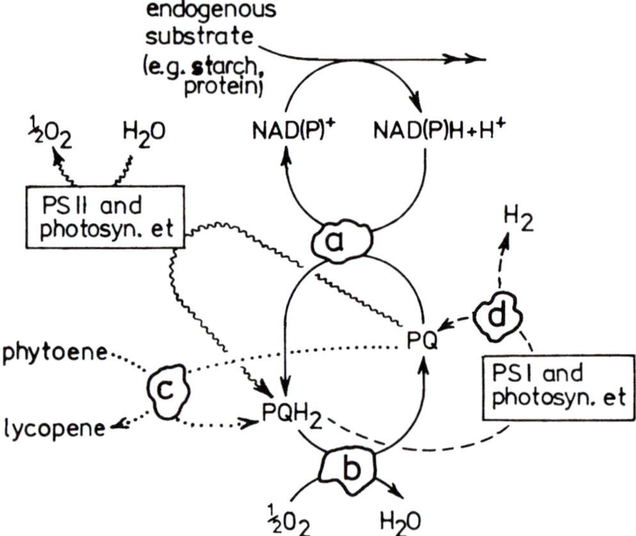

Fig. 1. Chlororespiration (*solid lines*) on the *Chlamydomonas* thylakoid membrane, and its connections with carotenoid biosynthesis (*dotted lines*), anaerobiosis-induced H_2 evolution (*dashed lines*), and a water-to-water cycle when photosystem I is missing (*squiggly lines*). *a* NAD(P)H-plastoquinone oxidoreductase; *b* plastid terminal oxidase; *c* phytoene desaturase; *d* hydrogenase. *photosyn. et* photosynthetic electron transport; *PQ* plastoquinone; *PS* photosystem. (Adapted from Cournac et al. 2000; Melis et al. 2000; Carol and Kuntz 2001)

but rather the carotenoid precursor phytoene, which undergoes four cycles of desaturation to generate lycopene and (after ring cyclisation) β-carotene (Bennoun 2001; Carol and Kuntz 2001). In the second, the reducing equivalents do not come from an organic molecule (and it is, therefore, strictly speaking, not a form of respiration), but rather from water which is photosynthetically split by photosystem II in the light (Grotjohann et al. 1999). Clearly, this process must be of low magnitude in normal cells (or else there would be no photosynthesis), but its capacity is considerable, as indicated by a *Chlamydomonas* mutant that is deficient in photosystem I and thus has nowhere else to deposit the reducing equivalents (Cournac et al. 2000). The third variant appears in certain green microalgae (and also certain cyanobacteria) that are capable of synthesising the enzyme hydrogenase. When these cells are placed in an anaerobic environment (thus precluding the transfer of the reducing equivalents from the plastoquinol to O_2; again, strictly speaking, the process is no longer a true respiration), they simply synthesise hydrogenase and use it to recombine the electron and proton parts of the reducing equivalents into H_2 gas (which is evolved). When this metabolism is linked in some way to the original release of hydrogen from water splitting by the photochemical apparatus of these cells (not an easy task because the hydrogenase is quite sensitive to inactivation by O_2, the other product of water splitting), then the application of photosynthesis to the production of a clean and renewable fuel (H_2) for humanity can be envisaged. Such an objective, long championed by Martin Gibbs, has recently come one step closer to reality (Melis et al. 2000).

The third form of respiration in photosynthetic cells, photorespiration, differs from the above two in that there is minimal involvement of cell membranes. As described in our earlier reviews, it is composed of the glycolate cycle (a series of eight reactions distributed in the soluble phases of chloroplasts, mitochondria and peroxisomes) connected to the Calvin cycle at the Rubisco reaction. A potential role of photorespiration is the consumption of excess ATP and NADPH and consequent avoidance of photoinhibition. More evidence supporting this role has come from a mutant barley with low glycine decarboxylase activity (Igamberdiev et al. 2001a), from a tobacco transgenic for low glycine decarboxylase activity (Yamaguchi and Nishimura 2000), and from plants under stress situations: tobacco exposed for 1 week to a low O_2 atmosphere (Migge et al. 1999), winter wheat exposed to high light at 20 °C (Savitch et al. 2000), and the rainforest plant *Arisaema heterophyllum* exposed to high light (Muraoka et al. 2000). A different facet of stress vis-à-vis photorespiration has been raised by Sivakumar et al. (2000) who found that the proline accumulated by plants in response to salt stress interacts with Rubisco in such a way as to counteract a salt-induced enhancement of the oxygenase activity of that enzyme.

Enzymes of the glycolate cycle are still attracting interest. Iwamoto and Ikawa (2000) remind us that two types of glycolate-oxidising enzymes (oxidase or dehydrogenase) exist in algae, sometimes even within the same algal group. Liepman and Olsen (2001) have DNA-based evidence that just one protein may accommodate the two transamination reactions required in the peroxisome. Finally, the tremendously high concentration and activity (five-times that of the Krebs cycle) of the four-enzyme complex that decarboxylates two glycines to one serine in the mitochondrial matrix has been re-emphasised by Douce et al. (2001). The sophistication of the complex is such that it contributes to the fractionation between ^{12}C and ^{13}C in photosynthetic cells (Igamberdiev et al. 2001b). This apparently happens despite the fact that most of the released CO_2 never gets to exit the leaf, but is returned to Rubisco (Loreto et al. 1999), an event that is intuitively intriguing but, of course, does not alter the basic premise that photorespiration can be viewed as a wasteful loss (as CO_2) of previously fixed CO_2.

d) Starch

The diurnal diversion of some of the product of the Calvin cycle into transitory starch in the chloroplast is achieved by the sequential action of hexose-P isomerase, P-glucomutase, ADP-glucose pyrophosphorylase, and a collection of starch synthases and associated enzymes that sculpture starch molecules. Indirect evidence for the indispensability of P-glucomutase (Tauberger et al. 2000; Fernie et al. 2001) has been complemented by the conclusion of Yu et al. (2000) that the hexose-P isomerase is also essential. Of course, the ADP-glucose pyrophosphorylase is essential because it forms the glucose-donor molecule (ADP-glucose). In leaves, all of this enzyme is in the chloroplast. It almost certainly influences the rate of starch synthesis through its regulatory properties (see our earlier reviews) and the overall amount present that, in turn, may be influenced by the mysterious sugar trehalose (Wingler et al. 2000). Another recent surprise has been the detection by Rodriguez-López et al. (2000) of an enzyme that destroys ADP-glucose by hydrolysing it into AMP and glucose-1-P, thereby potentially hindering starch synthesis. The role of this new enzyme is unclear. It might be envisioned that it could prevent starch biosynthesis from becoming excessive (Kleczkowski 2001), but its capacity to do this might be questioned given that chloroplasts are well known for their proclivity to accumulate excessive amounts of starch, e.g. after removal of sink organs (Nakano et al. 2000), or when the supply of CO_2 to the leaf is luxurious (Sawada et al. 2001). Such starch accumulation is also hard to reconcile with the proposal that the growth of the starch granule is kept under control by a refined regulatory system orchestrated by certain 14-3-3 proteins in the granule (Sehnke et al. 2001). [Note: "14-3-3" is a somewhat obscure term derived from the loci of migration of these proteins in a double chromatography system; the proteins are ubiquitous in nature, and usually enmeshed in intracellular molecular signalling and regulatory systems that involve the phosphorylation and dephosphorylation of other proteins (Voigt et al. 2001)].

Two points concerning the earlier surprise (see our 1998 review) that most ADP-glucose pyrophosphorylase is located in the cytosol, rather than the plastid, of non-photosynthetic starch-storing organs have come to light. The first is that this phenomenon is restricted to the seed endosperm of graminaceous species (Beckles et al. 2001a). An earlier report that it applies to developing tomato fruit has been challenged (Beckles et al. 2001b). The second is that, while this cytosolic form of the enzyme commonly lacks (Doan et al. 1999) the allosteric regulatory properties typical of the chloroplast, and particularly cyanobacterial (Gómez Casati et al. 2000) enzyme, this feature may not be universal: Sikka et al. (2001) report that the rice-endosperm enzyme is subject to allosteric regulation.

The manufacture of amylose and amylopectin by starch synthases and branching enzyme, and the reshaping of amylopectin by debranching enzyme, were described in our last review. The fate of the short chains of glucoses (termed maltooligosaccharides) released by the debranching enzyme can now be considered. One role is that they can be used as a start-up substrate by the granule-bound starch synthase, for elongation into amylose molecules (Denyer at al. 1999). A second, somewhat fascinating, role has been considered by Takaha et al. (1998), who propose that yet another enzyme, named disproportionating enzyme or "D-enzyme", can remove a piece of one maltooligosaccharide and add it to another. The latter (lengthened at the expense of the former) is then long enough to be recognised by a well-known enzyme of old, starch phosphorylase, that proceeds to cleave glucose-1-Ps from it. These are then able to re-enter starch synthesis via the ADP-glucose pyrophosphorylase reaction. The beauty of this scheme is that it explains how starch phosphorylase might participate in starch biosynthesis in chloroplasts and amyloplasts, an event long entertained by investigators of this enzyme (Albrecht et al. 2001; Yu et al. 2001). The likelihood of these events are strong, given that D-enzyme has been shown to be essential for the synthesis of normal starch in *Chlamydomonas reinhardtii* (Colleoni et al. 1999).

The nocturnal degradation of chloroplast starch begins with hydrolysis by endoamylases of the α-type, and possibly also of the β-type (Lao et al. 1999). Any released maltooligosaccharides too small to be recognised by the amylases are manipulated by D-enzyme (as described above) into suitably larger molecules (Critchley et al. 2001). Branch points in amylopectin are cleaved by debranching enzyme which, it seems, is regulated by the *h* form of thioredoxin (Cho et al. 1999).

e) Sucrose

The product of the chloroplast's CO_2 fixation is exported as triose-P to the cytosol where it is manufactured into sucrose (for subsequent translocation) in most species. Two challenges to this conventional scheme have appeared, one in which it is claimed that fumaric acid may have a role similar to that of sucrose (Chia et al. 2000), and the other reviving the old debate of whether or not sucrose might occur in chloroplasts as well as in the cytosol: Gerrits et al. (2001) provide evidence that substantial amounts of sucrose can be pulled into the chloroplast if it is then metabolised to an end product such as fructan.

Triose-P export is mediated by the well-known triose-P/Pi translocator ("phosphate translocator") of the chloroplast envelope's inner membrane. This translocator helps to determine how much photosynthetic product is directed to sucrose synthesis versus the alternative fates of transitory storage (as starch) in the chloroplast, and respiration by mitochondria (Häusler et al. 2000a), but it may not be the only player in this game: the chloroplast envelope's *outer* membrane has suddenly achieved prominence because it has been found to possess an anion-selective channel protein with the potential to control the direction of triose-P flow (Bölter et al. 1999), and a hexokinase (Wiese et al. 1999) that might energise glucose arriving from the inner membrane's glucose translocator during the night when the mobilisation of the transitory starch reserve takes place (Weber et al. 2000). Incidentally, a recent astonishing report concerning such glucose transporter proteins was that the marine alga *Phaeodactylum tricornutum* could be converted from photosynthesis to heterotrophy simply by inserting a single gene encoding the glucose transporter from human red blood cells (Zaslavskaia et al. 2001).

The roles of the cytosolic FBPase and sucrose-P synthase have been further investigated by decreasing the expression of their corresponding genes in *Arabidopsis thaliana*. In each case, sucrose synthesis was hindered, but only with lowered FBPase did a reciprocal rise in chloroplast starch occur (Strand et al. 2000). Confirmatory evidence that the FBPase, and also the enigmatic PPi-linked P-fructokinase, are subject to control by the regulatory sugar-P fructose-2,6-P_2 is provided by Scott et al. (2000) and Theodorou and Kruger (2001), and is complemented by a demonstration with transgenic *Arabidopsis* that this sugar-P has an influence on whether photosynthetic product is partitioned into cytosolic sucrose or chloroplast starch (Draborg et al. 2001). Finally, a full description of the ultimate enzyme of sucrose synthesis, sucrose-P phosphatase, has appeared (Lunn et al. 2000).

The movement of sucrose from mesophyll photosynthetic cells to phloem sieve tubes is a topic currently receiving intense scrutiny. At least five different sucrose transporter proteins associated with phloem-cell membranes have been identified, varying in their relative affinities for sucrose and capacities for transporting it. It is envisaged that most are engaged in loading sucrose into the phloem (Gottwald et al. 2000;

Meyer et al. 2000; Weise et al. 2000), but at least one is considered to be a sucrose sensor as well (Barker et al. 2000). In a unique report, Ludwig et al. (2000) describe how one, and possibly more, of these sucrose transporters can also transport the vitamin biotin. This may appear mystifying, but for those of us lucky enough to have had the opportunity to prepare culture media for growing marine microalgae, it is less mystical because biotin (and also thiamine and vitamin B_{12}) is included in the recipes for these media. Could an essential microalgal biotin transporter be the ancestor of the plant sucrose transporters? And to what extent are these transporters related to the mannitol transporter more recently identified (Noiraud et al. 2001) in the phloem of celery that translocates most photosynthetic product as mannitol rather than sucrose?

3 C_4 Photosynthesis and Crassulacean Acid Metabolism (CAM)

a) C_4 Photosynthesis

It has been more than 30 years since C_4 photosynthesis was discovered and, indeed, this topic may have passed its climax; during the past 2 years there has been little direct research on this marvellous metabolic extension of photosynthetic carbon metabolism. Great effort is being devoted to an indirect aspect, viz., the roles of DNA regulatory elements and associated proteins in bringing about the differential gene expression and distribution of enzymes between mesophyll and bundle-sheath cells, but this topic is beyond the scope of this review.

The tenant that C_4 photosynthesis is always associated with distinctive mesophyll and bundle-sheath cells (Kranz anatomy) is being shaken from three directions. Firstly, two aquatic angiosperms that display variants of C_4 photosynthesis lack Kranz anatomy (Casati et al. 2000; van Ginkel et al. 2001). Secondly, an obscure C_4 member of the Chenopodiaceae, *Borszczowia aralocaspica*, that has a strange leaf anatomy of one-layered chlorenchyma (Freitag and Stichler 2000), is being used by Voznesenskaya et al. (2001) to support the notion that Kranz anatomy is not essential for terrestrial C_4 photosynthesis. They describe how the two types of chloroplasts (and their associated enzymes) typical of mesophyll and bundle-sheath cells in conventional C_4 plants are positioned into discrete proximal and distal sections of the single type of photosynthetic cell in this plant. Finally, enzyme and ^{14}C-labelling experiments with the unicellular diatom *Thalassiosira weissflogii* under conditions that retard its CO_2-concentrating mechanism led Reinfelder et al. (2000) to claim that, under such conditions, the diatom performed a "unicellular C_4 photosynthesis". However, two types of chloroplasts were not identified, and the degree to which the data support C_4 photosynthesis in its traditional sense has been questioned by Johnston et al.

(2001); perhaps diatoms perform C_3 photosynthesis with a "C_4ish" tendency, akin to the aquatic angiosperms mentioned above.

More details concerning the activation upon sunrise of the C_4 pathway's initial enzyme, phosphoenolpyruvate (PEP) carboxylase, have emerged. As reviewed previously, the activation (now detected in one of the submersed aquatic species that perform C_4-like photosynthesis; Lara et al. 2001) results from a phosphorylation of the enzyme catalysed by a second enzyme appropriately named PEP carboxylase kinase (PEPCK). This PEPCK appears to be absent in darkness, or present in quite low amounts, with activity suppressed by yet a third protein, an inhibitor that binds to it (G.A. Nimmo et al. 2001). However, large amounts of PEPCK appear upon illumination when the light initiates an intracellular signalling system that culminates in the expression of its gene (H.G. Nimmo et al. 2001). The signalling system involves the ubiquitous second messengers inositol-P_3 (Coursol et al. 2000) and Ca^{2+} (Ogawa et al. 1998), but the nature of the (coloured?) molecule that detects light and sets off this chain of events is still elusive. The end result is an activated PEP carboxylase with, among other properties, an increased affinity for its substrate bicarbonate (Parvathi et al. 2000). This could be valuable because, at physiological levels of bicarbonate and Mg^{2+}, the enzyme operates below its potential (Tovar-Méndez et al. 2000).

b) Crassulacean Acid Metabolism

If C_4 photosynthesis has passed its 30th year, then crassulacean acid metabolism (CAM) must have passed its 40th. Nevertheless, the list of plants with CAM continues to lengthen, with more members of the Portulacaceae (Guralnick and Jackson 2001), and tropical epiphytic and lithophytic ferns (Holtum and Winter 1999) recently added. The principal benefit of CAM is that plant water is conserved when stomates are open only at night; CAM achieves nocturnal storage (as malate) of the CO_2 that enters these stomates. Next day this store is drawn upon for photosynthesis behind closed stomates. A not unexpected, but still impressive, observation concerning the CAM plant's water economy is that the plasma membrane and tonoplast of the CAM plant *Graptopetalum paraguayense* possess remarkably low levels (20 and 1% respectively) of the water-transporter protein, aquaporin, when compared to a representative C_3 plant, radish (Ohshima et al. 2001).

The distinctive nocturnal and diurnal metabolic activities of CAM plants are set into a circadian rhythm in the cells of these plants; wave fronts of the rhythm can be seen to propagate across leaves (Rascher et al. 2001). The master switch for the rhythm has not yet been confirmed, but a biophysical switch involving altered tonoplast tension is a strong contender (Lüttge 2000), especially since the activation of PEP carboxylase, which follows the turning-on of a PEPCK gene as described above for C_4 plants, follows a circadian rhythm determined by an oscillator that may be at the level of the tonoplast (Hartwell et al. 1999; Nimmo 2000; Taybi et al. 2000). Other points concerning PEP carboxylase relate to those plants for which CAM is an inducible phenomenon accompanied by an increase in the amount of the enzyme. Sometimes the extra enzyme is of a special form (Mazen 2000). The well-known induction by salinity in *Mesembryanthemum crystallinum* occurs only under high light (Miszalski et al. 2001). There is a surprise report that foliar application of gibberellic acid to this plant induces, like salinity, CAM and increased PEP carboxylase activity (Guralnick et al. 2001). Incidentally, another surprise arose when Malda et al. (1999) grew the CAM plant *Coryphantha minima* in vitro using plant tissue culture techniques. It showed net CO_2 uptake continuously in the light, as well as in the dark, and grew seven-fold larger than control plants.

Several other enzymes and membrane-located transporter proteins of CAM plants have been examined, including an unusual cytosolic NAD-linked malic enzyme with some function in night-time acidification (rather than daytime deacidification as established for its mitochondrial counterpart; Cuevas and Podestá 2000), a pyruvate Pi dikinase unexpectedly located in the cytosol (rather than in the chloroplast; Kondo et al. 2000), a putative tonoplast malate transporter (Lüttge et al. 2000), and chloroplast-envelope PEP/Pi and glucose-6-P/Pi translocators (Häusler et al. 2000b), all three of which are more plentiful when CAM is more prevalent. A notable conclusion from the latter study was that CAM chloroplasts might be more similar to non-photosynthetic plastids (Tauberger at al. 2000) than to other chloroplasts in that they can import glucose-6-P for starch synthesis.

4 Brief Notes on Carbon and Kyoto

Because the Kyoto Protocol makes reference to carbon sequestration in forests and soils, the current status of these systems and their responses in a world with extra CO_2 in its atmosphere are under intense investigation. The well-established stimulation of photosynthesis and growth by increased CO_2 is now even more strongly believed to be a characteristic and persistent attribute of trees (Curtis and Wang 1998; Saxe et al. 1998; Stylinski et al. 2000; Herrick and Thomas 2001; Idso and Kimball 2001) despite some down-regulation of the capacities of photosynthetic electron transport and Rubisco by some species (Medlyn et al. 1999; Griffin et al. 2000), and retardation of photosynthesis by excess starch accumulated in chloroplasts by others (Würth et al. 1998; see Sawada et al. 2001; Sicher and Bunce 2001). Another recent change in emphasis relates to respiration: it now seems that increased CO_2 does not reduce mitochondrial respiration, but rather has no effect on it (Amthor et al. 2001; Hamilton et al. 2001) when care is taken with measurement devices (Jahnke 2001). Of course, in quantifying the global carbon cycle, respiration below ground must also be factored in. This parameter may not be receiving proper attention in experiments with trees exposed to increased CO_2 (Luo 2001), and its importance is highlighted by the conclusion of Ekblad and Högberg (2001) that carbon photosynthetically fixed by tree leaves can be respired just 1–4 days later by the tree's roots or nearby soil microbes.

5 Genetic Manipulation of Photosynthetic Carbon Metabolism

The dream of bigger and/or faster growing crop plants has existed at least as far back as the times when the wonderful story of Gulliver's Travels and the impossible fantasy of Jack and the Beanstalk were penned (see quote above, and that preceding our 1994 review). Farmers, crop breeders and plant physiologists over the years have slowly but surely brought that dream to fruition (Richards 2000). However, dreams never die, and the prospect that the modern tools of molecular biology might be used to genetically engineer plants to be still bigger or still faster growing has encouraged many researchers in the past decade. Finally, there has appeared "the first report of the transgenic improvement of plant photosynthetic capacity" (Wood 2002). Encouraged, no doubt, by indications that the Calvin cycle enzyme sedoheptulose bisphosphatase (SBPase) plays a powerful part in determining the rate of photosynthetic carbon assimilation (Poolman et al. 2000; Harrison et al. 2001), and having on hand the gene for an unusual cyanobacterial SBPase that, among other odd properties, also has FBPase activity (Tamoi et al. 1998), Miyagawa et al. (2001) inserted this gene into tobacco chloroplasts and obtained transgenic plants with doubled SBPase activity and a photosynthetic rate increased by almost 25% (Fig. 2). One wonders whether an additional boost might be possible if extra transketolase could also be engineered into the chloroplast, given that Henkes et al. (2001) have discovered that the activity of this enzyme also has a dramatic influence on the rate of photosynthesis.

Whether or not the above was indeed the first case of a plant genetically engineered for faster photosynthesis might be debated. Three other cases deserve mention. In our 1998 review, we noted that Galtier et al. (1995) coaxed tomato plants to photosynthesise faster (the increase was about 20%) by transforming them with a maize sucrose-P synthase gene. More recently, Migge et al. (2000) reported a 20–30% increase in fresh weight of young tobacco transformed to contain a doubled activity of the chloroplast enzyme glutamine synthetase. This enzyme participates in the assimilation of ammonium, including that released by the mitochondrial four-enzyme complex that converts two glycines to one serine during photorespiration (Sect. 2.c). Finally, Paul et al. (2001) make reference to a relatively obscure publication pertaining to tobacco with photosynthesis increased by up to 30% following transformation with a

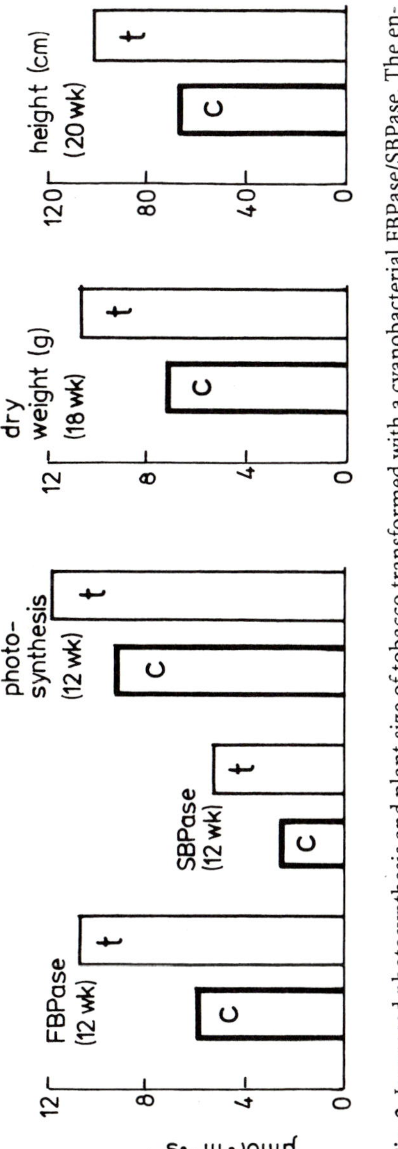

Fig. 2. Increased photosynthesis and plant size of tobacco transformed with a cyanobacterial FBPase/SBPase. The enzyme activities are also shown. *c* Control; *t* transformant TpFS-3. (Data from Miyagawa et al. 2001)

bacterial gene encoding trehalose-P synthase. Presumably, the trehalose alters sugar-signalling pathways such that the expression of the tobacco-leaf genes is altered.

Of course, the obliteration of photorespiration has been another goal of genetic engineers aiming to increase photosynthesis. Four approaches have been tried. One has been to alter the plant's native Rubisco so that it's oxygenase activity is removed, or at least altered. However, as noted in our earlier reviews, and as Lorimer (2001) has eloquently pointed out, this has come to naught after 30 years of effort (not to mention 2 billion years of evolution). A second approach recently considered and debated (see our last review) is to transform plants with an algal Rubisco that recognises O_2 (instead of CO_2) less often as a substrate. Whitney et al. (2001) now report that, when this was attempted, the two different subunits of the algal Rubisco were synthesised in the leaf chloroplasts in abundant amounts, but they failed to assemble into functional multi-subunit enzyme molecules. Unfortunately, the accumulated unassembled subunits clogged the chloroplast and photosynthesis consequently declined. A third idea from Barata et al. (2000) is to introduce the gene for the O_2-binding protein of legume nodules (leghaemoglobin) into chloroplasts so that the Rubisco no longer has access to free O_2. These authors attempted this, but so far gene expression has been too low to significantly lower the O_2 level. Finally, a popular approach to reduce photorespiration is to add characteristics of C_4 plants into C_3 plants (C_4 plants naturally avoid photorespiration by pumping CO_2 into bundle-sheath cells). To this end one, or sometimes two, of the genes for the C_4 enzymes PEP carboxylase, NADP-linked malate dehydrogenase, NADP-linked malic enzyme, PEP carboxykinase, and pyruvate Pi dikinase have been engineered into tobacco, potato or rice (Suzuki et al. 2000; Beaujean et al. 2001; Fukayama et al. 2001; Häusler et al. 2001). Gene expression was sometimes substantial (Fukayawa et al. 2001). However, in no case was a significant beneficial effect on the photosynthesis of the recipient plant observed. It could be argued that additional transformation with genes that direct the differentiation of cells into mesophyll and bundle-sheath types (possibly a daunting task) might be needed before a beneficial effect could ensue, but Voznesenskaya et al. (2001) argue that this is not necessarily so since they have discovered a C_4 plant that works well with only one type of cell (Sect. 3.a).

The only certainty concerning the genetic modification of photosynthesis to increase production is that a good deal more literature on the topic will be available for inclusion in the next review of this series in 2004. Meantime, optimism might be tempered by the comments from Horton (2000): "without extreme good fortune, manipulation of photosynthesis to increase yield is at present an unlikely prospect for the near future"; and from Richards (2000): "....photosynthesis does not limit grain yield and current crops have sufficient surplus photosynthetic capacity to respond to increases in yield potential". Nevertheless, we should be prepared for the unexpected. An example just reported in the media (but unconfirmed at the time of writing) is the production of a pig engineered to contain spinach genes. This first example of a mammal with plant genes will no doubt cause some of we older researchers of photosynthesis to squirm!

References

Albrecht T, Koch A, Lode A, Greve B, Schneider-Mergener J, Steup M (2001) Plastidic (Pho1-type) phosphorylase isoforms in potato (*Solanum tuberosum* L.) plants: expression analysis and immunochemical characterization. Planta 213:602–613

Amthor JS, Koch GW, Willms JR, Layzell DB (2001) Leaf O_2 uptake in the dark is independent of coincident CO_2 partial pressure. J Exp Bot 52:2235–2238

Atkin OK, Evans JR, Ball MC, Lambers H, Pons TL (2000) Leaf respiration of snow gum in the light and dark. Interactions between temperature and irradiance. Plant Physiol 122:915–923

Balmer Y, Schürmann P (2001) Heterodimer formation between thioredoxin *f* and fructose 1,6-bisphosphatase from spinach chloroplasts. FEBS Lett 492:58–61

Barata RM, Chaparro A, Chabregas SM, González R, Labate CA, Azevedo RA, Sarath G, Lea PJ, Silva-Filho MC (2000) Targeting of the soybean leghemoglobin to tobacco chloroplasts: effects on aerobic metabolism in transgenic plants. Plant Sci 155:193–202

Barker L, Kühn C, Weise A, Schulz A, Gebhardt C, Hirner B, Hellmann H, Schulze W, Ward JM, Frommer WB (2000) SUT2, a putative sucrose sensor in sieve elements. Plant Cell 12:1153–1164

Beaujean A, Issakidis-Bourguet E, Catterou M, Dubois F, Sangwan RS, Sangwan-Norreel BS (2001) Integration and expression of *Sorghum* C_4 phosphoenolpyruvate carboxylase and chloroplastic $NADP^+$-malate dehydrogenase separately or together in C_3 potato plants. Plant Sci 160:1199–1210

Beckles DM, Smith AM, ap Rees T (2001a) A cytosolic ADP-glucose pyrophosphorylase is a feature of graminaceous endosperms, but not of other starch storing organs. Plant Physiol 125:818–827

Beckles DM, Craig J, Smith AM (2001b) ADP-glucose pyrophosphorylase is located in the plastid in developing tomato fruit. Plant Physiol 126:261–266

Bennoun P (2001) Chlororespiration and the process of carotenoid biosynthesis. Biochim Biophys Acta 1506:133–142

Bernacchi CJ, Singsaas EL, Pimentel C, Portis AR, Long SP (2001) Improved temperature response functions for models of Rubisco-limited photosynthesis. Plant Cell Environ 24:253–259

Bölter B, Soll J, Hill K, Hemmler R, Wagner R (1999) A rectifying ATP-regulated solute channel in the chloroplastic outer envelope from pea. EMBO J 18:5505–5516

Büchel C, Zsíros O, Gareb G (1998) Alternative cyanide-sensitive oxidase interacting with photosynthesis in *Synechocystis* PCC6803. Ancestor of the terminal oxidase of chlororespiration? Photosynthetica 35:223–231

Cabrillac D, Cock JM, Dumas C, Gaude T (2001) The *S*-locus receptor kinase is inhibited by thioredoxins and activated by pollen coat proteins. Nature 410:220–223

Carol P, Kuntz M (2001) A plastid terminal oxidase comes to light: implications for carotenoid biosynthesis and chlororespiration. Trends Plant Sci 6:31–36

Casati P, Lara MV, Andreo CS (2000) Induction of a C_4-like mechanism of CO_2 fixation in *Egeria densa*, a submersed aquatic species. Plant Physiol 123:1611–1621

Cen Y-P, Turpin DH, Layzell DB (2001) Whole-plant gas exchange and reductive biosynthesis in white lupin. Plant Physiol 126:1555–1565

Chia DW, Yoder TJ, Reiter W-D, Gibson SI (2000) Fumaric acid: an overlooked form of fixed carbon in *Arabidopsis* and other plant species. Planta 211:743–751

Chiadmi M, Navaza A, Miginiac-Maslow M, Jacquot J-P, Cherfils J (1999) Redox signalling in the chloroplast: structure of oxidized pea fructose-1,6-bisphosphate phosphatase. EMBO J 18:6809–6815

Cho M-J, Wong JH, Marx C, Jiang W, Lemaux PG, Buchanan BB (1999) Overexpression of thioredoxin *h* leads to enhanced activity of starch debranching enzyme (pullulanase) in barley grain. Proc Natl Acad Sci USA 96:14641–14646

Colleoni C, Dauvillée D, Mouille G, Buléon A, Gallant D, Bouchet B, Morell M, Samuel M, Delrue B, d'Hulst C, Bliard C, Nuzillard J-M, Ball S (1999) Genetic and biochemical evidence for the involvement of α-1,4 glucanotransferases in amylopectin synthesis. Plant Physiol 120:993–1003

Cournac L, Redding K, Ravenel J, Rumeau D, Josse E-M, Kuntz M, Peltier G (2000) Electron flow between photosystem II and oxygen in chloroplasts of photosystem I-deficient algae is mediated by a quinol oxidase involved in chlororespiration. J Biol Chem 275:17256–17262

Coursol S, Giglioli-Guivarc'h N, Vidal J, Pierre J-N (2000) An increase in phosphoinositide-specific phospholipase C activity precedes induction of C_4 phosphoenolpyruvate carboxylase phosphorylation in illuminated and NH_4Cl-treated protoplasts from *Digitaria sanguinalis*. Plant J 23:497–506

Crafts-Brandner SJ, Law RD (2000) Effect of heat stress on the inhibition and recovery of the ribulose-1,5-bisphosphate carboxylase/oxygenase activation state. Planta 212:67–74

Crafts-Brandner SJ, Salvucci ME (2000) Rubisco activase constrains the photosynthetic potential of leaves at high temperature and CO_2. Proc Natl Acad Sci USA 97:13430–13435

Critchley JH, Zeeman SC, Takaha T, Smith AM, Smith SM (2001) A critical role for disproportionating enzyme in starch breakdown is revealed by a knock-out mutation in *Arabidopsis*. Plant J 26:89–100

Cuevas IC, Podestá FE (2000) Purification and physical and kinetic characterization of an NAD^+-dependent malate dehydrogenase from leaves of pineapple (*Ananas comosus*). Physiol Plant 108:240–248

Curtis PS, Wang X (1998) A meta-analysis of elevated CO_2 effects on woody plant mass, form, and physiology. Oecologia 113:299–313

Denyer K, Waite D, Motawia S, Møller BL, Smith AM (1999) Granule-bound starch synthase I in isolated starch granules elongates malto-oligosaccharides processively. Biochem J 340:183–191

Doan DNP, Rudi H, Olsen O-A (1999) The allosterically unregulated isoform of ADP-glucose pyrophosphorylase from barley endosperm is the most likely source of ADP-glucose incorporated into endosperm starch. Plant Physiol 121:965–975

Douce R, Bourguignon J, Neuburger M, Rébeillé F (2001) The glycine decarboxylase system: a fascinating complex. Trends Plant Sci 6:167–176

Draborg H, Villadsen D, Nielsen TH (2001) Transgenic *Arabidopsis* plants with decreased activity of fructose-6-phosphate-2-kinase/fructose-2,6-bisphosphatase have altered carbon partitioning. Plant Physiol 126:750–758

Ekblad A, Högberg P (2001) Natural abundance of ^{13}C in CO_2 respired from forest soils reveals speed of link between tree photosynthesis and root respiration. Oecologia 127:305–308

Feild TS, Nedbal L, Ort DR (1998) Nonphotochemical reduction of the plastoquinone pool in sunflower leaves originates from chlororespiration. Plant Physiol 116:1209–1218

Fernie AR, Roessner U, Trethewey RN, Willmitzer L (2001) The contribution of plastidial phosphoglucomutase to the control of starch synthesis within the potato tuber. Planta 213:418–426

Ferreira RB, Esquível MG, Teixeira AR (2000) An accurate method to quantify ribulose bisphosphate carboxylase content in plant tissue. Plant Cell Environ 23:1329–1340

Field CB (2001) Plant physiology of the "missing" carbon sink. Plant Physiol 125:25–28

Freitag H, Stichler W (2000) A remarkable new leaf type with unusual photosynthetic tissue in a central Asiatic genus of Chenopodiaceae. Plant Biol 2:154–160

Fukayama H, Tsuchida H, Agarie S, Nomura M, Onodera H, Ono K, Lee B-H, Hirose S, Toki S, Ku MSB, Makino A, Matsuoka M, Miyao M (2001) Significant accumulation of C_4-specific pyruvate, orthophosphate dikinase in a C_3 plant, rice. Plant Physiol 127:1136–1146

Galtier N, Foyer CH, Murchie E, Alred R, Quick P, Voelker TA, Thépenier C, Lascève G, Betsche T (1995) Effects of light and atmospheric carbon dioxide enrichment on photosynthesis and carbon partitioning in the leaves of tomato (*Lycopersicon esculentum* L.) plants over-expressing sucrose phosphate synthase. J Exp Bot 46:1335–1344

Gebbing T, Schnyder H (2001) ^{13}C labeling kinetics of sucrose in glumes indicates significant refixation of respiratory CO_2 in the wheat ear. Aust J Plant Physiol 28:1047–1053

Gerrits N, Turk SCHJ, van Dun KPM, Hulleman SHD, Visser RGF, Weisbeek PJ, Smeekens SCM (2001) Sucrose metabolism in plastids. Plant Physiol 125:926–934

Ghosh S, Bagchi S, Majumder AL (2001) Chloroplast fructose-1,6-bisphosphatase from *Oryza* differs in salt tolerance property from the *Porteresia* enzyme and is protected by osmolytes. Plant Sci 160:1171–1181

Gómez Casati DF, Aon MA, Iglesias AA (2000) Kinetic and structural analysis of the ultrasensitive behaviour of cyanobacterial ADP-glucose pyrophosphorylase. Biochem J 350:139–147

Gottwald JR, Krysan PJ, Young JC, Evert RF, Sussman MR (2000) Genetic evidence for the *in planta* role of phloem-specific plasma membrane sucrose transporters. Proc Natl Acad Sci USA 97:13979–13984

Griffin KL, Tissue DT, Turnbull MH, Whitehead D (2000) The onset of photosynthetic acclimation to elevated CO_2 partial pressure in field-grown *Pinus radiata* D. Don. after 4 years. Plant Cell Environ 23:1089–1098

Grotjohann N, Messdaghi D, Kowallik W (1999) Oxygen uptake during photosynthesis of isolated pea chloroplasts. Z Naturforsch 54c:209–219

Guralnick LJ, Jackson MD (2001) The occurrence and phylogenetics of crassulacean acid metabolism in the Portulacaceae. Int J Plant Sci 162:257–262

Guralnick LJ, Ku MSB, Edwards GE, Strand D, Hockema B, Earnest J (2001) Induction of PEP carboxylase and crassulacean acid metabolism by gibberellic acid in *Mesembryanthemum crystallinum*. Plant Cell Physiol 42:236–239

Häusler RE, Schlieben NH, Flügge U-I (2000a) Control of carbon partitioning and photosynthesis by the triose phosphate /phosphate translocator in transgenic tobacco plants (*Nicotiana tabacum*). II. Assessment of control coefficients of the triose phosphate/phosphate translocator. Planta 210:383–390

Häusler RE, Baur B, Scharte J, Teichmann T, Eicks M, Fischer KL, Flügge U-I, Schubert S, Weber A, Fischer K (2000b) Plastidic metabolite transporters and their physiological functions in the inducible crassulacean acid metabolism plant *Mesembryanthemum crystallinum*. Plant J 24:285–296

Häusler RE, Rademacher T, Li J, Lipka V, Fischer KL, Schubert S, Kreuzaler F, Hirsch H-J (2001) Single and double overexpression of C_4-cycle genes had differential effects on the pattern of endogenous enzymes, attenuation of photorespiration and on contents of UV protectants in transgenic potato and tobacco plants. J Exp Bot 52:1785–1803

Hamilton JG, Thomas RB, DeLucia EH (2001) Direct and indirect effects of elevated CO_2 on leaf respiration in a forest ecosystem. Plant Cell Environ 24:975–982

Harrison EP, Olcer H, Lloyd JC, Long SP, Paines CA (2001) Small decreases in SBPase cause a linear decline in the apparent RuBP regeneration rate, but do not affect Rubisco carboxylation capacity. J Exp Bot 52:1779–1784

Hartwell J, Gill A, Nimmo GA, Wilkins MB, Jenkins GI, Nimmo HG (1999) Phosphoenolpyruvate carboxylase kinase is a novel protein kinase regulated at the level of expression. Plant J 20:333–342

Haupt-Herting S, Klug K, Fock HP (2001) A new approach to measure gross CO_2 fluxes in leaves. Gross CO_2 assimilation, photorespiration, and mitochondrial respiration in the light in tomato under drought stress. Plant Physiol 126:388–396

Henkes S, Sonnewald U, Badur R, Flachmann R, Stitt M (2001) A small decrease of plastid transketolase activity in antisense tobacco transformants has dramatic effects on photosynthesis and phenylpropanoid metabolism. Plant Cell 13:535–551

Herrick JD, Thomas RB (2001) No photosynthetic down-regulation in sweetgum trees (*Liquidamber styraciflua* L.) after three years of CO_2 enrichment at the Duke Forest FACE experiment. Plant Cell Environ 24:53–64

Holtum JAM, Winter K (1999) Degrees of crassulacean acid metabolism in tropical epiphytic and lithophytic ferns. Aust J Plant Physiol 26:749–757

Horken KM, Tabita FR (1999) The "green" form I ribulose 1,5-bisphosphate carboxylase/oxygenase from the nonsulfur purple bacterium *Rhodobacter capsulatus*. J Bacteriol 181:3935–3941

Horton P (2000) Prospects for crop improvement through the genetic manipulation of photosynthesis: morphological and biochemical aspects of light capture. J Exp Bot 51:475–485

Idso SB, Kimball BA (2001) CO_2 enrichment of sour orange trees: 13 years and counting. Environ Exp Bot 46:147–153

Igamberdiev AU, Bykova NV, Lea PJ, Gardeström P (2001a) The role of photorespiration in redox and energy balance of photosynthetic plant cells: a study with a barley mutant deficient in glycine decarboxylase. Physiol Plant 111:427–438

Igamberdiev AU, Ivlev AA, Bykova NV, Threlkeld CN, Lea PJ, Gardeström P (2001b) Decarboxylation of glycine contributes to carbon isotope fractionation in photosynthetic organisms. Photosyn Res 67:177–184

Iwamoto K, Ikawa T (2000) A novel glycolate oxidase requiring flavin mononucleotide as the cofactor in the prasinophycean alga *Mesostigma viride*. Plant Cell Physiol 41:988–991

Jahnke S (2001) Atmospheric CO_2 concentration does not directly affect leaf respiration in bean or poplar. Plant Cell Environ 24:1139–1151

Jenks A, Gibbs AP (2000) Immunolocalization and distribution of form II Rubisco in the pyrenoid and chloroplast stroma of *Amphidinium carterae* and form I Rubisco in the symbiont-derived plastids of *Peridinium foliaceum* (Dinophyceae). J Phycol 36:127–138

Johnston AM, Raven JA, Beardall J, Leegood RC (2001) Photosynthesis in a marine diatom. Nature 412:40–41

Jung C-H, Hartman FC, Lu T-YS, Larimer FW (2000) D-Ribose-5-phosphate isomerase from spinach: heterologous overexpression, purification, characterization, and site-directed mutagenesis of the recombinant enzyme. Arch Biochem Biophys 373:409–417

Khan S, Andralojc PJ, Lea PJ, Parry MAJ (1999) 2′-Carboxy-D-arabinitol 1-phosphate protects ribulose 1,5-bisphosphate carboxylase/oxygenase against proteolytic breakdown. Eur J Biochem 266:840–847

Kleczkowski LA (2001) A new player in the starch field. Plant Physiol Biochem 39:759–761

Kondo A, Nose A, Yuasa H, Ueno O (2000) Species variation in the intracellular localization of pyruvate, Pi dikinase in leaves of crassulacean-acid-metabolism plants: an immunogold electron-microscope study. Planta 210:611–621

Krepinsky K, Plaumann M, Martin W, Schnarrenberger C (2001) Purification and cloning of chloroplast 6-phosphogluconate dehydrogenase from spinach. Eur J Biochem 268:2678–2686

Lao NT, Schoneveld O, Mould RM, Hibberd JM, Gray JC, Kavanagh TA (1999) An *Arabidopsis* gene encoding a chloroplast-targeted β-amylase. Plant J 20:519–527

Lara MV, Casati P, Andreo CS (2001) In vivo phosphorylation of phosphoenolpyruvate carboxylase in *Egeria densa*, a submersed aquatic species. Plant Cell Physiol 42:441–445

Liepman AH, Olsen LJ (2001) Peroxisomal alanine:glyoxylate aminotransferase (AGT1) is a photorespiratory enzyme with multiple substrates in *Arabidopsis thaliana*. Plant J 25:487–498

Loreto F, Delfine S, Di Marco G (1999) Estimation of photorespiratory carbon dioxide recycling during photosynthesis. Aust J Plant Physiol 26:733–736

Lorimer G (2001) Letter to the editor. Plant Physiol 127:3

Ludwig A, Stolz J, Sauer N (2000) Plant sucrose-H^+ symporters mediate the transport of vitamin H. Plant J 24:503–509

Lüttge U (2000) The tonoplast functioning as the master switch for circadian regulation of crassulacean acid metabolism. Planta 211:761–769

Lüttge U, Pfeifer T, Fischer-Schliebs E, Ratajczak R (2000) The role of vacuolar malate-transport capacity in crassulacean acid metabolism and nitrate nutrition. Higher malate-transport capacity in ice plant after crassulacean acid metabolism-induction and in tobacco under nitrate nutrition. Plant Physiol 124:1335–1347

Lunn JE, Ashton AR, Hatch MD, Heldt HW (2000) Purification, molecular cloning, and sequence analysis of sucrose-6^F-phosphate phosphohydrolase from plants. Proc Natl Acad Sci USA 97:12914–12919

Luo Y (2001) Transient ecosystem responses to free-air CO_2 enrichment (FACE): experimental evidence and methods of analysis. New Phytol 152:3–8

Malda G, Backhaus RA, Martin C (1999) Alterations in growth and crassulacean acid metabolism (CAM) activity of in vitro cultured cactus. Plant Cell Tissue Org Cult 58:1–9

Malhi Y, Grace J (2000) Tropical forests and atmospheric carbon dioxide. Trends Ecol Evol 15:332–337

Mazen AMA (2000) Changes in properties of phosphoenolpyruvate carboxylase with induction of crassulacean acid metabolism (CAM) in the C_4 plant *Portulaca oleracea*. Photosynthetica 38:385–391

Medlyn BE, Badeck F-W, De Pury DGG, Barton CVM, Broadmeadow M, Ceulemans R, De Angelis P, Forstreuter M, Jach ME, Kellomäki S, Laitat E, Marek M, Philippot S, Rey A, Strassemeyer J, Laitinen K, Liozon R, Portier B, Roberntz P, Wang K, Jarvis PG (1999) Effects of elevated [CO_2] on photosynthesis in European forest species: a meta-analysis of model parameters. Plant Cell Environ 22:1475–1495

Melis A, Zhang L, Forestier M, Ghirardi ML, Seibert M (2000) Sustained photobiological hydrogen gas production upon reversible inactivation of oxygen evolution in the green alga *Chlamydomonas reinhardtii*. Plant Physiol 122:127–135

Meyer S, Melzer M, Truernit E, Hümmer C, Besenbeck R, Stadler R, Sauer N (2000) *AtSUC3*, a gene encoding a new *Arabidopsis* sucrose transporter, is expressed in cells adjacent to the vascular tissue and in a carpel cell layer. Plant J 24:869–882

Michelis R, Gepstein S (2000) Identification and characterization of a heat-induced isoform of aldolase in oat chloroplast. Plant Mol Biol 44:487–498

Migge A, Kahmann U, Fock HP, Becker TW (1999) Prolonged exposure of tobacco to a low oxygen atmosphere to suppress photorespiration decreases net photosynthesis and results in changes in plant morphology and chloroplast structure. Photosynthetica 36:107–116

Migge A, Carrayol E, Hirel B, Becker TW (2000) Leaf-specific overexpression of plastidic glutamine synthetase stimulates the growth of transgenic tobacco seedlings. Planta 210:252–260

Miszalski Z, Niewiadomska E, Slesak I, Lüttge U, Kluge M, Ratajczak R (2001) The effect of irradiance on carboxylating/decarboxylating enzymes and fumarase activities in *Mesembryanthemum crystallinum* L. exposed to salinity stress. Plant Biol 3:17–23

Miyagawa Y, Tamoi M, Shigeoka S (2001) Overexpression of a cyanobacterial fructose-1,6-/sedoheptulose-1,7-bisphosphatase in tobacco enhances photosynthesis and growth. Nat Biotech 19:965–969

Mott KA, Woodrow IE (2000) Modelling the role of Rubisco activase in limiting non-steady-state photosynthesis. J Exp Bot 51:399–406

Muraoka H, Tang Y, Terashima I, Koizumi H, Washitani I (2000) Contributions of diffusional limitation, photoinhibition and photorespiration to midday depression of photosynthesis in *Arisaema heterophyllum* in natural high light. Plant Cell Environ 23:235–250

Nakano H, Muramatsu S, Makino A, Mae T (2000) Relationship between the suppression of photosynthesis and starch accumulation in the pod-removed bean. Aust J Plant Physiol 27:167–173

Nassoury N, Fritz L, Morse D (2001) Circadian changes in ribulose-1,5-bisphosphate carboxylase/oxygenase distribution inside individual chloroplasts can account for the rhythm in dinoflagellate carbon fixation. Plant Cell 13:923–934

Nimmo GA, Wilkins MB, Nimmo HG (2001) Partial purification and characterization of a protein inhibitor of phosphoenolpyruvate carboxylase kinase. Planta 213:250–257

Nimmo HG (2000) The regulation of phosphoenolpyruvate carboxylase in CAM plants. Trends Plant Sci 5:75–80

Nimmo HG, Fontaine V, Hartwell J, Jenkins GI, Nimmo GA, Wilkins MB (2001) PEP carboxylase kinase is a novel protein kinase controlled at the level of expression. New Phytol 151:91–97

Noiraud N, Maurousset L, Lemoine R (2001) Identification of a mannitol transporter, AgMaT1, in celery phloem. Plant Cell 13:695–705

Ogawa N, Yabuta N, Ueno Y, Izui K (1998) Characterization of a maize Ca^{2+}-dependent protein kinase phosphorylating phosphoenolpyruvate carboxylase. Plant Cell Physiol 39:1010–1019

Ohshima Y, Iwasaki I, Suga S, Murakami M, Inoue K, Maeshima M (2001) Low aquaporin content and low osmotic water permeability of the plasma and vacuolar membranes of a CAM plant *Graptopetalum paraguayense*: comparison with radish. Plant Cell Physiol 42:1119–1129

Padmasree K, Raghavendra AS (1999) Response of photosynthetic carbon assimilation in mesophyll protoplasts to restriction on mitochondrial oxidative metabolism: metabolites related to the redox status and sucrose biosynthesis. Photosyn Res 62:231–239

Padmasree K, Raghavendra AS (2001) Consequence of restricted mitochondrial oxidative metabolism on photosynthetic carbon assimilation in mesophyll protoplasts: decrease in light activation of four chloroplastic enzymes. Physiol Plant 112:582–588

Parvathi K, Bhagwat AS, Ueno Y, Izui K, Raghavendra AS (2000) Illumination increases the affinity of phosphoenolpyruvate carboxylase to bicarbonate in leaves of a C_4 plant, *Amaranthus hypochondriacus*. Plant Cell Physiol 41:905–910

Paul M, Pellny T, Goddijn O (2001) Enhancing photosynthesis with sugar signals. Trends Plant Sci 6:197–200

Paul MJ, Driscoll SP, Andralojc PJ, Knight JS, Gray JC, Lawlor DW (2000) Decrease of phosphoribulokinase activity by antisense RNA in transgenic tobacco: definition of the light environment under which phosphoribulokinase is not in large excess. Planta 211:112–119

Poolman MG, Fell DA, Thomas S (2000) Modelling photosynthesis and its control. J Exp Bot 51:319–328

Rascher U, Hütt M-T, Siebke K, Osmond B, Beck F, Lüttge U (2001) Spatiotemporal variation of metabolism in a plant circadian rhythm: the biological clock as an assembly of coupled individual oscillators. Proc Natl Acad Sci USA 98:11801–11805

Reinfelder JR, Kraepiel AML, Morel FMM (2000) Unicellular C_4 photosynthesis in a marine diatom. Nature 407:996–999

Richards RA (2000) Selectable traits to increase crop photosynthesis and yield of grain crops. J Exp Bot 51:447–458

Robinson JM (2000) Dark and daylight activity of glucose-6-phosphate dehydrogenase and 6-phosphogluconate dehydrogenase in the leaves of nitrogen-limited spinach and soybean plants. Int J Plant Sci 161:651–657

Rodriguez-López M, Baroja-Fernández E, Zandueta-Criado A, Pozueta-Romero J (2000) Adenosine diphosphate glucose pyrophosphatase: a plastidial phosphodiesterase that prevents starch biosynthesis. Proc Natl Acad Sci USA 97:8705–8710

Rogers A, Ellsworth DS, Humphries SW (2001) Possible explanation of the disparity between the in vitro and in vivo measurements of Rubisco activity: a study in loblolly pine grown in elevated pCO_2. J Exp Bot 52:1555–1561

Rokka A, Zhang L, Aro E-M (2001) Rubisco activase: an enzyme with a temperature-dependent dual function? Plant J 25:463–471

Salvucci ME, Osteryoung KW, Crafts-Brandner SJ, Vierling E (2001) Exceptional sensitivity of Rubisco activase to thermal denaturation in vitro and in vivo. Plant Physiol 127:1053–1064

Savitch LV, Massacci A, Gray GR, Huner NPA (2000) Acclimation to low temperature or high light mitigates sensitivity to photoinhibition: roles of the Calvin cycle and the Mehler reaction. Aust J Plant Physiol 27:253–264

Sawada S, Kuninaka M, Watanabe K, Sato A, Kawamura H, Komine K, Sakamoto T, Kasai M (2001) The mechanism to suppress photosynthesis through end-product inhibition in single-rooted soybean leaves during acclimation to CO_2 enrichment. Plant Cell Physiol 42:1093–1102

Saxe H, Ellsworth DS, Heath J (1998) Tree and forest functioning in an enriched CO_2 atmosphere. New Phytol 139:395–436

Schiermeier Q (2001) Cycle studies see carbon sinks rise to prominence. Nature 414:385

Schlesinger WH, Lichter J (2001) Limited carbon storage in soil and litter of experimental forest plots under increased atmospheric CO_2. Nature 411:466–469

Schlitter J, Wildner GF (2000) The kinetics of conformation change as determinant of Rubisco's specificity. Photosyn Res 65:7–13

Scott P, Lange AJ, Kruger NJ (2000) Photosynthetic carbon metabolism in leaves of transgenic tobacco (*Nicotiana tabacum* L.) containing decreased amounts of fructose 2,6-bisphosphate. Planta 211:864–873

Sehnke PC, Chung H-J, Wu K, Ferl RJ (2001) Regulation of starch accumulation by granule-associated plant 14-3-3 proteins. Proc Natl Acad Sci USA 98:765–770

Sharkey TD, Badger MR, von Caemmerer S, Andrews TJ (2001) Increased heat sensitivity of photosynthesis in tobacco plants with reduced Rubisco activase. Photosyn Res 67:147–156

Sicher RC, Bunce JA (2001) Adjustments of net photosynthesis in *Solanum tuberosum* in response to reciprocal changes in ambient and elevated growth CO_2 partial pressures. Physiol Plant 112:55–61

Sikka VK, Choi S-B, Kavakli IH, Sakulsingharoj C, Gupta S, Ito H, Okita TW (2001) Subcellular compartmentation and allosteric regulation of the rice endosperm ADPglucose pyrophosphorylase. Plant Sci 161:461–468

Sivakumar P, Sharmila P, Saradhi PP (2000) Proline alleviates salt-stress-induced enhancement in ribulose-1,5-bisphosphate oxygenase activity. Biochem Biophys Res Commun 279:512–515

Strand Å, Zrenner R, Trevanion S, Stitt M, Gustafsson P, Gardeström P (2000) Decreased expression of two key enzymes in the sucrose biosynthesis pathway, cytosolic fructose-1,6-bisphosphatase and sucrose phosphate synthase, has remarkably different consequences for photosynthetic carbon metabolism in transgenic *Arabidopsis thaliana*. Plant J 23:759–770

Stylinski CD, Oechel WC, Ganon JA, Tissue DT, Miglietta F, Raschi A (2000) Effects of lifelong [CO_2] enrichment on carboxylation and light utilization of *Quercus pubescens* Willd. examined with gas exchange, biochemistry and optical techniques. Plant Cell Environ 23:1353–1362

Suzuki S, Murai N, Burnell JN, Arai M (2000) Changes in photosynthetic carbon flow in transgenic rice plants that express C4-type phosphoenolpyruvate carboxykinase from *Urochloa panicoides*. Plant Physiol 124:163–172

Takaha T, Critchley J, Okada S, Smith SM (1998) Normal starch content and composition in tubers of antisense potato plants lacking D-enzyme (4-α-glucanotransferase). Planta 205:445–451

Tamoi M, Murakami A, Takeda T, Shigeoka S (1998) Acquisition of a new type of fructose-1,6-bisphosphatase with resistance to hydrogen peroxide in cyanobacteria: molecular characterization of the enzyme from *Synechocystis* PCC 6803. Biochim Biophys Acta 1383:232–244

Tauberger E, Fernie AR, Emmermann M, Renz A, Kossmann J, Willmitzer L, Trethewey RN (2000) Antisense inhibition of plastidial phosphoglucomutase provides compelling evidence that potato tuber amyloplasts import carbon from the cytosol in the form of glucose-6-phosphate. Plant J 23:43–53

Taybi T, Patil S, Chollet R, Cushman JC (2000) A minimal serine/threonine protein kinase circadianly regulates phosphoenolpyruvate carboxylase activity in crassulacean acid metabolism-induced leaves of the common ice plant. Plant Physiol 123:1471–1481

Theodorou ME, Kruger NJ (2001) Physiological relevance of fructose 2,6-bisphosphate in the regulation of spinach leaf pyrophosphate:fructose 6-phosphate 1-phosphotransferase. Planta 213:147–157

Tovar-Méndez A, Mújica-Jiménez C, Muñoz-Clares RA (2000) Physiological implications of the kinetics of maize leaf phosphoenolpyruvate carboxylase. Plant Physiol 123:149–160

van Ginkel LC, Bowes G, Reiskind JB, Prins HBA (2001) A CO_2-flux mechanism operating via pH-polarity in *Hydrilla verticillata* leaves with C_3 and C_4 photosynthesis. Photosyn Res 68:81–88

Voigt J, Liebich I, Kieß M, Frank R (2001) Subcellular distribution of 14-3-3 proteins in the unicellular green alga *Chlamydomonas reinhardtii*. Eur J Biochem 268:6449–6457

Voznesenskaya EV, Franceschi VR, Kiirats O, Freitag H, Edwards GE (2001) Kranz anatomy is not essential for terrestrial C_4 plant photosynthesis. Nature 414:543–546

Weber A, Servaites JC, Geiger DR, Kofler H, Hille D, Gröner F, Hebbeker U, Flügge U-I (2000) Identification, purification, and molecular cloning of a putative plastidic glucose transporter. Plant Cell 12:787–801

Weise A, Barker L, Kühn C, Lalonde S, Buschmann H, Frommer WB, Ward JM (2000) A new subfamily of sucrose transporters, SUT4, with low affinity/high capacity localized in enucleate sieve elements of plants. Plant Cell 12:1345–1355

Wendt UK, Wenderoth I, Tegeler A, von Schaewen A (2000) Molecular characterization of a novel glucose-6-phosphate dehydrogenase from potato (*Solanum tuberosum* L.). Plant J 23:723–733

Whitney SM, Andrews TJ (2001) The gene for the ribulose-1,5-bisphosphate carboxylase/oxygenase (Rubisco) small subunit relocated to the plastid genome of tobacco directs the synthesis of small subunits that assemble into Rubisco. Plant Cell 13:193–205

Whitney SM, Baldet P, Hudson GS, Andrews TJ (2001) Form I Rubiscos from non-green algae are expressed abundantly but not assembled in tobacco chloroplasts. Plant J 26:535–547

Wiese A, Gröner F, Sonnewald U, Deppner H, Lerchl J, Hebbeker U, Flügge U-I, Weber A (1999) Spinach hexokinase I is located in the outer envelope membrane of plastids. FEBS Lett 461:13–18

Wingler A, Fritzius T, Wiemken A, Boller T, Aeschbacher RA (2000) Trehalose induces the ADP-glucose pyrophosphorylase gene, *ApL3*, and starch synthesis in arabidopsis. Plant Physiol 124:105–114

Wood NT (2002) Cyanobacterial transgene paves the way to increased crop productivity. Trends Plant Sci 7:9

Würth MKR, Winter K, Körner C (1998) Leaf carbohydrate responses to CO_2 enrichment at the top of a tropical forest. Oecologia 116:18–25

Yamada S, Komori T, Hashimoto A, Kuwata S, Imaseki H, Kubo T (2000) Differential expression of plastidic aldolase genes in *Nicotiana* plants under salt stress. Plant Sci 154:61–69

Yamaguchi K, Nishimura M (2000) Reduction to below threshold levels of glycolate oxidase activities in transgenic tobacco enhances photoinhibition during irradiation. Plant Cell Physiol 41:1397–1406

Yu T-S, Lue W-L, Wang S-M, Chen J (2000) Mutation of *Arabidopsis* plastid phosphoglucose isomerase affects leaf starch synthesis and floral initiation. Plant Physiol 123:319–325

Yu Y, Mu HH, Wasserman BP, Carman GM (2001) Identification of the maize amyloplast stromal 112-kD protein as a plastidic starch phosphorylase. Plant Physiol 125:351–359

Zaslavskaia LA, Lippmeier JC, Shih C, Ehrhardt D, Grossman AR, Apt KE (2001) Trophic conversion of an obligate photoautotrophic organism through metabolic engineering. Science 292:2073–2075

Zhang N, Schürmann P, Portis AR (2001) Characterization of the regulatory function of the 46-kDa isoform of Rubisco activase from *Arabidopsis*. Photosyn Res 68:29–37

Dr. Grahame J. Kelly
Centre for Molecular Biotechnology
School of Life Sciences
Queensland University of Technology
Brisbane
Queensland 4000, Australia

e-mail: g.kelly@qut.edu.au

Light Sensory Responses in Lower Plants: Photomovement Versus Photoadaptation

Christoph Forreiter and Gottfried Wagner

Dedicated to Professor Silvia Braslavsky on the occasion of her 60th birthday

1 Introduction

"Plants usually adapt while animals move". This statement, often heard in student examinations, certainly contains some truth, but is not correct under all circumstances. Even seed plants move at certain stages of development and some of their prokaryotic ancestors do all the time (e.g., Taiz and Zeiger 1998 for basic information; Bhaya et al. 2001, and references therein). It is interesting to reflect on the two strategies of photobehavior and the molecular background in vegetative cells in particular when the life cycle is not interrupted by the complications of sexual reproduction. In vegetative cells with the capability of light-energy conversion, the major challenge in survival of both the individual organism and the whole organismic population is appropriate handling of the factor LIGHT. Survival of each single species in the natural environment is only guaranteed by each species finding an appropriate know-how to accomplish this predominant goal; the strategies of photomovement versus photoadaptation that have appeared leave room for an abundance of species to coexist. One chapter of this year's Progress in Botany will give a summary of prokaryotic photoadaptation. Complementary to this aspect of early plant evolution is the field of photomovement. This has lent the title to Volume 1 of a new Comprehensive Series in Photosciences, edited by Donat-Peter Häder and Giulio Jori (2001), and is taken as the basis for discussion of the antipode in photobehavior.

2 Photomovement

The current knowledge profile of photomovement is most comprehensively reflected in the multi-author encyclopedic work edited by Donat-P. Häder and Michael Lebert (2001). Key information from this volume,

Progress in Botany, Vol. 64

restricted to photomovement of single motile cells of eubacteria, archaea and vegetative eukaryotes, will represent this part of the review. Haupt (2001) starts with the subject in giving a clear-cut definition of the terminus technicus "Photomovement". Qualified as a mathematician, he talks with precision of the different categories of photosensory response and develops the field by scale of phylogenetic evolution. In Fig. 9 of his chapter, Foster (2001) has accepted the idea and shows an evolutionary tree of some of the photomovements described in the work of 940 pages in length. The molecular basis of the triggering of photomovement is described by Williams and Braslavsky (2001), i.e., events that occur upon excitation of the chromophores of the sensory photoreceptors. A general description is given of the processes of *cis-trans* isomerization and photoinduced electron transfer. Excursions are made also into fundamental biophysical themes. The chromophores undergoing *cis-trans* isomerization, which to the present knowledge are an open-chain tetrapyrrole in phytochromes, a *para*-hydroxycinnamyl anion in photoactive yellow protein, and retinal in archaean rhodopsins and of ciliate/flagellate algae, make up a good fraction of the nicely illustrated chapter. Blue-light photoreceptor chromophores such as flavins and the perylenquinone-type (e.g., hypericin derivatives in stentorin and blepharismin), which most likely undergo electron transfer as the primary photochemical process, are also discussed in detail. Foster's chapter (2001) on action spectroscopy of photomovement is a highlight in a set of excellent review papers to be appreciated by any scholar interested in photobiology. It is a detailed manual of more than 60 pages in length starting with precise definition of the subject (... "An action spectrum is a measurement as a function of wavelength or photon energy that is proportional to the activation cross section of a pigment that causes a biological response or effect"...). The exceptional benefit of quantitative action spectroscopy is developed from its historical roots [..."What has come to be known as the Stark-Einstein equivalence theory (Einstein 1912) means that if an electronic excitation occurs, then one photon absorbed has the possibility of leading to the excitation of a single molecule. This means that the sensitivity of response is proportional to the number of potentially absorbed photons rather than the total incident energy. All appropriate corrections should be made to remove any error due to cell screening, self-screening of pigments, the presence of other reactions, cell geometry, polarization, etc. The classical admonition was that one could do action spectroscopy when one had parallel logarithmic photon irradiance-response curves, the system is weakly absorbing and if only one primary reaction influenced the effect (Duysens 1970). One must take into account that these criteria are rarely met in photomovement studies"...]. The pros and cons of fairly common experimental situations are discussed, particularly the phototaxis population method and free swimming tracking. For the two techniques favored by Foster, a straight-line fit of the response is yielded with the log of irradiance. For this reason, the threshold can be determined from measurement of responses at only three or four irradiances above threshold. A complete photon irradiance-response curve is unnecessary (Smyth et al. 1988). Further, these straight-line fits need *not* be parallel as required in classical action spectroscopy.

This innovation put forward by Foster (2001) enables an accurate action spectrum of cellular behavior over a wide spectral range to be finished in a few hours or, in principle, even minutes with the proper equipment. Often misled by the experimental situation, plenty of false action spectra have been published in the literature. Such plots may roughly indicate regions of spectral sensitivity but the spectrum may greatly be disturbed by a side effect, and hence be difficult or even dangerous to interpret. Examples are given of distorted action spectra out of the family of rhodopsins and of flavoproteins/pterins, and measures of correction. Dichroic and photochromic receptors are also discussed, with special reference to recommendations given by Hartmann (1983) and Schäfer et al. (1983). A large part of the chapter is dedicated to the identification and characterization of rhodopsins and flavoproteins including their evolution from archaea (for rhodopsins) and euglenoids (for flavoproteins). The chapter concludes with practical advice on how to maximize results with minimum cost and time. After this seminar on photophysical principles, highly vivid in modern cell biology, the reader feels well prepared to focus on the present-day knowledge of biological systems moving with respect to light.

a) Bacteriochlorophyll

The chapter of Armitage (2001) on light responses in purple photosynthetic bacteria also starts from the historical perspective with special emphasis on the work of Christian Gottfried Ehrenberg (1838) on early microscopical observations almost 200 years ago in *Chromatium*. Fortunately enough, Engelmann (1883) isolated a similar *Chromatium* species from the river Rhine and conducted early photo- and chemobiology. He observed "breathing activity" and hunger responses, and concluded that bacteria must be animals with the same urges and needs as human beings, i.e., the behavior confirmed the "unity of organic nature". The current profile of physiology of chemotaxis in *Escherichia coli* and the far more complex system of chemo- and phototaxis in *Rhodobacter sphaeroides* is clearly described. The genomes of *Rhodobacter sphaeroides* and *Rhodopseudomonas palustris* have been sequenced (Choudhary et al. 1999; Pennisi 2001), suggesting four chemosensory gene loci and up to 29 receptors in *Rps. palustris*.

b) Archaean Rhodopsins

With respect to the four archaean rhodopsins in *Halobacterium salinarum*, it is amazing to see the evolutionary economy of nature. Each rhodopsin consists of seven transmembrane alpha-helices enclosing a retinal chromophore linked through a protonated Schiff base to a lysine residue in helix G. The sensory rhodopsins are complexed to their corresponding transducer proteins HtrI and HtrII (halobacterial transducers for sensory rhodopsins I and II), which have conserved methylation and histidine kinase-binding domains that modulate kinase activity which in turn controls flagellar motor switching through a cytoplasmic phosphoregulator. Spudich (2001) emphasizes that the structural similarity of sensory rhodopsin-I (SRI) and sensory rhodopsin-II (SRII) to the transport rhodopsins lends support to models in which the transport and signaling mechanisms derive from the same retinal-driven changes in protein conformation. A deprotonation/reprotonation cycle of the Schiff base in SRI is observed similar to that in the H^+-transport protein BR. In contrast, when SRI is complexed with HtrI, the proton transfer appears plugged within the complex (Haupts et al. 1995). In fact, no changes in the proton concentration are detected in the surrounding medium and the reactions are independent of external pH (Olson et al. 1992). The SRI protein forms a complex with HtrI both in the dark and in the light. Therefore, stimulus relay from the signaling states of SRI to HtrI does not involve protein association/dissociation, but rather structural changes within the complex, unlike signaling from visual rhodopsins to the G-protein transducin. The complete sequence of the *Halobacterium* genome reveals 16 Htr transducers in addition to HtrI and HtrII.

c) Photoactive Yellow Protein

For the photoactive yellow protein (PYP), rich detail concerning structure and function has been collected in recent years (Crielaard et al. 2001). Although the chromophore in PYP has a completely different chemical structure, its photocycle strongly resembles that of the archaean sensory rhodopsins, including *trans-cis* steps of isomerization and a relatively long-lived blue-shifted intermediate state.

PYP apoprotein can be seen as a structural prototype for the three-dimensional fold of the PAS domain superfamily (Pellequer et al. 1998 and references therein). Sequence similarities between the PAS repeats and PYP were identified by Lagarias and coworkers (1995), who probed a sequence database with a 43-residue consensus sequence constructed from the PAS-A and PAS-B domains of plant phytochromes. The PAS acronym was coined originally to describe the 270-residue region encompassing two direct sequence repeats (PAS-A and PAS-B) of 50 residues each that had been identified in the *Drosophila* Period clock protein (PER), the vertebrate dioxin receptor ARNT (Aryl hydrocarbon receptor nuclear translocator), and the *Drosophila* SIM (Single-minded). These three proteins are involved in regulation of circadian rhythms, activation of the xenobiotic response, and cell fate determination, respectively.

More recently, PAS domains have been found in many other proteins, including histidine-kinases, light receptor and regulator proteins, clock proteins, sensor proteins (oxygen-redox sensors), ion channels, and a Ser/Thr kinase with a putative redox-sensing or flavin-binding domain, in which PAS-like regions are named "LOV" (light, oxygen, or voltage). These PAS-containing proteins occur in a wide range of living organisms including: eubacteria, archaea, cyanobacteria, fungi, plants, insects, and mammals. PAS-containing proteins have been categorized into three functional subgroups: (1) transcription activators [DNA-binding proteins with both basic Helix-Loop-Helix (bHLH) and PAS sequence motifs], (2) sensor modules of two-component regulatory systems (oxygen sensor, nitrogen fixation, sensor kinase, etc.), and (3) ion channels (in eukarya).

d) Algal Rhodopsins

Following the nomenclature of Foster (2001), we now turn to *ciliate* green algae, in contrast to *flagellate* prokaryotes. Conspicuous eyespots are present in all major algal phylogenetic lineages. The eyespot in *sensu stricto* is located within a chloroplast, even though the origin of this organelle is, in contrast to that of chloroplasts, generally regarded as polyphyletic. Functional comparisons should therefore mainly be restricted to phylogenetically related phyla and care should be taken when generalizing signaling mechanisms. Kreimer (2001) has restricted his review to the best-studied group to date that is found in green algae. In most species here the globule layers within the eyespot are regularly spaced which is the most striking feature and one basis for contrast enhancement at the presumptive location of the photoreceptor in the plasma membrane patch overlying the globule layers.

In addition to the directivity already gained by its ultrastructure, the spectral properties of the whole optical system further increase the integral of directivity and sensitivity. Here, absorption, reflection and interference of phototactic active light work in harmony with the sensory pigment in dichroic orientation within the patch of cell membrane covering the eye spot. In *Haematococcus pluvialis* and *Chlamydomonas reinhardtii*, the chromophore is oriented almost parallel to the plane of the membrane (Sineshchekov 1991; Yoshimura 1994). In order to guarantee directivity, the strategy of photoreceptor dichroic orientation often is seen in lower plants, even complemented by change in transition moment upon photoexcitation of the photoreceptor involved. Jaffe and Haupt in their different systems noticed these phenomena in the second half of the last century and developed a concept which is known nowadays as the Jaffe–Haupt model (Wagner 2001).

In their thorough analysis of the optical principles of algal light antennas, Foster and Smyth (1980) were the first to postulate that green algal eyespots may take advantage of reflection to increase both light intensity and illumination time at the photoreceptor location. Analysis of eyespot reflection by confocal microscopy provided the first experimental support for the hypothesis of quarter-wave interference reflection in green algal eyespots. The fluence rate of reflected light at the plasma membrane dome spanning eyespots was found in multi-layered structures to be about twice that in single- or double-layered eyespots (Kreimer and Melkonian 1990).

Hegemann and Deininger (2001) describe the green algal rhodopsins as follows: (1) Chlamyrhodopsin contains an all-*trans*, 6-S-*trans* retinal chromophore very much like bacterial rhodopsins, whereas, in animal rhodopsins, 11-*cis* retinal or the derivatives 11-*cis*-3-hydroxy, 11-*cis*-4-hydroxy or 11-*cis*-3,4-dehydroretinal form the functional chromophore. (2) The algal chromophore undergoes a 13-*trans* to *cis* isomerization during illumination. The isomerization is converted into a conformational change of the protein via the 13-methyl group. (3) The retinal is in a planar conformation across the C6–C7 single bond, which links the *trans* conformation of the polyene chain to the ionone ring. (4) The ring is not a functionally essential component. It accelerates reconstitution, but a chromophore with only three conjugated double bonds plus methyl groups also restores photosensitivity.

Polyclonal antibodies against the purified chlamyopsin were used for the localization of the opsin in fixed and permeabilized cells. The opsin appeared as a sharply localized spot of 1 μm in diameter near the equatorial position where in living cells the eyespot is seen. Anti-chlamyopsin antibodies also identified the opsin in eyespot preparations of the chlorophycean alga *Spermatozopsis similis* (Calenberg et al. 1998). The similisopsin is of special interest because a light-regulated GTPase activity was discovered in eyespot preparations of this alga. Light-regulation is suppressed by addition of anti-chlamyopsin antibodies. Kreimer (2001) states that this is the only biochemical evidence so far that algal rhodopsins control a G-protein activity.

The algal opsins are unique in many respects. First of all, transmembrane segments can hardly be identified, and the topography of the algal rhodopsins remains unclear. No significant homology was found to archaean opsin sequences. However, algal opsins do show sequence homology to animal opsins, with a slightly higher relation to invertebrate than to vertebrate opsins. The motif EVEPSKKV is part of the microtubule-associated protein-2 from humans. The identical 8-amino-acid sequence is in the chlamyopsin part of the CL3-like domain and adjacent to the putative pore loop region. It might link the chlamyopsin to a microtubule-associated protein that attaches it to the microtubule rootlet and keeps the eyespot at its specific position. However, the abundant retinal protein of the *Chlamydomonas* eyespot is not the photoreceptor for phototaxis and photophobic responses (Fuhrmann et al. 2001). In opsin-deprived transformants, flash-induced photoreceptor currents are left unchanged. Moreover, photophobic responses as studied by motion analysis and phototaxis tested in a light-scattering assay were indistinguishable from the responses of untransformed wild-type cells.

The earliest step in the rhodopsins-mediated signal transduction chain in green flagellates is generation of the photoreceptor current (Sineshchekov and Govorunova 2001). Membrane depolarization induced by the photoreceptor current leads to the unbalanced motor response of the flagella, which is the basis for phototaxis. If depolarization exceeds a critical level, a voltage-gated flagellar current is triggered across the flagellar membrane, which gives rise to the photophobic response of the cell. Both the photoreceptor and the flagellar current are carried mostly by Ca^{2+} ions under physiological conditions.

e) Flavins and Pterins

Euglena gracilis is a unicellular, photosynthetic ciliate/flagellate which possesses a small, anterior bottle-like invagination, the reservoir. At the bottom of the reservoir, two flagella originate of which only one leaves the reservoir. Close to the contact point of the two flagella, a small organelle is found, i.e., the paraxonemal body (PAB). The electron dense material of the PAB is organized in a paracrystalline array (Piccinni and Mammi 1978 and references therein), while the carotenoid-rich stigma nearby does not show a distinct structure. As judged by the available structural information on the stigma, an interference reflection mechanism as in *Chlamydomonas* and other green algae seems unlikely in *Euglena*. This matches also the finding that the stigma of *Euglena* does not reflect light. On the other hand, while a whole body of indirect evidence points to the PAB as being the photosensory receptor, direct evidence, for example from local irradiation, is missing. Overwhelming indirect evidence exists for the hypothesis that flavin is the primary chromophore involved in photoreception of *Euglena* phototaxis. Pterins very likely act as antenna pigments for the flavins. However, recently, a discussion was raised about the possible role of rhodopsins in photoperception here (Gualtieri 2001). Lebert (2001) discusses the pros and cons in detail, as he does with respect to different methods used during the last century or so to determine tactic behavior.

f) Cryptophyte Phycobiliproteins and Diatom Unknown Pigments

Present knowledge says that cryptomonad algae (cryptophyte algae) originated from a eukaryotic host cell and a eukaryotic endosymbiont by secondary endosymbiosis. Cryptomonads contain phycobilin pigments, as do red algae and cyanobacteria. While organized here in "phycobilisomes" (see below), cryptomonades contain only a single type of phycobilin. These findings have given this taxonomic division an overwhelming significance in the analysis of plant ancestry. It may be expected that cryptomonads are of similar significance with respect to plant photosensory evolution. The striking similarity of the action spectra for phototaxis with that of photoaccumulation and light-induced membrane depolarization in the ciliate *Paramecium* suggests similarities between the still-unidentified photoreceptors (Watanabe and Erata 2001). Cyptomonads remain challenging organisms for future research.

Diatoms are beautiful and intricately shaped algae, known for their characteristic golden pigmentation and robust silica-based cell walls. Recently, diatoms have gained major interest in terms of biotechnology because of the phenomenon of bio-mineralization (Hampp and Noll 2000). It is less well known that one of the two groups, the pennates, is motile by segregation of cell material (mucilage) through the raphe, and recognize light/dark boundaries by photophobic response (step-up or step-down). Because diatoms and other microalgae are important contributors to the overall primary production of many aquatic communities, this cellular behavior is of major impact in terms of photoecology even though little is known of the photosensory physiology (see Cohn 2001; Garcia-Pichel and Castenholz 2001).

g) Stentorin and Blepharismin

Almost a century ago, *Stentor* was shown to avoid light which brings the cells to gather in the shady or dark areas of culture vessels (Holt and Lee 1901; Jennings 1904; Mast 1906). Similar to the case of *Stentor*, *Blepharisma* has been shown to accumulate in shaded regions as a result of step-up photophobic responses, negative phototaxis and positive photokinesis. The photoreceptive function of the two chromophores, stentorin and blepharismin, has been concluded on the basis of the resemblance of the action spectra to the absorption spectra of the whole cells and of the extracted pigment. The action spectra for the membrane receptor potentials are also consistent with the absorption spectra of the pigments (Lenci et al. 2001; Wood 2001). Stentorin and blepharismin do not exhibit a photochemical transformation cycle; a photo-induced change in pigment configuration seems thereby ruled out. Both pigments of the parent compound hypericin (Petrich 2001) can be good electron donors as well as electron acceptors, depending on the redox potentials of the donor/acceptor pairs present in solution. The photomovement responses in the two ciliates are greatly pH-dependent. Upon addition of protonophores, the photoresponses in *Stentor* were shown to decrease significantly. A whole body of supporting information leads to the hypothesis of the sensory transduction chain being based on intracellular pH change, including membrane-limited Ca^{2+} fluxes. Furthermore, pharmacological evidence points to the involvement of heterotrimeric G-protein as a signal transducer in *Stentor* and *Blepharisma* (Lenci et al. 2001).

3 Photoadaptation

Alternative strategies for sessile photosynthetic organisms to respond appropriately to changing light environment are various intracellular operations. Many of them respond to light by metabolic adaptation. The most prominent example is phycobilisome (PBS) assembly in cyanobacteria. Some of these blue-green algae are able to adjust the ratio of their photosynthetic pigments according to environmental conditions. This change in response to light quality was termed *chromatic adaptation*. In addition to photomovement, chromatic adaptation in cyanobacteria turned out to be a valuable model system to study light-driven physiological responses. Since considerable progress has been achieved in understanding the underlying mechanisms, we want to present an update on recent findings regarding photoreceptor and signal transduction, which are strikingly similar to effects observed also in photomovement.

Already a century ago, Engelman and others described that photosynthetic organisms can acclimate to their light environment by changing their pigmentation. Generally, low light intensities stimulate the synthesis of PBS and the rods increase in length. However, many cyanobacteria can alter the composition of PBS even in response to a given light quality. This ability varies considerably between species: Some cannot adapt to light at all, some alter levels of phycoerythrin (PE) only, whereas certain organisms modulate both, PE and phycocyanin (PC) levels. In these organisms, PC and PE are modulated by different light qualities in opposite directions, a process termed *complementary chromatic adaptation* (CCR). This effect can be demonstrated when PCB composition is compared after growth in red light (RL) and green light (GL). In RL, *Fremyella diplosiphon*, a cyanobacteria used for most of the experiments performed so far, accumulates high levels of the blue pigment PC and small amounts of the red pigment PE. In GL, the organism has low levels of PC and high levels of PE. This makes sense, since PC can efficiently absorb RL and PE can efficiently absorb GL, and not vice versa (comprehensively reviewed by Grossman et al. 1993).

Considerable effort was made to identify the genes involved in chromatic adaptation. It turned out that most of them encode structural components of PBS. These genes have been characterized with respect to their sequences and relative levels of their transcripts in cells grown in RL and GL. Generally, abundance of all PC and PE coding mRNAs reflect PBS polypeptide composition under a given light condition (Oelmüller et al. 1988a,b; Mazel et al. 1991). *Fremyella diplosiphon* contains three PC gene sets, two of them are important for chromatic adaptation. The mRNA coding for these two genes accumulates only in cells maintained in RL (Conley et al. 1985). The other mRNA accumulates constitutively. Interestingly, transcription of linker polypeptides is also affected by light. This is not surprising, since most of the coding PBS-assembling linkers are organized in an operon with overlapping sequences, together with the corresponding PC or PE genes (Conley et al. 1988) and genes which attach the tetrapyrrole chromophore to the biliprotein (Fairchild and Glazer 1994). It has been suggested that transcription results in a primary mRNA that is further processed to the appropriate transcript coding either for the linker or the chromoprotein.

To understand the regulatory elements responsible for the light-driven transcription of PC, PE and the corresponding linker peptides, 5′ ends of several transcripts have been mapped. One observation was that a transcript coding for one particular linker element of PE has an mRNA leader of 187 bp. This leader sequence contains several small overlapping open reading frames (ORFs), which are 14 to 21 amino acids long and are probably associated with potential ribosome-binding sites. Although their significance with respect to light regulation of this transcript is not known yet, such ORFs play a role in translational control in other bacteria. However, light-regulated genes should contain additional sequences besides general common regulatory motifs within the promoter region. They should not be present in genes coding for constitutively expressed PCB genes and may be similar for genes that are regulated in a similar manner. One candidate for such an element was pinpointed 83 bp upstream of the transcription start site of the PE coding gene and a second one 195 bp upstream of the transcription start site of the corresponding linker operon. Although both operons respond similarly to light conditions, the function of this element in controlling GL-regulated transcription of PE mRNA remains questionable since it resembles repetitive sequences found in other non-light-regulated genes in *F. diplosiphon* (Mazel et al. 1990). Up to now, too few light-regulated genes have been identified and characterized in these algae.

Many different types of experiments have also been performed to elucidate the nature of the photoreceptor. It has been shown that the action maximum for PE synthesis in *F. diplosiphon* (Haury and Bogorad 1977; Vogelman and Scheibe 1978) and *Tolypothrix tenuis* (Diakoff and Scheibe 1973) is between 540 and 550 nm. Maximum for PC synthesis is between 650 and 660 nm. These action spectra indicate that the photoreceptor involved in chromatic adaptation is most probably a phycobiliprotein (Schäfer and Briggs 1986). Some early lines of evidence indicate that the photoreceptor that controls chromatic adaptation is allophycocyanin (Scheibe 1972; Björn and Björn 1980; Ohad et al. 1981). Cell extracts containing allophycocyanin (AP) exhibit photoreversible absorption changes, as expected for a photoreceptor. Additionally, accumulation of PC is promoted by short-wavelength light of 360 nm (Vogelman and Scheibe 1978; Ohki et al. 1982). However, photoreversibility is a common artifact of biliprotein under mild denaturing conditions (Ohki and Fujita 1979a,b, 1981) and it is not known so far whether the short-wavelength response is mediated by the same regulatory elements as the RL/GL photoreversible response. Other photobiological characterizations revealed that transcription of both PC and PE genes can be triggered by a pulse of inductive light followed by darkness (Oelmüller et al. 1988a,b; 1989). For both gene sets, transcript population increases immediately following transfer to inductive light and reaches a maximum within 2 h at 25 °C. The fluence required for altered transcription from two gene sets is different, suggesting that two distinct photoreceptors regulate expression of the two different phycobiliprotein gene sets. Alternatively, a complex signal transduction chain between photoperception and the control of transcriptional activity might result in the different light responsiveness observed.

It was speculated that the GL/RL control system involves a bilin-associated photochromic photoreceptor resembling a phytochrome. However, some differences to a classical phytochrome-mediated response are obvious: The maxima of the action spectrum for the cyanobacterial photoreceptor are 540 nm (GL) and 640 nm (RL), whereas those for plant phytochromes are at 660 nm (red) and 730 nm (far red). Secondly, it seems that both forms of the cyanobacterial photoreceptor are biologically active, whereas plant phytochromes are physiologically impotent in far red. Thirdly, in the plant phytochrome system there is an "escape" from photocontrol, which is not seen in the cyanobacterial system. Transcription from PC or PE operons remains fully on or off immediately following the terminal irradiation and it is possible to fully reverse the effect with a pulse of complementary light, even after several hours of darkness following the terminal light pulse.

To study this more closely, mutants deficient in chromatic adaptation were created and subsequently complemented by plasmids carrying the wild-type genetic fragments (Cobley and Miranda 1983; Tandeu de Marsac 1983; Bruns et al. 1989). Cobley and Miranda (1983) were the first to report the characterization of pigment mutants after treatment of wild-type cells with UV-C irradiation. In their work they define three classes of mutants dependent on the pigmentation of the cells. In "green" mutants synthesis of PE neither occurred in GL nor RL whereas PC synthesis was normal under both light conditions. In "blue" mutants, both photoinduction of PE and photorepression of PC synthesis were impaired. In "black" mutants, PE was partially induced and PC partially repressed in RL. Hence, black mutants responded to light quality in the opposite way to the wild-type response. Six classes of PBS regulatory mutants were described by Tandeu de Marsac (1983). These strains, isolated after exposure of wild-type cells to nitrosoguanidine, were postulated to represent six of eight possible mutant phenotypes in which PC and PE are abnormally regulated. In the process of optimizing conditions for electroporation, Bruns and coworkers (1989) isolated numerous mutants and focused on their characterization as a primary means of examining the regulatory mechanisms governing chromatic adaptation in *F. diplosiphon*. They defined and characterized three mutant classes (red, blue, and green). A comprehensive analysis, including spectral analyses of cells and cell extracts, determination of phycobiliprotein including linker polypeptide composition together with a quantification of transcripts from PC and PE genes, allowed them to distinguish mutants with lesions in the structural genes of the PBS and/or assembly processes from regulatory mutants. This established two general features of chromatic adaptation. First, a group of single-lesion mutants showed altered transcription of both PE- and PC-coding genes. This suggests the presence of common elements in the chain of events governing transcription of PCB. A second class of mutants, however, revealed phenotypes indicating distinct events that specifically control the expression of either PC or PE genes. Indeed, results from several groups (Oelmüller et al. 1989; Sobczyk et al. 1993) suggest that more than one regulatory element may be involved in controlling the expression of each of the biliprotein genes. Further insight has been obtained by introduction of wild-type DNA into the mutant strains to complement these regulatory mutants (Chiang et al. 1992a).

Using this approach, several genes encoding putative components of light perception and the downstream signal transduction pathway have been isolated. The first thoroughly characterized gene, designated *rcaC*, was localized on a large plasmid rescued from a single complemented colony of a red mutant (Chiang et al. 1992b). Sequence analysis of the complementing fragment revealed that the *rcaC* gene encodes a protein that has strong sequence identity with the *Bacillus subtilis* PhoP protein (Seki et al. 1987). It is a 73-kDa polypeptide with sequence similarities to response regulators of two-component regulatory systems (Chiang et al. 1992a). It is, however, much larger than most response regulators but has two characteristically conserved, aspartate-containing receiver domains, one N-terminal at position 51 and the other near the C-terminus. The Asp-51 residue is likely to be phosphorylated in RL-grown cells leading to enhanced PC and repression of PE synthesis. Under GL conditions, wild-type cells probably dephosphorylate the aspartate residue, thus triggering elevated PE synthesis and depressing PC synthesis. Adjacent to the N-terminal receiver domain of RcaC is a sequence predicted to bind DNA (Appleby et al. 1996).

The "black" mutants proved to be particular interesting. They failed to respond to light changes at all. However, they could be complemented by a *rcaE*, which encodes a polypeptide of 74 kDa (Kehoe and Grossmann 1997). The C-terminal region of RcaE has motifs typical of bacterial sensor kinases. The N-terminal half of the polypeptide has a domain of about 140 amino acids with similarity to a part of the phytochrome sensor domain, although not to the attachment site of the chromophore itself. Similar to plant phytochromes, the central region of the protein contains a PAS domain (see above; Taylor and Zhulin 1999). They are involved in protein–protein interactions or binding of a redox-active prosthetic group. Recently, Grossmann and Kehoe (1997) reported that RcaE binds covalently to a linear tetrapyrrole chromophore within the phytochrome-like domain. The phenotype of these mutants and the similarity of RcaE to sensor kinases and eukaryotic phytochrome photoreceptors suggest that the *rcaE* gene product encodes a photoreceptor in chromatic adaptation (Grossman and Kehoe 1997; Kehoe and Grossman 1997).

Some of the "red" mutants could not be complemented by *rcaC* (Kehoe and Grossman 1997), whereas one was complemented by *rcaE*. Interestingly, a second was rescued by *rcaF*, a gene located downstream of *rcaE* encoding a small response regulator. This is similar to the situation described for *Synechocystis*, in which Cph1, a prokaryotic phytochrome histidine kinase of the two-component type, lies upstream of its response regulator Rcp1 (Hughes and Lamparter 1997; Hughes et al. 1997; Yeh et al. 1997; Lamparter et al. 2001). RcaF may act as an intermediate in the phosphorelay pathway controlling chromatic adaptation and facilitate phosphate transfer from its sensor (presumably RcaE) to other response regulators such as RcaC. These findings would be in line with the observation made by the analysis of several other bacterial phytochrome-like histidine kinases (summarized by Vierstra and Davis 2000). In this context it may be of particular interest that, in contrast to these findings, in *Rhodobacter sphaeroides*, a photosynthetic bacteria, adaptation of the photosynthetic apparatus is under control of a blue-light receptor system (G. Klug and S. Baatsch, pers. comm.). Not surprisingly, other regulatory components turned out to be involved in controlling chromatic adaptation. A novel class of mutants in *F. diplosiphon* only affecting PE expression has been identified (designated turquoise, FdTq). These mutants exhibit normal downregulation of PC but cannot activate PE gene transcription in GL. Complementation of the FdTq strains uncovered two genes, designated *trqA* and *trqB*, encoding polypeptides which were related to protein phosphatases. This finding is interesting, especially because phosphorylation of the putative regulatory protein RcaA has been implicated in the control of PE gene expression (Sobczyk et al. 1993).

Summarizing the above data, Grossman and coworkers (2001) deduced a model of how regulation of complementary chromatic adaptation could be achieved (Fig. 1): In RL neither a repressor of PC nor an activator of PE is functioning. This results in increased transcription of PC and repressed transcription of PE in RL. After shifting to GL, transcriptional regulators become active, leading to enhanced transcription of PE and repression of PC. The three regulatory elements controlling complementary chromatic adaptation are RcaE, RcaF and RcaC. Although these polypeptides have features of bacterial two-component regulators, the phosphorelay system responsible for chromatic adaptation is unique, because it includes five potential Pi-acceptor domains. RcaE, the putative photoreceptor, perceives the light signal. RL causes RcaE to undergo an autophosphorylation followed by transfer of the phosphoryl groups to the response regulator RcaF. In the absence of RcaE, RcaF might interact with other phosphoryl donors. RcaF then might transfer phosphoryl groups to the conserved histidine within RcaC, which can pass it (probably) to either the N- or C-terminal receiver domain. The N-terminal receiver domain of RcaC is critical, whereas the role of the C-terminal receiver is still unclear. In GL, RcaE acts as a phosphatase or can block phosphotransfer by binding to RcaF, causing activation of PE synthesis and suppression of PC.

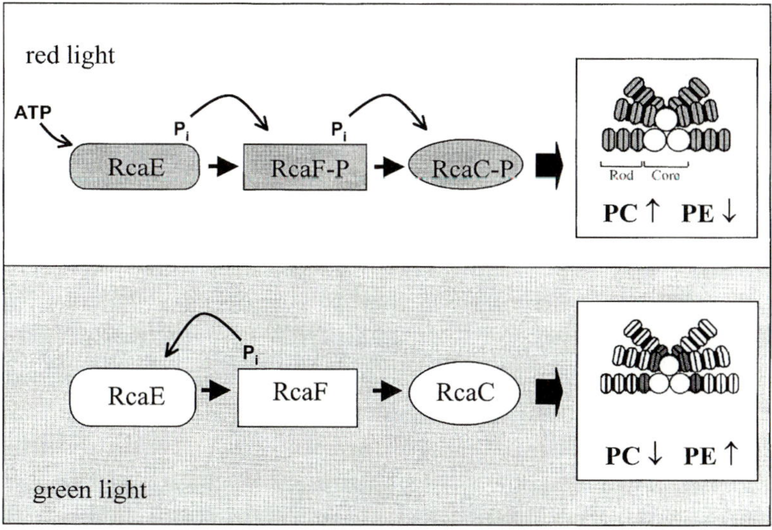

Fig. 1. Phosphorelay model explaining complementary chromatic adaptation in cyanobacteria. Red light stimulates phosphorylation of components of the signal transduction pathway thus activating transcription for inducible PC genes and suppression of PE genes. Dephosphorylation of signal transduction components in green light results in suppression of PC and activation of PE synthesis. (Modified after Grossman et al. 2001)

4 Résumé and Outlook

The survival strategies of photomovement versus photoadaptation have in recent years reached levels of clear-cut genetic analysis, as evidenced for lower plants in this review. Not always seen, the field of photoperception has been molecularly minded from the very beginning because quantitative action spectroscopy allows nothing but molecular conclusions. Photoreceptor families were identified in the last century long before biochemistry came into play. Conclusions on actual gene products have been left to more recent years based on techniques of, e.g., successful cell transformation. Interestingly enough, we learn from those experiments the concept of bulk members of the sensory photoreceptor families discussed so far that are not essential to the particular photobehaviour observed. Thus, biology in the matter of survival has built up a large stock of opportunities to choose from, and simple economy is not the only directive, even in lower plant organisms.

Acknowledgments. We would like to thank Jon Hughes and Frank Landgraf for carefully reading the manuscript and helpful discussions.

References

Appleby JL, Parkinson JS, Bourret RB (1996) Signal transduction via the multi-step phosphorelay: not necessarily a road less traveled. Cell 86:845–848

Armitage JP (2001) Light responses in purple photosynthetic bacteria. In: Häder DP, Lebert M (eds) Photomovement. Elsevier, Amsterdam, pp 117–150

Bhaya D, Takahasi A, Grossman AR (2001) Light regulation of type IV pilus-dependent motility by chemosensor-like elements in *Synechocystis* PCC6803. Proc Natl Acad Sci USA 98:7540–7545

Björn L0, Björn GS (1980) Photochromic pigments and photoregulation in blue-green algae. Photochem Photobiol 32:849–852

Bruns B, Briggs WR, Grossman AR (1989) Molecular characterization of phycobilisome regulatory mutants in *Fremyella diplosiphon.* J Bacteriol 171:901–908

Calenberg M, Brohsonn U, Zedlacher M, Kreimer G (1998) Light and Ca^{2+}-modulated heterotrimeric GTPases in the eyespot apparatus of a flagellate green alga. Plant Cell 10:91–103

Chiang GG, Schaefer MR, Grossman AR (1992a) Complementation of a red-light indifferent cyanobacterial mutant. Proc Natl Acad Sci USA 89:9415–9419

Chiang GG, Schaefer MR Grossman (1992b) Transformation of the filamentous cyanobacterium *Fremyella diplosiphon* by conjugation or electroporation. Plant Physiol Biochem 30:315–325

Choudhary M, Mackenzie C, Mouncey NJ, Kaplan S (1999) RsGDB, the *Rhodobacter sphaeroides* genome database. Nucleic Acids Res 27:61–62

Cohn SA (2001) Photo-stimulated effects on diatom motility. In: Häder DP, Lebert M (eds) Photomovement. Elsevier, Amsterdam, pp 375–401

Cobley JG, Miranda RD (1983) Mutations affecting chromatic adaptation in the cyanobacterium *Fremyella diplosiphon.* J Bacteriol 153:1486–1492

Conley PB, Lemaux PG, Grossman AR (1985) Cyanobacterial light-harvesting complex subunits encoded in two red light-induced transcripts. Science 230:550–553

Conley PB, Lemaux PG, Grossman AR (1988) Molecular characterization and evolution of sequences encoding light harvesting components in the chromatically adaptive cyanobacterium *Fremyella diplosiphon*. J Mol Biol 199:447–465

Crielaard W, Kort R, Hellingwerf KJ (2001) Photoactive yellow protein, a photoreceptor from purple bacteria. In: Häder DP, Lebert M (eds) Photomovement. Elsevier, Amsterdam, pp 179–192

Diakoff S, Scheibe S (1973) Action spectra for chromatic adaptation in *Tolypothrix tenuis*. Plant Physiol 51:382–385

Duysens LNM (1970) Photobiological principles and methods. In: Halldal P (ed) Photobiology of microorganisms. Wiley-Interscience, New York, pp 1–16

Einstein A (1912) Thermodynamische Begründung des photochemischen Äquivalentgesetzes. Ann D Phys 37:832–838

Ehrenberg CG (1838) Die Infusionsthierchen als vollkommene Organismen. Ein Blick in das tiefere organische Leben der Natur, L. Voss, Leipzig,15 pp

Engelmann TW (1883) *Bakterium photometricum*. Ein Beitrag zur vergleichenden Physiologie des Licht- und Farbensinnes. Pfuegers Arch Gesamte Physiol Menschen Tiere 42:183–186

Fairchild CD, Glazer AN (1994) Oligomeric structure, enzyme kinetics, and substrate specificity of the phycocyanin alpha subunit phycocyanobilin lyase. J Biol Chem 269:8686–8694

Foster KW (2001) Action spectroscopy of photomovement. In: Häder DP, Lebert M (eds) Photomovement. Elsevier, Amsterdam, pp 51–115

Foster KW, Smyth RD (1980) Light antennas in phototactic alga. Microbiol Rev 44:572–630

Fuhrmann M, Stahlberg A, Govorunova E, Rank S, Hegemann P (2001) The abundant retinal protein of the *Chlamydomonas* eye is not the photoreceptor for phototaxis and photophobic responses. J Cell Sci 114:3857–3863

Garcia-Pichel F, Castenholz RW (2001) Photomovement of microorganisms in benthic and soil microenvironments. In: Häder DP, Lebert M (eds) Photomovement. Elsevier, Amsterdam, pp 403–420

Grossman AR, Kehoe DM (1997) Phosphorelay control of phycobilisome biogenesis during complementary chromatic adaptation. Photosynth Res 53:95–108

Grossman AR, Schaefer MR, Chiang GG, Collier JL (1993) The phycobilisome, a light-harvesting complex responsive to environmental conditions. Microbiol Rev 57:725–749

Grossman AR, Bhaya D, He Q (2001) Tracking the light environment by cyanobacteria and the dynamic nature of light harvesting. J Biol Chem 276:11449–11452

Gualtieri P (2001) Rhodopsin-like-proteins: light detection pigments in *Leptolyngbya*, *Euglena*, *Ochromonas*, *Pelvetia*. In: Häder DP, Lebert M (eds) Photomovement. Elsevier, Amsterdam, pp 281–295

Häder DP, Lebert M (eds) (2001) Photomovement. In: Häder DP, Jori G (eds) Comprehensive series in photosciences, vol 1. Elsevier, Amsterdam

Häder DP, Jori G (eds) (2001) Comprehensive series in photosciences, vols 1–3. Elsevier, Amsterdam

Hampp N, Noll F (2000) Nanobionics – from molecules to applications (238. WEH-Seminar). Phys Bl 10:75

Hartmann KM (1983) Action spectroscopy. In: Hoppe W, Markl H, Ziegler H (eds) Biophysics. Springer, Berlin Heidelberg New York, pp 115–144

Haupt W (2001) Photomovement: past and future. In: Häder DP, Lebert M (eds) Photomovement. Elsevier, Amsterdam, pp 1–13

Haupts U, Haupts C, Oesterhelt D (1995) The photoreceptor sensory rhodopsin I as a two-photon-driven proton pump. Proc Natl Acad Sci USA 92:3834–3838

Haury JF, Bogorad L (1977) Action spectra for phycobiliprotein synthesis in a chromatically adapting cyanophyte. Plant Physiol 60:835–839

Hegemann P, Deininger W (2001) Algal eyes and their rhodopsin photoreceptors. In: Häder DP, Lebert M (eds) Photomovement. Elsevier, Amsterdam, pp 229–243

Holt EB, Lee FS (1901) The theory of phototactic response. Am J Physiol 4:460–481

Hughes JE, Lamparter T (1997) Prokaryotes and phytochrome. The connection to chromophores and signaling. Plant Physiol 121:1059–1068

Hughes JE, Lamparter T, Mittmann F, Hartmann E, Gärtner W, Wilde A, Börner T (1997) A prokaryotic phytochrome. Nature 386:663

Jennings HS (1904) Contributions to the study of the behaviour of lower organisms. Pub Carnegie Inst Washington, no 16, pp 1–256

Kehoe DM, Grossman AR (1997) New classes of mutants in complementary chromatic adaptation provide evidence for a novel four-step phosphorelay system. J Bacteriol 179:3914–3921

Kreimer G (2001) Light perception and signal modulation during photoorientation of flagellate green algae. In: Häder DP, Lebert M (eds) Photomovement. Elsevier, Amsterdam, pp 193–227

Kreimer G, Melkonian M (1990) Reflection confocal laser scanning microscopy of eyespots in flagellated green algae. Eur J Cell Biol 53:101–111

Lagarias DM, Wu SH, Lagarias JC (1995) Atypical phytochrome gene structure in the green alga *Mesotaenium caldariorum*. Plant Mol Biol 29:1127–1142

Lamparter T, Esteban B, Hughes J (2001) Phytochrome Cph1 from the cyanobacterium *Synechocystis* PCC6803. Purification, assembly and quaternary structure. Eur J Biochem 268:4720–4730

Lebert M (2001) Phototaxis of *Euglena gracilis* – flavins and pterins. In: Häder DP, Lebert M (eds) Photomovement. Elsevier, Amsterdam, pp 297–341

Lenci F, Ghetti F, Song P-S (2001) Photomovement in ciliates. In: Häder DP, Lebert M (eds) Photomovement. Elsevier, Amsterdam, pp 475–503

Mast SO (1906) Light reactions in lower organisms. I. *Stentor coeruleus*. J Exp Zool 3:359–399

Mazel D, Houmard J, Tandeau de Marsac N (1990) Highly repetitive DNA sequences in cyanobacterial genome. J Bacteriol 172:2755–2761

Mazel D, Bernard C, Schwarz A, Castets M, Houmard J, Tandeau de Marsac N (1991) Characterization of two insertion sequences IS701 and IS702 from the cyanobacterium *Calothrix* sp. PCC 7601. Mol Microbiol 5:2165–2170

Oelmüller R, Conley PB, Federspiel N, Briggs WR, Grossman AR (1988a) Changes in accumulation and synthesis of transcripts encoding phycobilisome components during acclimation of *Fremyella diplosiphon* to different light qualities. Plant Physiol 88:1077–1083

Oelmüller R, Grossman AR, Briggs WR (1988b) Photoreversibility of the effect of red and green light pulses on the accumulation in darkness of mRNAs coding for phycocyanin and phycoerythrin in *Fremyella diplosiphon*. Plant Physiol 88:1084–1091

Oelmüller R, Grossman AR, Briggs WR (1989) Role of protein synthesis in regulation of phycobiliprotein mRNA abundance by light quality in *Fremyella diplosiphon*. Plant Physiol 90:1486–1499

Ohad I, Schneider HAW, Gendel S, Bogorad L (1981) Light-induced changes in allophycocyanin. Plant Physiol 65:6–12

Ohki K, Fujita J (1979a) In vivo transformation of phycobiliproteins during bleaching of *Tolyphotrix tenuis* to forms active in photoreversible absorption changes. Plant Cell Physiol 20:1341–1347

Ohki K, Fujita J (1979b) Photoreversible absorption changes of guanidine-HCl-treated phycocyanin and allophycocyanin isolated from the blue-green alga *Tolyphotrix tenuis*. Plant Cell Physiol 20:483–490

Ohki K, Fujita J (1981) On the relationship of photocontrol of phycoerythrin formation and the photoreversible pigment of Scheibe. Plant Cell Physiol 22:347–357

Ohki K, Watanabe M, Fujita J (1982) Action of near UV and blue light on the photocontrol of phycobiliprotein formation. A complementary chromatic adaptation. Plant Cell Physiol 23:651–656

Olson KD, Deval P, Spudich JL (1992) Absorption and photochemistry of sensory rhodopsin-I: pH effects. Photochem Photobiol 56:1181–1187

Pellequer JL, Wager-Smith KA, Kay SA, Getzoff ED (1998) Photoactive yellow protein: a structural prototype for the three-dimensional fold of the PAS domain superfamily. Proc Natl Acad Sci USA 95:5884–5890

Pennisi E (2001) Microbial genomes. Sequences reveal borrowed genes. Science 294:1634–1635

Petrich J (2001) Hypericin and its analogs. Photochem Photobiol 74:iii–iv

Piccinni E, Mammi M (1978) Motor apparatus of *Euglena gracilis*: ultrastructure of the basal portion of the flagellum and the paraflagellar body. Boll Zool 45:405–414

Schäfer E, Briggs WR (1986) Photomorphogenesis from signal perception to gene expression. Photobiochem Photobiophys 12:305–320

Schäfer E, Fukshansky L, Shropshire W Jr (1983) Action spectroscopy of photoreversible pigment systems. In: Shropshire W Jr, Mohr H (eds) Photomorphogenesis. Springer, Berlin Heidelberg New York, pp 39–67

Scheibe J (1972) Photoreversible pigment: occurrence in blue-green algae. Science 176:1037–1039

Seki T, Yoshikawa H, Takahashi H, Saito H (1987) Cloning and nucleotide sequence of *phoP*, the regulatory gene for alkaline phosphatase und phosphodiesterase in *Bacillus subtilis*. J Bacteriol 169:2913–2916

Sineshchekov OA (1991) Electrophysiology of photomovements in flagellated algae. In: Lenci F, Ghetti F, Colombetti G, Häder DP, Song PS (eds) Biophysics of photoreceptors and photomovements in microorganisms. Plenum Press, New York, pp 191–202

Sineshchekov OA, Govorunova G (2001) Electrical events in photomovement of green flagellated algae. In: Häder DP, Lebert M (eds) Photomovement. Elsevier, Amsterdam, pp 245–280

Smyth RD, Saranak J, Foster KW (1988) Algal visual systems and their photoreceptor pigments. Prog Phycol Res 6:255–286

Sobczyk A, Schyns G, Tandeau de Marsac N, Houmard J (1993) Transduction of the light signal during complementary chromatic adaptation in the cyanobacterium *Calothrix* sp. PCC 7601: DNA binding proteins and modulation by phosphorylation. EMBO J 12:997–1004

Spudich JL (2001) Color-sensitive vision by haloarchaea. In: Häder DP, Lebert M (eds) Photomovement. Elsevier, Amsterdam, pp 151–178

Taiz L, Zeiger E (1998) Plant physiology. Sinauer Assoc, Sunderland, pp 1–792. Homepage: http://www.sinauer.com

Tandeau de Marsac N (1983) Phycobilisomes and complementary adaptation in cyanobacteria. Bull Inst Pasteur 81:201–254

Taylor BL, Zhulin IB (1999) PAS domains: internal sensors of oxygen, redox potential, and light. Microbiol Mol Biol Rev 63:479–506

Vierstra RD, Davis SJ (2000) Bacteriophytochromes: new tools for understanding phytochrome signal transduction. Cell Dev Biol 11:511–521

Vogelman TC, Scheibe J (1978) Action spectrum from chromatic adaptation in the blue-green alga *Fremyella diplosiphon*. Planta 143:233–239

Wagner G (2001) Phytochrome as an algal photoreceptor. In: Häder DP, Lebert M (eds) Photomovement. Elsevier, Amsterdam, pp 897–925

Watanabe M, Erata M (2001) Yellow-light sensing phototaxis in cryptomonad algae. In: Häder DP, Lebert M (eds) Photomovement. Elsevier, Amsterdam, pp 343–373

Williams RM, Braslavsky SE (2001) Triggering of photomovement-molecular basis. In: Häder DP, Lebert M (eds) Photomovement. Elsevier, Amsterdam, pp 15–50

Wood DC (2001) Electrophysiology and light responses in *Stentor* and *Blepharisma*. In: Häder DP, Lebert M (eds) Photomovement. Elsevier, Amsterdam, pp 505–518
Yeh KC, Wu SH, Murphy JT, Lagarias JC (1997) A cyanobacterial phytochrome two-component light sensory system. Science 277:1505–1508
Yoshimura K (1994) Chromophore orientation in the photoreceptor of *Chlamydomonas* as probed by stimulation with polarised light. Photochem Photobiol 60:594–597

Christoph Forreiter
Gottfried Wagner
Pflanzenphysiologie, Justus-Liebig-Universität
Senckenbergstr. 3
35390 Giessen, Germany

e-mail: christoph.forreiter@bot3.bio.uni-giessen
e-mail: gottfried.wagner@bio.uni-giessen.de

Circadian Rhythmicity: Is the "Biological Clock" Hardware or Software?

Ulrich Lüttge

1 Historical Reminiscence and Current Questions

Awareness of the influence of external rhythms such as seasons, tides and the solar day on organisms, plants and animals including humans, dates back to the early advent of human culture (Waterhouse 2001). One of the earliest demonstrations of an intrinsic or endogenous free-running daily rhythm in organisms is often acknowledged to be that of the nyctinastic day/night movements of the pinnulae of leaves of the higher plant *Mimosa* by De Mairan (1729). Pioneers of higher-plant chronobiology are Wilhelm Pfeffer (1845–1920) and Erwin Bünning (1906–1990). Notwithstanding much work producing a wealth of evidence, even in the latter half of the twentieth century, it often occurred – as occasionally mentioned by Erwin Bünning – that the notion of a "biological clock" was associated with parascience. As recently as 1989 one could still read that biorhythmicity and the "biological clock" belonged to the field of parabiology together with fire-walking, the divining rod, and the like (Resch 1989). Conversely, international societies and journals of chronobiology were founded, and indeed occurrence of endogenous "daily" rhythmicity in organisms including green organisms such as the prokaryotic cyanobacteria, eukaryotic unicellular and multicellular algae and higher plants is now a solid fact.

Endogenous "daily" rhythms corresponding to the solar day of 24 h are always circadian, i.e. approximately (*circa*) a day (*dies*) with period lengths of somewhat shorter or longer than 24 h. Conversely, there is also a wealth of rhythms in plants which are not related to the solar day, with considerably shorter or longer periods. Although much can also be learned from their physiological, biochemical, molecular and theoretical analysis for the understanding of circadian rhythms, they must remain excluded from this review to keep its volume manageable.

To the extent this is possible with the space allowed, the phenomenology of circadian rhythmicity, i.e. output of the "biological clock", in terms of overt rhythms will be reviewed. However, with the bewildering number of reports in the literature this must remain limited, and we must ask for conceptual progress. This is best characterised by the major

Progress in Botany, Vol. 64

questions currently asked. These emerge from important recent reviews (Millar 1999; Somers 1999; Barak et al. 2000; McClung 2000; Staiger 2000; 2002; Golden and Strayer 2001), marking the forefront of research:

1. Is there one "central clock", or are there individual oscillators on the structural level and/or the level of genes?
2. What is the interaction of external control parameters (*zeitgeber)* with signal transduction networks?
3. How can we assign observations to input pathways, oscillator elements and output pathways, respectively, of circadian rhythms?
4. What are hierarchical links of oscillating elements?
5. What is the role of synchronisation and desynchronisation and noise?
6. What is the "hardware" and the "software", respectively, in the oscillators?
7. Is there a biologically adaptive value of circadian timing?

However, some of these questions are not even new (1 and 2). They were already posed in early books by Bünning (1973) and Sweeney (1969, 1987); evergreens indeed on the reading list of chronobiologists.

Almost unanimously, the quoted recent reviews (1999–2001, see above) present a graphic scheme of

Input (external control parameters) — input pathways → oscillator — output pathways → output (overt rhythms)

to establish the order of events; sometimes the oscillator is explicitly called a "central oscillator" (Somers 1999; McClung 2000). However, it is also noted that this model is insufficient and inadequate (McClung 2000). An essential point of view here is that we must distinguish between the structural molecular and cytological hardware and the software of transmission of information, where the latter needs both theoretical enlightenment and actual experimental documentation.

2 One "Central Clock" or Individual Oscillators

a) The Morphological and Anatomical Level

Although it is hard to be sure, it may well be that the conception of a central clock originates from the localisation of respective morphological structures in highly developed animals. For instance, nerve dissection experiments in a cockroach have shown that the central circadian oscillator in relation to the external control parameter light must reside in the *lobus opticus* (Czihak et al. 1990). In mammals including man the central circadian oscillator is accepted to be the suprachiasmatic nucleus of the hypothalamus in the brain. In plants, we cannot delineate such neurological centres. However, even in higher animals, insects, vertebrates, and mammals, morphological evidence for one central oscillator is inconclusive. The cockroach experiment cannot be generalised, and among the insects there is no evidence for a single central clock for all circadian functions (Czihak et al. 1990). In fish, there are independent oscillators, which also function ex vivo in tissue and organ cultures (Whitmore et al. 2000), and in mammals there are peripheral oscillators that may operate independently of the suprachiasmatic nucleus (Cermakian and Sassone-Corsi 2001). Thus, even in the highest animals, it is as Czihak et al. wrote as early as 1990, that most likely all highly differentiated animals have a system of several circadian oscillators. This is not the place to review the animal literature, but it does not provide a basis for assuming a morphologically defined central oscillator.

In plants, there are also examples of different circadian rhythms in different organs, e.g. photoperiodic flowering and leaf movements in *Pharbitis nil* (Bollig 1977). When the two cotyledons of tobacco seedlings were entrained so that they showed phases offset by 180° against each other, they remained in antiphase under continuous light indicating independent clocks (Thain et al. 2000). We even have evidence that there is at least one oscillator in every cell (Millar 1998, 1999) and "that these oscillators act autonomously and independently of any centralised pacemaker" (Barak et al. 2000). There may be different oscillators within a given cell providing different overt outputs (Roenneberg and Morse 1993; Morse et al. 1994), but this then is a matter of cell function rather than the morphological and anatomical level (see Sect. 5). When the different types of output rhythms are separated in phase but still have the same period length, of course this is not yet very strong evidence for different oscillators driving them. A stronger indication for the existence of separate oscillators is the occurrence of different rhythms with different circadian periods, as for example found in bean plants. In the same leaves a rhythm of stomatal opening with a circadian period length of approximately 24 h and a rhythm of pulvinar leaf movements with a period length of approximately 27 h have been observed (Hennessey et al. 1992).

b) The Cellular Level

In the early days of biorhythm research it was assumed that the organisation of the eukaryotic cell and that of the genome in chromosomes was prerequisite for circadian rhythmicity. Thus, the observation of a circadian advancement of a culture of *Escherichia coli* in a "race tube" containing nutrients was an important finding (Rogers and Greenbank 1930). Transplantation experiments of nuclei between cells of two cultures of the giant cells of *Acetabularia* which were kept under conditions of circadian photosynthetic rhythmicity separated in phase by 180° showed that the nucleus determined the phase of the rhythm (Schweiger et al. 1964a,b). Conversely, rhythmicity also persisted for days and longer than a month in enucleated cells (Schweiger et al. 1964a,b). However, although as recently as 1986, it was still proposed that circadian rhythmicity was a domain of eukaryotic cells, the 80S ribosomes of which being responsible for functioning of periodicity (Jerebzoff 1986), a wealth of overt circadian phenomena in procaryotes especially in cyanobacteria (Huang et al. 1994; Golden et al. 1997; Ishiura et al. 1998; Iwasaki and Kondo 2000) has now clearly made this question obsolete as a basic problem. There is no reason to assume the nucleus of eukaryotic cells to be the "master-clock" (Bünning 1973), and at any rate it would be the quest for identifying a gene or genes as master-clock(s) that we must move on to in the next section.

Another question, of course, is the involvement of the cell cycle of eukaryotic cells (Doerner 1994; Francis and Halford 1995) in circadian rhythms of cell division (Edmunds and Tamponnet 1990; Tamponnet and Edmunds 1990), growth and development, where the question was asked if the cell cycle was the driving oscillator or, if conversely, there was an oscillator driving the cell cycle which was taken to be more likely (Makarov et al. 1995; Titlyanov et al. 1996).

c) The Gene Level

The first genetic experiments with crossings affecting period length of endogenous leaf movements were performed by Bünning (1935) using *Phaseolus multiflorus*. Currently, genetics and studies with mutants and molecular biology produce new phenomena of oscillating genes and gene products at a breathtaking speed (see also Sect. 3.a). The major "model" systems have become *Drosophila* (insects), *Neurospora* (fungi), *Synechococcus* (cyanobacteria) and *Arabidopsis* (higher plants).

From such work the most compelling argument for the possible existence of central oscillators arises, namely by the identification of genes that appear to master an array of different overt rhythmic outputs. Mutations of the *period* (*PER*) locus in *Drosophila* affect period length of the circadian rhythmicity of eclosion and locomoter activity and the very short ultradian rhythm of the male courtship song in the same way, i.e. per^s shortens and per^l increases period length (Dunlap 1993). There may have been at least some conservation of such clock genes during evolution as the clock gene *frequency* (*FRQ*) of the fungus *Neurospora* (Loros 1995) shares a sequence element with the *PER* gene of *Drosophila* (McClung et al. 1989), although such conservation may not be very strong (Millar 1999; McClung 2000). In *Arabidopsis* such a central clock gene may be *APRR1/TOC1* (from timing of chlorophyll *a*, *b*-binding light harvesting complex b expression; Somers et al. 1998b; Barak et al. 2000; McClung 2000; Golden and Strayer 2001; Makino et al. 2001).

Often organisms, plants and animals, and even single cells, show different overt circadian rhythms at the same time, which also may be separated in phase. In *Kalanchoë* species with crassulacean acid metabolism (CAM), the circadian rhythms of CAM and nitrate reductase (NR) can be separated. They are arrested at opposite phases when plants are transferred from continuous light to darkness (Lillo et al. 2001). In the unicellular *Gonyaulax polyedra*, photosynthesis, bioluminescence and mitosis oscillate with different phases (see Czihak et al. 1990; for more examples, see Sect. 5).

If such different rhythms were governed by the same gene product, one might well arrive at considering it based on a central oscillator gene. For example, the genes *LHY* (late elongated hypocotyl) and *CCA1* (circadian clock associated) appear to control overt rhythmicity of leaf movement, hypocotyl elongation and other genes, the expression of which peaks with different phases (Barak et al. 2000). Indeed, there are many clock-controlled gene products (CCGs) that may act as "slaves" or suboscillators (Somers 1999; Barak et al. 2000). We may have hierarchies of genes just like in development or pattern formation in space, e.g. with the maternal master genes, gap genes, pair rule genes, homeotic genes and segment polarity genes in the early development of *Drosophila*. For rhythmicity, or pattern formation in time, the *PER* gene of *Drosophila* and for plants *TOC1*, *CCA1* and *LHY* of *Arabidopsis* may actually have such master functions.

Perhaps the closest advance to reconstruction of a molecular oscillator currently is achieved for cyanobacteria (*Synechococcus elongatus* PCC 7942). However, this is not the result of a single gene activity but a complex highly regulated arrangement of gene products, i.e. a three-gene cluster *kai A*, *B* and *C*, where *kai C* encodes an ATP-binding peptide, *cik A* (circadian input kinase) a bilin-type phytochrome-related photoreceptor at the input side and the family of group 2 sigma factors, RpoD2, RpoD3, RpoD4 and SigC at the output side (Ishiura et al. 1998; Katayama et al. 1999; Iwasaki et al. 2000; Schmitz et al. 2000).

The overwhelming multitude of genes with transcriptional oscillation (oscillating mRNA) currently identified, and where new ones are added at high rate, makes it difficult to envisage a single gene as the decisive hardware of a central oscillator (see Sect. 6). The existence of a single-gene "central" clock is far from being a proven fact. If a "central oscillator" was a discrete function of a given gene, one might also expect much larger homologies between the various so-called master genes identified in the "clock mutants" of *Drosophila*, mammals, *Neurospora*, *Arabidopsis* and cyanobacteria, as is the case for genes of many other basic functions of living cells. The important plant genes have no sequence homology with the clock genes of *Drosophila* or mammals (Somers 1999; Staiger and Heintzen 1999), and CCGs of cyanobacteria have no homologies with CCGs of eukaryotes (Ishiura et al. 1998). The extent to which the central clock concept is attended by difficulty and ambiguity is also documented by observations like the same authors in one publication strongly referring to the "central clock" (Tóth et al. 2001) and in a simultaneous different publication writing that there is "evidence for multiple copies of self-sustained and entrainable clocks in both plants and animals. The clocks can be desynchronized experimentally" (Hall et al. 2001).

3 The Output of Overt Rhythms

The overt rhythms that we see as output of the oscillators constitute a vast phenomenology. We observe circadian rhythms on a wide scope of scaling, from molecules and individual enzymes to metabolic pathways, from membranes and organelles to cells, organs and whole plants.

a) The Molecular and Enzyme Level

Circadian oscillations occur at all points on the way from the gene to the protein, i.e. in transcription, transcript abundance, translation and post-translational processing (Somers 1999). Circadian gene expression in the procaryotic *Synechococcus* appears to be an universal phenomenon, while in the eukaryotic *Chlamydomonas reinhardtii* many but not all genes are under circadian clock regulation (Jacobshagen et al. 2001) and in *Arabidopsis thaliana* only 2–6% of the genes are circadian (Harmer et al. 2000; Schaffer et al. 2001). Observations have accumulated in the last decade and new discoveries are added at increasing pace. This is now much facilitated by the break-through technique of using bioluminescence (firefly luciferase) transcriptional reporter constructs to monitor gene activity and thus obtain long and densely spaced time-series of measurements without painstaking analytical efforts. Another new approach accelerating acquisition of molecular data on large amounts of genes with oscillating activities is the microarray technique. In fact, this technique produces data sets where not only the complexitiy is bewildering, but simply the sheer amount of data piled up. However, with the appropriate sampling protocols, this contains spatiotemporal information which directly merges into the consideration of spatial as well as relational networks (see Sect. 5).

A detailed account of the observations would by far exceed the space limits given here. By way of overview, and in a summarizing and fairly

Table 1. Gene products, their functions and their genes exhibiting circadian oscillations (+ oscillating; – not oscillating)

Protein/enzyme/function	Circadian oscillation		Plant	References
	Protein: amount, activity	Gene expression, mRNA		
Photosynthesis				
Rubisco-activase		+	Tomato, *Arabidopsis*	Martino-Catt and Ort (1992); Pilgrim and McClung (1993); Liu et al. (1996)
Rubsico, small subunit		+	Barley, *Arabidopsis*	Beator et al. (1992); Pilgrim and McClung (1993)
Rubisco, large subunit		+	*Chlamydomonas*	Salvador et al. (1993)
Chlorophyll a/b-binding	+	+	Tomato, *Arabidopsis*, barley, wheat	Piechulla and Riesselmann (1990); Beator et al. (1992); O'Neill et al. (1994); Millar and Kay (1996); Kellmann et al. (1999); Millar (1999); Piechulla (1999); Sai and Johnson (1999); Pott et al. (2000)
Light-harvesting protein of PS II	+	+	Bean	Tavladoraki et al. (1989); Paraskevopoulou et al. (1995)
Oxygen-evolving enzyme		+	Tomato	Giuliano et al. (1988)
Ferredoxin/thioredoxin pathway		+	*Chlamydomonas*	Lemaire et al. (1999)
δ-Aminolevulonic acid synthesis	+	+	Barley	Beator et al. (1992); Beator and Kloppstech (1993); Kruse et al. (1997)
ATP-synthase		+	*Chlamydomonas*	Salvador et al. (1993)

Table 1 (continued)

Protein/enzyme/function	Circadian oscillation		Plant	References
	Protein: amount, activity	Gene expression, mRNA		
Photorespiration				
Catalase	+	+	*Arabidopsis*, maize	Zhong et al. (1994, 1997); McClung (1997); Scandalios et al. (1997)
Serine hydroxymethyltrans-ferase		+	*Arabidopsis*	McClung et al. (2000)
Photoprotection				
Zeaxanthin epoxidase		+	Tomato	Thompson et al. (2000)
Crassulacean acid metabolism				
Phosphoenolpyruvate carboxylase (PEPC)	+		*Mesembryanthemum crystallinum*	Kusumi et al. (1994); Nimmo (2000); Taybi et al. (2000)
PEPC-kinase	+	+	*M. crystallinum*	Carter et al. 1991; Hartwell et al. (1996); Nimmo (1998, 2000); Taybi et al. (2000)
V-ATPase subunit c		+	*M. crystallinum*	Rockel et al. (1997)
Various metabolic functions				
3-Phosphoglycerate kinase	+		*Chlorella*	Walla et al. (1994)
Sucrose-phosphate synthase	+		Tomato, soybean	Rufty et al. (1983); Jones et al. (1998)
Sucrose-phosphate phosphatase	+		Tomato	Jones and Ort (1997)
Starch synthase		+	Snapdragon	Mérida et al. (1999)
Amylolytic activity	+		Spinach	Pongratz and Beck (1978)
Adenylate kinase	+		*Chenopodium rubrum*	Frosch et al. (1973)

NAD(P)-glyceraldehyde-3-phosphatases	+		*Chenopodium rubrum*	Frosch et al. (1973)
NAD-kinase, NADP-phosphatase	+		*Euglena*	Edmunds and Tamponnet (1990); Laval-Martin et al. (1990)
Carbonic anhydrase		+	*Chlamydomonas*	Fujiwara et al. (1996)
Bioluminescence				
Luciferase binding protein	+	–	*Gonyaulax*	Johnson et al. (1984); Schröder-Lorenz and Rensing (1987); Mittag et al. (1994); Mittag and Hastings (1996)
Mineral nutrition: nitrogen				
Nitrate reductase	+	+	*Chenopodium rubrum*, wheat, *Gonyaulax*, tomato, *Arabidopsis* , maize, tobacco	Cohen and Cumming (1974); Upcroft and Done (1976); Lillo and Ruoff (1989); Deng et al. (1990); Pilgrim et al. (1993); McClung and Kay (1994); Ramalho et al.(1995); Jones et al. (1998); Lillo et al. (2001)
Nitrate transport	+		Tomato	Ono et al. (2000)
Nitrogenase	+		*Synechococcus RF-1*	Huang et al. (1994)

Table 1 (continued)

Protein/enzyme/function	Circadian oscillation		Plant	References
	Protein: amount, activity	Gene expression, mRNA		
Development, flowering; hormones and regulation				
Gigantea gene		+	*Arabidopsis*	Park et al. (1999)
Flower induction	+	+	*Sinapis, Arabidopsis, Pharbitis nil*	Heintzen et al. (1994 a,b, 1997); O'Neill et al. (1994); Kreps and Simon (1997); Sage-Ono et al. (1998); Staiger and Heintzen (1999); Covington et al. (2001); Hicks et al. (2001); Liu et al. (2001)
1-Aminocyclopropane-1-carboxylic acid oxidase (ethylene biosynthesis)	+	+	*Stellaria longipes, Sorghum*	Kathiresan et al. (1996); Finlayson et al. (1999)
S-Adenosylmethionine decarboxylase (polyamine biosynthesis)		+	*Pharbitis nil*	Yoshida et al. (1999)
Serine/threonine phosphoprotein phosphatases	+		*Gonyaulax*	Comolli et al. (1996)
Glycine-rich RNA-binding proteins		+	*Pelargonium × hortum*	Clark et al. (1999)

generalizing vein, the most outstanding examples are compiled in Table 1, where proteins, enzymes and functions are given rather than genes, and information is added whether protein amount and activity and/or gene expression and mRNA levels were observed to oscillate in the various plants listed. (More details and additional references are found in the reviews cited at the end of Section 1 and in the references given in the table.) In summary, we note that the vast majority of circadian oscillations described at the molecular and enzyme level are directly or indirectly (photorespiration, crassulacean acid metabolism) related to photosynthesis (for plastid genes, see Piechulla and Gruissem 1987; Piechulla 1988). Oscillations of functions of nitrogen metabolism also deserve a special mention. The field of development and regulation is probably one of the most promising areas of further excitement in plant chronobiology in the future.

b) Membranes and Transport

Membrane functions and transport processes are long known to be essential components of biological oscillators. Experiments with valinomycin interfering with K^+-transport involved in the leaf movements of *Phaseolus* led Bünning and Moser (1972) to suggest that membrane transport is the pacemaker of circadian rhythms. Njus et al. (1974) and Sweeney (1974) also proposed membrane models for circadian rhythmicity. Georg Schweiger developed a coupled translation - membrane model of endogenous rhythmicity with synthesis of a peptide that is incorporated into a membrane, where it has negative feedback on its own synthesis, while the membrane devoid of the peptide has positive feedback on synthesis. Thus, the membrane can function as a hysteresis switch or oscillator (see Fig. 28-6, p. 568 in Lüttge et al. 2002; Wiedemann et al. 1992).

A rather elaborate membrane model was suggested for the *PER* oscillator in *Drosophila* (Konopka and Orr 1980). The PER protein is taken to be an ion pump which builds up an ion gradient regulating a gated ion channel which is also photosensitive. The PER-channel protein exerts negative feedback on its own synthesis (transcription, translation) and the photosensitivity of the ion channel confers a light-inhibited circadian rhythmicity in continuous darkness. (A more complex comprehensive model scheme for *Drosophila* has now been developed; see Millar 1999; McClung 2000.)

Cyanobacteria (*Synechococcus* RF-1) show a circadian rhythm in amino acid uptake (Chen et al. 1991, 1996; Huang et al. 1994). In *Neurospora crassa*, light perceived by a flavin-associated photoreceptor stimulates plasma membrane H^+-ATPase activity and shifts the phase of circadian rhythmicity of conidiation (Nakashima 1982). Circadian changes in the cell surface affect stickiness of cells of *Chlamydomons reinhardtii* to glass (Straley and Bruce 1979), i.e. a kind of circadian electrical Tanada effect (Jaffe 1968), which is also under phytochrome control. Novak and Greppin (1979) observed a circadian rhythm of electrical membrane potential in spinach. In the duckweed *Lemna gibba* K^+-uptake shows circadian rhythmicity and is regulated by Na^+, Li^+ and factors affecting protein synthesis, where supply of amino acid analogues leads to altered forms of two proteins which appear to be involved in timing of the circadian oscillator (Kondo 1983, 1984, 1989). Rhythmic variations in the levels of a fatty acid, oleic acid, in phospholipids, phoshatidylcholine and phosphatidylethanolamine were correlated with rhythmic expansion growth of stems of *Chenopodium rubrum* and possibly linked to the endomembrane flow in relation to the rate of growth (Lecharny and Wagner 1984; Lecharny et al. 1990). (For a circadian oscillator based on tonoplast membrane functions, see also Section 3c.β.)

c) Metabolism

α) C_3- and C_4-Photosynthesis and Stomata

The recognition of circadian oscillations in C_3-photosynthesis is based on early work with unicellular algae, dinoflagellates (*Glenodinium* sp., *Ceratium furca*, *Gonyaulax polyedra*: Prézelin and Sweeney 1977; Prézelin et al. 1977) and *Euglena gracilis* (Lonergan and Sargent 1979). In *Euglena*, non-cyclic electron transport was rhythmic while the electron transport rates of PS II and PS I remained constant. Thus, it was assumed that electron flow between the two photosystems was subject to circadian control (Lonergan and Sargent 1979). Conversely, in *Gonyaulax polyedra*, electron flow through both photosystems and also PS II alone was rhythmic and only electron flow through PS I, including or excluding the plastoquinone pool, was constant with time in the circadian cycle (Samuelsson et al. 1983). Organisation of the pigment-protein complexes plays an essential role in expression of rhythmicity in this organism (Knoetzel and Rensing 1990). Enzymes such as carbonic anhydrase and glyceraldehyde-3-phosphate dehydrogenase (*Elodea*; Lonergan and Sargent 1978), Rubisco and Ru-5-P-kinase (*Phaseolus vulgaris*; Fredeen et al. 1991) did not appear to be involved in regulation of photosynthetic rhythmicity.

In view of the observations that the circadian rhythms of bioluminescence in *Gonyaulax polyedra* (Sweeney 1969, 1974) and of the CAM cycle in *Kalanchoë daigremontiana* (Lüttge and Beck 1992) in continuous light only persist at low light intensities, it is interesting to note contrasting results found among C_3-plants. In *Euglena gracilis*, photosynthesis was rhythmic at high light intensities (Lonergan and Sargent 1979), in three leguminous species in dim light (but only in the presence of uncouplers; Lonergan 1981), in *Ectocarpus siliculosus* not at irradiances below saturation (Schmid and Dring 1992; Schmid et al. 1992), and in *Phaseolus* there was rhythmicity at limiting conditions of either light or CO_2 (low light/high CO_2 and high light/low CO_2, respectively; Hennessey et al. 1993). Light regulation of C_3-photosynthesis rhythmicity must be complex, and light quality may also be involved (see Sect. 4).

Circadian rhythmicity of photosynthesis was also shown for a C_4-plant, the grass *Sorghum bicolor* (Britz et al. 1987).

With rhythms of photosynthesis in higher plants it is, of course, always important to distinguish photosynthesis itself from stomatal rhythms. Therefore, the interpretation is more straightforward for algal systems: *Euglena*, Dinoflagellates, *Ectocarpus* (see above), and also the green alga *Ulva lactuca* (Mishkind et al. 1979), the red macroalga *Kappaphycus alvarezii* (Granbom et al. 2001) and the brown alga *Laminaria saccharina* (Klenell et al. 2001). Endogenous circadian stomatal rhythms have been known for a long time (Stålfelt 1963; Kana and Miller 1977; Gorton et al. 1989, 1993; Gorton 1990; Assmann 1993; Hennessey et al. 1993; Kaiser and Kappen 1997; Webb and Hetherington 1997; Webb 1998). They appear to be under the control of the *TOC1* gene (Somers et al. 1998b). Circadian stomatal rhythms also occur in isolated epidermal strips independent of photosynthesis in the mesophyll, and oscillations directly residing in the guard cells may therefore affect rhythmicity of photosynthesis (Gorton et al. 1989; Meidner and Willmer 1993). They are related to the varying ability of the guard cells to accumulate or retain both Cl^- and K^+ ions (Snaith and Mansfield 1986). In *Phaseolus vulgaris*, Freeden et al. (1991) and Hennessey and Field (1991) detected both stomatal and non-stomatal contributions. They found that under constant light an external temperature cycle entrained the rhythm of stomatal conductance but not carbon assimilation, which indicates that the oscillators driving the two rhythms differ in their sensitivity to external stimuli.

β) Crassulacean Acid Metabolism (CAM)

It has long been known that crassulacean acid metabolism shows endogenous circadian rhythmicity (Wilkins 1992). Circadian rhythms of gas exchange were first observed in continuous darkness and CO_2-free air (Wilkins 1959, 1960) but then also in continuous light and normal air (Nuernbergk 1961; Lüttge and Ball 1978; Wilkins 1984). With studies mainly in species of *Kalanchoë*, CAM is now rapidly developing into a foremost model system for investigating circadian rhythmicity of a metabolic cycle. It has been shown that under continuous conditions

CAM is only rhythmic within certain limits of light intensities and temperature beyond which it reversibly changes to arrhythmicity (Buchanan-Bollig 1984; Buchanan-Bollig et al. 1984; Anderson and Wilkins 1989a,b; Wilkins, 1992; Lüttge and Beck 1992; Grams et al. 1996, 1997; Rascher et al. 1998). The transitions between rhythmic and arrhythmic behaviour at the critical levels of light intensity and temperature are very sharp (Lüttge and Beck 1992). It is most noteworthy that a similar response to light intensity *and* temperature is also observed with the circadian rhythm of bioluminescence in *Gonyaulax* (Njus et al. 1977). Responses of the circadian CAM rhythm to complex external thermoperiodic perturbations reveal interesting transitions between free running oscillations, full synchronisation to the external driver and arrhythmic behaviour (Lüttge et al. 1996; Bohn et al. 2001), and circadian time windows, i.e. gating (for gating, see Sect. 4), of the responses are indicated (Bohn et al. 2002).

The tonoplast membrane, which mediates energy-dependent vacuolar malate accumulation and passive malate efflux, and the enzyme phosphoenolpyruvate carboxylase (PEPC), serving nocturnal CO_2 fixation and malate synthesis, have both been considered most seriously as the possible candidates for the cellular, physiological or molecular hysteresis switch or beat oscillator (Wilkins 1984, Anderson and Wilkins 1989a,b; Lüttge and Beck 1992; Grams et al. 1997; see Lüttge 2000; Nimmo 2000). PEPC activity is regulated by phosphorylation/dephosphorylation, and this is under circadian gene control because the PEPC-kinase expression shows circadian rhythmicity (Nimmo et al. 1987, 2001; Carter et al. 1991, 1995a, 1996; Kusumi et al. 1994; Hartwell et al. 1996, 1999). However, studies of gas exchange and malate accumulation/remobilisation at the tonoplast (Grams et al. 1997; Borland et al. 1999) as well as model simulations (see Sect. 6) have stressed the concept of a tonoplast membrane oscillator in CAM rhythmicity, switching between net accumulation in and net remobilization of malate from the vacuole. It turned out that PEPC-kinase activity cannot be regarded as an essential driving force of the circadian CAM rhythm (Carter et al. 1995b). Rather, the circadian rhythm is under metabolic control (Lüttge 2000; Nimmo 2000; Nimmo et al. 2001) because "... the circadian control of kinase mRNA and activity can be influenced by metabolic status, specifically by treatments that affect the content or compartmentation of malate ..." (Borland et al. 1999). Thus, PEPC-kinase oscillations are a control point downstream to the tonoplast oscillator, which nature may have selected to stabilise a complex reaction network. Indeed, model simulations also show that this stabilises rhythmicity against perturbations (Blasius 1997). Thus, as far as CAM is concerned, it appears to be at least one clear system where a primary oscillatory element is not at the gene and molecular, but at a membrane level.

On the other hand, Boxall et al. (2001) have recently identified a large number of circadianly oscillating gene transcripts in the C_3-photosynthesis/CAM-intermediate plant *Mesembryanthemum crystallinum*; in addition to PEPC-kinase (see also above), transcripts of other CAM-associated enzymes, of enzymes involved in photosynthesis, in starch and sucrose metabolism, in glycolysis, in nitrate assimilation and in housekeeping functions. (These results, unpublished except for the cited poster of Boxall et al. 2001, are not incorporated in Table 1.) In the CAM state more genes were oscillatory than in the C_3 state. It remains to be seen how these findings integrate in oscillator regulation-chains. Certainly, unravelling hierarchical links needs more attention in future research (Roenneberg and Merrow 1999). However, systems may turn out to be so complex that we must move from linear signalling systems or open networks with cause-effect linkages to circular regulation networks where hierarchies and cause-effect linkages are obsolete (Watts 1999; see Sect. 5). This may then even reconcile the dispute on priority of genetic, metabolic or membrane control in oscillations such as that of CAM.

γ) Non-photosynthetic Energy Turnover Including Circadian Rhythmicity of Respiration

It was found that a circadian rhythm in O_2 uptake was correlated with the energy demand of pulvinar movements (see Sect. 3.d.α) of leaves (Satter et al. 1979). In a set of studies, E. Wagner and colleagues have shown endogenous rhythmicity in reduced and oxidised NAD(P) levels, adenylate kinase, NAD(P)-glyceraldehyde-3-phosphate dehydrogenase and energy charge in *Chenopodium rubrum* (Deitzer et al. 1974; Frosch and Wagner 1973a,b; Frosch et al. 1973; Wagner and Frosch 1974; Wagner et al. 1974a,b). Period length was 9–12 h (Wagner et al. 1974a). Wolff and Künne (2000), however, found a light-regulated circadian activity of respiration in mutants of *Euglena gracilis* that lacked chloroplasts, and whether the energy was supplied by photosynthesis or respiration depended on the *Euglena* cells tested. They suggested that the demand for energy is under light and circadian control.

d) Organs

α) Pulvini

Wilhelm Pfeffer (1907, 1915) performed his seminal studies in plant chronobiology on the leaf movements of bean. Ever since then, movements of pulvini have remained one of the major overt circadian rhythms in plant studies (Satter et al. 1990; Engelmann and Johnson 1998): "Leaf movements: rosetta stone of plant behavior?" (Satter and Galston 1973). Leaf movements are now also known to include movements of cotyledons of the widely studied model plant *Arabidopsis thali-*

ana (Engelmann et al. 1992; Engelmann and Johnsson 1998; Dowson-Day and Millar 1999). However, these are not pulvinar motor-tissue movements but asymmetric growth reactions.

Pulvinar movements were last reviewed in this series 15 years ago (Hensel 1987). They comprise a broad array of physiological functions in their operation: (1) involvement of mainly K^+ and Cl^- but also other inorganic ions, as well as malate and sucrose as osmotica in swelling and shrinking of pulvinar extensor and flexor cells (e.g. refs. recent to Hensel 1987: Erath 1998; Freudling et al. 1988; Starrach and Mayer 1989; Zucker Lowen and Satter 1989; Irving et al. 1997), which also show circadian oscillations (e.g. Lee and Satter 1987). (2) Involvement of transmembrane electrical potenials and pH gradients (e.g. Racusen and Satter 1975; Freudling et al. 1980; Lee and Satter 1989) related to activities of the plasmalemma H^+-transporting ATPase (Satter et al. 1987; Mayer et al. 1994; Fleurat-Lessard et al. 1997a,b) and K^+ channels (Moran et al. 1988; Kim et al. 1992, 1993; 1996; Moshelion and Moran 2000; Suh et al. 2000) in generating circadian periodicity. (3) Involvement of aquaporins in the plasma membrane (Moshelion et al. 2002) and tonoplast (Harmer et al. 2000) with their circadian regulation. (4) Involvement of phytochrome, blue and red light as external control parameters of signalling networks (Kim et al. 1992, 1993; Koller et al. 1995; Nishizaki et al. 1997; Okazaki et al. 2000; Suh et al. 2000). (5) Involvement of various phytohormones (Białczyk and Lechowski 1987; Everat-Bourbouloux et al. 1990). (6) Involvement of the Ca^{2+}/phosphoinositides/G-protein complement of regulation networks (Coté et al. 1989; Moysset and Simon 1989; Roblin et al. 1989; Kim et al. 1996; Kayali et al. 1997; Mayer et al. 1997).

We do not know what conducts the concerted action of all these elements. Applewhite et al. (1973) wrote that protein synthesis is involved.

β) Roots, Shoots and Flowers

Rhythmic outputs are observed in all plant organs. We have already noted that circadian leaf movements may also be non-pulvinar growth reactions (Sect. 3.d.α). Even in gymnosperms circadian leaf movements have been observed (Brinker et al. 2001). Circadian petal movements are long known in *Kalanchoë* (Bünsow 1953; Oltmanns 1960). The emission of volatile components, scents and terpenoids by flowers may also be circadian (Helsper et al. 1998; Kolosova et al. 2001; Pott et al. 2002).

There are diurnal rhythms of metabolism and transport in roots (Couderchet and Girard 1976; Knight and Weissman 1982; Macduff and Dhanoa 1996; Henzler et al. 1999). It is not so clear if they also run endogenously. An endogenous rhythm of amino acids and organic acids in the xylem of stems of *Lagerstroemia indica*, mainly glutamine, in darkness and terminating in light was described by Andersen et al. (1993).

Circadian effects on morphogenesis are reflected in induction of flowering (Bünning 1936; *Pharbitis nil*: Zheng et al. 1993; *Sinapis alba*:

Heintzen et al. 1994a,b; *Lolium temulentum*: Perilleux et al. 1996; *Arabidopsis thaliana*: e.g. Liu et al. 2001) and in stem (Lecharny and Wagner 1984) and hypocotyl elongation. After the discovery of the circadian early flowering gene *ELF3* in *Arabidopsis* this has become a vigorous field of circadian-rhythm research in plants (Kreps and Kay 1997; Schaffer et al. 1998; Dowson-Day and Millar 1999; Millar 1999; Somers 1999; McClung 2000, Reves and Coupland 2000; Covington et al. 2001; Hicks et al. 2001; Liu et al. 2001).

4 External Control Parameters, Receptors and Internal Signalling Systems

Environmental factors can be input factors (*zeitgeber*) and act as external control parameters in circadian rhythmicity. They exert their actions via receptors (*zeitnehmer*) and the elements of internal signalling networks. As for the genes and enzymes (Sect. 3.a), again only a general survey of the information available is possible here (Table 2).

The major external control parameters for circadian rhythmicity in plants are light and temperature and, for rhythms related to photosynthesis, most likely also CO_2. Their action usually is that of a *zeitgeber* eliciting phase shifting and entrainment of rhythms. Often their action and efficiency is determined by the phase of the rhythm when they are acting, an effect called "circadian gating", i.e. implying that gates for their action are only open at certain circadian times (Edmunds and Tamponnet 1990; Rikin 1991; Millar and Kay 1996; Millar 1999). Thus, sensitivity of the *zeitnehmer* of oscillators is altered circadianly (circadian gates) which is expressed by phase shifts, which may be delays or advances, or the total absence of a response as a function of the circadian phase when stimuli (or *zeitgeber*) are given (early example: Fig. 3.22 in Sweeney 1969; Johnson 1992; McWatters et al. 2000; Covington et al. 2001).

While some rhythms in *Drosophila* and *Neurospora* only run in darkness, rhythmicity in plants is mainly restricted to continuous light (Millar 1999). Light intensity may play different roles. In some cases, high light intensities are required, while in others circadian rhythmicity only persists under non-saturating continuous illumination (see Sect. 3.c.α). The latter does not exclude that light is acting via photosynthetic energy supply (Sweeney 1969). For example, light intensity as a photosynthetic resource parameter may regulate reversible shifts between circadian rhythmicity and arrhythmicity of the CAM cycle (Lüttge and Beck 1992). Actually, the absence of rhythmicity in continuous darkness in plants in most cases apears to be due to a dampening effect

Table 2. External control parameters, receptors and elements of internal signalling systems involved in circadian rhythmicity

Control parameter, receptors, elements of signaling systems	Plants, organs, enzymes and functions	References
Light and darkness		
Only in darkness	locomotor activity, *Drosophila*; conidiation, *Neurospora*	Millar (1999)
Mainly in light	Plants	Millar (1999)
Light intensity	*Gonyaulax, Kalanchoë daigremontiana, Euglena, Ectocarpus, Phaseolus*	Sweeney (1974); Lonergan and Sargent (1979); Lüttge and Beck (1992); Schmid and Dring (1992); Schmid et al. (1992); Hennessey et al. (1993)
Red light (phytochrome) and blue light (cryptochrome)	Pulvini	Moysset and Simon (1989); Roblin et al. (1989); Kim et al. (1992, 1993); Koller et al. (1995); Nishizaki et al. (1997); Okazaki et al. (2000); Suh et al. (2000)
	Arabidopsis	Millar et al. (1995); Johnson et al. (1998); Somers et al. 1(998a); Somers (1999); Hall et al. (2001); Liu et al. (2001); Tóth et al. (2001)
	Chlorophyll a/b binding	Pott et al. (2000)
	Catalase	Zhong et al. (1997)
	Gonyaulax	Deng and Roenneberg (1997); Roenneberg and Deng (1997)
	Ethylene production, *Sorghum*	Finlayson et al. (1998, 1999)
	S-Adenosylmethionine decarboxylase (polyamine biosynthesis)	Yoshida et al. (1999)
Temperature	*Kalanchoë daigremontiana* (crassulacean acid metabolism)	Wilkins (1962, 1983); Anderson and Wilkins (1989a,b); Lüttge and Beck (1992); Grams et al. (1995, 1996, 1997); Bohn et al. (2002)
	Ethylene production, *Sorghum*	Finlayson et al. (1998)
	Conidiation, *Neurospora*	Gooch van et al. (1994); Goto et al. (1994); Liu et al. (1997)

Phytohormones		
Ethylene	Barley, wheat, *Stellaria*, *Sorghum*, rye, *Chenopodium rubrum*	Ievinsh and Kreicbergs (1992); Máchacková et al. (1997); Morgan et al. (1997); Finlayson et al. (1998, 1999)
Auxin	*Arabidopsis*	Jouve et al. (1998, 1999)
Gibberellic acid	*Sorghum*	Foster and Morgan (1995)
Auxin, gibberellic acid, kinetin, abscisic acid, fusicoccin	Pulvini	Bialczyk and Lechowski (1987); Everat-Bourbouloux et al. (1990)
Ca^{2+}	*Euglena*, tobacco, *Arabidopsis*	Edmunds and Tamponnet (1990); Tamponnet and Edmunds (1990); Johnson et al. (1995); Sai and Johnson (1999); Wood et al. (2001)
	Pulvini	Moysset and Simon (1989); Roblin et al. (1989); Kayali et al. (1997); Mayer et al. (1997); Moshelion and Moran (2000)
	C_3-photosynthesis, chloroplasts (*Arabidopsis*)	Lonergan (1990); Johnson et al. (1995)
Phosphoinositides	Pulvini	Coté et al. (1989); Kim et al. (1996); Mayer et al. (1997)
cAMP	*Chlorella*	Edmunds and Tamponnet (1990)

in the absence of photosynthetic energy supply, and the dissipative systems of overt rhythms cease oscillating while the basic circadian oscillator continues (Johnson et al. 1998; Strayer et al. 2000; Johnson 2001; Xu and Johnson 2001).

Conversely, Hall et al. (2001) and Tóth et al. (2001) underline the signalling function of light. The qualities of signalling light (input parameters) are red and blue light. The receptors are phytochrome and cryptochrome, respectively. References for both are listed together in Table 2 because phytochrome and cryptochrome effects are often closely related. Circadian rhythms and phytochrome were reviewed by Lumsden (1991). In *Arabidopsis*, the promoter activity of all crytochrome and phytochrome genes (except *PHYC*) is rhythmic in both continuous light and darkness (Tóth et al. 2001), although this does not aply to all higher plants (Hall et al. 2001). The dampening in continuous darkness is thought to be due to low levels of the far-red form (pfr) of phytochrome (Hall et al. 2001). Leaf excision prevents dampening suggesting that the interaction between organs of the plant is involved possibly via signal transmission. The oscillator may go on but may be masked by other processes (Hall et al. 2001). Ongoing oscillator activity coupling to outputs may be under genetic control. Xu and Johnson (2001) identified a new gene (*ZGT*) in tobacco that is involved in coupling oscillator and outputs. It is most noteworthy that cryptochrome, which was postulated to occur specifically in plants a long while ago (Rau 1967; Ahmad and Cashmore 1993), also plays a role in circadian rhythmicity of insects (*Drosophila*) and mammals (Barak et al. 2000; Golden and Strayer 2001).

Ever since it was originally reported that the day/night pulvinar movements of leaflets (see Sect. 3.d.α) are under phytochrome control, pulvini have remained a major model system for the study of phytochrome and cryptochrome as light receptors in circadian rhythmicity, although with the superior molecular approaches available in *Arabidopsis* this is now very rapidly catching up.

Temperature compensation keeping period length stable is one of the basic essentials and characteristics of circadian rhythmicity (Somers et al. 1998b); this is also supported by theoretical modelling (Leloup and Goldbeter 1977). In *Neurospora* this is under the control of FRQ protein (Liu et al. 1997). Sometimes there is overcompensation in that period length is somewhat extended rather than shortened at higher temperatures (Lüttge and Beck 1992). However, temperature may also have direct regulatory functions in circadian rhythmicity (McClung 2000), although no temperature sensor equivalents of photoreceptors have been isolated (Barak et al. 2000). In CAM, temperature appears to be a key element for functioning of the hysteresis switch at the tonoplast (Sects. 3.c.β and 6). In cotton, chilling resistance and temperature-induced phase shifting have been shown to be under circadian control

(Rikin 1991). Heat shock effects interacted with rhythmicity in barley (Beator et al. 1992).

With phytohormones, the standard elements of signalling networks, there is some evidence for the involvement of auxin and gibberellic acid in circadian rhythmicity and even more for ethylene, so that at least three of the major phytohormones are involved (McClung 2000).

Much more information is available on the role of the ubiquitous cytosolic secondary messenger Ca^{2+} ("le calcium, c'est la vie", Trewavas 1999). Oscillations of cytosolic Ca^{2+} levels are correlated with circadian rhythmicity (Table 2). Cytosolic Ca^{2+} regulation is typically related to the membrane-associated phosphoinositide cycle with inositol-tris-phosphate and diacylglycerol as secondary messengers. Evidence for their involvement has been obtained in studies with pulvini. Another secondary messenger particularly well known from animal physiology, i.e. cAMP, may play a role in *Euglena*.

5 Synchronisation/Desynchronisation

If there is a central oscillator or biological clock in each organism, this would make dealing with the multitude of overt circadian rhythms we observe (Sect. 3) rather comfortable. Observations as to why we need to reject this more facile conception and renounce more precisely defined targets for research have been presented in Section 2. Here, we must now consider synchronisation/desynchronisation. Individual rhythms may live their own life independent of each other, but they may also interact.

In tobacco, different tissues and cells generate different circadian phases of cytoplasmic Ca^{2+} concentrations, and it was asked whether there are different oscillators or whether each cell or tissue type interprets the same fundamental oscillator in different ways (Wood et al. 2001). Even in the same cell there are different circadian rhythms. In the prokaryotic *Synechococcus elongatus*, class 1 genes with expression peaks at "subjective" dusk and class 2 genes with expression peaks ca. 12 h out of phase to the class 1 genes are distinguished (Min and Golden 2000). In eukaryotes, examples of different rhythms within one cell are the rhythms of phototaxis, chemotaxis, stickiness to glass, cell division and release of daughter cells and ammonia uptake in the unicellular *Chlamydomonas* (Straley and Bruce 1979, Johnson et al. 1992); the rhythms of bioluminescence, photosynthesis and cell division in *Gonyaulax polyedra* that have their peaks at different circadian times (Roenneberg and Morse 1993; Morse et al. 1994; Roenneberg 1996); the expression of two catalase genes in *Arabidopsis* with opposite (ca. 180°) circadian phase (Zhong and McClung 1996); and the rhythms of phototaxis and gravitaxis in *Euglena* (Lebert et al. 1999). In *Arabidopsis*, chlorophyll a/b binding (*LHCb* promoter gene) and free cytosolic Ca^{2+} oscillate independent of each other (Sai and Johnson 1999), although the 19 *LHC* genes are synchronously expressed (Kellmann et al. 1993).

Such rhythms evidently are not synchronised, and it is likely that they follow separate oscillating systems. Synchronisation is required when the same overt output rhythm needs input by different control parameters, especially if there is different circadian gating for the control parameters: Circadian ethylene production in *Sorghum* needs both phytochrome-activated and temperature signals (Finlayson et al. 1998). The circadian system of *Gonyaulax* receives light signals via two distinct input pathways (Roenneberg 1996). There are different phase responses for red and blue light (phytochrome and cryptochrome), respectively (Deng and Roenneberg 1997; Roenneberg and Deng 1997). The authors think there may be two circadian oscillators. Or are there complex synchronisations? A blue light input pathway is linked to xanthine oxidase (inhibited by allopurinol) while a second blue and red light input pathway is not (Deng and Roenneberg 1997). In tomato it was found that circadian and phytochrome control act at different promotor regions of the chlorophyll a/b binding gene (*LHCa3*). This requires synchronisation at the promoter level because there are no indications that one regulatory element leads to convergence (Pott et al. 2000).

We have absolutely no idea at the moment of how such synchronisation at the genetic level may occur, although this certainly is one of the major new questions arising when the central-clock-gene concept is rejected. A recent exciting observation in *Arabidopsis* may result in a breakthrough. There are circadian oscillations of expression of genes of a gene-quintett (*APRR1/TOC1* family), which are induced by a red light pulse presumably perceived by phytochrome (Makino et al. 2000, 2001; Matsushika et al. 2000). These oscillations follow each other in 2 to 3 h intervals in a given 24 h photoperiod. A possibility discussed for gene expression in a series by Waterhouse (2001) is that "the protein produced by the first gene is transported back to the nucleus to initiate" [transcription and] "translation of the next gene in the sequence, and so on. Since the last protein then re-activates the first gene, a cycle is produced. The length of this cycle might be determined by the number of base pairs in the genes, as well as by the time taken for the gene product to be manufactered and transported ...". When we understand if and how the information inherent in these oscillations following each other in time and expressing themselves as waves in the spatial system may lead to synchronised outputs, we shall have much advanced our understanding of functioning of the circadian-rhythmicity information-software.

With input and signalling systems (Sect. 4) we are already used to talking of networks because we know branching and anastomosing signal pathways. Networks are characterised by vertices and edges with connecting pathways, including shortcuts, between them for cross-talk of their elements (Watts 1999). Individual oscillatory systems can be considered as modules when we advance "from molecular to modular cell biology" (L.H. Hartwell et al. 1999). Individual oscillatory systems or modules are complex and themselves may be considered as networks, e.g. the genetic systems described in Sections 2.c and 3.a and clusters of genes (Staiger 2002) or the CAM-tonoplast beat-oscillator discussed in Section 3.c.β. Amzallag (2001) and Genoud et al. (2001) underline the robustness, adaptive plasticity and flexibility given by using the genome as network-like systems (NELI) in contrast to deterministic undimensional cause-effect linkage (DUCE) or linear signalling systems. Individual oscillatory systems as individual networks may be connected or coupled in super-networks (Watts 1999). Such connectivity evidently does not necessarily require physical contact in space. We may distinguish between spatial networks, when indeed physical contacts are effective, and relational networks, where connectivity is rather based on the software of information flow. Networks also offer a conceptual integration of modules of such different physical characteristics as molecular and non-molecular oscillators, e.g. as in the rhythm of CAM with its tonoplast-membrane oscillator and gene oscillations (Sect. 3.c.β).

In the above discussion we considered synchronisation/desynchronisation of different types of oscillators, i.e. different systems of input and reactions or relational networks. There is also the case, however, of individual cells of spatial networks in a tissue where all cells have the same oscillator for the same overt output rhythm and that they can synchronise and desynchronise. This has been shown for cultured cells of the two suprachiasmic nuclei at the base of the hypothalamus of the brain of mammals (Waterhouse 2001). It is seen in leaves, where individual stomata may show desynchronised oscillations. Even adjacent stomata may oscillate by more than 2 h out of phase with each other (Kaiser and Kappen 1997). This has been studied in much detail in the leaves of *Kalanchoë daigremontiana* in relation to the circadian rhythm of CAM in continuous light. The leaves are of uniform tightly packed spherical cells. Model simulations (see Sect. 6) had suggested that patches of cells may oscillate separately (desynchronised) and also may become synchronised. The circadian CO_2-exchange J_{CO2} rhythm of entire leaves becomes arrhythmic above a certain threshold of temperature. This is reversible, but only when temperature is reduced again abruptly and not when this occurs gradually. A strong signal or *zeitgeber* apparently is needed for synchronisation (Rascher et al. 1998). Such patchiness was then observed using the non-intrusive approach of photographic imaging of chlorophyll fluorescence during the circadian rhythm of J_{CO2}.It was not related to stomatal patchiness (last review in this series, Beyschlag and Eckstein 1997) but to patchy efficiency of photosystem II, and it was most likely due to restricted diffusion of CO_2 between the densely packed cells of the leaves having the role of the synchronising signal. Quantified patchiness (using a cellular automaton and nearest

nearest neighbour algorithm; Hütt and Neff 2001) showed rhythmicity itself and changed during long-lasting oscillations of J_{CO2} in CAM. It was also clearly seen that small patches of 5–10 mm diameter (approximately 30–60 cells) at close distance to each other (approximately 15–30 mm) desynchronised at falling values of J_{CO2} and synchronised again in the following phase of increasing J_{CO2} (Rascher 2001; Rascher et al. 2001). This shows that circadian rhythmicity is not only a pattern developed in time, but requires spatiotemporal organisation and flow of information.

6 The Software: Theory and Simulation Models

Any system, even if composed only of a very small number of elements, can move in a limit cycle or oscillate as long as it has (1) feedback loop(s) and (2) a hysteresis switch or beat oscillator (Winfree 1990; Schuster 1995). This holds for inorganic (technical systems or systems in inorganic chemistry) as well as for organic living systems. It may produce periods of any length. This is only a matter of the capacity or size of elements or pools and the time constants of changes or flows of or between them. This also regulates timing of the trinity of the states excitable–excited–refractory of both inorganic excitable media and excitable living systems. Much can be learned here for the understanding of circadian rhythmicity especially from ultradian short period free running rhythms of single isolated enzymes, reaction sequences (e.g. glycolysis), pattern formation (e.g. *Dictyostelium*), which due to space limitations cannot be reviewed here.

Many of these systems oscillate without a direct involvement of genes and their function and we need formal approaches (Friesen et al. 1993) and theoretical systems analysis to get to grips with them: *"Doch um die Systemeigenschaften der belebten Welt zu verstehen, braucht man ... Anschauung, experimentelle Beobachtungen und Daten, eine geeignete begriffliche Erfassung der erfahrenen Wirklichkeit und – nicht zuletzt – angemessene mathematische Konzepte ... Trotz aller Verdienste der molekularen Biologie – nicht jede Erklärung muß ausschließlich oder in erster Linie molekular sein." ... "Es gab und gibt die Meinung, die grundlegenden Probleme seien gelöst, und um komplexe biologische Vorgänge zu verstehen, müßten eigentlich immer nur die beteiligten Moleküle ermittelt werden. Das stimmt natürlich nicht ..."* (Gierer 1998).

Indeed, most models now trying to explain circadian rhythms based on the activity of genes have adopted such approaches with interlocked feedback loops (Merrow et al. 1999; Somers 1999; Gonze et al. 2000), be that Schweiger's simple coupled translation–membrane model (Sect. 3.b), or the more sophisticated model of *Drosophila* (McClung 2000). Staiger (2000) has reduced it to the simple formula that proteins switch on and off their own transcription in a 24-h rhythm.

However, theoretical software of simulation models may also work assuming hardware other than genes. This has been demonstrated for the circadian rhythm of CAM. The experimental evidence for circadian rhythmicity of transcription and translation of a gene (PEPC-kinase) being a module downstream to a tonoplast-membrane oscillator has been summarised in Sect. 3.c.β. A model for computer simulation of the CAM cycle was first developed by Nungesser et al. (1984). It has been continuously refined (Grams et al. 1996, 1997; Blasius et al. 1997, 1998, 1999) to a minimal or skeleton model with only the three pools of internal CO_2, cytoplasmic malate and vacuolar malate, the control parameters light, temperature and external CO_2 and the hysteresis switch at the tonoplast (based on a theoretical thermodynamic membrane model; Neff et al. 1998). It simulates many facets of the endogenous CAM rhythm astonishingly precisely (refs. above and Lüttge 2000). It confirmed the primary role of a tonoplast hysteresis switch in the endogenous rhythm of CAM. Furthermore, it allowed the function of noise to be demonstrated. The model reproduces well the experimental observation of threshold levels of temperature for rhythmicity, where at a critical lower level rhythmicity is lost with a state of "full vacuole" (with respect to malate) and at a critical higher level it is lost with a state of "empty vacuole" (Grams et al. 1997; see Lüttge 2000). In the simulations, corresponding fixed points are reached at the lower and higher temperatures, with limit cycles (oscillations) in between. The model clearly shows a constructive regulatory role of purely stochastic noise. Close to critical temperature, where the system would spiral into a fixed point, noise can regularly kick it out of the fixed point and force it into a quasi-periodic trajectory with an oscillatory output (Beck et al. 2001; see Lüttge 2000). "The model ... is unlike the orthodox transcription/translation models that nearly everyone in the field of circadian rhythms uses ... [it] ... provides another perspective of how rhythmicity can be generated ..." (anonymous reviewer of Beck et al. 2001). Periodic output of the model is optimal at intermediate noise intensity. We call this stochastic resonance.

There is no experimental evidence for such stochastic resonance in the CAM cycle yet, and it is very difficult to design the appropriate experiments with a given background of the usual "disturbing noise" of instruments and other parameters. However, other evidence for the ordering role that noise may have in functions of living systems is gradually accumulating (Wiesenfeld and Moss 1995; Hänggi 2001).

Together with the observations of critical threshold levels of temperature and light beyond which rhythmicity reversibly changes to arrhythmicity (Sect. 3.c.β), this may present a kind of warning for identification of arrhythmicity mutants (Somers 1999). Clearly, a wide array of conditions must be tested before the negative result of arrhythmicity can be (more or less) reliably ascertained. Model simulations may also be suitable to facilitate assignment of elements of rhythmicity to input and output pathways and the oscillator, respectively.

More fundamentally, theory and simulation models teach us that it is the software, viz. information and communication, that makes systems oscillate. The situation of pattern formation in time discussed here is very similar to that of pattern formation in space (development). The search for a central developmental control gene is just as frustrating as the search for a central circadian oscillator gene (Theißen and Saedler 1997). In both cases, theoretical systems analysis is required.

7 Biological Sense and Adaptive Value

The question is often asked what adaptive value photoperiodic timing may have. There is a widely adopted saying that endogenous timing provides readiness, alertness and preparedness in advance to changing conditions of day and night. Indeed, common sense suggests that anticipation is a decisive advantage. An ecological advantage of circadian stomatal rhythmicity has been discussed (Kaiser and Kappen 1997). On the other hand, Williams and Gorton (1998) compared a linear and a rhythmic simulation model and found that rhythmicity does not have advantages for CO_2 acquisition in the field. Generally, there is little solid evidence that the circadian clock is of adaptive value.

There is one exception though, i.e. work with mutants of the cyanobacterium *Synechococcus elongatus* PCC7942. There are mutants/strains with period lengths of circadian rhythmicity of 23 h (short), 25 h (wild type) and 30 h (long). In pure culture all strains grow equally well under normal dark/light cycles or continuous illumination. If they are mixed, however, the strain which is closest to its endogenous period very rapidly outcompetes the other two strains, i.e. the short-period strain (23 h) at D/L of 11:11 h, the wild-type strain (25 h) at D/L of 12:12 h and the long-period strain (30 h) at D/L of 15:15 h (Ouyang et al. 1998; Johnson and Golden 1999). The result of this experiment is interesting with respect to the theory of Daido (2001), which suggests that an endogenous period equalling that of the milieu is most advantageous in adaptation to the environment but leads to the most intense competition. For *S. elongatus*, the phase relationship, i.e. living in synchrony with the environment, seems to be the more important trait as compared with having an endogenous period at the edge of entrainment avoiding the most direct competition. The observation of increased fitness of the *S. elongatus* strain with the (nearly) correct circadian period in the in vitro experiments with mixed mutant strains (Johnson et al. 1998; Yan et al. 1998) provides encouragement for further research in the quest for a biological meaning and adaptive value of circadian rhythmicity (McClung 2000).

However, this experiment does not rule out yet that endogenous circadian rhythmicity is only an inescapable side-product of evolution of life under the continuous entrainment by the natural environmental rhythm of days and nights, which would have forced the time-constants of the turnover of elements in signalling networks to evolve in a way that such networks may display circadian oscillations. The problem of driving force or product of evolution remains an open question.

8 Conclusions and Outlook: Hardware, Software and Multidisciplinary Approaches

The most exciting developments in biology currently are (1) "functional molecular biology" bringing "genomics", "proteomics" and "matabolomics" to life by relating them to physiological functions and (2) the assessment of biological functions within the theoretical framework of non-linear dynamics that has already revolutionised our understanding of the inorganic world.

Molecular biology and genetics are essential in three domains of biorhythm research, namely: (1) the identification of clock mutants and putative "oscillator" genes, (2) the phenomenology of clock-controlled genes (CCGs), and (3) the identification and phenomenology of receptors and signal transduction networks. So far it has not revealed the material existence of "central oscillators" which may be an idea rather than a materialised fact. Conversely, the theory of non-linear dynamics producing software of simulation models, which can be parameterised and checked against experimental reality, can reproduce oscillations.

Software is a language document, while hardware is a material basis. Information is laid down in both the hardware (structure of organs, cells, membranes, nucleic acids and proteins) as well as in the software. The hardware itself is not oscillating; it does not constitute oscillators (Sects. 2.a–c). The software needs the hardware to produce oscillations. In the simulations (Sect. 6), software is essentially the leading and determining element that can cause the computer hardware to produce oscillations. The analogy with the communication software regulating the concerted action of the elements of the material hardware of organisms may not be so farfetched.

Clearly, to ultimately unravel the nature of circadian rhythmicity, an intimate multidisciplinary approach of biology, non-linear dynamics of theoretical physics and information science is required.

Acknowledgements. Professor Dr. Birgit Piechulla, Dr. Marc-Thorsten Hütt and Dr. Uwe Rascher are thanked for reading the manuscript. Their critical reviews greatly contributed to improving the text.

References

Ahmad M, Cashmore AR (1993) Seeing blue: the discovery of cryptochrome. Plant Mol Biol 30:851–861

Amzallag GN (2001) Data analysis in plant physiology: are we missing the reality? Plant Cell Environ 24:881–890

Anderson CM, Wilkins MB (1989a) Period and phase control by temperature in the circadian rhythm of carbon dioxide fixation in illuminated leaves of *Bryophyllum fedtschenkoi*. Planta 177:456–469

Anderson CM, Wilkins MB (1989b) Phase resetting of the circadian rhythm of carbon dioxide assimilation in *Bryophyllum* leaves in relation to their malate content following brief exposure to high and low temperatures, darkness and 5% carbon dioxide. Planta 180:61–73

Andersen PC, Brodbeck BV, Mizell RF (1993) Diurnal variations of amino acids and organic acids in xylem fluid from *Lagerstroemia indica*: an endogenous circadian rhythm. Physiol Plant 89:783–790

Applewhite PB, Satter RL, Galston AW (1973) Protein synthesis during endogenous rhythmic leaflet movement in *Albizzia*. J Gen Physiol 62:707–713

Assmann SM (1993) Signal-transduction in guard cells. Annu Rev Cell Biol 9:345–375

Barak S, Tobin EM, Andronis C, Sugano S, Green RM (2000) All in good time: the *Arabidopsis* circadian clock. Trends Plant Sci 5:517–522

Beator J, Kloppstech K (1993) The circadian oscillator coordinates the synthesis of apoproteins and their pigments during chlorophyll development. Plant Physiol 103:191–196

Beator J, Pötter E, Kloppstech K (1992) The effect of heat shock on morphogenesis in barley. Coordinated circadian regulation of mRNA levels for light-regulated genes and of the capacity for accumulation of chlorophyll protein complexes. Plant Physiol 100:1780–1786

Beck F, Blasius B, Lüttge U, Neff R, Rascher U (2001) Stochastic noise interferes coherently with a model biological clock and produces specific dynamic behaviour. Proc R Soc B 268:1307–1313

Beyschlag W, Eckstein J (1997) Stomatal patchiness. Progress in Botany 59. Springer, Berlin Heidelberg New York, pp 283–298

Białczyk J, Lechowski Z (1987) The effect of abscisic acid and fusicoccin on malic acid concentration in pulvini of *Phaseolus coccineus* L. New Phytol 105:469–475

Blasius B (1997) Rhythmus und CAM, Zeitreihenanalyse und Modellierung regulärer und irregulärer Photosyntheseoszillationen bei CAM-Pflanzen. PhD Thesis, TU Darmstadt, Darmstadt, Germany

Blasius B, Beck F, Lüttge U (1997) A model for photosynthetic oscillations in crassulacean acid metabolism (CAM). J Theor Biol 184:345–351

Blasius B, Beck F, Lüttge U (1998) Oscillatory model of crassulacean acid metabolism: structural analysis and stability boundaries with a discrete hysteresis switch. Plant Cell Environ 21:775–784

Blasius B, Neff R, Beck F, Lüttge U (1999) Oscillatory model of crassulacean acid metabolism with a dynamic hysteresis switch. Proc R Soc B 266:93–101

Bohn A, Geist A, Rascher U, Lüttge U (2001) Responses to different external light rhythms by the circadian rhythm of crassulacean acid metabolism in *Kalanchoë daigremontiana*. Plant Cell Environ 24:811–820

Bohn A, Rascher U, Hütt MT, Kaiser F, Lüttge U (2002) Responses of a plant circadian rhythm to thermoperiodic perturbations with asymmetric temporal patterns and the rate of temperature change. Biol Rhythm Res (in press)

Bollig I (1977) Different circadian rhythms regulate photoperiodic flowering response and leaf movement in *Pharbitis nil* (L.) Choisy. Planta 135:137–142

Borland AM, Hartwell J, Jenkins GI, Wilkins MB, Nimmo HG (1999) Metabolite control overrides circadian regulation of phosphoenolpyruvate carboxylase kinase and CO_2 fixation in crassulacean acid metabolism. Plant Physiol 121:889–896

Boxall SF, Bohnert HJ, Cushman JC, Nimmo HG, Hartwell J (2001) The circadian clock and crassulacean acid metabolism in *Mesembryanthemum crystallinum*. Poster S18-D12. 12th International Congress of Photosynthesis, Brisbane

Brinker M, Engelmann W, Piechulla B (2001) Circadian rhythms of leaf movement in Gymnosperm species. Biol Rhythm Res 32:467

Britz SJ, Hungerford WE, Lee DR (1987) Rhythms during extended dark periods determine rates of net photosynthesis and accumulation of starch and soluble sugars in subsequent light periods of *Sorghum*. Planta 171:339–345

Buchanan-Bollig IC (1984) Circadian rhythms in *Kalanchoë*: effects of irradiance and temperature on gas exchange and carbon metabolism. Planta 160:264–271

Buchanan-Bollig IC, Fischer A, Kluge M (1984) Circadian rhythms in *Kalanchoë*: the pathway of $^{14}CO_2$ fixation during prolonged light. Planta 161:71–80

Bünning E (1935) Zur Kenntnis der erblichen Tagesperiodizität bei den Primärblättern von *Phaseolus multiflorus*. Jahrb Wiss Bot 81:411–418

Bünning E (1936) Die endogene Tagesrhythmik als Grundlage der photoperiodischen Reaktion. Ber Dtsch Bot Ges 54:590–607

Bünning E (1973) The physiological clock. Circadian rhythms and biological chronometry. The English Universities Press Ltd, London; Springer, Berlin Heidelberg New York

Bünning E, Moser I (1972) Influence of valinonmycin on circadian leaf movements of *Phaseolus*. Proc Natl Acad Sci USA 69:2732–2733

Bünsow R (1953) Endogene Tagesrhythmik und Photoperiodismus bei *Kalanchoë blossfeldiana*. Planta 42:220–252

Carter PJ, Nimmo HG, Fewson CA, Wilkins MB (1991) Circadian rhythms in the activity of a plant protein kinase. EMBO J 10:2063–2068

Carter PJ, Wilkins MB, Nimmo HG, Fewson CA (1995a) Effects of temperature on the activity of phosphoenolpyruvate carboxylase and on the control of CO_2 fixation in *Bryophyllum fedtschenkoi*. Planta 196: 375–380

Carter PJ, Wilkins MB, Nimmo HG, Fewson CA (1995b) The role of temperature in the regulation of the circadian rhythm of CO_2 fixation in *Bryophyllum fedtschenkoi*. Planta 196:381–386

Carter PJ, Fewson CA, Nimmo GA, Nimmo HG, Wilkins MB, (1996) Roles of circadian rhythms, light and temperature in the regulation of phosphoenolpyruvate carboxylase in crassulacean acid metabolism. In: Winter K, Smith JAC (eds) Crassulacean acid metabolism: biochemistry, ecophysiology and evolution. Ecological Studies, vol 114. Springer, Berlin Heidelberg New York, pp 46–52

Cermakian N, Sassone-Corsi P (2001) Rythmes biologiques: les secrets d'une horloge. La Recherche 338:38–42

Chen HM, Chien CY, Huang TC (1996) Regulation and molecular structure of a circadian oscillating protein located in the cell membrane of the prokaryote *Synechococcus* RF-1. Planta 199:520–527

Chen TH, Chen TL, Hung LM, Huang TC (1991) Circadian rhythm in amino acid uptake by *Synechococcus* RF-1. Plant Physiol 97:55–59

Clark DG, Richards C, Brown KM (1999) Characterization of circadian-regulated mRNAs encoding glycine-rich RNA-binding proteins in *Pelargonium × hortorum*. Physiol Plant 106:409–414

Cohen AS, Cumming BG (1974) Endogenous rhythmic activity of nitrate reductase in a selection of *Chenopodium rubrum*. Can J Bot 52:2351–2360

Comolli J, Taylor W, Rehman J, Hastings JW (1996) Inhibitors of serine/threonine phosphoprotein phosphatases alter circadian properties in *Gonyaulax polyedra*. Plant Physiol 111:285–291

Coté GG, DePass AL, Quarmby LM, Tate BF, Morse MJ, Satter RL, Crain RC (1989) Separation and characterisation of inositol phospholipids from the pulvini of *Samanea saman*. Plant Physiol 90:1422–1428

Couderchet J, Girard D (1976) Débit et teneur en potassium de la sève émise par des racines excisées d'*Helianthus annuus* L. C R Acad Sci Paris 282 Série D:173–176

Covington MF, Panda S, Liu XL, Strayer CA, Wagner DR, Kay SA (2001) ELF3 modulates resetting of the circadian clock in *Arabidopsis*. Plant Cell 13:1305–1315

Czihak G, Langner H, Ziegler H (1990) Biologie, 4. Aufl. Springer, Berlin Heidelberg New York

Daido H (2001) Why circadian rhythms are circadian: competitive population dynamics of biological oscillators. Phys Rev Lett 87: 048101-1 -048101-4

Deitzer GF, Kempf O, Fischer S, Wagner E (1974) Endogenous rhythmicity and energy transduction. IV. Rhythmic control of enzymes involved in tricarboxylic-acid cycle and the oxidative pentose-phosphate pathway in *Chenopodium rubrum* L. Planta 117:29–41

De Mairan J (1729) Observation botanique. In: Histoire de l'Academie Royale des Sciences. pp 35–36

Deng TS, Roenneberg T (1997) Photobiology of the *Gonyaulax* circadian system. II. Allopurinol inhibits blue-light effects. Planta 202:502–509

Deng MD, Moureaux T, Leydecker MT, Caboche M (1990) Nitrate-reductase expression is under the control of a circadian rhythm and is light-inducible in *Nicotiana tabaccum* leaves. Planta 180:257–261

Doerner PW (1994) Cell cycle regulation in plants. Plant Physiol 106:823–827

Dowson-Day MJ, Millar AJ (1999) Circadian dysfunction causes aberrant hypocotyl elongation patterns in *Arabidopsis*. Plant J 17:63–71

Dunlap JC (1993) Genetic analysis of circadian clocks. Annu Rev Physiol 55:683–728

Edmunds LN, Tamponnet C (1990) Oscillator control of cell division cycles in *Euglena*: role of calcium in circadian time-keeping. In: O'Day DH (ed) Calcium as an intracellular messenger in eucaryotic microbes. Am Soc Microbiol, Washington, DC, pp 97–123

Engelmann W, Johnsson A (1998) Rhythms in organ movement. In: Lumsden PJ, Millar AJ (eds) Biological rhythms and photoperiodism in plants. BIOS Scientific, Oxford, pp 35–50

Engelmann W, Simon K, Phen CJ (1992) Leaf movement rhythm in *Arabidopsis thaliana*. Z Naturforsch 47c:925–928

Erath F, Ruge WA, Mayer WE, Hampp R (1988) Isolation of functional extensor and flexor protoplasts from *Phaseolus coccineus* L. pulvini: potassium induced swelling. Planta 173:447–452

Everat-Bourbouloux A, Fleurat-Lessard P, Roblin G (1990) Comparative effects of indole-3-acetic acid, abscisic acid, gibberellic acid and 6-benzylaminopurine on the dark- and light-induced pulvinar movements in *Cassia fasciculata* Michx. J Exp Bot 41:315–324

Finlayson SA, Lee IJ, Morgan PW (1998) Phytochrome B and the regulation of circadian ethylene production in sorghum. Plant Physiol 116:17–25

Finlayson SA, Lee IJ, Mullet JE, Morgan PW (1999) The mechanism of rhythmic ethylene production in sorghum. The role of phytochrome B and simulated shading. Plant Physiol 119:1083–1089

Fleurat-Lessard P, Bouché-Pillon S, Leloup C, Bonnemain JL (1997a) Distribution and activity of the plasma membrane H^+-ATPase in *Mimosa pudica* L. in relation to ionic fluxes and leaf movements. Plant Physiol 113:747–754

Fleurat-Lessard P, Frangne N, Maeshima M, Ratajczak R, Bonnemain JL, Martinoia E (1997b) Increased expression of vacuolar aquaporin and H^+-ATPase related to motor cell function in *Mimosa pudica* L. Plant Physiol 114:827–834

Foster KR, Morgan PW (1995) Genetic regulation of development in *Sorghum bicolor.* IX. The ma^R_3 allele disrupts diurnal control of gibberellin biosynthesis. Plant Physiol 108:337–343

Francis D, Halford NG (1995) The plant cell cycle. Physiol Plant 93:365–374

Fredeen AL, Hennessey TL, Field CB (1991) Biochemical correlates of the circadian rhythm in photosynthesis in *Phaseolus vulgaris.* Plant Physiol 97:415–419

Freudling C, Mayer WE, Gradmann D (1980) Electrical membrane properties and circadian rhythm in extensor cells of the laminar pulvini of *Phaseolus coccineus* L. Plant Physiol 65:966–968

Freudling C, Starrach N, Flach D, Gradmann D, Mayer WE (1988) Cell walls as reservoirs of potassium ions for reversible volume changes of pulvinar motor cells during rhythmic leaf movements. Planta 175:193–203

Friesen WO, Block GD, Hocker CG (1993) Formal approaches to understanding biological oscillations. Annu Rev Physiol 55:661–681

Frosch S, Wagner E (1973a) Endogenous rhythmicity and energy transduction. II. Phytochrome action and the conditioning of rhythmicity of adenylate kinase, NAD- and NADP-linked glyceraldehyde-3-phosphate dehydrogenase in *Chenopodium rubrum* by temperature and light intensity cycles during germination. Can J Bot 51:1521–1528

Frosch S, Wagner E (1973b) Endogenous rhythmicity and energy transduction. III. Time course of phytochrome action in adenylate kinase, NAD- and NADP-linked glyceraldehyde-3-phosphate dehydrogenase in *Chenopodium rubrum.* Can J Bot 51:1529–1535

Frosch S, Wagner E, Cumming BG (1973) Endogenous rhythmicity and energy transduction. I. Rhythmicity in adenylate kinase, NAD- and NADP-linked glyceraldehyde-3-phosphate dehydrogenase in *Chenopodium rubrum.* Can J Bot 51:1355–1367

Fujiwara S, Ishida N, Tsuzuki M (1996) Circadian expression of the carbonic anhydrase gene, *Cah1*, in *Chlamydomonas reinhardtii.* Plant Mol Biol 32:745–749

Genoud T, Trevino Santa Cruz MB, Métraux JP (2001) Numeric simulation of plant signaling networks. Plant Physiol 126:1430–1437

Gierer A (1998) Im Spiegel der Natur erkennen wir uns selbst. Wissenschaft und Menschenbild. Rowohlt, Hamburg

Giuliano G, Hoffman NE, Ko K, Scolnik PA, Cashmore AR (1988) A light-entrained circadian clock controls transcription of several plant genes. EMBO J 7:3635–3642

Golden SS, Strayer C (2001) Time for plants. Progress in plant chronobiology. Plant Physiol 125:98–101

Golden SS, Ishiura M, Johnson CH, Kondo T (1997) Cyanobacterial circadian rhythms. Annu Rev Plant Physiol Plant Mol Biol 48:327–354

Gonze D, Leloup JC, Goldbeter A (2000) Theoretical models for circadian rhythms in *Neurospora* and *Drosophila.* C R Acad Sci Paris Sci de la vie 323:57–67

Gooch van D, Wehseler RA, Gross CG (1994) Temperature effects on the resetting of the phase of the *Neurospora* circadian rhythm. J Biol Ryhthm 9:83–94

Gorton HL (1990) Stomates and pulvini: a comparison of two rhythmic, turgor-mediated movement systems. In: Satter RL, Gorton HL et al. (eds) The pulvinus: motor organ for leaf movement. The American Society of Plant Physiologists, Rockville, MD, pp 223–237

Gorton HL, Williams WE, Binns ME, Gemmell CN, Leheny EA, Shepherd AC (1989) Circadian stomatal rhythms in epidermal peels from *Vicia faba.* Plant Physiol 90:1329–1334

Gorton HL, Williams WE, Assmann SM (1993) Circadian rhythms in stomatal responsiveness to red and blue light. Plant Physiol 103:399–406

Goto R, Kane R, Morishita M, Nakashia H (1994) Effect of temperature on the circadian conidiation rhythm of temperature-sensitive mutants of *Neurospora crassa.* Plant Cell Physiol 35:613–618

Grams TEE, Kluge M, Lüttge U (1995) High temperature adapted plants of *Kalanchoë daigremontiana* show changes in temperature dependence of the endogenous CAM rhythm. J Exp Bot 46:1927–1929

Grams TEE, Beck F, Lüttge U (1996) Generation of rhythmic and arrhythmic behaviour of crassulacean acid metabolism in *Kalanchoë daigremontiana* under continuous light by varying the irradiance or temperature: measurements in vivo and model simulations. Planta 198:110–117

Grams TEE, Borland AM, Roberts A, Griffiths H, Beck F, Lüttge U (1997) On the mechanism of reinitiation of endogenous crassulacean acid metabolism rhythm by temperature changes. Plant Physiol 113:1309–1317

Granbom M, Pedersén M, Lüning K (2001) A circadian rhythm of photosynthesis in the red macroalga *Kappaphycus alvarezii*. Biol Rhythm Res 32:460

Hänggi P (2001) Stochastische Resonanz: Rauschen macht sensibel. Physikalische Blätter 57:15–16

Hall A, Kozma-Bognár L, Tóth R, Nagy F, Millar AJ (2001) Conditional circadian regulation of phytochrome A gene expression. Plant Physiol 127:1808–1818

Harmer SL, Hogenesch JB, Straume M, Chang HS, Zhu T, Wang X, Kreps JA, Kay SA (2000) Orchestrated transcription of key pathways in *Arabidopsis* by the circadian clock. Science 290:2110–2113

Hartwell J, Smith LH, Wilkins MB, Jenkins GI, Nimmo HG (1996) Higher plant phosphoenolpyruvate carboxylase kinase is regulated at the level of translatable mRNA in response to light or a circadian rhythm. Plant J 10:1071–1078

Hartwell J, Gill A, Nimmo GA, Wilkins MB, Jenkins GI, Nimmo HG (1999) Phosphoenolpyruvate carboxylase kinase is a novel protein kinase regulated at the level of expression. Plant J 20:333–342

Hartwell LH, Hopfield JJ, Leibler S, Murray AW (1999) From molecular to modular cell biology. Nature 402 Supp:C47-C52

Heintzen C, Fischer R, Melzer S, Kappeler S, Apel K, Staiger D (1994a) Circadian oscillation of a transcript encoding a germin-like protein that is associated with cell walls in young leaves of the long-day plant *Sinapis alba* L. Plant Physiol 106:905–915

Heintzen C, Melzer S, Fischer R, Kappeler S, Apel K, Staiger D (1994b) A light-entrained and temperature-entrained circadian clock controls expression of transcripts encoding nuclear proteins with homology to RNA-binding proteins in meristematic tissue. Plant J 5:799–813

Heintzen C, Nater M, Apel K, Staiger D (1997) AtGRP7, a nuclear RNA-binding protein, as a component of a circadian-regulated negative feedback loop in *Arabidopsis thaliana*. Proc Natl Acad Sci USA 94:8515–8520

Helsper JPFG, Davies JA, Bouwmeester HJ, Krol AF, van Kampen MH (1998) Circadian rhythmicity in emission of volatile compounds by flowers of *Rosa hybrida* L. cv. Honesty. Planta 207:88–95

Hennessey TL, Field CB (1991) Circadian rhythms in photosynthesis. Oscillations in carbon assimilation and stomatal conductance under constant conditions. Plant Physiol 96:831–836

Hennessey TL, Freeden AL, Field CB (1992) Evidence of multiple circadian oscillators in bean plants. J Biol Rhythms 7:105–113

Hennessey TL, Freeden AL, Field CB (1993) Environmental effects of circadian rhythms in photosynthesis and stomatal opening. Planta 189:369–376

Hensel W (1987) Movements of pulvinated leaves. Progress in Botany49. Springer, Berlin Heidelberg New York, pp 171–180

Henzler T, Waterhouse RN, Smyth AJ, Carvajal M, Cooke DT, Schäffner AR, Steudle E, Clarkson DT (1999) Diurnal variations in hydraulic conductivity and root pressure can be correlated with the expression of putative aquaporins in the roots of *Lupinus japonicus*. Planta 210:50–60

Hicks KA, Albertson TM, Wagner DR (2001) *Early flowering3* encodes a novel protein that regulates circadian clock function and flowering in *Arabidopsis*. Plant Cell 13:1281–1292

Huang TC, Chen HM, Pen SY, Chen TH (1994) Biological clock in the prokaryote *Synechococcus* RF-1. Planta 193:131–136

Hütt MT, Neff R (2001) Quantification of spatiotemporal phenomena by means of cellular automata techniques. Physica A 289:498–516

Ievinsh G, Kreicbergs (1992) Endogenous rhythmicity of ethylene production in growing intact cereal seedlings. Plant Physiol 100:1389–1391

Irving MS, Ritter S, Tomos AD, Koller D (1997) Phototropic response of the bean pulvinus: movement of water and ions. Bot Acta 110:118–126

Ishiura M, Kutsuna S, Aoki S, Iwasaki H, Andersson CR, Tanabe A, Golden SS, Johnson CH, Kondo T (1998) Expression of a gene cluster *kaiABC* as a circadian feedback process in cyanobacteria. Science 281:1519–1523

Iwasaki H, Kondo T (2000) The current state and problems of circadian clock studies in cyanobacteria. Plant Cell Physiol 41: 1013–1020

Iwasaki H, Williams SB, Kitayama Y, Ishiura M, Golden SS, Kondo T (2000) A KaiC-interacting sensory histidine kinase, SasA, necessary to sustain robust circadian oscillation in cyanobacteria. Cell 101:223–233

Jacobshagen S, Whetstine JR, Boling JM (2001) Many but not all genes in *Chlamydomonas reinhardtii* are regulated by the circadian clock. Plant Biol 3:592–597

Jaffe MJ (1968) Phytochrome-mediated bioelectric potentials in mung bean seedlings. Science 162:1061–1067

Jerebzoff S (1986) Cellular circadian rhythms in plants: recent approaches to their molecular bases. Physiol Vég 24:367–376

Johnson CH (1992) Phase response curves: What can they tell us about circadian clocks? In: Hiroshige T, Honma K (eds) Circadian clocks from cell to human. Hokkaido University Press, Sapporo, Japan, pp 209–249

Johnson CH (2001) Endogenous timekeepers in photosynthetic organisms. Annu Rev Physiol 63:695–728

Johnson CH, Golden SS (1999) Circadian programs in cyanobacteria: adaptiveness and mechanism. Annu Rev Microbiol 53:389–409

Johnson CH, Roeber JF, Hastings JW (1984) Circadian changes in enzyme concentration account for rhythm of enzyme activity in *Gonyaulax*. Science 223:1428–1430

Johnson CH, Kondo T, Goto K (1992) Circadian rhythms in *Chlamydomonas*: In: Hiroshige T, Homna K (eds) Circadian clocks from cell to human. Hokkaido Press, Sapporo, Japan, pp 139–155

Johnson CH, Knight MR, Kondo T, Masson P, Seedbrook J, Haley A, Trewavas A (1995) Circadian oscillations of cytosolic and chloroplastic free calcium in plants. Science 269:1863–1865

Johnson CH, Knight M, Trewavas A, Kondo T (1998) A clockwork green: circadian programs in photosynthetic organisms. In: Lumsden P, Millar A (eds) Biological rhythms and photoperiodism in plants. BIOS Scientific, Oxford, pp 1–34

Jones TL, Ort DR (1997) Circadian regulation of sucrose phosphate synthase activity in tomato by protein phosphatase activity. Plant Physiol 113:1167–1175

Jones TL, Tucker DE, Ort DR (1998) Chilling delays circadian pattern of sucrose phosphate synthase and nitrate reductase activity in tomato. Plant Physiol 118:149–158

Jouve L, Greppin H, Agosti RD (1998) *Arabidopsis thaliana* floral stem elongation: evidence for an endogenous circadian rhythm. Plant Physiol Biochem 36:469–472

Jouve L, Gaspar T, Kevers C, Greppin H, Agosti RD (1999) Involvement of indole-3-acetic acid in the circadian growth of the first internode of *Arabidopsis*. Planta 209:136–142

Kaiser H, Kappen L (1997) In situ observations of stomatal movements in different light-dark regimes: the influence of endogenous rhythmicity and long term adjustments. J Exp Bot 48:1583–1589

Kana TM, Miller JH (1977) Effect of photoperiod on stomatal opening in *Vicia faba*. Plant Physiol 60:803–804

Katayama M, Tsinoremas NF, Kondo T, Golden SS (1999) *cpmA*, a gene involved in an output pathway of the cyanobacterial circadian system. J Bacteriol 181:3516–3524

Kathiresan A, Reid DM, Chinnappa CC (1996) Light- and temperature-entrained circadian regulation of activity and mRNA accumulation of 1-aminocyclopropane-1-carboxylic acid oxidase in *Stellaria longipes*. Planta 199:329–335

Kayali S, Greppin H, Agosti RD (1997) Effect of EGTA on the diurnal leaf movement of *Phaseolus vulgaris*. Plant Physiol Biochem 35:915–922

Kellmann JW, Merforth N, Wiese M, Pichersky E, Piechulla B (1993) Concerted circadian oscillations in transcript levels of nineteen (*Lha/b*)(*cab*) genes in *Lycopersicon esculentum* (tomato). Mol Gen Genet 237:439–448

Kellmann JW, Hoffrogge R, Piechulla B (1999) Transcriptional regulation of oscillating steady-state *Lhc* mRNA levels: characterization of two *Lhca* promoter fragments in transgenic tobacco plants. Biol Rhythm Res 30:264–271

Kim HJ, Coté GG, Crain RC (1992) Effects of light on the membrane potential of protoplasts from *Samanea saman* pulvini. Involvement of K^+ channels and the H^+-ATPase. Plant Phsiol 99:1532–1539

Kim HJ, Coté GG, Crain RC (1993) Potassium channels in *Samanea saman* protoplasts controlled by phytochrome and the biological clock. Science 260:960–962

Kim HJ, Coté GG, Crain RC (1996) Inositol 1,4,5-triphosphate may mediate closure of K^+ channels by light and darkness in *Samanea saman* motor cells. Planta 198:279–287

Klenell M, Pedersén M, Lüning K (2001) A circadian rhythm of photosynthesis in the brownalga *Laminaria saccharina*: dependence on light intensity and spectral quality. Biol Rhythm Res 32:461

Knight TJ, Weissman GS (1982) Rhythms in glutamine synthetase activity, energy charge, and glutamine in sunflower roots. Plant Physiol 70:1683–1688

Knoetzel J, Rensing L (1990) Characterization of the photosynthetic apparatus from the marine dinoflagellate *Gonyaulax polyedra*. II. Circadian rhythmicity of photosynthesis and the supramolecular organization of pigment-protein complexes. J Plant Physiol 136:280–288

Koller D, Björkman O, Ritter S (1995) Role of pulvinar chloroplasts in light-driven leaf movements of the trifoliate leaf of bean (*Phaseolus vulgaris* L.). J Exp Bot 46:1215–1222

Kolosova N, Gorenstein N, Kish CM, Dudareva N (2001) Regulation of circadian methyl benzoate emission in diurnally and nocturnally emitting plants. Plant Cell 13:2333–2347

Kondo T (1983) Phase shift in the potassium uptake rhythm of the duckweed *Lemna gibba* G3 caused by an azide pulse. Plant Physiol 73:605–608

Kondo T (1984) Removal by a trace of sodium of the period lengthening of the potassium uptake rhythm due to lithium in *Lemna gibba* G3. Plant Physiol 75:1071–1074

Kondo T (1989) Comparison of phase shifts of the circadian rhythm of K^+ uptake in *Lemna gibba* G3 by various amino acid analogs. Plant Physiol 90:1600–1608

Konopka RJ, Orr D (1980) Effects of a clock mutation on the subjective day – implications for a membrane model of the *Drosophila* circadian clock. In: Siddiqi O, Babu P, Hall LM (eds) Development and neurobiology of *Drosophila*. Plenum, New York, pp 409–416

Kreps JA, Kay SA (1997) Coordination of plant metabolism and development by the circadian clock. Plant Cell 9:1235–1244

Kreps JA, Simon AE (1997) Environmental and genetic effects on circadian clock-regulated gene expression in *Arabidopsis*. Plant Cell 9:297–304

Kruse E, Grimm B, Beator J, Kloppstech K (1997) Developmental and circadian control of the capacity for δ-aminolevulinic acid synthesis in green barley. Planta 202:235–241

Kusumi K, Arata H, Iwasaki I, Nishimura M (1994) Regulation of PEP-carboxylase by biological clock in a CAM plant. Plant Cell Physiol 35:233–242

Laval-Martin DL, Carré IA, Barbera SJ, Edmunds LN (1990) Rhythmic changes in the activities of NAD kinase and NADP phosphatase in the achlorophyllous ZC mutant of *Euglena gracilis* Klebs strain Z. Arch Biochem Biophys 276:433–441

Lebert M, Porst M, Häder DP (1999) Circadian rhythm of gravitaxis in *Euglena gracilis.* J Plant Physiol 155:344–349

Lecharny A, Wagner E (1984) Stem elongation rate in light-grown plants. Evidence for an endogenous circadian rhythm in *Chenopodium rubrum.* Physiol Plant 60:437–443

Lecharny A, Tremolières A, Wagner E (1990) Correlation between the endogenous circadian rhythmicity in growth rate and fluctuations in oleic acid content in expanding stems of *Chenopodium rubrum* L. Planta 182:211–215

Lee Y, Satter RL (1987) H^+-uptake and release during circadian rhythmic movements of excised *Samanea* motor organs. Plant Physiol 83:856–862

Lee Y, Satter RL (1989) Effects of white, blue, red light and darkness on pH of the apoplast in the *Samanea* pulvinus. Planta 178:31–40

Leloup J-C, Goldbeter A (1997) Temperature compensation of circadian rhythms: control of the period in a model for circadian oscillations of the PER protein in *Drosophila.* Chronobiol Int 14:511–520

Lemaire SD, Stein M, Issakidis-Bourguet E, Keryer E, Benoit V, Pineau B, Gérard-Hirne C, Miginiac-Maslow M, Jacquot JP (1999) The complex regulation of ferredoxin/thioredoxin-related genes by light and the circadian clock. Planta 209:221–229

Lillo C, Ruoff P (1989) An unusually rapid light-induced nitrate reductase mRNA pulse and circadian oscillations. Naturwissenschaften 76:526–528

Lillo C, Meyer C, Ruoff P (2001) The nitrate reductase circadian system. The central clock dogma contra multiple oscillatory feedback loops. Plant Physiol 125:1554–1557

Liu XL, Covington MF, Fankhauser C, Chory J, Wagner DR (2001) *ELF3* encodes a circadian clock-regulated nuclear protein that functions in an *Arabidopsis PHYB* signal transduction pathway. Plant Cell 13:1293–1304

Liu Y, Garceau NY, Loros JJ, Dunlap JC (1997) Thermally regulated translational control of FRQ mediates aspects of temperature responses in the *Neurspora* circadian clock. Cell 89:477–486

Liu Z, Taub CC, McClung CR (1996) Identification of an *Arabidopsis thaliana* ribulose-1,5-bisphosphate carboxylase/oxygenase activase (RCA) minimal promoter regulated by light and the circadian clock. Plant Physiol 112:43–51

Lonergan TA (1981) A circadian rhythm in the rate of light-induced electron flow in three leguminous species. Plant Physiol 68:1041:1046

Lonergan TA (1990) Steps linking the photosynthetic light reactions to the biological clock require calcium. Plant Physiol 93:110–115

Lonergan TA, Sargent ML (1978) Regulation of the photosynthesis rhythm in *Euglena gracilis.* I. Carbonic anhydrase and glyceraldehyde-3-phosphate dehydrogenase do not regulate the photosynthesis rhythm. Plant Physiol 61:150–153

Lonergan TA, Sargent ML (1979) Regulation of the photosynthesis rhythm in *Euglena gracilis.* II. Involvement of electron flow through both photosystems. Plant Physiol 64:99–103

Loros J (1995) The molecular basis of the *Neurospora* clock. Sem Neurosci 7:3–13

Lüttge U (2000) The tonoplast functioning as the master switch for circadian regulation of crassulacean acid metabolism. Planta 211:761–769

Lüttge U, Ball E (1978) Free running oscillations of transpiration and CO_2 exchange in CAM plants without a concomitant rhythm of malate levels. Z Pflanzenphysiol 90:69–77

Lüttge U, Beck F (1992) Endogenous rhythms and chaos in crassulacean acid metabolism. Planta 188:28–38

Lüttge U, Grams TEE, Hechler B, Blasius B, Beck F (1996) Frequency resonances of the circadian rhythm of CAM under external temperature rhythms of varied period lengths in continuous light. Bot Acta 109:422–426

Lüttge U, Kluge M, Bauer G (2002) Botanik. 4th edn., VCH, Weinheim

Lumsden PJ (1991) Circadian rhythms and phytochrome. Annu Rev Plant Physiol Plant Mol Biol 42:351–371

Macduff JH, Dhanoa MS (1996) Diurnal and ultradian rhythm in K^+ uptake by *Trifolium repens* under natural light patterns: evidence for segmentation at different root temperatures. Physiol Plant 98:298–308

Màchackovà I, Chauvaux N, Dewitte W, van Onckelen H (1997) Diurnal fluctuations in ethylene formation in *Chenopodium rubrum*. Plant Physiol 113:981–985

Makarov VN, Schoschina EV, Lüning K (1995) Diurnal and circadian periodicity of mitosis and growth in marine macroalgae. I. Juvenile sporophytes of Laminariales (Phaeophyta). Eur J Physiol 30:261–266

Makino S, Kiba T, Imamura A, Hanaki N, Nakamura A, Suzuki T, Taniguchi M, Ueguchi C, Sugiyama T, Mizuno T (2000) Genes encoding pseudo-response regulators: insight into his-to-asp phosphorelay and circadian rhythm in *Arabidopsis thaliana*. Plant Cell Physiol 41:791–803

Makino S, Matsushika A, Kojima M, Oda Y, Mizumo T (2001) Light response of the circadian waves of the APRR1/TOC1 quintet: when does the quintet start singing rhythmically in *Arabidopsis*? Plant Cell Physiol 42:334–339

Martino-Catt S, Ort DR (1992) Low temperature interrupts circadian regulation of transcriptional activity in chilling-sensitive plants. Proc Natl Acad Sci USA 89:3731–3735

Matsushika A, Makino S, Kojima M, Mizumo T (2000) Circadian waves of expression of the APRR1/TOC1 family of pseudoresponse regulators in *Arabidopsis thaliana:* insight into the plant circadian clock. Plant Cell Physiol 41:1002–1012

Mayer WE, Betz S, Schöffel S (1994) Are K^+ channels and H^+-ATPases of the plasma membrane involved in the control and generation of circadian rhythmicity in the pulvinar motor cells of *Phaseolus*? Biol Rhythm Res 25:301–314

Mayer WE, Hohloch C, Kalkuhl A (1997) Extensor protoplasts of the *Phaseolus* pulvinus: light-induced swelling may require extracellular Ca^{2+} influx, dark-induced shrinking inositol 1,4,5-triphosphate-induced Ca^{2+} mobilization. J Exp Bot 48:219–228

McClung CR (1997) The regulation of catalase in *Arabidopsis*. Free Rad Biol Med 23:489–496

McClung CR (2000) Circadian rhythms in plants: a millenial view. Physiol Plant 109:359–371

McClung CR, Kay SA (1994) Circadian rhythms in the higher plant, *Arabidopsis thaliana*. In: Somerville CR, Meyerowitz EM (eds) *Arabidopsis thaliana*. Cold Spring Harbor Laboratory Press, Cold Spring Harbor, NY, pp 615–637

McClung CR, Fox BA, Dunlap JC (1989) The *Neurospora* clock gene *frequency* shares a sequence element with the *Drosophila* clock gene *period*. Nature 339:558–562

McClung CR, Hsu M, Painter JE, Gagne JM, Karlsberg SD, Salomé PA (2000) Integrated temporal regulation of the photorespiratory pathway. Circadian regulation of two *Arabidopsis* genes encoding serine hydroxymethyltransferase. Plant Physiol 123:381–391

McWatters HG, Bastow RM, Hall A, Millar AJ (2000) The *ELF3 zeitnehmer* regulates light signaling to the circadian clock. Nature 408:716–720

Meidner H, Willmer CM (1993) Circadian rhythms of stomatal movements in epidermal strips. J Exp Bot 44:1649–1652

Mérida A, Rodríguez-Galán JM, Vincent C, Romero JM (1999) Expression of the granule-bound starch synthase I (*waxy*) gene from snapdragon is developmentally and circadian clock regulated. Plant Physiol 120:401–409

Merrow M, Brunner M, Roenneberg T (1999) Assignment of circadian function of the *Neurospora* clock gene *frequency*. Nature 399:584–586

Millar AJ (1998) The cellular organization of circadian rhythms in plants: not one but many clocks. In: Lumsden PJ, Millar AJ (eds) Biological rhythms and photoperiodism in plants. BIOS Scientific, Oxford, pp 51–68

Millar AJ (1999) Biological clocks in *Arabidopsis thaliana.* New Phytol 141:175–197

Millar AJ, Kay SA (1996) Integration of circadian and phototransduction pathways in the network controlling *CAB* gene transcription in *Arabidopsis.* Proc Natl Acad Sci USA 93:15491–15496

Millar AJ, Straume M, Chory J, Chua NH, Kay SA (1995) The regulation of circadian period by phototransduction pathways in *Arabidopsis.* Science 267:1163–1166

Min H, Golden SS (2000) A new circadian class 2 gene, *opcA*, whose product is important for reductant production at night in *Synechococcus elongatus* PCC7942. J Bacteriol 182:6214–6221

Mishkind M, Mauzerall D, Beale SI (1979) Diurnal variation in situ of photosynthetic capacity in *Ulva* is caused by a dark reaction. Plant Physiol 64:896–899

Mittag M, Hastings JW (1996) Exploring the signaling pathway of circadian bioluminescence. Physiol Plant 96:727–732

Mittag M, Lee DH, Hastings JW (1994) Circadian expression of the luciferin-binding protein correlates with the binding of a protein to the 3′ untranslated region of its mRNA. Proc Natl Acad Sci USA 91:5257–5261

Moran N, Ehrenstein G, Iwasa K, Mischke C, Bare C, Satter RL (1988) Potassium channels in motor cells of *Samanea saman.* A patch-clamp study. Plant Physiol 88:643–648

Morgan PW, Finlayson SA, Lee IJ, Childs KL, He CJ, Creelman RA, Drew MC, Mullet JE (1997) Regulation of a circadianly rhythmic ethylene production by phytochrome B in sorghum. In: Kanellis AK, Chang C, Kende H, Grierson D (eds) Biology and biotechnology of the plant hormone ethylene. Kluwer, Dordrecht, pp 105–111

Morse D, Hastings JW, Roenneberg T (1994) Different phase responses of the two circadian oscillators in *Gonyaulax.* J Biol Rhythms 9:263–274

Moshelion M, Moran N (2000) Potassium-efflux channels in extensor and flexor cells of the motor organ of *Samanea saman* are not identical. Effects of cytosolic calcium. Plant Physiol 124:911–919

Moshelion M, Becker D, Biela A, Uehelin N, Hedrich R, Otto B, Levi H, Moran N, Kaldenhoff R (2002) Plasmamembrane aquaporins in the motor cells of *Samanea saman*: diurnal and circadian regulation. Plant Cell 14:727–739

Moysset L, Simon E (1989) Role of calcium in phytochrome-controlled nyctinastic movements of *Albizzia lophanta* leaflets. Plant Physiol 90:1108–1114

Nakashima H (1982) Effects of membrane ATPase inhibitors on light-induced phase shifting of the circadian clock in *Neurospora crassa.* Plant Physiol 69:619–623

Neff R, Blasius B, Beck F, Lüttge U (1998) Thermodynamics and energetics of the tonoplast membrane operating as a hysteresis switch in an oscillatory model of crassulacean acid metabolism. J Membr Biol 165:37–43

Nimmo GA, Wilkins MB, Fewson CA, Nimmo HG (1987) Persistent circadian rhythm in the phosphorylation state of phosphoenolpyruvate carboxylase from *Bryophyllum fedtschenkoi* leaves and its sensitivity to inhibition by malate. Planta 170:408–415

Nimmo HG (1998) Circadian regulation of a plant protein kinase. Chronobiol Int 15:109–118

Nimmo HG (2000) The regulation of phosphoenolpyruvate carboxylase in CAM plants. Trends Plant Sci 5:75–80

Nimmo HG, Fontaine V, Hartwell J, Jenkins GI, Nimmo GA, Wilkins MB (2001) PEP carboxylase kinase is a novel protein kinase controlled at the level of expression. New Phytol 151:91–97

Nishizaki Y, Kubota M, Yamamiya K, Watanabe M (1997) Action spectrum of light pulse-induced membrane depolarization in pulvinar motor cells of *Phaseolus.* Plant Cell Physiol 38:526–529

Njus D, Sulzmann FM, Hastings JW (1974) Membrane model for the circadian clock. Nature 248:116–120

Njus D, McMurry L, Hastings JW (1977) Conditionality of circadian rhythmicity synergistic action of light and temperature. J Comp Physiol 117:335–344

Novak B, Greppin H (1979) High-frequency oscillations and circadian rhythm of the membrane potential in spinach leaves. Planta 144:235–240

Nuernbergk EL (1961) Endogener Rhythmus und CO_2-Stoffwechsel bei Pflanzen mit diurnalem Säurestoffwechsel. Planta 56:28–70

Nungesser D, Kluge M, Tolle H, Oppelt W (1984) A dynamic computer model of the metabolic and regulatory processes in crassulacean acid metabolism. Planta 162:204–214

Okazaki Y, Azuma K, Nishizaki Y (2000) A pulse of blue light induces a transient increase in activity of apoplastic K^+ in laminar pulvinus of *Phaseolus vulgaris* L. Plant Cell Physiol 41:230–233

Oltmanns O (1960) Über den Einfluß der Temperatur auf die endogene Tagesrhythmik und die Blühinduktion bei der Kurztagspflanze *Kalanchoë blossfeldiana*. Planta 54:233–264

O'Neill SD, Zhang XS, Zheng CC (1994) Dark and circadian regulation of mRNA accumulation in the short-day plant *Pharbitis nil*. Plant Physiol 104:569–580

Ono F, Frommer WB, von Wirén N (2000) Coordinated diurnal regulation of low- and high-affinity nitrate transporters in tomato. Plant Biol 2:17–23

Ouyang Y, Andersson CR, Kondo T, Golden SS, Johnson CH (1998) Resonating circadian clocks enhance fitness in cyanobacteria. Proc Natl Acad Sci USA 95:8660–8664

Paraskevopoulou T, Anastassiou R, Argyroudi-Akoyunoglou JH (1995) Circadian expression of the light-harvesting protein of photosystem II in etiolated bean leaves following a single red light pulse: coordination with the capacity of the plant to form chlorophyll and the thylakoid bound protease. Photosyn Res 44:93–106

Park DH, Somers DE, Kim YS, Choy YH, Lim HK, Soh MS, Kim HJ, Kay SA, Nam HG (1999) Control of circadian rhythms and photoperiodic flowering by the *Arabidopsis GIGANTEA* gene. Science 285:1579–1582

Perilleux C, Ongena P, Bernier G (1996) Changes in gene-expression in the leaf of *Lolium temulentum* L. Ceres during the photoperiodic induction of flowering. Planta 200:32–40

Pfeffer W (1907) Abh Math Phys Kl Kgl Sächs Ges Wiss 30 III:259–472

Pfeffer W (1915) Abh Math Phys Kl Kgl Sächs Ges Wiss 34 I:1–154

Piechulla B (1988) Plastid and nuclear mRNA fluctuations in tomato leaves – diurnal and circadian rhythms during extended dark and light periods. Plant Mol Biol 11:345–353

Piechulla B (1999) Circadian expression of the light-harvesting complex protein genes in plants. Chronobiol Int 16:115–128

Piechulla B, Gruissem W (1987) Diurnal mRNA fluctuations of nuclear and plastid genes in developing tomato fruits. EMBO J 6:3593–3599

Piechulla B, Riesselmann S (1990) Effect of temperature alterations on the diurnal expression pattern of the chlorophyll *a/b* binding proteins in tomato seedlings. Plant Physiol 94:1903–1906

Pilgrim ML, McClung CR (1993) Differential involvement of the circadian clock in the expression of genes required for ribulose-1,5-bisphosphate carboxylase/oxygenase synthesis, assembly, and activation in *Arabidopsis thaliana*. Plant Physiol 103:553–564

Pilgrim ML, Caspar T, Quail PH, McClung CR (1993) Circadian and light regulated expression of nitrate reductase in *Arabidopsis*. Plant Mol Biol 23:349–364

Pongratz P, Beck E (1978) Diurnal oscillations of amylolytic activity in spinach chloroplasts. Plant Physiol 62:687–689

Pott MB, Kellmann JW, Piechulla B (2000) Circadian and phytochrome control act at different promoter of the tomato *Lhca3* gene. J Plant Physiol 157:449–452

Pott MB, Pichersky E, Piechulla B (2002) Nocturnal oscillations of methyl salicylate emission, SAMT enzyme activity, and SAMT mRNA in flowers of *Stephanotis floribunda*. J Plant Physiol (in press)

Prézelin BB, Sweeney BM (1977) Characterization of photosynthetic rhythms in marine dinoflagellates. II. Photosynthesis-irradiance curves and in vivo chlorophyll a fluorescence. Plant Physiol 60:388–392

Prézelin BB, Meeson BW, Sweeney BM (1977) Characterization of photosynthetic rhythms in marine dinoflagellates. I. Pigmentation, photosynthetic capacity and respiration. Plant Physiol 60:384–387

Racusen RH, Satter RL (1975) Rhythmic and phytochrome-regulated changes in transmembrane potential in *Samanea* pulvini. Nature 225:408–410

Ramalho CB, Hastings JW, Colepicolo P (1995) Circadian oscillations of nitrate reductase activity in *Gonyaulax polyedra* is due to changes in cellular protein levels. Plant Physiol 107:225–231

Rascher U (2001) Der endogene CAM-Ryhthmus von *Kalanchoë daigremontiana* als nichtlineares Modellsystem zum Verständnis der raum-zeitlichen Dynamik einer biologischen Uhr. Der Andere Verlag, Osnabrück

Rascher U, Blasius B, Beck F, Lüttge U (1998) Temperature profiles for the expression of endogenous rhythmicity and arrhythmicity of CO_2 exchange in the CAM plant *Kalanchoë daigremontiana* can be shifted by slow temperature changes. Planta 207:76–82

Rascher U, Hütt MT, Siebke K, Osmond B, Beck F, Lüttge U (2001) Spatiotemporal variation of metabolism in a plant circadian rhythm: the biological clock as a an assembly of coupled individual oscillators. Proc Natl Acad Sci USA 98:11801–11805

Rau W (1967) Untersuchungen über die lichtabhängige Carotinbiosynthese. I. Das Wirkungsspektrum von *Fusarium aquaeductum*. Planta 72:14–28

Resch A (1989) Paranormologie: Geschichte und Fachgebiete. Universitas 44:310–320

Reves PH, Coupland G (2000) Response of plant development to environment: control of flowering by daylength and temperature. Curr Opin Plant Biol 3:37–42

Rikin A (1991) Temperature-induced phase shifting of circadian rhythms in cotton seedlings as related to variations in chilling resistance. Planta 185:407–414

Roblin G, Fleurat-Lessard P, Bonmort J (1989) Effects of compounds affecting calcium channels on phytochrome- and blue pigment-mediated pulvinar movements of *Cassia fasciculata*. Plant Physiol 90:697–701

Rockel B, Blasius B, Beck F, Ratajczak R, Lüttge U (1997) Endogenous oscillations of the transcript amounts of subunit-c of the V-ATPase of *Mesembryanthemum crystallinum* with harmonic frequency resonances under continuous illumination. Cell Mol Biol Lett 2:69–76

Roenneberg T (1996) The complex circadian system of *Gonyaulax polyedra*. Physiol Plant 96:733–737

Roenneberg T, Deng TS (1997) Photobiology of the *Gonyaulax* circadian system. I. Different phase response curves for red and blue light. Planta 202:494–501

Roenneberg T, Merrow M (1998) Molecular circadian oscillators: an alternative hypothesis. J Biol Rhythms 13:167–179

Roenneberg T, Morse D (1993) Two circadian oscillators in one cell. Nature 362:362–364

Rogers LA, Greenbank GR (1930) The intermittent growth of bacterial cultures. J Bacteriol 19:181–190

Rufty TW, Kerr PS, Huber SC (1983) Characterization of diurnal changes in activities of enzymes involved in sucrose biosynthesis. Plant Physiol 73:428–433

Sage-Ono K, Ono M, Harada H, Kamada H (1998) Accumulation of a clock-regulated transcript during flower-inductive darkness in *Pharbitis nil*. Plant Physiol 116:1479–1485

Sai J, Johnson CH (1999) Different circadian oscillators control Ca^{2+} fluxes and *Lhcb* gene expression. Proc Natl Acad Sci USA 96:11659–11663

Salvador ML, Klein U, Bogorad L (1993) Light-regulated and endogenous fluctuations of chloroplast transcript levels in *Chlamydomonas*. Regulation by transcription and RNA degradation. Plant J 3:213–219

Samuelsson G, Sweeney BM, Mattick HA, Prézelin BB (1983) Changes in photosystem II account for the circadian rhythm in photosynthesis in *Gonyaulax polyedra*. Plant Physiol 73:329–331

Satter RL, Galston AW (1973) Leaf movements: Rosetta stone of plant behavior? Bioscience 23:407–416

Satter RL, Hatch AM, Gill MK (1979) A circadian rhythm in oxygen uptake by *Samanea* pulvini. Plant Physiol 64:379–381

Satter RL, Xu Y, De Pass A (1987) Effects of temperature on H^+-secretion and uptake by excised flexor cells during dark-induced closure of *Samanea* leaflets. Plant Physiol 85:850–855

Satter RL, Gorton HL, Vogelmann TC (1990) The pulvinus: motor organ for leaf movement. Am Soc Plant Physiol, Rockville, MD

Scandalios JG, Guan L, Polidoros AN (1997) Catalases in plants: gene structure, properties, regulation and expression. In: Scandalios JG (ed) Oxidative stress and the molecular biology of antioxidant defenses. Cold Spring Harbor Laboratory Press, Plainville, NY, pp 343–406

Schaffer R, Ramsay N, Samach A, Corden S, Putterill J, Carré IA, Coupland G (1998) The late elongated hypocotyl mutation of *Arabidopsis* disrupts circadian rhythms and the photoperiodic control of flowering. Cell 93:1219–1229

Schaffer R, Landgraf J, Accerbi M, Simon V, Larson M, Wisman E (2001) Microarray analysis of diurnal and circadian-regulated genes in *Arabidopsis*. Plant Cell 13:113–123

Schmid R, Dring MJ (1992) Circadian rhythm and fast responses to blue light of photosynthesis in *Ectocarpus* (Phaeophyta, Ectocarpales). I. Characterization of the rhythm and the blue-light response. Planta 187:53–59

Schmid R, Forster R, Dring MJ (1992) Circadian rhythm and fast responses to blue light of photosynthesis in *Ectocarpus* (Phaeophyta, Ectocarpales). II. Light and CO_2 dependence of photosynthesis. Planta 187:60–66

Schmitz O, Katayama M, Williams SB, Kondo T, Golden SS (2000) CikA, a bacteriophytochrome that resets the cyanobacterial circadian clock. Science 289:765–768

Schröder-Lorenz A, Rensing L (1987) Circadian changes in protein-synthesis rate and protein phosphorylation in cell-free extracts of *Gonyaulax polyedra*. Planta 170:7–13

Schuster HG (1995) Deterministic chaos, 3rd edn. VCH, Weinheim

Schweiger E, Wallraff HG, Schweiger HG (1964a) Über tagesperiodische Schwankungen der Sauerstoffbilanz kernhaltiger und kernloser *Acetabularia mediterranea*. Z Naturforsch 196:499–505

Schweiger E, Wallraff AG, Schweiger HG (1964b) Endogenous circadian rhythm in cytoplasm of *Acetabularia*: influence of the nucleus. Science 146:656–659

Snaith PJ, Mansfield TA (1986) The circadian rhythm of stomatal opening: evidence for the involvement of potassium and chloride fluxes. J Exp Bot 37:188–199

Somers DE (1999) The physiology and molecular bases of the plant circadian clock. Plant Physiol 121:9–19

Somers DE, Devlin PF, Kay SA (1998a) Phytochromes and cryptochromes in the entrainment of the *Arabidopsis* circadian clock. Science 282:1488–1490

Somers DE, Webb AAR, Pearson M, Kay SA (1998b) The short-period mutant, *toc1–*, alters circadian clock regulation of multiple outputs throughout development in *Arabidopsis thaliana*. Development 125:485–494

Staiger D (2000) Biologische Zeitmessung bei Pflanzen. Biol Unserer Zeit 30:76–79

Staiger D (2002) Circadian rhythms in *Arabidopsis*: time for nuclear proteins. Planta 214:334–344

Staiger D, Heintzen C (1999) The circadian system of *Arabidopsis thaliana.* Chronobiol Int 16:1–16

Stålfelt MG (1963) Diurnal dark reactions in the stomatal movements. Physiol Plant 16:756–766

Starrach N, Mayer WE (1989) Changes of the apoplastic pH and K^+ concentration in the *Phaseolus pulvinus* in situ in relation to rhythmic leaf movements. J Exp Bot 40:865–873

Straley SC, Bruce VG (1979) Stickiness to glass. Circadian changes in the cell surface of *Chlamydomonas reinhardtii.* Plant Physiol 63:1175–1181

Strayer C, Oyama T, Schultz TF, Raman R, Somers DE, Más P, Panda S, Kreps JA, Kay SA (2000) Cloning of the *Arabidopsis* clock gene *TOC1*, an autoregulatory response regulator homolog. Science 289:768–771

Suh S, Moran N, Lee Y (2000) Blue light activates potassium-efflux channels in flexor cells from *Samanea saman* motor organs via two mechanisms. Plant Physiol 123:833–843

Sweeney BM (1969) Rhythmic phenomena in plants. Academic Press, London

Sweeney BM (1974) A physiological model for circadian rhythms derived from the *Acetabularia* rhythm paradoxes. Int J Chronobiol 2:25

Sweeney BM (1987) Rhythmic phenomena in plants, 2nd edn. Academic Press, San Diego

Tamponnet C, Edmunds LN (1990) Entrainment and phase shifting of the circadian rhythm of cell division by calcium in synchronous cultures of the wild-type Z strain and of the ZC achlorophyllous mutant of *Euglena gracilis.* Plant Physiol 93:425–431

Tavladoraki P, Kloppstech K, Argyroudi-Akoyunoglou JH (1989) Circadian rhythm in the expression of the mRNA coding for the apoprotein of the light-harvesting complex of photosystem II. Phytochrome control and persistent far red reversibility. Plant Physiol 90:665–672

Taybi T, Patil S, Chollet R, Cushman JC (2000) A minimal serine/threonine protein kinase circadianly regulates phospho*enol*pyruvate carboxylase activity in crassulacean acid metabolism-induced leaves of the common ice plant. Plant Physiol 123:1471–1481

Thain SC, Hall A, Millar AJ (2000) Functional independence of circadian clocks that regulate plant gene expression. Curr Biol 10:951–956

Theißen G, Saedler H (1997) Molecular architects of plant body plans. Progress in Botany 59. Springer, Berlin Heidelberg New York, pp 227–256

Thompson AJ, Jackson AC, Parker RA, Morpeth DR, Burbidge A, Taylor IB (2000) Abscisic acid biosynthesis in tomato: regulation of zeaxanthin epoxidase and 9-cis-epoxycarotenoid dioxygenase mRNAs by light/dark cycles, water stress and abscisic acid. Plant Mol Biol 42:833–845

Titlyanov EA, Titlyanova TV, Lüning K (1996) Diurnal and circadian periodicity of mitosis and growth in marine macroalgae. II. The green alga *Ulva pseudocurvata.* Eur J Phycol 31:181–188

Tóth R, Kevei E, Hall A, Millar AJ, Nagy F, Kozma-Bognár L (2001) Circadian clock-regulated expression of phytochrome and cryptochrome genes in *Arabidopsis.* Plant Physiol 127:1607–1616

Trewavas A (1999) Le calcium, c'est la vie: calcium makes waves. Plant Physiol 120:1–6

Upcroft JA, Done J (1976) Circadian rhythm of nitrate reductase (NADH) activity in wheat seedlings grown in continuous light. Aust J Plant Physiol 3:421–428

Wagner E, Frosch S (1974) Endogenous rhythmicity and energy transduction. VI. Rhythmicity in reduced and oxidized pyridine nucleotide levels in seedlings of *Chenopodium rubrum.* J Interdiscip Cycle Res 5:230–239

Wagner E, Frosch S, Kempf O (1974a) Endogenous rhythmicity and energy transduction. VII. Phytochrome-modulated rhythms in pyridine nucleotide levels in seedlings of *Chenopodium rubrum.* Plant Sci Lett 3:43–48

Wagner E, Tetzner J, Haertlé U, Deitzer GF (1974b) Endogenous rhythmicity and energy transduction. VIII. Kinetic in enzyme activity, redox state and energy charge as related to photomorphogenesis in seedlings of *Chenopodium rubrum* L. Ber Dtsch Bot Ges 87:291–302

Walla OJ, de Groot EJ, Schweiger M (1994) On the molecular mechanism of the circadian clock. The 412,000 M(r) clock protein of *Chlorella* was identified as 3-phosphoglycerate kinase. J Cell Sci 107:719–726

Waterhouse J (2001) Time in biology with particular reference to humans. Eur Rev 9:31–42

Watts DJ (1999) Small worlds. The dynamics of networks between order and randomness. Princeton University Press, Princeton, NJ

Webb AAR (1998) Stomatal rhythms. In: Lumsden PJ, Millar AJ (eds) Biological rhythms and photoperiodism in plants. BIOS Scientific, Oxford, pp 69–79

Webb AAR, Hetherington AM (1997) Convergence of the ABA, CO_2 and extracellular calcium signal transduction pathways in stomatal guard cells. Plant Physiol 114:1157–1560

Whitmore D, Foulkes NS, Sassone-Corsi P (2000) Light acts directly on organs and cells in culture to set the vertebrate circadian clock. Nature 404:87–91

Wiedemann I, de Groot EJ, Schweiger M (1992) On the molecular mechanism of the circadian clock: the 64,000-Mr protein of *Chlamydomonas reinhardtii* might be related to the biological clock. Planta 186:593–599

Wiesenfeld K, Moss F (1995) Stochastic resonance and the benefits of noise: from ice ages to cray fish and squids. Nature 373:33–36

Wilkins MB (1959) An endogenous rhythm in the rate of carbon dioxide output of *Bryophyllum*. I. Some preliminary experiments. J Exp Bot 10:377–390

Wilkins MB (1960) An endogenous rhythm in the rate of carbon dioxide output of *Bryophyllum*. II. The effects of light and darkness on the phase and period of the rhythm. J Exp Bot 11:269–288

Wilkins MB (1962) An endogenous rhythm in the rate of carbon dioxide output of *Bryophyllum*. II. The effects of light and darkness on the phase and period of the rhythm. J Exp Bot 11:269–288

Wilkins MB (1983) The circadian rhythm of carbon-dioxide metabolism in *Bryophyllum*: the mechanism of phase-shift induction by thermal stimuli. Planta 157:471–480

Wilkins MB (1984) A rapid circadian rhythm of carbon-dioxide metabolism in *Bryophyllum fedtschenkoi*. Planta 161:381–384

Wilkins MB (1992) Circadian rhythms: their origin and control. New Phytol 121:347–375

Williams WE, Gorton HL (1998) Circadian rhythms have insignificant effects on plant gas exchange under field conditions. Physiol Plant 103:247–256

Winfree AT (1990) The geometry of biological time. Springer, Berlin Heidelberg New York

Wolff D, Künne A (2000) Light-regulated, circadian respiration activity of *Euglena gracilis* mutants that lack chloroplasts. J Plant Physiol 156:52–59

Wood NT, Haley A, Viry-Moussaid M, Johnson CH, van der Luit AH, Trewavas AJ (2001) The calcium rhythms of different cell types oscillate with different circadian phases. Plant Physiol 125:787–796

Xu Y, Johnson CH (2001) A clock- and light-regulated gene that links the circadian oscillator to *LHCB* gene expression. Plant Cell 13:1411–1425

Yan OY, Andersson CR, Kondo T, Golden SS, Johnson CH (1998) Resonating circadian clocks enhance fitness in cyanobacteria. Proc Natl Acad Sci USA 95:8660–8664

Yoshida I, Yamagata H, Hirasawa E (1999) Blue- and red-light regulation and circadian control of gene expression of *S*-adenosylmethionine decarboxylase in *Pharbitis nil*. J Exp Bot 50:319–326

Zheng CC, Bui AQ, O'Neill SD (1993) Abundance of an mRNA encoding a high mobility group DNA-binding protein is regulated by light and an endogenous rhythm. Plant Mol Biol 23:813–823

Zhong HH, McClung CR (1996) The circadian clock gates expression of two *Arabidopsis* catalase genes to distinct and opposite circadian phases. Mol Gen Genet 251:196–203

Zhong HH, Young JC, Pease EA, Hangarter RP, McClung CR (1994) Interactions between light and the circadian clock in the regulation of CAT2 expression in *Arabidopsis*. Plant Physiol 104:889–898

Zhong HH, Resnick AS, Straume M, McClung CR (1997) Effects of synergistic signaling by phytochrome A and cryptochrome on circadian clock-regulated catalase expression. Plant Cell 9:947–955

Zucker Lowen C, Satter RL (1989) Light-promoted changes in apoplastic K^+ activity in the *Samanea saman* pulvinus, monitored with liquid membrane microelectrodes. Planta 179:421–427

Prof. Dr. Ulrich Lüttge
Institut für Botanik
Technische Universität Darmstadt
Schnittspahnstrasse 3–5
64287 Darmstadt, Germany

e-mail: luettge@bio.tu-darmstadt.de

Systematics

Systematics of the Pteridophytes

Stefan Schneckenburger

1 Systematics

The modes and mechanisms of speciation in pteridophytes were impressively outlined by Haufler (1997). In his excellent article he discusses the basic challenges that lie in the high age of the lineages and in the combination of speciation (with difficulties in determining character homologies within young clades) and extinction during this time. Furthermore, the problem of two independent generations requiring different environments and the high chromosome numbers of some groups pose additional difficulties. He presents three speciation modes from which the following two coincide with known models: primary (divergence of diploid populations to the level of species), and secondary speciation (origin of new species involving hybridization or polyploidy within or between existing species). A divergence among populations of secondary species is postulated and described as tertiary speciation. The mechanisms of these modes in respect to pteridophytes are discussed in detail. Pryer et al. (2001a) presented a phylogenetic analysis of combined data (from morphology and from four genes) for 35 representatives from all the main lineages of land plants. They showed that there are three monophyletic groups of extant vascular plants: lycophytes, seed plants, and a clade including equisetophytes, psilophytes and all eusporangiate and leptosporangiate ferns. The horsetails and ferns proved to be the closest relatives to seed plants.

The sequences of the chloroplast genome *rbc*L were initially used for phylogenetic analyses on higher levels (usually interfamilial) and achieved excellent results (see last report within this series; vol. 60, 1999). Recently, its use for intergeneric and intrageneric analyses has impressively increased. For ferns, in particular, the diversity in *rbc*L is very high, even in morphologically similar species. This provides the chance to clarify relationships in poorly discriminated lower taxa. A lot of these studies were carried out in recent years. In many cases, resulting 'species' are defined only by molecular means. It remains questionable whether such differentiations based solely on one molecular marker

Progress in Botany, Vol. 64

justify the definition of new taxa. Before creating them it should be decided whether the examination of further markers supports this decision.

The phylogeny of **Lycopodiaceae** was studied by Wikström and Kenrick (1997). *rbc*L analyses clearly support the monophyly of this family. Within the family, a *Huperzia-Phylloglossum* clade is strongly supported and is sister to a *Lycopodium-Lycopodiella* group. These results provide the first clear evidence for the relationships of the enigmatic *Phylloglossum drummondii*. As in the past, differences in life cycle and morphology between this monotypic genus and other Lycopodiaceae are interpreted in terms of pedomorphosis and are viewed as adaptive responses to drought and bush fire. In the past, numerous attempts have been made to identify subgeneric groups within *Huperzia* and it has proven difficult to find discrete morphological features. By analyzing *rbc*L sequences (Wikström and Kenrick 1999, 2000), *Huperzia* species are partitioned into neotropical and paleotropical clades, incongruent with traditional morphology-based taxonomy. The study reveals that the diversification of epiphytic *Huperzia* is a comparatively recent phenomenon (Upper Cretaceous-Tertiary), and there has been morphological convergence in epiphytes throughout the tropics at a great rate. Furthermore, it documents a single origin of epiphytism prior to the final rifting of South America and Africa and multiple reversals to a terrestrial habitat in the Neotropics. In the Andes, it has been shown that alpine terrestrial species evolved from montane epiphytes, which correlates well with regional orogenesis during the Miocene.

The phylogeny of **Aspleniaceae** was studied by Murakami et al. (1999b) by analyzing *rbc*L of 27 taxa of *Asplenium*. It was demonstrated that leaf shape is not congruent with *rbc*L phylogeny, whereas rhizome morphology (erect-ascending or creeping) reflects this. The study revealed that naturally occurring hybrids are generated only between closely related species and reflect *rbc*L phylogeny. *rbc*L and isozyme studies were used by Murakami et al. (1999a) to discriminate the three Japanese species of *Asplenium* sect. *Thamnopteris* ('*Asplenium-nidus*-complex'), which are only poorly defined morphologically. The Japanese populations of *A. australasicum* have a very different *rbc*L sequence from the 'normal' form from Australia and the South Pacific Islands. Based only on *rbc*L a new species was proposed and described, closely allied to *A. nidus*. The species cannot be separated morphologically; they occur parapatrically and have different habitat preferences with no mixed populations. *A. nidus* is one of the most common epiphytic fern species in the Old World. Its simple leaves and the lack of any particular appendages hinder a good species recognition. Material from 25 individuals – all belonging to *A. nidus*, following Holttum's most recent monograph (1974) – were collected in a Javanese national park (Murakami et al. 1999c). Three different types of *rbc*L sequences were found, together with correlation among these types and some morphological characters and ecological traits. It remains questionable how far these *rbc*L types represent true biological units, as postulated by the authors, especially in respect to the small number of samples. *rbc*L analyses were used to clarify the relations within the *Asplenium* sect. *Hymenasplenium* by Murakami et al. (1998). Within the Asiatic *Hymenasplenium obliquissimum*, three clades were detected. One differs in having a two-cell-thick translucent lamina. The two other clades, having three to four cell layers, cannot be distinguished morphologically.

A new treatment of monogeneric **Cheiropleuriaceae** was presented by Kato et al. (2001). Both morphological and nucleotide differences in the *rbc*L gene support the separation of three species in *Cheiropleuria*. Their range extends from Borneo and SE Asia to Japan, Taiwan, and possibly China.

Thomson (2000a,b) reports on a collection of sporophytes of *Pteridium* representatives of the genus worldwide, grown under standardized conditions at Sydney Botanical Garden. With this material, morphostructural and DNA fingerprinting studies of relationships were carried out. Clearly demarcated from other genera of **Dennstaedtiaceae**, nomenclature within the genus is still uncertain and contentious. Reasons are the paucity of suitable characters and the strong environmental influence on bracken, including effects on expression of characters, and the existence of various morphological intermediates between currently recognized subspecific taxa. Some decades ago, Tryon had proposed 12 varieties within the genus, considered as monotypic. The studies of Thomson revealed five well-demarcated groups, identical to formerly described varieties that should best be treated as species. Two more of Tryon's varieties fell in one of these species; another two showed to be of hybridogenous origin. Further studies are needed to clarify both the relationships of the European *aquilinum* complex and the status of the Central American '*feei*-group'. It can be summarized that *Pteridium* contains about seven distinct species. Spores of 20 species of *Lindsaea* were studied by scanning electron microscopy by Lin et al. (1999). The two-layered perispore shows differentiations interspecifically and intraspecifically. The variations in ornamentations among sections or species are obvious. Therefore, spores provide a useful tool in species differentiation within this complex genus.

The lady fern group (Physematieae=Athyrieae) is one of the five tribes of **Dryopteridaceae** sensu Kramer and contains about 700 species. Nucleotide sequences of the chloroplast gene *rbc*L from 42 species were analyzed by Sano et al. (2000a) and provided some insights into the puzzling inter- and intrageneric relationships and the generic circumscriptions of the group. For example, *Athyrium* proved to be polyphyletic, whereas *Deparia* seems to be monophyletic. There is significant evidence for a clade, containing *Athyrium*, *Cornopteris*, *Pseudo-*

cystopteris, and *Anisocampium* and another with *Diplaziopsis* and *Homalosorus*, which is isolated from the other genera of the tribe. These *rbc*L data, together with morphological and cytological arguments, require the shifting of several *Diplazium* species to *Deparia* (Sano et al. 2000b,c). In her revisional series on **Grammitidaceae** with the aim of creating natural units, Parris (1998) described the genus *Chrysogrammitis* with two species in SE Asia and Malesia. It was segregated from *Ctenopteris*, *Grammitis*, and *Xiphiopteris*. The monophyly of **Hymenophyllaceae** was confirmed convincingly by Pryer et al. (2001b) by using *rbc*L data. They provided strong support for a basic split in two monophyletic units, *Hymenophyllum* s.l. and *Trichomanes* s.l. Some monotypic genera such as *Cardiomanes*, e.g., are convincingly included in *Hymenophyllum*. Whereas *rbc*L data provide a well-supported phylogenetic estimate of *Trichomanes*, they are inadequate for resolving relationships within *Hymenophyllum*, which requires data from further sources. Nucleotide sequences from *rbc*L were used to clarify relationships of **Hymenophyllopsidaceae** and **Lophosoriaceae** (Wolf et al. 1999). Both families include only one genus. Whereas the large terrestrial ferns of *Lomariopsis* are widely distributed in tropical America, the tiny species of *Hymenophyllopsis* are narrow endemics of the Roraima formation of Venezuela, Guyana, and northernmost Brazil. The studies provided strong evidence that both families are part of a well-supported tree fern clade. Within this clade, *Lophosoria* seems to be more closely allied to *Dicksonia*, and *Hymenophyllopsis* to the scaly tree ferns (as *Cyathea* and *Cibotium*). Further studies are urgently needed to enlarge our knowledge, especially of *Hymenophyllopsis*.

The neotropical species of *Lomariopsis* (**Lomariopsidaceae**) were revised by Moran (2000). 15 species are recognized; all are illustrated and keyed. In the Neotropics the genus occurs from southern Florida, the Antilles, and Mexico to Bolivia, and southern Brazil. On the basis of heteroblastic leaf series, two groups can be recognized. Beneath this, three Antillean species are unusual by abortion of the rachis apex and terminalization of the distalmost lateral pinna. **Loxsomataceae** are a relictual and taxomically isolated fern family with two genera: *Loxsoma*, which is restricted to New Zealand, and *Loxsomopsis*; the second genus was revised by Lehnert et al. (2001). It proved to be monotypic with one variable species. This morphological variation seems to be the result of the distribution of the species in small isolated populations in ephemeral early successional habitats from Costa Rica to Peru and Bolivia. The phylogeny of **Marsileaceae** was studied by Pryer (1999). The once disputed hypothesis of the monophyly of heterosporous ferns was confirmed earlier (see previous report). By several analyses of *rbc*L and morphological characters, Pryer's studies revealed a single and completely congruent topology for extant heterosporous leptosporangiate ferns: (*Marsilea*, (*Regnellidium*, *Pilularia*)), (*Azolla*, *Salvinia*). The phylogenetic position of the fossil *Hydropteris* is viewed as unresolved. The **Matoniaceae** were revised by Kato within the Flora Malesiana series (see Kalkman and Nooteboom 1998). Two genera are distinguished, *Matonia* and *Phanerosorus*, with two species each. Phylogenetic analyses based on *rbc*L sequences support the treatment of these two genera as monophyletic units (Kato and Setoguchi 1999). While adult leaves differ very strongly, the young leaves are similar. Together with paleobotanical evidence, this may suggest that the pinnate leaves of *Phanerosorus* are derived - perhaps via pedomorphosis - from the pedate leaves like those of fossil Matoniaceae and extant *Matonia* species.

Fossil specimens of *Osmunda claytoniites* (**Osmundaceae**) from the Triassic of Antarctica showed virtual identity with modern *O. claytoniana* in all vegetative and generative features (Phibbs et al. 1998); similar results were obtained by Serbet and Rothwell (1999). They were able to trace back *O. cinnamomea* to the Upper Cretaceous - this implies a minimum age of 70 million years for that species. Both demonstrate the evolutionary stasis of this ancient genus which can be traced back to the late Paleozoic. Yatabe et al. (1999) studied the phylogeny of this family using molecular methods. *rbc*L analyses of 11 of 15 extant species showed that *Osmunda* is not monophyletic, because *Todea* and *Leptopteris* are positioned within *Osmunda. O. cinnamomea* is the most basally positioned species in this family, which can be called a 'living fossil'. Conspecific samples of *O. claytoniana* and *O. cinnamomea* from the USA and Japan showed a greater nucleotide variation. Therefore, each of these species may comprise more than two biologically differentiated species.

The **Plagiogyriaceae** were revised by Zhang and Nooteboom (1998). The only genus *Plagiogyria* consists of 11 species only. With the exception only of *P. pectinata* (New World), all species are restricted to the Old World with one species and one variety reaching the southern Pacific (Solomon Islands). An interesting study (Li and Haufler 1999) deals with the Hawaiian endemic *Polypodium pellucidum* which belongs to the primarily temperate *P.-vulgare* complex (**Polypodiaceae**). Several varieties had been named; the infraspecific delimitations remained unclear. Isozyme analyses showed that the levels of genetic variability were high for an island endemic with rare outcrossing between genetically differentiated gametophytes. The high genetic variability together with low morphological differentiation contrasts with the endemic spermatophytes, which show great morphological diversity, containing only low genetic variations. The studies indicate that after a single introduction with subsequent diversification *P. pellucidum* has been a resident of the archipelago for a long time, perhaps longer than the oldest of the current islands and has migrated from island to island. The varieties may eventually become separate species. Wollenweber and Schneider (2000) gave a comprehensive overview of lipophilic exudates of **Pteridaceae.** 80

pteridophyte species out of 14 genera of only this family produce farinose waxes, mostly located on the lower leaf surface. Mostly, these exudates are flavonoids, although sometimes diterpenoids and triterpenoids are found. The paper surveys the chemical composition of these exudates and their occurrence within the genera. Their scattered distribution suggests the polyphyletic origin of farinose waxes within the family and her subfamilies (only Adiantoideae, Cheilanthoideae, Taenitoideae). From a chemosystematic point of view, the flavonoid patterns proved to be significant at the species and also the populational level, whereas they seem to be of less value at the generic level. They may have high value in taxonomic and phylogenetic studies of genera with frequent occurrence of farinose waxes (e.g., *Argyrochosma*, *Chrysochosma*, *Pityrogramma*). The cheilanthoids, a difficult group of Pteridaceae of xeric regions, show an unsettled taxonomy. Two analyses were carried out by Gastony and Rollo (1998): data from nucleotide sequences of the maternally inherited chloroplast-encoded *rbc*L gene from 57 species were compared with those based on ITS sequences of biparentally inherited nuclear ribosomal DNA. These two data sets yield remarkably congruent topologies: the monotypic Mexican *Llavea* is rejected from cheilanthoids. Traditional *Cheilanthes*, *Notholaena*, and *Pellaea* proved to be polyphyletic, the segregations of *Argyrochosma*, *Aspidotis*, *Astrolepis*, and *Pentagramma* are supported. There is further need to incorporate data from less rapidly evolving nrDNA regions. A small group of the neotropical *Adiantum* species, *A. gracile* and its relatives, was revised by Lellinger and Prado (2001), a contribution to the much-needed monographic treatment of this huge genus.

Yatabe et al. (1998) used *rbc*L data for analyses on infraspecific levels of the Japanese *Stenogramma pozoi* ssp. *mollissima* (**Thelypteridaceae**). Species and subspecies recognition of ferns on leaf morphology can only be fine-tuned by molecular features.

Hecistopteris, a neotropical genus of **Vittariaceae** with members of inconspicuous size, was thought to be monotypic until 1995; but now a third species with multi-furcate leaves of length less than 2 cm from Kaietur National Park (Guyana) has been described by Kelloff and McKee (1998). *Callipteris* is a pantropical fern genus of **Woodsiaceae**, defined by rhizome scales with dark castaneous borders and bifid marginal teeth. A group of neotropical species with anastomosing veins was revised by Pacheco and Moran (1999); 15 species belong to this group, occurring from Guatemala to Bolivia, northern Brazil, and the Lesser Antilles. Three of them were recognized as new, 12 were placed in *Diplazium* formerly. The two centers of species richness are the Chocó region (western Andes in Colombia and Ecuador; with eight species, five of them endemics) and the mountains of Costa Rica and Panama (six species, three of them endemics). Only one (endemic) species occurs in the Lesser Antilles, the four species of the eastern Andes represent at least two separate dispersal events from the western side. The intraspecific differentiations within the Japanese *Diplazium doederleinii* with respect to cytology, morphology, genetics and molecular phylogenetics were studied by Takamiya et al. (2001). Two apomictic cytotypes (triploid and tetraploid) were found. The possible origin of these two cytotypes and evolution within the species were discussed.

2 Bibliography, Collections, Nomenclature

Over 2000 references on spores and gametophytes of pteridophytes with full bibliographical details and published between 1699 and 1996 have been compiled by Pérez-García and Riba (1998). Even fossil gametophytes were included. Each of the 2195 entries of this important source is annotated with reference to all categories that relate in respect to 33 research areas. A comprehensive bibliography for the pteridophyte flora of Macaronesia has been published by Horn and Wells (1998).

3 Floristics

a) Asia, Australia, Pacific

Two excellent books present the pteridophyte flora of two Indian regions: Dixit and Sinha (2001) deal with the Andaman and Nicobar Islands (126 species within a land area of 8290 km^2, 7 endemics) and Borthakur et al. (2001) with Assam. This eastern Himalayan region is richest in plant diversity with a highly diversified pteridophyte flora of 221 species in 87 genera. Both books are illustrated and present treatments on geography, geology, climate, and vegetation of the areas. They contain keys on all taxonomic levels; especially interesting is the chapter on the useful pteridophytes of the Andaman and Nicobar Islands with detailed information on the various uses as nutrients or medicines. The vertical distribution of pteridophytes of Nepal Himalaya in relation to their different ecological habitats of different zones was comprehensively outlined by Gurung (1997). After the recent treatment of Australia's oceanic islands, the pteridophyte flora (456 species in 112 genera) of this continent was completely revised in the Flora of Australia series (Flora of Australia 1998). Psilotopsida are represented by two genera with eight species of which two are endemic. Lycopodiaceae, Selaginellaceae, and Isoetaceae contain 47 species in 6 genera. A remarkable rate of endemics shows *Isoetes* with 14 of the 15 Australian species. The only endemic genera are found within true ferns. About 56 of the 106 genera are broadly pantropical or cosmopolitan. A further 37 genera occur widely in the Old World tropics and 21 of these extend to Africa. The remainder has a more restricted distribution, seven are predominantly southern temperate and only four are endemic (with five species). In total, 390 species of true ferns in 103 genera are known, 141 of them (36%) are endemic. This is astonishingly low in comparison with the whole vascular flora that shows an endemism rate of 90%. The highest rates on the species level can be found in *Blechnum* (10 of 18), *Cyathea* (8 of 12), *Lastraeopsis* (12 of 15), and *Marsilea* (8 of 12). The excellent flora contains keys on all taxonomic levels, various figures, 157 photographs of pteridophyte species, and distribution maps.

As the seventh installment of the Malesian pteridoflora, an outstanding contribution was edited by Kalkman and Nooteboom (1998). The volume contains taxonomic revisions of seven families as the large Polypodiaceae (183 species in 18 genera), Davalliaceae (31 species in three genera), Azollaceae (a single species in one genus), Cheiropleuriaceae (monogeneric), Equisetaceae (one single subspecies), Matoniaceae (two species in two genera), Plagiogyriaceae (seven species in one genus) for Malesia. In the case of Matoniaceae and Cheiropleuriaceae (but see Kato et al. 2001), which are restricted to the Malesian region, the treatments are monographs of the families. In particular, the treatments of Davalliaceae for their horticultural importance and Polypodiaceae for their complexity are extremely useful. They are recommended to anyone who is interested in understanding the modern generic classification especially of the last mentioned family and its Old World species. In the case of some revisions (esp. from Polypodiaceae and Davalliaceae), it seems that they are overwhelmingly based on herbarium material and are backed up by apparently little field study. So, some of the lumping seems to be rather extreme. The Malesian species of *Blechnum* (Blechnaceae) were revised by Chambers and Farrant (2001). In the past, confusion was caused by a bad collection status (esp. for New Guinea). Only 20 taxa (17 species, 3 subspecies) were distinguished; 43 species and 7 varieties were put into synonymy. Many species that were thought to be endemics were shown to have a much wider range. The morphology and taxonomy of the Malay-Pacific species of the polypodiaceous genera *Selliguea*, *Crypsinus*, *Holcosorus*, *Phymatopsis*, *Grammatopteris*, *Pycnoloma*, *Oleandropsis*, and *Crysinopsis* have been discussed in detail by Hovenkamp (1998). All these genera were merged in *Selliguea*; 52 species were revised and described.

b) Africa, Macaronesia

Ferns and fern allies of the Cape Verde Islands were revised by Lobin et al. (1998). 34 species are mentioned, described and illustrated. It was not possible to confirm the presence of two species that were mentioned by earlier authors. Many species show a worldwide or paleotropical distribution; only six demonstrate relationships to the neighboring Canary

Islands. 21 species (65.6%) are classified as extinct or threatened. A revised version of the Red List of Capeverdian pteridophytes is presented.

Several groups of ferns and fern allies were revised for the 'Flora of Tropical East Africa' series (Verdcourt 1999a,b,c,d,e,f,g,h; Actiniopteridaceae, Davalliaceae, Equisetaceae, Marattiaceae, Osmundaceae, Parkeriaceae, Psilotaceae, Vittariaceae). In the majority there are very common species (distribution nearly worldwide or within the Old World tropics); only within the Vittariaceae (9 species in the region) does one find one species only known from Tanzania. These data reflect the poorness of Africa in pteridophyte species and endemics. A second series dealt with Dennstaedtiaceae, Gleicheniaceae, and Salviniaceae (Verdcourt 2000a,b,c). The first groups mentioned above show endemics of the region covered by this flora: Gleicheniaceae (one species) and one variety of a total of four species in three genera and Dennstaedtiaceae with a slightly higher amount. The treatment lists 19 species in 9 genera, from which 2 are newly described; 5 are endemic to eastern Africa and another 10 are confined to the African/Madagascan/Mascarene region.

c) Europe

The first volume of Flora Nordica, covering Denmark, Norway, Finland, Sweden, Iceland, and Svalbard, has been published which contains, beneath other groups, the pteridophytes (Jonsell 2000). The flora describes 67 naturally occurring species in 27 genera with an additional 6 species from 4 genera which escaped from gardens. Keys and distribution maps are given; figures of critical taxa are very helpful. Most species occur in remaining Europe (with only three exemptions). Only one variety seems to be an endemic of the region (*Dryopteris expansa* var. *willeana* in western Norway). These facts reflect the low age of the flora of these northern areas that were covered by ice until 10,000 years ago.

An illustrated guide to pteridophytes of Switzerland and adjacent countries was presented by Kopp and Schneebeli-Graf (1998). 88 taxa were treated, keyed and illustrated by a mixture of silhouettes of plants and fronds and line drawings of details. There are additional discussions of taxonomy and conservation status.

Cystopteris dickieana currently receives legal protection in Britain on the basis that it is extremely rare and endemic to Scotland. Its taxonomic status within the *Cystopteris fragilis* complex is uncertain. This was reviewed by Dyer et al. (2000). The study of isozymes, spore sculptures, and frond variation in Scottish populations of *C. dickieana* (including the type population) and *C. fragilis* (Parks et al. 2000) gave no grounds for recognizing *C. dickieana* or other members of the complex as separate species. The former opinion that *C. dickieana* represents a populational variant in the differentiating polyploid *C. fragilis* is strongly supported.

d) America

The Flora of Cuba started with the treatments of pteridophytes: initially, Hymenophyllaceae were presented by Sánchez (2000). Only *Hymenophyllum* und *Trichomanes*, the first with 19, the second with 30 species, occur in Cuba. Many of them are neotropical species; within the two genera, only two species are endemics of Cuba. From the Flora of México series, two volumes, treating Marsileaceae (Pérez-García et al. 1999; 2 genera with 10 species) and Dryopteridaceae (Riba and Pérez-García 1999; 10 genera with 16 species; 8 of them restricted to México), were published. A regional revision of *Isoetes*, largely unknown in Alaska and the Aleutians until recent years, was presented by Britton et al. (1999). Three species and two hybrids were recognized and their cytology, spore morphology, and distribution were documented. A third interspecific hybrid is expected, but missed until now.

The Lycopodiaceae of Colombia have been outlined by Murillo-Pulido and Murillo-Aldana (1999). They consist of 55 species in 3 genera; the most diversified is *Huperzia*. Most species are found at altitudes between 3000 and 3500 m. The contribution contains a key to genera and an annotated checklist. Smith et al. (1999) added 145 new records of pteridophytes to the flora of Bolivia. Many of the species could be expected from their occurrence in adjacent regions, but some 35 species show remarkable disjunctions. Most surprising is the occurrence of two species only known from southern Chile and Argentina (*Hypolepis poeppigii*, *Blechnum blechnoides*). All these records show that this megadiverse country is also rich in pteridophytes and botanically extremely neglected.

The first part of the Flora Patagónica series deals with the pteridophytes (de la Sota et al. 1998). Patagonia is one of the three major areas for pteridophytes in Argentina. 79 species in 33 genera are treated (all figured); about 50% are endemics of the southern part of the subcontinent (Argentina and Chile). Most species occur in the region of the Andean-Patagonian forests. Mainly, they are terrestrial species; some of them (Hymenophyllaceae and members of *Grammitis* and *Polypodium*) are epiphytes. Narrow relations exist to the pteridophyte floras of New Zealand and the islands of the southern Atlantic and Pacific Ocean. The checklist of Trinidad and Tobago (Baksh-Comeau 2000) contains 302 species in 27 families and 77 genera. The families richest in species are Pteridaceae (36), Hymenophyllaceae (29), and Polypodiaceae (24). The series on pteridophytes of the Mato Grosso State (Brazil) was continued by Windisch (1997, 1998).

4 Geography, Ecology, and Conservation

Lectures given at the Pteridophyte Biogeography Symposium at the International Botanical Congress in St. Louis, Missouri, 1999, were published in Brittonia 53(2) 2001. There are papers concerned with ancient floras, about times and places when certain groups of pteridophytes first evolved (Cainozoic ferns: Collinson 2001; Mesozoic ferns: Skog 2001). The authors compared the ancient distributions with the present-day ones. The floristic similarities between the Neotropics and Africa-Madagascar were analyzed by Moran and Smith (2001). The reasons for the similarities (esp. continental drift in the geologically old Schizaeaceae, mainly long-distance dispersal; in most cases from the Neotropics to Africa-Madagascar) are discussed. Brownsey (2001) examines the origin of the New Zealand fern flora. It comprises 194 native species, of which 89 (46%) are endemic. Several reasons indicate that most pteridophytes first appeared after separation from Gondwana by long-distance dispersal. Parris (2001) discussed the circum-Antarctic patterns and their origins. Four major patterns could be found; long-distance dispersal, rather than continental drift, is a likely explanation for them. Dassler and Farrar (2001) discussed how gemmae on the gametophytes had affected the distribution of tropical epiphytic ferns having these structures. They proposed that gemmae significantly aid long-distance colonization of outbreeding species because they allow gametophytes to exploit available niches through their dispersal and through clonal expansion and persistence of the gametophytes. Therefore, sexual reproduction is facilitated by providing the opportunity for sperm and antheridiogen transfer and by providing new sources of tissues for antheridia formation.

A very interesting comparison of ecologically similar but phylogenetically independent biotic communities in tropical montane regions has been given by Kessler et al. (2001). They studied the pteridophyte floras of Mount Kinabalu, Borneo, and Parque Nacional Carrasco, Bolivia. The numbers of species, genera, and families recorded on Mt Kinabalu was 14–23% higher than PN Carrasco, whereas species richness per 400-m^2 plot was somewhat higher at the latter. There was a remarkable similarity in the elevational distribution of species numbers, of pteridophyte families, and of life forms. A pronounced peak of species richness at 1500 m on Mt. Kinabalu, considerably higher species numbers of *Elaphoglossum* in South America and of Grammitidaceae in SE Asia were recorded. An interesting contribution (Vogel et al. 1999) deals with the importance of pteridophytes in reconstructing glacial refugia in Europe. Today, most of Europe has been colonized by polyploid *Asplenium* species, while the diploids that gave rise to them are more-or-less confined to the Mediterranean Basin as refugial region. The arguments are taken from genetic diversity, breeding types, and colonization behavior.

Tyson et al. (1999) showed that bracken takes up (by fronds, roots, and even rhizome tips), retains (concentrating in developing organs) and recycles radiocesium. Concentration in meristems could potentially cause genetic and phenotypic changes in clones.

A very useful book has been published by Chin (1998). He gives an introduction to the ferns of the tropics with special emphasis on Southeast Asia. Besides an introduction to ferns and their allies it contains a cross section of fern diversity in this region, together with many superb color photographs. The western Ghats (South India) is one of the most important centers of plant diversity and richness in India. About 260 of a total of 1200 species of ferns and fern allies are known from this region. Manickam and Dominic Rajkumar (1999) studied the polymorphism of 10 fern species, analyzing more than 10,000 specimens and taking into account numerous morphological and cytological characters. Variations proved to be great and continuous with numerous intermediates (morphotypes), so that botanists without field knowledge would always be tempted to place them as different species. As causative agents, environmental (climate, soil, exposure, altitude, temperature, humidity, wind, and fire) and genetic factors (intraspecific polyploidy, hybridization, genetic imbalance) were discussed exhaustively. Using this impressive broad basis and data collection, further cytomorphological, cytogeographical and molecular studies would surely give deep insights into speciation processes in pteridophytes.

The distribution of Cyatheaceae across the successional mosaic in an Andean cloud forest was studied by Arens and Sánchez-Baracaldo (1998). Species richness was highest in secondary forest. Conservation of tree fern diversity requires the maintenance of a variety of successional habitats. Tree ferns are important components of species that colonize abandoned open sites in Andean forests. They may be useful in restoration strategies in cloud forests. An interesting sense-and-response mechanism that allows reaction to overtopping by woody plants was discovered by Arens and Sánchez Baracaldo (2000). *Cyathea caracasana* is a common open habitat tree fern in the Andes with a high growth rate. After 10–15 years the individuals are regularly overtopped by spermatophyteous trees. As a reaction, *Cyathea* produces nearly vertical fronds with very long stipes. Stipes and blades were shortest in open habitats and longest in low canopy forests; ferns in high canopy had intermediate measurements. Because per-frond pinnae numbers did not differ among the habitats, the elongation cues must be received late in the development of the fronds. This cue seems to be a low-red/far-red wavelength ratio of the light received by the apical meristem. A very interesting tree fern of the genus *Cyathea* from Columbia has been described by Arens and Smith (1998). It possesses stems, creeping above the ground. Their buds give rise to upright stems of a height of about 1 m with three to four twice pinnate-pinnatified fronds. *Cyathea planadae* is a common element of mid-elevation undisturbed Andean cloud forests. Ants inhabit aerophore pads located at the base of the pinnae and possibly inside the rhachis. Further studies should be carried out to search for nectar-extruding structures and to clarify the nature of this association. The Australian tree fern *Sphaeropteris cooperi* is an invasive in Hawaiian wet forests where it displaces native tree ferns (esp. *Cibotium* species). Durand and Goldstein (2001) compared annual growth rates, fertile frond production, and leaf traits. Annual height increase and fertile frond production were significantly higher in *Sphaeropteris*, whereas leaf mass was lower and leaf life span was shorter than that of native *Cibotium*. 'Costs' of leaf construction of the invasive species seems to be profitably low.

A comparative study of the feedback between forest understory and canopy dynamics in a temperate hardwoods-conifer forest has been carried out by Hill and Silander (2001). *Dennstaedtia punctilobula* and *Thelypteris noveboracensis* were studied, considering light response, soil moisture, and seasonal changes. Molecular studies (inter-simple sequence repeats) by Camacho and Liston (2001) evaluated the role of asexual propagation by subterranean sporophytic gemmae in *Botrychium pumicola*. They conclude that long-distance dispersal by gemmae is at best a rare event.

Bremer and Smit (1999) studied the colonization of polder woodland plantations with ferns in the Netherlands. They showed that these processes are more rapid in coniferous forests whereas deciduous woodland stands support more species with time. Furthermore, the presence of trenches promoted species diversity.

Conservation

From a central European point of view, the most important work in this field was published by Bennert (1999) in a lavishly illustrated book. 45 rare or endangered pteridophytes were studied. A wealth of ecological data concerning soil factors, light intensity and quality, population data, and genetic variability within the populations was included. Germination experiments were carried out in situ and ex situ with special reference to conservation measurements like ex situ cultivation or reintroduction of plants grown in botanical gardens. In situ data of gas exchange revealed that many pteridophytes show very low photosynthesis rates (sometimes lower than mosses!) and are weak competitors. The data are combined in so-called histograms that allow a quick overview.

Conservation biology, population structure, and habitat characteristics of *Trichomanes speciosum* have been studied in detail by Rumsey et al. (1998, 1999). Gametophytes are widespread, far beyond the present range of the sporophyte, which is rare and vulnerable. If one considers the gametophyte, the species is neither threatened with extinction, nor does it appear to face the danger of marked genetic erosion, because the long-lived gametophyte stage contains all of the genetic variability present in the area and can be regarded as a valuable 'seed bank'. The gametophytes of *T. speciosum* thrive in extremely deep shade over a long period of time. Their morphology and ultrastructure have been studied by Makgomol and Sheffield (2001). The success of the fern is attributed to a low metabolic rate at low temperatures and inability of other species to cope with extreme low light. It was demonstrated (Johnson et al. 2000) that there is little or no protection of the photosynthetic apparatus from light-induced damage and an extremely efficient use of what little light is available (<0.01% of full sunlight).

Two articles contribute to the role and possible control of a fern species as a dangerous weed (Pemberton 1998; Pemberton and Ferrier 1998). *Lygodium microphyllum*, a native of the warm and wet regions of the Old World, was first detected in southeastern Florida in 1965. In the meantime it has developed into a dangerous aggressive invader in many different habitats with rapid spreading. At the moment, no effective method of control exists. Tests with herbicides show the damage to the nativ plant species as well. According to the wide range and the isolated taxonomic position of this fern, specialized natural enemies are predicted.

5 Morphology and Anatomy

A very impressive and highly recommended series of articles dealing with spermatozoid structure and ontogenesis, reproductive innovations and evolution of early land plants has been presented by Renzaglia and collaborators (Maden et al. 1997; Renzaglia et al. 1998, 1999, 2001; Renzaglia and Maden 2000, especially recommended as conspectus Renzaglia and Garbary 2001 and as overview Renzaglia et al. 2000). The contributions include fine illustrations and elucidating reconstructions. These gametes and their ontogenesis may serve as a very useful informative system in approaching fundamental questions relating to cellular differentiation and motility. Cladistic analyses show that hornworts are supported as the earliest divergent embryophyte clade with a liverwort/moss clade sister to vascular plants. Lycophytes are monophyletic among pteridophytes and an assemblage containing ferns, horsetails and psilophytes are sister to seed plants.

An important paper was presented by Wilson (1999) who studied the ontogeny of the sporangia of *Sphaeropteris cooperi*. They have a distinctly different developmental sequence from that known in the higher leptosporangiate ferns, although both types originate from a single superficial cell. Similar appearing structures, such as the annulus, have different origins and should be compared with each other with great caution only. Therefore, additional ontogenetic studies of fern sporangia (especially in basal groups) seem to be needed urgently. The dogma of an one-type ontogeny of the leptosporangium of higher ferns must be given up because it is based on incomplete, unclear and incorrect observations. Relationships between marginal and submarginal sori were outlined by Schölch (2000a,b) in detailed and precise studies. She defines the type of the 'basipetal marginal sorus' (ridge-, cushion-, or rod-shaped receptacle, basipetal sequence of sporangia initiation, and presence of a lower and an upper indusium). By comparative morphological arguments she concluded that Hypolepidaceae and Dennstaediaceae should not be combined in a single family and that the primitiveness of Dicksoniaceae, Hymenophyllaceae, and Dennstaedtiaceae (the letter with a highly complex soral construction) is not supported. Furthermore, Hymnophyllaceae cannot be regarded as closely related either to Dennstaedtiaceae or to Dicksoniaceae.

Karrfalt (1999) reported the existence of apogamic embryos deeply embedded within the somatic tissue of the megagametophytes in *Isoetes andicola*. The nuclei of these embryos were about twice the size and contained about twice as much DNA as those in adjacent gametophytic cells. The form of apogamy is not known. The archegonia were lacking neck canals. A very rare phenomenon in ferns are storage roots. In addition to the three known cases (2 species of *Afropteris*, 1 species of *Adiantum*), Schneider (1999) reported storage roots from *Cheilanthes bolborrhiza*, where the bases of the shoot-borne roots are thickened. The cells of the pericyle contain starch, differing to the cases mentioned above, where the cortex cells bear starch. He discussed the relationship between the habitat with periodical dryness and fires and the storage roots which replace the leaf bases which act as storage organs in ferns with tiny rhizomes normally. Schneider (2000) studied the various constructions of the root systems of the filmy ferns of the tribe Trichomaneae (=*Trichomanes* s.l.) corresponding to their different growth forms. Most terrestrial species possess a short erect shoot with many thick roots, whereas epiphytic species show long creeping rhizomes with few thin roots. The loss of roots seems to be a secondary simplification as an adaptive trait in epiphytic plants. Rootless species occur in two monophyletic groups (subgen. *Crepidomanes* and subgen. *Didymoglossum*). Consequences for the classification of the group are discussed. The loss of roots is partly balanced out by the development of specialized transformed structures as root-like shoots and adhesive hairs. Hopefully, future studies of rootless filmy ferns (e.g., embryology, morphological characters) may give better insights into the evolution of this highly interesting group. Furthermore, it may help to understand comparable cases of loss of complex structures, the transfer of functions to other organs and the blurring of organ distinction.

Problems of speciation between pairs of rheophytes and closely related dryland species were summarized and impressively discussed by Imaichi and Kato (1997). In this case, speciation seems to be necessarily associated with distinct morphological changes (esp. stenophyllization) that enable plants to invade new ecological niches as periodically flooded localities. This is true for adult sporophytes; however, the gametophytes and sporelings are more or less similar. *Stromatopteris moniliformis*, an endemic New Caledonian fern with doubtful affinities (Gleicheniaceae), is a fern with subterranean mycorrhizal gametophytes. Whittier and Pintaud (1999) studied the germination of the spores, which takes place in the dark. The pattern of cell division for early gametophyte development is that of Gleicheniaceae. No similarities could be seen with the type of germination observed in primitive groups as was postulated formerly.

Occurrence of spores in tetrads at the dispersal times is very rare in Bryophyta and Pteridophyta. Among the latter, they are only known from megaspores of *Selaginella*. For the first time, formation of microspore polyads has been observed in the Indian *S. intermedia* by Mukhopadhyay and Bandhari (1999). Many tetrads are clumped together in a ball-like or chain-like fashion connected by sticky threads. Possibly, cross-fertilization is ensured by a transfer of massed microspores.

6 Ethnobotany, Uses, Techniques

A very useful overview of ferns as experimental systems for morphogenesis has been presented by Cheema (1997). He takes into account sporophyte morphogenesis, apogamy, and apospory as well as gametophyte morphogenesis. Pence (2000) studied the survival of chlorophyllous and nonchlorophyllous fern spores after exposure to liquid nitrogen. She demonstrated that air-dried spores showed no inhibition of germination. Therefore, they are candidates for long-term germplasm storage at low temperatures in connection with ex situ conservation measurements.

An interesting contribution to the knowledge of the use of pteridophytes in Central America under commercial and conservational aspects was given by Thomas (1999). Large-scale cultivation exists for *Ruhmora adiantiformis* whose cut fronds are exported to the USA and Europe. Some species from the *Phlebodium aureum* group are used as medicine (skin complaints, urinary and liver disorders) and to produce an extract which is imported to the USA, to Spain and France for a similar purpose. Alarming is the large-scale use of wild-collected tree fern trunks as pots for ornamentals. In many cases, orchids or cycads – in both cases collected from the wild – are sold in such pots. A review of the pteridophyte ethnobotany of Uttar Pradesh (India) is presented by Joshi (1997). 44 species of pteridophytes are used by the local inhabitants. Beneath 21 species used as ornamentals, 16 are used as food, 9 as medicine and 2 for magico-religious purposes. The role of *Azolla* and its nitrogen-fixing symbiotic relation with *Anabaena* as well as its use as fertilizer are discussed within the treatment of the family in the Flora Malesiana series (see Kalkman and Nooteboom 1998).

The International Bracken Group (IBG) has published a comprehensive volume with contributions to a conference 'Bracken fern: toxicity, biology and control' (Taylor and Smith 2000; overview: Smith 2000). It deals with bracken taxonomy (Speer 2000; Thomson 2000b, see Speer and Hilu 1999; Speer et al. 1999, too), phytochemistry, ecology, control and management, as well as toxicity and animal and human health. As well as several animal diseases such as neoplasia of the urinary bladder, human stomach cancer seems to be a problem in regions where cows are fed a lot of bracken. The carcinogenic substances (ptaquiloside, illudane-type sesquiterpenes; Castillo et al. 2000) are transferred via milk (Alonso-Amelot et al. 2000) and possibly the local water supply. The molecular mechanisms of carcinogenesis are discussed (Shahin et al. 2000) and synergisms between papillomavirus and bracken in carcinogenesis of the intestinal tract in cattle and humans were uncovered (Campo et al. 2000). Although reports of health risks have been published in the past, young croziers, known to be highly carcinogenic, are eaten in Japan (Morrow 2000) and Brazil (Marlière et al. 2000). Potter (2000) demonstrated that some of the compounds are similar to substances with pharmaceutical activity (e.g., antibiotics). With respect to risks to cattle and man, steps must obviously be taken to eradicate the fern in regions where no alternative grazing or land use is possible. The context of bracken expansion and the influence of environmental pollution are discussed – altogether a most important contribution not only for 'brackenologists'. Various aspects of bracken perception and bracken control in Great Britain were published in a foregoing publication of the IBG (Taylor 1999). Simán et al. (2000) give an overview of human health risks from ferns, especially from fern spores, together with a comprehensive list of literature concerning this field. The most important fern genus as a health hazard is *Pteridium*. The authors summarize the effects of various parts of *Pteridium* on humans and livestock and evaluate the role of spores as a risk to human health. Most reports of effects caused by spores relate to allergy-like symptoms, but in the case of *Pteridium* it is probable that the carcinogenic substances (ptaquilosides) detected in vegetative parts may be contained in spores, too. However, attempts to extract and identify them from spores had not been made until now. Because many stands of bracken in Europe are infertile, the danger does not seem to be very high. Precaution is necessary when passing fertile stands or nearby them, whereas the risk cannot be quantified at the moment. Handling of fertile fronds by sensitive persons should be accompanied by the use of a face mask, especially when working with bracken. Simán et al. (1999) showed that spores and spore extracts of a range of species (*Pteridium aquilinum*, *Anemia phyllitidis*, *Pteris vittata*, *Sadleria pallida*, *Dicksonia antarctica*) can cause DNA lesions both in vivo and in vitro. No effects were observed with spores of *Osmunda regalis*.

An encyclopedic treatment of 700 species from 124 genera of ferns and their allies can be found in the second edition of the 'Fern grower's manual' (Hoshizaki and Moran 2001). All are described in detail, including cultural requirements, hardiness and special use. Each species is illustrated by line drawings and frond silhouettes - an important horticultural publication, not only for pteridologists.

An old and horticulturally much used fern is the Barbados Farley Fern, *Adiantum tenerum* 'Farleyense'. It was discovered on a horticulturally sophisticated Barbados sugar plantation in the mid-nineteenth century. Rogers (1998) confirmed that it is a cultivar of *Adiantum tenerum* and summarized the fascinating history of this well-known fern. A neotype was selected and a standard for the cultivar name was given.

References

Alonso-Amelot ME, Castillo UF, Avendaño, Smith BL, Lauren DR (2000) Milk as a vehicle for the transfer of ptaquiloside, a bracken carcinogen. In: Taylor JA, Smith RT (eds) Bracken fern: toxicity, biology, and control. Int Bracken Group Spec Publ 4. Aberystwyth, pp 86–90

Arens NC, Sánchez Baracaldo P (1998) Distribution of tree ferns (Cyatheaceae) across the successional mosaic in an Andean cloud forest, Nariño, Colombia. Am Fern J 88:60–71

Arens NC, Sánchez Baracaldo P (2000) Variation in tree fern stipe length with canopy height: tracking preferred habitat through morphological change. Am Fern J 90:1–15

Arens NC, Smith AR (1998) *Cyathea planadae*, a remarkable new creeping tree fern from Columbia, South America. Am Fern J 89:49–59

Baksh-Comeau YS (2000) Checklist of the pteridophytes Trinidad and Tobago. Fern Gaz 16:11–122

Bennert HW (1999) Die seltenen und gefährdeten Farnpflanzen Deutschlands. Biologie, Verbreitung, Schutz. Bundesamt für Naturschutz, Bonn

Borthakur SK, Deka P, Nath KK (2001) Illustrated manual of ferns of Assam. Bishen Singh Mahendra Pal Singh, Dehra Dun

Bremer P, Smit A (1999) Colonization of polder woodlands with particular reference to the ferns. Fern Gaz 15:289–308

Britton DM, Brunton DF, Talbot SS (1999) *Isoetes* in Alaska and the Aleutians. Am Fern J 89:133–141

Brownsey PJ (2001) New Zealand's pteridophyte flora - plants of ancient lineage but recent arrival? Brittonia 53:284–303

Camacho FJ, Liston A (2001) Population structure and genetic diversity of *Botrychium pumicola* (Ophioglossaceae) based on inter-simple sequence repeats (ISSR). Am J Bot 88:1065–1070

Campo MS, Beniston RG, Conolly JA, Grindlay GJ (2000) Synergism between papillomavirus and bracken fern in carcinogenesis of the upper gastrointestinal tract in cattle and humans: quercetin and cell transformation. In: Taylor JA, Smith RT (eds) Bracken fern: toxicity, biology, and control. Int Bracken Group Spec Publ 4. Aberystwyth, pp 116–122

Castillo UF, Ojika M, Sakagami Y, Wilkins AL, Lauren DR, Alonso-Amelot ME, Smith BL (2000) Isolation and structural determination of three new toxic illudane-type sesquiterpene glucosides and a new protoilludane sesquiterpene from *Pteridium*. In: Taylor JA, Smith RT (eds) Bracken fern: toxicity, biology, and control. Int Bracken Group Spec Publ 4. Aberystwyth, pp 55–59

Chambers TC, Farrant PA (2001) Revision of *Blechnum* (Blechnaceae) in Malesia. Blumea 46:283–350

Cheema HK (1997) Ferns as an excellent experimental system for morphogenesis: an overview. Indian Fern J 14:1–9

Chin WY (1998) Ferns of the tropics. Timber Press, Portland

Collinson ME (2001) Cainozoic ferns and their distribution. Brittonia 53:173–235

Dassler CL, Farrar DR (2001) Significance of gametophyte form in long-distance colonization by tropical epiphytic ferns. Brittonia 53:352–369

De la Sota ER, Ponce M, Morbelli MA, Cassá de Pazos L (1998) Pteridophyta. In: Correa MN (ed) Flora Patagónica, parte I. Colección Científica del INTA, Tomo VIII, Buenos Aires

Dixit RD, Sinha BK (2001) Pteridophytes of Andaman and Nicobar Islands. Bishen Singh Mahendra Pal Singh, Dehra Dun

Durand LZ, Goldstein G (2001) Growth, leaf characteristics, and spore production in native and invasive tree ferns in Hawaii. Am Fern J 91:25–35

Dyer AF, Parks JC, Lindsay S (2000) Historical review of the uncertain taxonomic status of *Cystopteris dickieana* R.Sim (Dieckie's Bladder Fern). Edinb J Bot 57:71–81

Flora of Australia (1998) Vol 48: ferns, gymnosperms, and allied groups. ABRS/CISRO, Melbourne, Australia

Gastony GJ, Rollo DR (1998) Cheilanthoid ferns (Pteridaceae: Cheilanthoideae) in the southwestern United States and adjacent Mexico – a molecular phylogenetic reassessment of generic lines. Aliso 17:131–144

Gurung VL (1997) Ecology and distribution of pteridophytes of Nepal Himalaya. Indian Fern J 14:51–88

Haufler CH (1997) Modes and mechanisms of speciation in pteridophytes. In: Iwatsuki K, Raven PH (eds) Evolution and diversification of land plants. Springer, Berlin Heidelberg New York, pp 291–307

Hill JD, Silander JA Jr (2001) Distribution and dynamics of two ferns: *Dennstaedtia punctilobula* (Dennstaedtiaceae) and *Thelypteris noveboracensis* (Thelypteridaceae) in a northeast mixed hardwoods-hemlock forest. Am J Bot 88:894–902

Horn K, Welss W (1998) Bibliography for the pteridophyte flora of Macaronesia. Vieraea 25:89–101

Hoshizaki BJ, Moran RC (2001) Fern grower's manual: revised and expanded edition. Timber Press, Portland

Hovenkamp P (1998) An account of the Malay-Pacific species of *Selliguea* (Polypodiaceae). Blumea 43:1–108

Imaichi R, Kato M (1997) Speciation and morphological evolution in rheophytes. In: Iwatsuki K, Raven PH (eds) Evolution and diversification of land plants. Springer, Berlin Heidelberg New York, pp 309–318

Johnson GN, Rumsey FJ, Headley AD, Sheffield E (2000) Adaptations to extreme low light in the fern *Trichomanes speciosum*. New Phytol 148:423–431

Jonsell B (ed) (2000) Flora Nordica 1. The Bergius Foundation. The Royal Academy of Sciences, Stockholm

Joshi P (1997) Ethnobotany of hilly districts of Uttar Pradesh, India. Indian Fern J 14:14–18

Kalkman C, Nooteboom HP (eds) (1998) Flora Malesiana. Series II - fern and fern allies, vol 3. Rijksherbarium/Hortus Botanicus, Leiden

Karrfalt E (1999) Some observations on the reproductive anatomy of *Isoetes andicola.* Am Fern J 89:198–203

Kato M, Setoguchi H (1999) An *rbc*L-based phylogeny and heteroblastic leaf morphology of Matoniaceae. Syst Bot 23:391–400

Kato M, Yatabe Y, Sahashi N, Murakami N (2001) Taxonomic studies of *Cheiropleuria* (Dipteridaceae). Blumea 46:513–525

Kelloff CL, McKee GS (1998) A new species of *Hecistopteris* from Guyana, South America. Am Fern J 88:155–157

Kessler M, Parris BS, Kessler E (2001) A comparison of the tropical montane pteridophyte floras of Mount Kinabalu, Borneo, and Parque Nacional Carrasco, Bolivia. J Biogeogr 28:611–622

Kopp E, Schneebeli-Graf R (1998) Illustrierter Leitfaden zum Bestimmen der Farne und farnverwandten Pflanzen der Schweiz und angrenzender Gebiete. Schweizerische Farnfreunde, Luzern

Lehnert M, Mönnich M, Pleines T, Schmidt-Lebuhn A, Kessler M (2001) The relictual fern genus *Loxsomopsis.* Am Fern J 91:25–35

Lellinger DB, Prado J (2001) The group of *Adiantum gracile* in Brazil and environs. Am Fern J 91:1–8

Li J, Haufler CH (1999) Genetic variation, breeding systems, and patterns of diversification in Hawaiian *Polypodium* (Polypodiaceae). Syst Bot 24:339–355

Lin S-L, Kato M, Iwatsuki K (1999) Spore morphology of the fern genus *Lindsaea.* J Jpn Bot 74:353–366

Lobin W, Fischer E, Ormonde J (1998) The ferns and fern-allies (Pteridophyta) of the Cape Verde Islands, West Africa. Nova Hedwigia Beih 115:1–115

Maden AR, Whittier DP, Garbary DJ, Renzaglia KS (1997) Ultrastructure of the spermatozoid of *Lycopodiella lateralis* (Lycopodiaceae). Can J Bot 75:1728–1738

Makgomol K, Sheffield E (2001) Gametophyte morphology and ultrastructure of the extremely deep shade fern, *Trichomanes speciosum.* New Phytol 151:243–255

Manickam VS, Dominic Rajkumar S (1999) Polymorphic ferns of the western Ghats, South India. Bishen Singh Mahendra Pal Singh, Dehra Dun

Marlière CA, Wathern P, Freitas SN, Castro MCFM, Galvao MAM (2000) Bracken fern (*Pteridium aquilinum*) consumption and oesophagal and stomach cancer in the Ouro Preto region, Minas Gerais, Brazil. In: Taylor JA, Smith RT (eds) Bracken fern: toxicity, biology, and control. Int Bracken Group, Spec Publ 4. Aberystwyth, pp 144–149

Moran RC (2000) Monograph of the neotropical species of *Lomariopsis* (Lomariopsidaceae). Brittonia 52:55–111

Moran RC, Smith AR (2001) Phytogeographic relationships between neotropical and African-Madagascan pteridophytes. Brittonia 53:304–351

Morrow MF (2000) Bracken as food: a Japanese perspective. In: Taylor JA, Smith RT (eds) Bracken fern: toxicity, biology, and control. Int Bracken Group, Spec Publ 4. Aberystwyth, pp 204–205

Murakami N, Yokoyama J, Cheng X, Iwasaki H, Imaichi R, Iwatsuki K (1998) Molecular α-taxonomy of *Hymenasplenium obliquissimum* complex (Aspleniaceae) based on *rbc*L sequence comparisons. Plant Species Biol 13:51–56

Murakami N, Watanabe M, Yokoyama J, Yatabe Y, Iwasaki H, Serizawa S (1999a) Molecular taxonomic study and revision of three Japanese species of *Asplenium* sect. *Thamnopteris.* J Plant Res 112:15–25

Murakami N, Nogami S, Watanabe M, Iwatsuki K (1999b) Phylogeny of Aspleniaceae inferred from *rbc*L nucleotide sequences. Am Fern J 89:232–243

Murakami N, Yatabe Y, Iwasaki H, Darnaedi D, Iwatsuki K (1999c) Molecular α-taxonomy of a morphologically simple fern *Asplenium nidus* complex from Mt. Halimum National Park, Indonesia. In: Kato M (ed) The biology of biodiversity. Springer, Berlin Heidelberg New York

Mukhopadhyay R, Bandhari JB (1999) Occurrence of microspore polyads in *Selaginella intermedia* (Bl.) Spring. Phytomorphology 49:75–78

Murillo-Pulido MT, Murillo-Pulido J (1999) Pteridófitos de Colombia I. Composición y distribución de las Lycopodiaceae. Rev Acad Colomb Cienc 23:19–38

Pacheco L, Moran RC (1999) Monograph of the neotropical species of *Callipteris* with anastomosing veins (Woodsiaceae). Brittonia 51:343–388

Parks JC, Dyer AF, Lindsay S (2000) Allozyme, spore, and frond variation in some Scottish populations of the ferns *Cystopteris dickieana* and *Cystopteris fragilis*. Edinb J Bot 57:83–105

Parris BS (1998) *Chrysogrammitis*, a new genus of Grammitidaceae. Kew Bull 53:909–918

Parris BS (2001) Circum-Antarctic continental distribution patterns in pteridophyte species. Brittonia 53:270–283

Pemberton RW (1998) The potential biological control to manage Old World climbing fern (*Lygodium microphyllum*), an invasive weed in Florida. Am Fern J 88:176–182

Pemberton RW, Ferrier AP (1998) Old World climbing fern *(Lygodium microphyllum)*, a dangerous invasive weed in Florida. Am Fern J 88:165–175

Pence VC (2001) Survival of chlorophyllous and nonchlorophyllous fern spores through exposure to liquid nitrogen. Am Fern J 90:119–126

Pérez-García B, Riba R (1998) Bibliographía sobre gametofitos de helechos y plantas afines. Monographs in systematic botany 7. Missouri Botanical Garden Press, St. Louis, pp 1699–1996

Pérez-García B, Riba R, Johnson DM (1999) Marsileaceae. – Flora de México, vol 6, no 5. Consejo Nacional de la Flora de Mexico, México, DF

Phibbs CJ, Taylor TN, Taylor EL, Cúneo NR, Boucher LD, Yao X (1998) *Osmunda* (Osmundaceae) from the Triassic of Antarctica: an example of evolutionary stasis. Am J Bot 85:888–895

Potter DM (2000) The Pteridaceae as a source of compounds with pharmaceutical activity. In: Taylor JA, Smith RT (eds) Bracken fern: toxicity, biology, and control. Int Bracken Group, Spec Publ 4. Aberystwyth, pp 60–67

Pryer KM (1999) Phylogeny of marsileaceous ferns and relationships of the fossil *Hydropteris pinnata* reconsidered. Int J Plant Sci 160:931–954

Pryer KM, Schneider H, Smith AR, Cranfill R, Wolf PG, Hunt JS, Sipes SD (2001a) Horsetails and ferns are a monophyletic group and the closest living relatives to seed plants. Nature 409:618–622

Pryer KM, Smith AR, Hunt JS, Dubuisson J-Y (2001b) *rbc*L data reveal two monophyletic groups of filmy ferns (Filicopsida: Hymenophyllaceae). Am J Bot 88:1118–1130

Renzaglia KS, Garbary DJ (2001) Motile gametes of land plants: diversity, development, and evolution. Crit Rev Plant Sci 20:107–213

Renzaglia KS, Maden AR (2000) Microtubule organizing centers and the origin of centrioles during spermatogenesis in the pteridophyte *Phylloglossum*. Micros Res Tech 49:496–505

Renzaglia KS, Dengate SB, Bernhard DL (1998) Architecture of the spermatozoid of *Selaginella australiensis*. Am Fern J 88:1–16

Renzaglia KS, Bernhard DL, Garbary DJ (1999) Developmental ultrastructure of the male gamete of *Selaginella*. Int J Plant Sci 160:14–28

Renzaglia KS, Duff RJ, Nickrent DL, Garbary DJ (2000) Vegetative and reproductive innovations of early land plants: implications for a unified phylogeny. Philos Trans R Soc Lond B 355:769–793

Renzaglia KS, Johnson TH, Gates HD, Whittier DP (2001) Architecture of the sperm cell of *Psilotum.* Am J Bot 88:1151–1163

Riba R, Pérez-García B (1999) Dryopteridaceae. Flora de México, vol 6, no 4. Consejo Nacional de la Flora de Mexico, México, DF

Rogers G (1998) The history and classification of the farley fern, *Adiantum tenerum* 'Farleyense'. Am Fern J 88:32–46

Rumsey FJ, Jermy AC, Sheffield E (1998) The independent gametophytic stage of *Trichomanes speciosum* Willd. (Hymenophyllaceae), the Killarney fern and its distribution in the British Isles. Watsonia 22:1–19

Rumsey FJ, Vogel JC, Russell SJ, Barrett JA, Gibby M (1999) Population structure and conservation biology of the endangered fern *Trichomanes speciosum* Willd. (Hymenophyllaceae) at its northern distributional limit. Biol J Linn Soc 66:333–344

Sánchez C (2000) Hymenophyllaceae. In: Flora de la Republica de Cuba, Ser A, fasc. 4. Koeltz, Königstein

Sano R, Takamiya M, Ito M, Kurita S, Hasebe M (2000a) Phylogeny of the lady fern group, tribe Physematieae (Dryopteridaceae), based on chloroplast *rbc*L gene sequences. Mol Phylogenet Evol 15:403–413

Sano R, Takamiya M, Kurita S, Ito M, Hasebe M (2000b) *Diplazium subsinuatum* and *Di. tomitaroanum* should be moved to *Deparia* according to molecular, morphological, and cytological characters. J Plant Res 113:157–163

Sano R, Ito M, Kurita S, Hasebe M (2000c) *Deparia formosana* (Rosenst.) R. Sano, a new name for *Diplazium formosanum* Rosenst. Acta Phytotax Geobot 51:17–20

Schneider H (1999) Yet another fern with storage roots – *Cheilanthes bolborrhiza* Mickel and Beitel (Pteridaceae: Pteridophyta) from Mexico and El Salvador. Fern Gaz 15:269–272

Schneider H (2000) Morphology and anatomy of roots in the filmy fern tribe Trichomaneae H. Schneider (Hymenophyllaceae, Filicatae) and evolution of rootless taxa. Bot J Linn Soc 132:29–46

Schölch A (2000a) Relations between submarginal and marginal sori in ferns. I. The sori of selected Hypolepidaceae and Dennstaedtiaceae. Plant Syst Evol 220:161–183

Schölch A (2000b) Relations between submarginal and marginal sori in ferns. I. The sori of selected Dicksoniaceae and Hymenophyllaceae. Plant Syst Evol 220:185–198

Serbet R, Rothwell GW (1999) *Osmunda cinnamomea* (Osmundaceae) in the Upper Cretaceous of western North America: additional evidence for exceptional species longevity among filicalean ferns. Int J Plant Sci 160:425–433

Shahin M, Seawright AA, Smith BL, Prakash AS (2000) Molecular mechanism of bracken carcinogenesis. In: Taylor JA, Smith RT (eds) Bracken fern: toxicity, biology, and control. Int Bracken Group, Spec Publ 4. Aberystwyth, pp 91–95

Simán SE, Povey AC, Sheffield E (1999) Human health risks from fern spores. A review. Fern Gaz 15:275–288

Simán SE, Povey AC, O'Connor PJ, Ward TH, Margison GP, Sheffield E (2000) The genotoxicity of fern spores. In: Taylor JA, Smith RT (eds) Bracken fern: toxicity, biology, and control. Int Bracken Group, Spec Publ 4. Aberystwyth, pp 99–105

Skog JE (2001) Biogeography of Mesozoic leptosporangiate ferns related to extant ferns. Brittonia 53:236–269

Smith RT (2000) The spread of bracken: an end-of-century assessment of factors, risks, and land use realities. In: Taylor JA, Smith RT (eds) Bracken fern: toxicity, biology, and control. Int Bracken Group, Spec Publ 4. Aberystwyth, pp 2–8

Smith AR, Kessler M, Gonzales J (1999) New records of pteridophytes from Bolivia. Am Fern J 89:244–266

Speer WD (2000) A systematic assessment of British and North American *Pteridium* cpDNA gene sequences. In: Taylor JA, Smith RT (eds) Bracken fern: toxicity, biology, and control. Int Bracken Group, Spec Publ 4. Aberystwyth, pp 37–42

Speer WD, Hilu KW (1999) Relationships between two infraspecific taxa of *Pteridium aquilinum* (Dennstaedtiaceae). I. Morphological evidence. Syst Bot 23:305–312

Speer WD, Werth CR, Hilu KW (1999) Relationships between two infraspecific taxa of *Pteridium aquilinum* (Dennstaedtiaceae). II. Isozyme evidence. Syst Bot 23:312–325

Takamiya M, Ohta N, Yatabe Y, Murakami N (2001) Cytological, morphological, genetic, and molecular phylogenetic studies on intraspecific differentiations within *Diplazium doederleinii* (Woodsiaceae: Pteridophyta). Int J Plant Sci 162:625–636

Taylor J (ed) (1999) Bracken perceptions and bracken control in the British Uplands. Int Bracken Group, Spec Publ 3. Aberystwyth

Taylor JA, Smith RT (eds) (2000) Bracken fern: toxicity, biology, and control. Int Bracken Group, Spec Publ 4, Aberystwyth

Thomas BA (1999) Some commercial uses of pteridophytes in Central America. Am Fern J 89:101–105

Thomson JA (2000a) Morphological and genomic diversity in the genus *Pteridium* (Dennstaedtiaceae). Ann Bot (suppl B):77–99

Thomson JA (2000b) New perspectives on taxonomic interrelationships in *Pteridium*. In: Taylor JA, Smith RT (eds) Bracken fern: toxicity, biology, and control. Int Bracken Group, Spec Publ 4. Aberystwyth, pp 15–34

Tyson MJ, Sheffield E, Callaghan TV (1999) Uptake, transport and seasonal recycling of ^{134}Cs applied experimentally to bracken *Pteridium aquilinum* (L.) Kuhn. J Environ Radioact 46:1–14

Verdcourt B (1999a) Actiniopteridaceae. In: Beentje HJ (ed) Flora of tropical East Africa. AA Balkema, Rotterdam

Verdcourt B (1999b) Davalliaceae. In: Beentje HJ (ed) Flora of tropical East Africa. AA Balkema, Rotterdam

Verdcourt B (1999c) Equisetaceae. In: Beentje HJ (ed) Flora of tropical East Africa. AA Balkema, Rotterdam

Verdcourt B (1999d) Marattiaceae. In: Beentje HJ (ed) Flora of tropical East Africa. AA Balkema, Rotterdam

Verdcourt B (1999e) Osmundaceae. In: Beentje HJ (ed) Flora of tropical East Africa. AA Balkema, Rotterdam

Verdcourt B (1999f) Parkeriaceae. In: Beentje HJ (ed) Flora of tropical East Africa. AA Balkema, Rotterdam

Verdcourt B (1999g) Psilotaceae. In: Beentje HJ (ed) Flora of tropical East Africa. AA Balkema, Rotterdam

Verdcourt B (1999h) Vittariaceae. In: Beentje HJ (ed) Flora of tropical East Africa. AA Balkema, Rotterdam

Verdcourt B (2000a) Dennstaedtiaceae. In: Beentje HJ (ed) Flora of tropical East Africa. AA Balkema, Rotterdam

Verdcourt B (2000b) Gleicheniaceae. In: Beentje HJ (ed) Flora of tropical East Africa. AA Balkema, Rotterdam

Verdcourt B (2000c) Salviniaceae. In: Beentje HJ (ed) Flora of tropical East Africa. AA Balkema, Rotterdam

Vogel JC, Rumsey FJ, Schneller JJ, Barrett JA, Gibby M (1999) Where are the glacial refugia in Europe? Evidence from pteridophytes. Biol J Linn Soc 66:23–37

Whittier DP, Pintaud J-C (1999) Spore germination and early gametophyte development in *Stromatopteris*. Am Fern J 89:142–148

Wikström N, Kenrick P (1997) Phylogeny of Lycopodiaceae (Lycopsida) and the relationships of *Phylloglossum drummondii* Kunze based on *rbc*L sequences. Int J Plant Sci 158:862–871

Wikström N, Kenrick P (1999) Epiphytism and terrestrialization in tropical *Huperzia* (Lycopodiaceae). Plant Syst Evol 218:221–243

Wikström N, Kenrick P (2000) Phylogeny of epiphytic *Huperzia* (Lycopodiaceae): paleotropical and neotropical clades corroborated by *rbc*L sequences. Nord J Bot 20:165–171

Wilson KA (1999) Ontogeny of the sporangia of *Sphaeropteris cooperi*. Am Fern J 89:204–214

Windisch PG (1997) Pteridófitas do Estado do Mato Grosso: Psilotaceae. Bradea 8:57–60

Windisch PG (1998) Pteridófitas do Estado do Mato Grosso: Osmundaceae. Bradea 8:107–109

Wolf PG, Sipes SD, White MR, Martines ML, Pryer KM, Smith AR, Ueda K (1999) Phylogenetic relationships of the enigmatic fern families Hymenophyllopsidaceae and Lophosoriaceae: evidence from *rbc*L nucleotide sequences. Plant Syst Evol 219:263–270

Wollenweber E, Schneider H (2000) Lipophilic exudates of Pteridaceae – chemistry and chemotaxonomy. Biochem Syst Ecol 28:751–777

Yatabe Y, Takamiya M, Murakami N (1998) Variation in *rbc*L sequence of *Stegogramma pozoi* subsp. *mollissima* (Thelypteridaceae) in Japan. J Plant Res 111: 557–564

Yatabe Y, Nishida H, Murakami N (1999) Phylogeny of Osmundaceae inferred from *rbc*L nucleotide sequences and comparison to the fossil evidence. J Plant Res 112:397–404

Zhang XC, Nooteboom HP (1998) A taxonomic revision of Plagiogyriaceae. Blumea 43:401–469

Dr. Stefan Schneckenburger
Botanischer Garten
Technische Universität Darmstadt
Schnittspahnstraße 3–5
64287 Darmstadt, Germany

e-mail: schneckenburger@bio.tu-darmstadt.de

Ecology

Impact of Ozone on Trees: an Ecophysiological Perspective

Rainer Matyssek and Heinrich Sandermann, Jr

1 Ozone as a Pollutant and Potential Risk Factor in Trees

Photooxidants have been recognized since the 1950s as gaseous agents that are potentially harmful to plants (Lefohn 1992). Early observations in the Los Angeles area had established links between vegetation damage and high photooxidant levels which were generated, in the presence of sunlight, from photochemical reactions of nitrogen oxides and organic compounds (as released from industrial and other anthropogenic sources into the atmosphere, Middleton et al. 1950; Haagen-Smit et al. 1952). Studies on tobacco then clarified that the typical spot-like visible symptoms were due to the O_3 component of photochemical smog (Heggestad and Middleton 1959). During recent decades, high O_3 regimes have spread across most major urban areas around the world, and enhanced O_3 concentrations are encountered even in rural regions (Stockwell et al. 1997). Towards the beginning of the twenty-first century, ozone has become a pollutant of great concern, regarding its impact on trees and forests, although the role of this agent in forest decline of the eastern USA and Europe has remained controversial (Matyssek and Innes 1999).

What do we know about the action of ozone in forest trees? We are inclined to believe that there is ample information, given the intense research on forest decline phenomena during the 1980s and 1990s (Sandermann et al. 1997). And indeed, the database elaborated to date is appreciably large. However, there are doubts about the extent to which this database is applicable to actual forest sites and stands with adult trees.

The following account will investigate what we really know about the O_3 sensitivity of trees, and what is lacking *prior to* achieving a functional basis for risk assessments of O_3 impact at forest sites. In doing so, the issues discussed will aim at pursuing an ecophysiological perspective, i.e., evaluate, wherever possible, the extent to which findings are relevant to prevalent site conditions. The account will integrate review articles recently published by the two authors of this present paper (e.g., Sandermann et al. 1998; Matyssek and Innes 1999; Polle et al. 2000;

Progress in Botany, Vol. 64

Sandermann 2000a,b; Kolb and Matyssek 2001; Langebartels et al. 2001; Matyssek 2001a). After briefly addressing the chemistry of ozone, trends of O_3 regimes and approaches for quantifying O_3 exposure, emphasis will be on the mechanisms of O_3 impact in trees, viewing consistencies and prerequisites for scaling responses across internal levels of tree functioning and to adult trees under stand conditions. The analysis of O_3 impact on trees in the field will be highlighted, directing focus on experimentation rather than modelling. Conclusions will be drawn about the extent to which ozone may represent a risk factor for trees and forests during the decades to come.

a) The Process of O_3 Formation

A certain portion of tropospheric ozone originates from downward flow of stratospheric ozone; however, most of the tropospheric ozone is attributed to the anthropogenic emission of ozone precursors. The basic reaction of tropospheric ozone formation consists of the light-induced oxygen transfer from nitrogen dioxide (NO_2) to oxygen (O_2), resulting in formation of ozone (O_3) and nitrogen monoxide (NO). This reaction is reversible so that increases in NO can destroy ozone (see Sect. 1.c). Excess ozone formation is generated by a shift of the described reversible reaction due to the occurrence of hydroxyl radicals and hydrocarbon peroxy radicals. Tropospheric hydrocarbons are due in part to anthropogenic (e.g., car traffic, industry) and in part to biogenic emissions (e.g., of isoprene and monoterpenes from tree foliage). Hydrocarbon double bonds can consume ozone, while hydrocarbon-derived radicals strongly enhance ozone formation. Due to these double-sided reactivities, the overall reactions between oxygen, ozone, nitrogen oxides and hydrocarbons are complex and non-linear. Modelling studies and practical measures of ozone reduction are thereby made difficult. The reader is referred to the excellent review of Stockwell et al. (1997) for more detailed information.

b) Trends in O_3 Regimes

Recordings during the last century reveal that the surface O_3 concentrations at mid to high latitudes of the northern hemisphere have increased by a factor of 2–5, as pre-industrial O_3 levels appear to have ranged between 10 and 20 nl l^{-1}; peak concentrations can distinctly exceed 100–200 nl l^{-1} in urban areas (Musselman et al. 1994), and, during recent decades, O_3 levels have increased on average by 1–2% each year (Staehelin and Schmid 1991; Stockwell et al. 1997). Although this trend may fluctuate to some extent, modelling predicts substantial further increase during the course of the twenty-first century (Fowler et al. 1999), which is reflected not only across the northern, but increasingly also in the southern hemisphere. Amongst the developing regions, China and central Africa, in particular, may become new "hot spots" of high O_3 regimes (Shao et al. 2000; Emberson et al. 2001).

In the temperate climate zone, high O_3 levels typically occur during spring and summer, being favored under sunny and dry weather conditions. In particular, the concentrations of NO_2 and reactive volatile organic compounds (VOCs) involved in O_3 formation (see Sect. 1.a) are responsible for the contrasting diurnal time courses of ozone in rural and urban areas. In the latter case, concentrations of VOCs tend to limit the formation of O_3, whereas, in rural areas, O_3 production tends to be limited by NO_x concentrations (Chameides et al. 1992). High NO release from traffic emissions leads in cities to a distinct decay of ozone during the early afternoon. Tree plantations can lower the O_3 levels to a minor but detectable extent in cities (Nowak et al. 2000). Rural surroundings contrast by the lack of NO release (because of less traffic emissions) in that O_3 levels stay high throughout the afternoons and evenings. Also, long-distance advection from urban areas can contribute to elevated O_3 regimes during the second half of the day in rural areas (Lefohn and Jones 1986). Such time courses resemble those at high-altitude sites (Wieser and Havranek 1994; Matyssek et al. 1997a) – basically for similar reasons, but also because of episodic O_3 incursions at high elevation from the stratosphere.

Such O_3 incursions are typically associated with periods of high pressure, when atmospheric conditions are relatively stable. Such episodes may also occur in spring and contribute to the high O_3 regimes that are often observed at that time in Scandinavia (e.g., Davies and Schuepbach 1994; Staehelin et al. 1994; Laurila and Tuovinen 1996; Skärby and Karlsson 1996). More typically, however, O_3 episodes occur primarily downwind of industrialized areas in central and northern Europe (Proyou et al. 1991; Grennfelt and Beck 1994). A peculiar situation is encountered in the Mediterranean region, where anticyclones in the west and low-pressure systems in the east create a boundary of relevance for air quality (Sanz and Millan 1998). This large-scale scenario is modified by local conditions, e.g., mountain ranges and their effects on sea breezes and up-slope winds, as found in the western Mediterranean. Stacked air layers (2–3 km deep, about 300 km wide) occur at the coastline (which provides substantial NO_x sources) through up and downward movement of air masses along the mountain slopes, which leads to the build-up of high O_3 concentrations (Millan et al. 1996, 1997). A similar mechanism is reported from the Los Angeles area (Stockwell et al. 1997).

When considering rural regions that typically harbor the forested areas, a major deficit is encountered in that monitoring stations of ozone are located in urban rather than rural surroundings, given the primary aim of assessing human health risks. Such stations in cities have little relevance to the O_3 regimes at rural sites and forests (see above).

c) How to Determine Constraints by Ozone on Trees

Ozone can act on plants both in chronic and acute ways. In the latter case, episodes of short duration (e.g., half an hour) and rather high O_3 concentration (100 to above 200 nl l^{-1}) may cause sudden and irreversible, physiological and macroscopic injury. The proposed initial event is membrane destruction (Heath 1980). Depending on the location and climate, such O_3 episodes may be rare events. However, at many sites of the northern hemisphere, the mean seasonal O_3 exposure, as occurring on a long-term scale, is significantly enhanced above the pre-industrial level (see Sect. 1.b). This kind of O_3 impact is termed "chronic". Plant responses to chronic exposure can distinctly differ from those to acute impact, even though the accumulating O_3 doses may be similar (Reich 1987). Contrasting with acute effects, responses to chronic impact may reflect acclimation, i.e., metabolic regulation, to O_3 stress including enhanced defense and repair capacities, but also endogenous "burst" induction (see Sect. 2.a). Also, such latter effects may eventually become irreversible. Under most field scenarios, even in southern California, chronic rather than acute O_3 regimes appear to be ecologically meaningful for the long-term development of trees. Therefore, experiments that have employed acute O_3 regimes may have little relevance for interpreting plant performance, in particular of long-lived individuals like trees, under the prevailing, chronic O_3 scenarios of given field sites (Reich 1987). For example, chamber experiments did induce acute O_3 injury in the youngest needles of Ponderosa pine, whereas it was the older needle ages that showed chronic O_3 symptoms (yellowish "mottling") at O_3-affected field sites (Miller and McBride 1999).

Different "exposure indices" have been defined for quantifying the O_3 impact on plants, aiming at creating "biologically meaningful" measures of O_3 stress. However, most of these measures are "dose surrogates" (Stockwell et al. 1997), as they cannot express the actual O_3 uptake into the plant during a given time interval (i.e., the metabolically relevant O_3 dose). Rather, the exposure indices reflect the "external O_3 dose", and there have been numerous kinds of such index definitions (Stockwell et al. 1997). For example, one approximation is the seasonal means of the average of daily O_3 concentrations, as assessed during the daylight hours (7- or 12-h intervals). When all hourly mean O_3 concentrations are summed up while neglecting threshold levels, the index is called "SUM0". A particular definition is "W126", as integrals of O_3 levels are adapted to weighting functions. This relates to the issue of whether O_3

concentrations below thresholds, being regarded ineffective in plant response, may be discarded, and high O_3 levels, distinctly exceeding the mean exposure, may be accentuated in index definition. Observations exist that peak concentrations rather than the extended impact of an enhanced mean O_3 exposure may be injurious (Lefohn 1992, Musselman et al. 1994). Accentuating high O_3 levels, "SUM06", "SUM07" or "SUM08", have been introduced that represent O_3 sums only of levels equal to or greater than 60, 70 or 80 nl O_3 l^{-1}, respectively.

Another approach is AOT40 (i.e., accumulated O_3 exposure - calculated as the sum of 1-h means - over a threshold of 40 nl l^{-1}) as proposed by UNECE (Fuhrer and Achermann 1994, 1999), which (1) neglects O_3 levels <40 nl l^{-1} while (2) integrating the differences of higher levels to the threshold of 40 nl l^{-1}. With respect to trees, it is believed that only daylight hours between April and September are relevant for characterizing the O_3 impact in this way (Käremlampi and Skärby 1996). This concept being applied in Europe provides a "Critical Level for Ozone" below which O_3 effects are regarded as insignificant. AOT40 has been set to 10 µl l^{-1} h, as this "Critical Level" of an external O_3 dose is expected to cause an annual growth reduction of 10%. However, this exposure threshold was derived from a limited number of chamber experiments with young beech plants and neglects potential O_3 effects on belowground growth (Matyssek and Innes 1999).

AOT40 is intended to provide an index for indicating O_3 risks on plants rather than quantifying O_3 impacts (Fuhrer et al. 1997). Calculations of AOT40 suggest large forested areas in Europe to be at risk from O_3 injury. However, visible injury does not seem to be a widespread phenomenon, and evidence of growth reductions in trees is scarce (Spiecker et al. 1996). Overall, the consistency between AOT40 and the apparent forest condition is low (Matyssek and Innes 1999). Major shortcomings of AOT40 are that it does not account for the actual species and site-dependent O_3 uptake and growth periods, focuses on the aboveground production, and neglects nightly O_3 exposure, apart from being derived from a limited database. The threshold of 10 µl l^{-1} h has turned out to be too high for sensitive, and too low for less sensitive plants (Skelly et al. 1999; Vanderheyden et al. 2001). In some cases, AOT40=30 µl l^{-1} h appeared to be appropriate (Skärby 1994; Pääkkönen et al. 1996), and in the USA, SUM06 has become a widely used index (Matyssek and Innes 1999). In addition, the relationships between macroscopic leaf injury, defoliation and growth have turned out to be very variable (Innes et al. 1997; Somers et al. 1998). Given the shortcomings in the risk assessment of trees, the situation is more complex with respect to forest ecosystems. However, observations of O_3 sensitivity exist for other wild plants besides trees (Davison and Barnes 1998), and the chronic O_3 impact on sensitive genotypes, potentially leading to their loss, may be significant for the long-term development of forest ecosystems (Karnosky 1981; Berrang et al. 1989, 1991; Miller and McBride 1999).

More meaningful than exposure indices is the O_3 flux into the leaves, i.e., the actual O_3 uptake, which represents the physiologically relevant O_3 dose (Reich 1987). The main pathway of uptake is through the stomata, and, therefore, all factors which modify stomatal width will affect O_3 influx (Guderian 1985). The internal O_3 dose is not necessarily proportional to the external O_3 dose (Emberson et al. 2000a). Variations in stomatal conductance and defense metabolites introduce non-linearity, and in addition, active oxygen species in the cells are produced by endogenous "burst" reactions. Such reactions may also occur in the chloroplasts and mitochondria (Pell et al. 1997, Pellinen et al. 1999). Further non-linear effects arise from the involvement of second messenger molecules (active oxygen species, ethylene, salicylic acid, jasmonic acid, Ca^{2+} ions, see Sect. 2.a).

Ozone influx can be assessed through the "water vapor surrogate method", converting stomatal conductance for water vapor (g_{H2O}) into g_{O3} (Matyssek et al. 1995a), and through evidence that the O_3 concentration in the intercellular space within the mesophyll is close to zero (Laisk et al. 1989). Influx assessment is complicated, however, by numerous diffusion resistances that depend on atmospheric turbulences and boundary layers within canopies and around leaves (Bortier et al. 2000). Research is being intensified, at present, in quantifying and modelling the O_3 flux from the free atmosphere towards the leaf (while distinguishing stomatal from non-stomatal components in O_3 deposition), and in preparing new flux-orientated concepts to derive critical levels for ozone (Emberson et al. 2000a,b; Mikkelsen et al. 2000; Fowler et al. 2001; Grünhage et al. 2001). Karlsson et al. (2000) found "memory effects" of drought on O_3 uptake but less variation between years than compared with the AOT40 approach. Such flux-orientated concepts are aimed at overcoming the conventional exposure indices towards a quantitative risk assessment of O_3 injury in trees and forests, as based on actual cause-effect relationships (Fuhrer and Achermann 1999). At present, no database exists that would translate, in a thoroughly mechanistic way, O_3 uptake into the complex metabolic response networks of trees.

d) The Kind of Knowledge Available on the O_3 Sensitivity of Trees

The impact of chronic O_3 exposure throughout decades is likely to have effects on long-lived organisms like adult forest trees (Reich 1987; Pye 1988). However, the available information is rather limited, although in the USA and Europe levels of ozone regularly exceed the various international threshold levels. In the field, the typical multi-factorial impact may bias any tree response to ozone (Arndt and Seufert 1990), and whole-tree analyses are complicated due to logistic reasons. Hence, most of our knowledge on O_3 effects derives from young trees exposed to controlled O_3 treatments (Matyssek and Innes 1999). Typically, seedlings or saplings have been grown in pots in various kinds of fumigation chambers, often under non-limiting conditions except for ozone, and under the exclusion of competition and parasites. Comparisons with control plants exposed to low O_3 levels or O_3-free air enabled the detection of even minor O_3 effects, in particular, if clonal plants were used to minimize the statistical variability in individual plant responses. In this way, principles of O_3 action have been revealed, also proving ozone to be an injurious pollutant. However, such findings are, *a priori*, not representative for forest sites, as they are more or less biased by the kind of experimental exposure (Chappelka and Chevone 1992), and because juvenile trees differ in physiological terms from adult individuals (Kolb et al. 1997). Given the lack of validation for forest sites and the uncertainties about the long-term acclimation and sensitivity of adult trees to ozone, the current knowledge about O_3 effects must largely rely on findings from young trees under controlled chamber exposures. Therefore, the following account will first comprehend the knowledge from controlled experiments with young trees and then examine to what extent this database is consistent with observations and experimentation on adult forest trees.

2 Mechanisms of O_3 Impact in Trees

According to the above conclusions, the following account must mainly refer to findings on young trees from chamber experiments of relevance, to some extent at least, for chronic O_3 regimes.

a) O_3 Responses at the Different Levels of Tree Functioning

α) Cellular Level

After passage through the stomata, the primary attack sites of ozone (and its oxidative derivatives) are the apoplast and the adjacent plasmalemma of leaf mesophyll cells. The level of apoplastic ascorbate may be increased under chronic O_3 impact and serve as a measure of "front line" oxidant scavenging and, thus, O_3 detoxification (Polle et al. 2000). Further components like glutathione, glutathione reductase and peroxidase are active as well and correspond with enhanced levels of ascorbate and glutathione in the symplast. Interaction of defense metabolites between apoplast and symplast is required, as the regeneration of the apoplastic ascorbate (i.e., reduction of dehydroascorbate) is mediated through metabolic processing in the cytoplasm, involving the turnover of glutathione (see Polle 1998 for details). The cell membranes, however, may suffer injury that results in disturbed functionality regarding permeability and ion gradients, so that, for example, Ca^{2+} levels can increase in the cytoplasm (Heath 1999). This increase displayed two kinetic peaks (Clayton et al. 1999). The second peak carried the information for inducing a GST stress gene. Also, wound-type responses and those of pathogen infection may be induced by ozone (Heath and Taylor 1997).

With regard to the incipient pathway of O_3 action, the view pursued up to the mid-1980s was focused on the chemical reactivity of ozone (i.e., with membrane lipids and proteins) and subsequent physiological injury (Heath 1980). More recent research has indicated that such direct mechanisms of ozone action are restricted to the acute dose range. Under chronic O_3 regimes, an elicitor-like action of ozone is more typical (Sandermann et al. 1990, 1998; Sandermann 1996). An amplifying ozone receptor has been proposed (Sandermann 1996) but is so far not experimentally identified. Still, ozone mimics pathogens like fungi and viruses that act through receptor molecules and induce ethylene emis-

Table 1. Ozone-induced gene expression in deciduous and conifer tree species. (Adapted from Langebartels et al. 2001)

Gene	Tree species
Phenylalanine ammonia-lyase Pathogenesis-related protein PR10 Mitochondrial phosphate translocator	*Betula pendula* Roth.
Extensin	*Fagus sylvatica* L.
Phenylalanine ammonia-lyase	Hybrid poplar (*Populus maximowizii* × *P. trichocarpa*)
O-Methyltransferase Pathogenesis-related protein PR1 Wound-induced gene *WIN*3.7	
Cinnamyl alcohol dehydrogenase Extensin Short-chain alcohol dehydrogenase Metallothionin, Porin	*Picea abies* (L.) Karst
Stilbene synthase Cinnamyl alcohol dehydrogenase Extensin 3-Hydroxymethylglutaryl CoA-synthase Polyubiquitin Ozone-related proteins, unknown function	*Pinus sylvestris* L.

Table 2. Ozone-induced stress metabolites in deciduous and conifer tree species. (Adapted from Langebartels et al. 1997)

Metabolite	Tree species
Total phenolics	*Pinus sylvestris* L.
Kaempferol 3-*O*-glucoside	*Picea abies* (L.) Karst
Pinosylvin, pinosylvin methyl ether	*Pinus sylvestris* L.
Catechin	*Picea abies* (L.) Karst *Pinus sylvestris* L.
3,3′,4,4′-Tetramethoxybiphenyl	*Fagus sylvatica* L.
Putrescine, putrescine conjugates	*Picea abies* (L.) Karst
Tyramine conjugates	*Fagus sylvatica* L.
Ethylene, ACC	*Pinus sylvestris* L. *Picea abies* (L.) Karst *Fagus sylvatica* L.

sion, gene transcription and major plant defence systems, such as programmed cell death and systemic acquired resistance (Sandermann 1996). The genes shown to be induced in tree species by ozone are listed in Table 1. Stress metabolites triggered in an elicitor-like fashion in trees are summarized in Table 2. These results indicate that many reactions of plant secondary metabolism are increased by ozone. In addition to second messenger molecules such as ethylene, salicylic acid and jasmonic acid, protein kinases of the MAP type are induced by ozone in analogy to many other biotic and abiotic stressors (Overmyer et al. 2000; Rao and Davis 2001; Samuel et al. 2000). Besides mimicking the effects of pathogens (Sandermann et al. 1990; Sandermann 1996), ozone can also mimick other environmental stressors, such as heat shock, UV-B and heavy metal ions (Sandermann et al. 1998; Utriainen et al. 1998).

There must be signal transduction between the apoplast and the chloroplast (Sandermann 1996), but up to now this signal chain has not been identified. Acclimation of chloroplasts to oxidative stress is assumed to proceed via H_2O_2 (Karpinski et al. 2001). The same is true for induced cellular resistance on one hand and lesion formation and injury on the other. H_2O_2 is formed directly or via dismutation of superoxide anion from O_3 or probably in much higher amounts from the oxidative burst (Schraudner et al. 1998; Pellinen et al. 1999; Rao and Davis 2001).

Most of the induced responses summarized in Tables 1 and 2 are transient and decline within hours or days. However, in certain cases, such as the transcript for the lignin-biosynthetic enzymes cinnamyl alcohol dehydrogenase (Zinser et al. 1998, 2000) and the metabolite catechin (Langebartels et al. 1998), longer lasting induction has been shown. A number of soluble plant metabolites, such as kampferol-3-*O*-glucoside, tyramine, stilbenes (Lange et al. 1994) and indole-3-carboxylic acid (Hagemeier et al. 2001), are known to be incorporated into plant cell walls. The changed plant cell wall structure is expected to have a long lifetime and thereby play a role in the growth and "memory" effects of ozone (see below). Another level where the elicitor-like action of ozone may change the plant permanently is by down-regulating photosynthetic transcripts and enzyme activities (Pell et al. 1997). Photosynthesis as well as translocation of photosynthates are inhibited. Free glucose that may arise by this inhibition is a strong inducer of protective genes and of protein kinases (Smeekens 2000). Free glucose could be an ecophysiological signal molecule because normally sucrose and not glucose is the predominant carbohydrate in plants. The elicitor-like actions of ozone typically lead to ecophysiological consequences such as changed growth behavior and changed susceptibility to pathogens and herbivores. The induced parameters may also serve as biomarkers for ozone action. Some of the best biomarkers known are the stilbene phytoalexins induced in pine needles (Rosemann et al. 1991). These antimicrobial metabolites occur in healthy pine trees only as constitutive components of heartwood. Ozone leads to rapid induction of the corresponding biosynthetic genes and enzymes as well as the stilbene metabolites (Rosemann et al. 1991; Zinser et al. 1998, 2000). Remarkably, the transcripts can have a lifetime longer than the metabolites because the latter tend to rapidly incorporate into cell walls (Lange et al 1994). Methylated stilbenes of heartwood possess higher fungitoxicity and are also induced by ozone from zero levels in pine needles (Rosemann et al. 1991). More recently, the gene for a pine stilbene methyltransferase has been cloned (Chiron et al. 2000a), and its stress regulation has been characterized at the levels of transcripts and of promotor elements (Chiron et al. 2000b). Stilbene needle content was correlated with ozone symptoms in the Californian San Bernardino Mountains (Sandermann 2000b). The grapevine stilbenes are also strongly induced by ozone (Schubert et al. 1997). In conclusion, stilbene induction is one of the few cases where an ozone-induced change in disposition to fungal attack (Manning and von Tiedemann 1995; Sandermann 1996) is understood at the molecular level.

O_3-induced symptoms and loss of spruce and pine needles occurred only in the year following treatment (Langebartels et al. 1998). Certain O_3-responsive secondary compounds remained elevated over many months so that the concept of a "memory" of conifers for ozone was generated (Sandermann et al. 1990; Langebartels et al. 1998). Beech trees also showed carry-over or "memory" effects into the year after O_3 treatment (delayed bud break, increased lammas shoot growth, Pearson and Mansfield 1994; Lippert et al. 1998). Similar carry-over effects have also been found for pine species (Andersen et al. 1997) and birch (Oksanen and Saleem 1999). These various observations indicate that trees may sense a long-lasting cumulative O_3 dose. It has been assumed that the O_3

sensitivity of deciduous and conifer trees becomes very similar when O_3-lifetime dose per leaf or needle is considered (Reich 1987), although experimental validation is still lacking. In conclusion, memory and long-lasting carryover effects are typical elements of chronic rather than acute mechanisms of O_3 impact. Such memory effects are at the center of a recently proposed epidemiological model for the analysis of ozone/biotic disease interactions (Sandermann 2000b). EDU injections into trees as protectants against O_3 injury is a suitable tool for studying O_3 action (Bortier et al. 2001), but has no relevance for practice or the ecophysiology of trees in the field.

β) Leaf Level

The metabolic perturbations addressed above appear to be reflected by structural reactions induced by ozone in mesophyll tissue. The outer surfaces of cell walls that face the intercellular space can develop droplet-like protrusions (indicated as "unetchable exudates" under low-temperature scanning electron microscopy: Matyssek et al. 1991; Günthardt-Goerg et al. 1993) that spread from the substomatal cavities, the sites of O_3 uptake, across the remainder of the leaf apoplast. Histological analyses proved these protrusions to mainly consist of pectates containing calcium and to be distinguishable from structural responses to other kinds of stress (Günthardt-Goerg et al. 1997). Given increased Ca^{2+} levels in cells under O_3 stress (Lock and Price 1994; Clayton et al. 1999), perhaps these wall structures represent a mechanism involved in calcium regulation.

Accompanying the changes in the wall surface, the wall width can be increased under O_3 stress (Günthardt-Goerg et al. 1993) and accompanied by enhancements in the leaf mass/area ratio (LMA) and stomatal density, while the leaf size is reduced. However, it was also found that stomatal density was lowered under chronic O_3 exposure (Matyssek et al. 1993a). Stimulated lignin biosynthesis (Sandermann et al. 1998) may contribute to the enhanced LMA, although reduced lignification was found in the stomatal apparatus of spruce needles (Maier-Maercker 1998). Inhibited cell extension growth may be conducive both to high LMA and reduced leaf size. Since ozone has the potential to directly or indirectly perturb membrane integrity (Heath and Taylor 1997; Schraudner et al. 1998), increased leakiness may lower the cell turgor during leaf differentiation and constrain tissue extension (cf. Matyssek et al. 1988). Loss of cell fluid appears to be the cause of advancing leaf injury, as shrinkage eventually leads to the collapse of mesophyll cells (Günthardt-Goerg et al. 1993) – this decline being associated with "programmed cell death" (Sandermann et al. 1998; Schraudner et al. 1998; Pellinen et al. 1999; Rao and Davis 2001). In this way, the intercellular space is enhanced as long as the epidermal cell layers stay intact. The necrotic decline of leaves typically proceeds from the sub-stomatal cavity across the mesophyll, before the epidermis and - eventually - the guard cells are destroyed by ozone (Günthardt-Goerg et al. 2000).

The structural changes affect the gas exchange of the leaves. The photosynthetic performance becomes restricted by both the collapsing mesophyll cells (Matyssek et al. 1991) and preceding decreases in Rubisco activity and gene-regulated de novo synthesis (Pell et al. 1997; Saurer et al. 1995; see Sect. 2.a). The weakened CO_2 fixation is paralleled by changes in the chlorophyll fluorescence, e.g., lowered effective quantum yield of photosystem II (ΔF/Fm' of PS II) and increased non-photochemical quenching (NPQ, Shavnin et al. 1999), whereas the optimal quantum yield of PS II, Fv/Fm, can stay rather stable (Maurer et al. 1997). Photosynthesis is initially affected only close to the stomatal pores (Leipner et al. 2001). Both the light and CO_2-saturated capacity of photosynthesis and the carboxylation efficiency can decline so that the CO_2 concentration in the intercellular space of the mesophyll (c_i) may increase. Given the fact that ozone hardly reaches the chloroplasts (Laisk et al. 1989; Urbach et al. 1989), their structural (Fink 1989; Mikkelsen et al. 1996; Manninen et al. 1999) and functional decline is meditated through mechanisms of signal transduction arising in the plasmalemma, as addressed above. The increase in c_i may provide the stimulus for some extent of stomatal closure that can override the effect of increased stomatal density on leaf transpiration. O_3-induced narrowing of stomatal pores was confirmed by low-temperature scanning electron microscopy (Matyssek et al. 1991, Frey et al. 1996). However, some studies did not find O_3 effects on stomatal conductance or even reported increases (cf. Grams et al. 1999). Partial stomatal closure typically does not prevent, however, a decline in the water-use efficiency of photosynthesis (WUE, Sasek and Richardson 1989; Schweizer and Arndt 1990; Matyssek et al. 1991; Lippert et al. 1996; Maurer et al. 1997). Macroscopically, the leaf-internal decline is accompanied by discoloration, often progressing in angiosperm trees from chlorotic stippling through a stage of "bronzing" towards the formation of large necroses, before the leaves may prematurely be shed (cf. Matyssek 1998). O_3 symptoms can vary strongly, however, among species and, in particular, between angiosperm and conifer trees, as the latter may develop a specific, chlorotic mottling pattern in the needles (Hanisch and Kilz 1990; Hartmann et al. 1995; Skelly et al. 1999).

The structural changes probably affect the translocation of assimilates within the leaves, given the disruption of symplastic pathways due to cell collapse. Phloem loading and transport may be disturbed (Rennenberg et al. 1996) as indicated by starch accumulation along leaf veins (Matyssek et al. 1992; Günthardt-Goerg et al. 1993). Such an accumulation can be paralleled by increased sugar levels. The reduction in sucrose phosphate synthase activity and stimulation of sucrose synthase and invertase may reflect, however, the degradation rather than synthesis of sucrose (Einig et al. 1997; Landolt et al. 1997). The metabolic changes associated with the impeded assimilate export out of the leaves may exert inhibitory effects on the photosynthesis (cf. Stitt and Schulze 1994), with the internal carbon flux being directed to glycolytic and anaplerotic processes that feed detoxification and repair of O_3 injury. This appears to be consistent with an enhanced respiratory activity and stimulation of the CO_2-binding PEP-carboxylase (PEPC, Lüthy-Krause et al. 1990; Gerant et al. 1996; Landolt et al. 1997; Maurer et al. 1997; Lütz et al. 2000), this latter enzyme fuelling carbon into the citric cycle (cf. Wiskich and Dry 1985). Regarding the O_3-induced PEPC activity, it is the low ^{13}C discrimination of this enzyme rather than a limited CO_2 uptake through the stomata that can raise the level of $\delta^{13}C$ in the tree biomass (Saurer et al. 1995). Given a decline in WUE despite partial stomatal closure (Matyssek et al. 1992), the response of $\delta^{13}C$ contrasts with suggestions by Farquhar et al. (1989) about the O_3 effect on this isotopic ratio.

γ) Whole-Tree Level

As the assimilate transport out of the leaves can be impeded under O_3 impact (Rennenberg et al. 1996), the C flux is curtailed into the non-green tree organs (Spence et al. 1990; Friend and Tomlinson 1992; Kelly et al. 1993; Smeulders et al. 1995; Takemoto et al. 2001). This limitation can be exacerbated by the enhanced C demand of the O_3-exposed leaves for detoxification and repair, which is indicated by the raised respiration and altered primary metabolism of the leaves (Skärby et al. 1987; Adams et al. 1990; Saurer et al. 1995; Einig et al. 1997; Landolt et al. 1997; Maurer et al. 1997), but may depend on the developmental status of the tissue (Reich 1983). Also, compensatory growth of new leaves may occur (Tjoelker and Luxmoore 1991; Bortier et al. 2000), making use of nutrient and C retranslocation from declining leaves or reserve storage (Beyers et al. 1992; Greitner et al. 1994; Pell et al. 1994; Temple and Riechers 1995). Such repair processes are likely mediated by phytohormones or second messengers of ozone action (see above).

According to the "set-point argument" by Mooney and Winner (1991) that claims a balance to be sustained between the C and nutrient flux through the plant, compensatory processes (new leaf growth, repair, detoxification) do counteract the O_3-caused reduction in C availability and reflect "acclimation" to ozone. Except for ethylene that initiates the stress response to ozone, the plant's sensitivity to other phytohormones (auxins, cytokinins, ABA) or second messenger molecules and their signalling between shoot and root under O_3 impact for sustaining the

C/nutrient balance is largely unclear (Lucas and Wolfenden 1996; Sandermann et al. 1998; Rao and Davis 2001).

Given the "extra demands" for C under O_3 stress and the change in allocation, the organ most limited has often turned out to be the root (Andersen et al. 1991; Coleman et al. 1995). C partitioning between root diameter classes may remain stable (Maurer and Matyssek 1997), although increased fine-root production has been found in tree seedlings under O_3 stress (Kelly et al. 1995). In parallel, the above-ground allocation can also be affected by ozone. Typically, the radial rather than longitudinal stem growth, as well as the lateral branching and the weight and size of individual leaves, have been found to be reduced (Schier et al. 1990; Matyssek et al. 1992, 1993a,b; Samuelson et al. 1996; Broadmeadow and Jackson 2000). Limitation in radial stem growth can be accompanied by structural decline in the cortex and phloem tissue (Günthardt-Goerg et al. 1993; Polle et al. 2000).

The longitudinal growth which is fuelled to some extent from the previous-year reserve storage (Kozlowski and Pallardy 1997; Kolb and Matyssek 2001), however, may be overestimated under O_3 stress, as exposure experiments with young trees were often preceded by growth conditions under reduced (or absent) O_3 impact. Hence, stems may display a similar number of nodes and leaves along their axes irrespective of the O_3 stress (Maurer 1995), which in such a case indicates the O_3-induced changes in biomass partitioning to not merely result from retarded ontogeny (Matyssek et al. 1998; Heath 1999; cf. Walters et al. 1993). Reduced branching and leaf size along with premature leaf loss were shown to more distinctly limit the biomass production than does the decline in leaf photosynthesis, and the efficiency of the whole-plant production, as based on the foliage area which was actually formed during the entire growing season, can be reduced by more than 50% (Matyssek et al. 1992; Matyssek 2001a).

As the respiratory costs can be high under O_3 exposure, the WUE of whole-plant production may be lowered, which was found to be consistent with the decline in WUE at the single-leaf level (Maurer and Matyssek 1997; Polle et al. 2000). In addition to the inhibited assimilate translocation, premature leaf loss can limit the increments of stem and root and the delivery of storage compounds (cf. Dickson and Isebrands 1991). The O_3-caused reduction in C flux and re-adjustments in allocation may eventually curtail fructification (Drogoudi and Ashmore 2000) and weaken the whole-tree defense status (Sandermann 1996).

Whenever crown and root architecture are modified, plant competitiveness may be affected, because the efficiency between space sequestration and resource acquisition is influenced (Küppers 1994; Tremmel and Bazzaz 1995). The significance of O_3 exposure for plant competition has been demonstrated in semi-natural plant communities (Nebel and Fuhrer 1995; Barbo et al. 1998), but information is rather scarce for woody species (Grams et al. 2002). The risks caused by O_3 through a changing resource allocation should not be overlooked relative to limitations of plant productivity, given the potential effects on competitiveness, reserve storage and fructification as well as whole-tree defense status. These latter plant functions altogether make up the individual fitness of a plant (Herms and Mattson 1992).

b) Scaling of O_3 Effects Within the Tree

As far as young trees have been studied under controlled chamber conditions, principles of O_3 action have been unravelled to an extent that allows response scaling across the internal levels of cells, leaves and the entire plant (Kolb and Matyssek 2001). Nevertheless, the validation for "realistic" site conditions is still lacking. The conceptual scaling scheme of Fig. 1 reflects the consistency that exists, under controlled experimental conditions, in the O_3 response pattern across the whole (juvenile) tree.

The ozone being taken up through the stomata, the oxidative impact on the apoplast and the cell interior, as mediated through signal chains, can lead to chloroplast decline and membrane injury and induce various kinds of gene expression (Fig. 1: cell level). While membrane injury may perturb the cell metabolism and leaf morphology (e.g., reduced area), gene expression can initiate protein degradation, programmed cell death (as a means of defense) and the formation of a broad array of secondary compounds and stress proteins. The latter are of significance for defense, either through lignin and callose formation (with effects on leaf morphology, e.g., LMA) or metabolites that are relevant for repair and detoxification. The enhanced defense activity is associated with raised respiratory demands, the involvement of anaplerotic PEPC stimulation and, as a consequence, an increase in tissue ^{13}C content (for comprehensive review on the C metabolism of O_3-stressed trees, see Dizengremel 2001).

As the leaf area stays reduced, LMA and stomatal density may increase, and transpiration decreases because of partial stomatal closure (Fig. 1: leaf level). The latter may be linked with photosynthesis that decreases via biochemical signalling from the impacted plasmalemma through reduced activity and gene expression of Rubisco. Mesophyll collapse contributes to the decline of photosynthesis, can lower WUE, and finally leads to premature leaf loss and a diminished foliage area. In addition, developmental processes are likely to respond to ozone-induced changes in phytohormones and second messenger molecules.

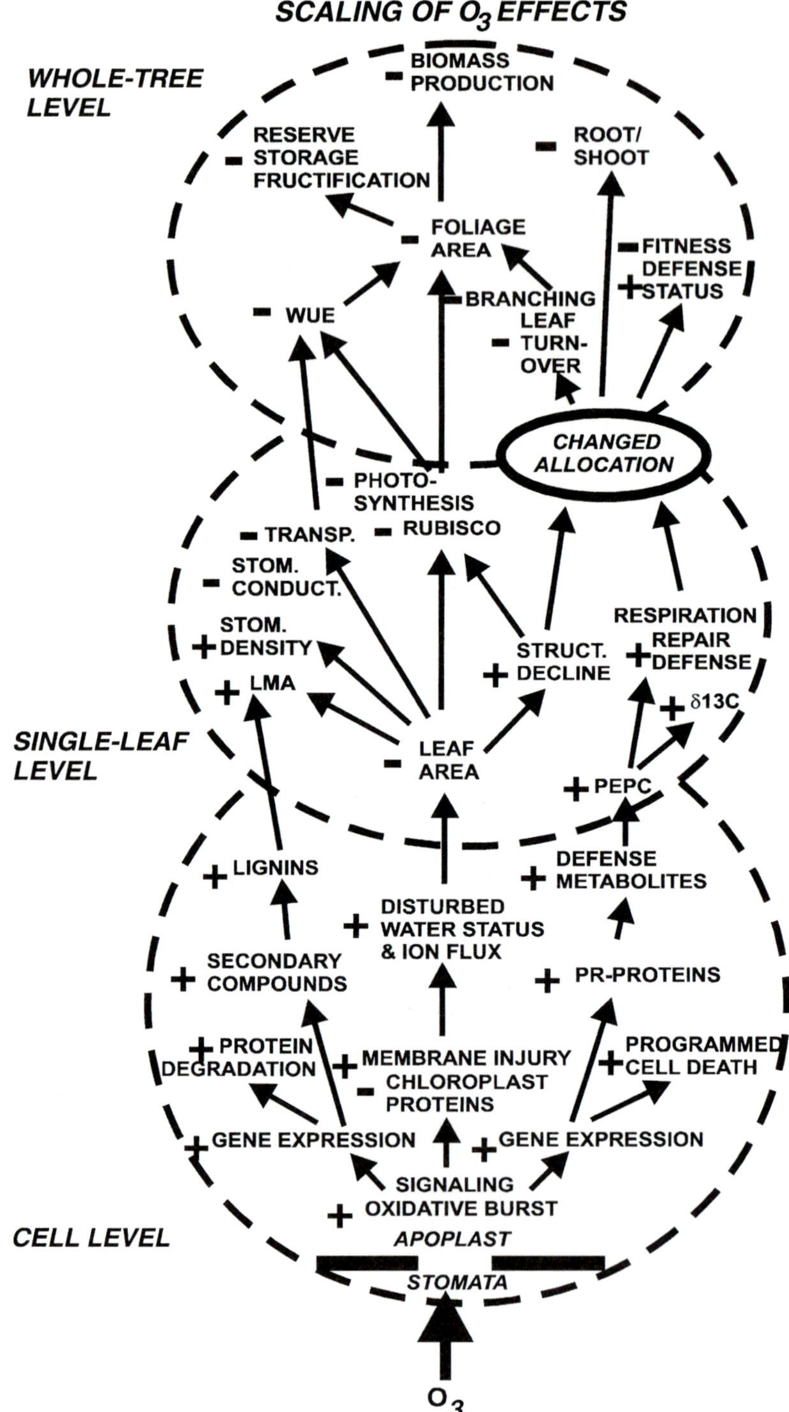

SCALING OF O_3 EFFECTS
WHOLE-TREE LEVEL
- BIOMASS PRODUCTION
- RESERVE STORAGE FRUCTIFICATION
- ROOT/ SHOOT
- FOLIAGE AREA
- FITNESS DEFENSE
+ STATUS
- WUE
- BRANCHING LEAF
- TURN-OVER
CHANGED ALLOCATION
- PHOTO-SYNTHESIS
- TRANSP.
- RUBISCO
- STOM. CONDUCT.
+ STOM. DENSITY
+ LMA
+ STRUCT. DECLINE
RESPIRATION
+ REPAIR DEFENSE
+ δ13C
SINGLE-LEAF LEVEL
- LEAF AREA
+ PEPC
+ LIGNINS
+ DEFENSE METABOLITES
+ DISTURBED WATER STATUS & ION FLUX
+ SECONDARY COMPOUNDS
+ PR-PROTEINS
+ PROTEIN DEGRADATION
+ MEMBRANE INJURY
- CHLOROPLAST PROTEINS
+ PROGRAMMED CELL DEATH
+ GENE EXPRESSION
+ GENE EXPRESSION
SIGNALING
+ OXIDATIVE BURST
CELL LEVEL
APOPLAST
STOMATA
O_3

Leaf loss along with disrupted resource translocation and enhanced respiratory demand alter whole-tree C allocation and decreases the increment of the root to a larger extent than of the shoot (Fig. 1: whole-tree level). Even though the root/shoot ratio stays low, biomass production declines because of decreases in leaf size and branching, and increase in leaf loss – altogether accounting for a decreased foliage area. In consistency with the lowered single-leaf WUE, biomass production occurs at an inefficient transpiratory water use, and shortcomings on reserve storage, fructification, above- and belowground space sequestration and overall defense status may weaken the fitness of the individual tree. The impact of ozone on the reproductive development of plants has recently been comprehensively reviewed (Black et al. 2000).

One has to be aware that the scaling scheme of Fig. 1 does not necessarily follow linear interrelationships and may be distinctly modified by the genotype, tree ontogeny, prevailing site conditions and temporal variability (see Sects. 2.d, f), which, in particular, "bias" the molecular and biochemical processes (Baldocchi 1993). Nevertheless, knowledge about these processes is indispensable for understanding the ecophysiological O_3 responses at the organ and whole-tree level (Waring 1993).

c) Does O_3 Stress Accelerate Senescence?

Plant senescence is driven by a declining capacity to cope with oxidants that are released from metabolic processes, and, in addition, this capacity becomes strained by the extent of oxidant production through ozone (Pell and Dann 1991). Tree species vary in the degree of ozone-induced acceleration of foliar senescence (Pell et al. 1999). Hence, senescence – in particular of leaves – is often regarded to be "accelerated" by O_3 stress. There are many similarities between ozone damage and natural senescence, e.g., the decline of Rubisco enzyme activity and its rbcS transcript (Pell et al. 1997, 1999). However, important differences have been identified at the biochemical level. At the level of the chloroplast, ozone may cause a decline in the maximum quantum yield of PS II (F_V/F_M, Grams et al. 1999). This ratio is maintained in naturally senescent leaves (Lu and Zhang 1998; Quirino et al. 2000). Regarding chloroplast pigments, chlorophylls decline through senescence as well as under ozone stress. However, β-carotin was hardly affected by ozone, although xanthophyll cycle

◄

Fig. 1. Conceptual scaling scheme of O_3 action across the functional levels of cells, leaves and the whole (juvenile) tree (adapted from Kolb and Matyssek 2001). Interrelationships are not necessarily linear, as may be suggested by the vectors bridging between responses. + denotes stimulation, – inhibition by ozone. *WUE* Water-use efficiency of net CO_2 uptake; *LMA* leaf mass per leaf area; *PR* pathogen-related; *RUBISCO* ribulose-bisphosphate carboxylase/oxygenase; *PEPC* phosphoenolpyruvate carboxylase

pigments were always elevated (Langebartels et al. 1997). Several senescence-associated (SAG) genes were induced by ozone, but SAG 17 (metallothionein) and SAG 12 (protease) were not induced (Miller et al. 1999, as shown in *Arabidopsis thaliana*). In conclusion, ozone acts on molecular and biochemical processes in a way that is partially similar to natural senescence, but does not trigger the complete senescence program.

Equating effects of O_3 stress with natural senescence becomes even more questionable, when moving from the cell towards the organ level. Obviously, O_3-caused discoloration of leaves distinctly differs from that of autumnal degradation (Matyssek 1998). Natural senescence is a gene-regulated process (Thomas and Stoddart 1980) that requires metabolic and structural integrity to ensure controlled leaf degradation. This control is reflected in, e.g., stable WUE of the leaf gas exchange and efficient nitrogen retranslocation from the leaves during autumnal senescence (Schulze and Hall 1982; Matyssek 1986), while the leaf structure stays intact. However, WUE may decline and the nitrogen concentration stay rather stable in O_3-exposed leaves before the co-occurring tissue destruction leads to premature abscision (Matyssek et al. 1991; Günthardt-Goerg et al. 1999). The intensity of O_3 impact versus the defense capacity appears to be crucial in determining the rapidity of the breakdown of leaves, and this probably decides to what extent mechanisms of natural senescence may become effective under O_3 stress. Further research on gene expression and hormonal relationships (cf. Lucas and Wolfenden 1996; Miller et al. 1999) appears to be rewarding for distinguishing O_3-specific degradation from senescence.

d) Modifications of O_3 Response Through Factorial Interactions

Pursuing an ecophysiological perspective, the action of ozone must be evaluated in concert with other environmental impacts that altogether determine the tree's sensitivity to stress (Mooney et al. 1991; Sandermann 1996; Skärby et al. 1998; Matyssek and Innes 1999). The principles of O_3 action highlighted above may be moderated, perhaps even masked, under the multi-factorial scenarios of actual forest sites – and still, our judgment must largely rely on findings from young trees under controlled chamber conditions.

α) O_3/Light

The injurious potential of ozone is typically increased by high irradiance, because the chloroplast degradation proceeds under high energy capture, as this energy uptake eventually turns, via formation of singlet oxygen, into photoinhibitory bleaching of chlorophylls (Heath and Taylor 1997). In addition, undisturbed photosynthesis produces oxidants that require detoxification. As the latter is warranted by the photosynthetic energy supply (Menser 1964; Schupp and Rennenberg 1988), the question arises about the effect of additional oxidative stress as caused by ozone when occurring under light limitation (cf. Foyer et al. 1991). There is evidence that low-light conditions can raise the sensitivity to ozone (Volin et al. 1993; Tjoelker et al. 1995; Oren and Schoeneberger in Matyssek et al. 1995a; Kolb et al. 1997; Günthardt-Goerg et al. 1997). In birch (Matyssek et al. 1995b), the same O_3 exposure at night (including dawn and dusk) led to more distinct declines in the biomass production and allometric relationships than during the daylight hours. Stomata were open at night, allowing O_3 influx, although the stomatal conductance (and O_3 uptake) was only about 50% of the levels during daylight. Stomata have the potential to stay open at night (Tobiessen 1982; Wieser and Havranek 1993), and there are a number of conditions (e.g., chronic O_3 impact itself) that favor such a stomatal behavior (Matyssek and Innes 1999). It is unclear whether the light-demanding character of a tree species (like birch) predisposes to high O_3 sensitivity under light limitation (Kolb and Matyssek 2001). Current AOT40 definitions might be reconsidered, however, as they only take into account daylight conditions (see Sect. 1.c).

High O_3 sensitivity of shade leaves (Tjoelker et al. 1995) may be related to their low palisade/spongy mesophyll ratio and less compact mesophyll structure that alleviates O_3 diffusion within the intercellular space and increases the exposure of palisade cells relative to sun leaves (Bennett et al. 1992). Shaded leaves may in addition have a higher ratio of O_3 versus CO_2 uptake, given the light limitation of photosynthesis (Fredericksen et al. 1996a,b). Perhaps, ratios relating agent uptake to assimilatory or defense capacities provide estimates of O_3 tolerance (Kolb 2001, Wieser et al. 2002) and reflect light limitation on detoxification.

β) O_3/Temperature

This interaction has comprehensively been accounted for by Skärby et al. (1998), with special emphasis on frost temperatures. Tree injury during winter is believed to be mediated in the preceding summer, through O_3 effects that may weaken subsequent frost hardening (Chappelka et al. 1990), but evidence is still ambiguous. Affected may be the synthesis of lipids and the proportion of unsaturated fatty acids in membranes to prevent freezing injury (Heath 1980; Wolfenden and Mansfield 1991) as well as the formation of glycoproteins and carbohydrates to act as cryoprotectants. Such demands relate to the tree's carbon budget that may have been curtailed through the O_3 impact. Starch has to be converted into raffinose and sucrose in order to ensure depression of the solution freezing-point (Sheppard 1994), and, in fact, Alscher et al. (1989) observed delay in raffinose formation under O_3 exposure. Although reduced accumulation of starch and glucose was reported by Barnes et al. (1995), no O_3 effects were found on sucrose. Ozone increased pinitol but decreased inositol levels in Scots pine (Landolt et al. 1989), both compounds being relevant as cryoprotectants. Some studies, however, did not find O_3 effects on starch formation (Gorissen et al. 1994), or reported differences in sugar availability between species or needle age in O_3-exposed conifers (Tingey et al. 1976; Wellburn et al. 1996). Also, desiccation and photoinhibition during winter can relate to preceding O_3 impact in summer (Mikkelsen and Ro-Poulsen 1995). Acclimation to frost temperatures can be delayed, and frost tolerance reduced (Cape et al. 1990; Edwards et al. 1990), although evidence of O_3-induced freezing injury is scarce. "Memory effects" of ozone were described (Langebartels et al. 1990), when needle injury and changed metabolism were observed only towards the beginning of the subsequent growing season. It is an open issue of whether repeated O_3 exposure in summer and, through this, delayed or lowered acclimation to frost can impose a long-term risk to trees. Risks by temperature increase and its interaction with enhanced O_3 regimes, as predicted for global-change scenarios, are uncertain as well, regarding the susceptibility to spring and autumn frosts during prolonged growing seasons or to forests pests which may profit from mild winter conditions (Karnosky et al. 2001a).

γ) O_3/Moisture

Temperature increase due to global change may be coupled with alterations in water availability, in some regions increasing the probability of drought (Scarascia-Mugnozza et al. 2001). Partial stomatal closure caused by moisture deficits of the air or soil can limit the O_3 uptake so that during sunny, dry days with high O_3 levels the influx of O_3 may be less than that under lower O_3 regimes during overcast, humid days (Wieser and Havranek 1993). However, significant exclusion of ozone is accompanied by restricted CO_2 uptake (Panek and Goldstein 2001). On the other hand, ozone can influence stomatal sensitivity and predispose trees to drought stress, being indicated by sluggishness in stomatal movements and inefficient control of transpiration (Keller and Häsler 1984; Barnes et al. 1990a,b; Pearson and Mansfield 1993; Karlsson et al. 1995; Götz 1996). The changed stomatal performance may be caused by altered "mechanics" in the stomatal apparatus due to reduced cell wall lignification, or disturbed osmotic control of the guard cells because of membrane impairment (Heath and Taylor 1997; Maier-Maercker 1998).

Changes in stomatal sensitivity to ABA under O_3 stress have not been clarified yet; however, O_3-induced dysfunction in the phloem transport out of the leaves may lead to enhanced ABA levels (Neals and McLeod 1992) and, through this, alleviate stomatal closure. In birch, drought and O_3 stress were additive in reducing leaf number, foliage area and starch formation in the mesophyll cells, but increasing epidermal cell wall width and tannin-like depositions in the vacuoles - these two latter effects being regarded as defense reactions (Pääkkönen et al. 1998a). Drought counteracted O_3 effects on leaf differentiation by decreasing the stomatal density; however, both stresses lowered the stomatal widths. Only severe drought tended to protect from O_3 impact (cf. Retzlaff et al. 2000), whereas mild soil moisture deficit rather was conducive to O_3 injury (Pääkkönen et al. 1998b). Also, "memory effects" of drought on stomatal conductance and O_3 uptake have been observed (Karlsson et al. 2000).

δ) O_3/Nutrition

Nutrition relates to the water relations, as in particular low nitrogen availability tends to ensure high sensitivity in stomatal control (Schulze 1994). Chronic O_3 exposure lowered the stomatal conductance in birch only at low nutrition (Maurer et al. 1997), restricting O_3 influx. In general, O_3 effects on nutrition have been found to be inconsistent and may be confounded by processes of senescence and retranslocation or depend on the specific nutrient under consideration (Matyssek et al. 1995b; Polle et al. 2000). Retranslocation of nutrients from injured leaves can stimulate the photosynthetic performance of new compensatory leaf growth under O_3 stress (Beyers et al. 1992; Greitner et al. 1994; Pell et al. 1994; Temple and Riechers 1995). High nutrient supply does not necessarily enhance O_3 tolerance (Maurer and Matyssek 1997; Matyssek et al. 1997b) and may foster the action of ozone in limiting root growth and promoting leaf loss (Grulke and Balduman 1999). In birch, low nutrition enhanced the antioxidative defense capacity against ozone and delayed premature leaf loss (Fredericksen et al. 1995; Maurer et al. 1997; Polle et al. 2000; Matyssek 2001a), while the carbon metabolism was changed accordingly (Saurer et al. 1995; Einig et al. 1997; Landolt et al. 1997). Also in this case, root growth was limited rather than stimulated (cf. Mooney and Winner 1991).

The maintenance of O_3-injured foliage can substantially raise the respiratory costs in the whole-tree carbon balance and, by this, lower WUE of the biomass production (Maurer and Matyssek 1997). However, low nutrition does not necessarily prevent O_3 injury (Bielenberg et al. 2001), which contrasts with hypotheses by Weinstein et al. (1991). The role of nutrition in driving leaf turnover, i.e., replacement of O_3-injured leaves, may be determined by the seasonal dynamics of the different patterns in shoot growth (determinate vs indeterminate, Tjoelker and Luxmoore 1991; Laurence et al. 1994; Matyssek et al. 1997b; Kolb and Matyssek 2001). If N is the only nutrient to vary, high supply tends to decrease O_3 injury and premature leaf loss (Pääkkönen and Holopainen 1995; cf. Greitner and Winner 1989), and high N deposition is believed to ameliorate O_3 effects (Takemoto et al. 2001).

ε) O_3/CO_2

As opposed to chronic O_3 impact, elevated atmospheric CO_2 concentration may stimulate tree growth (Ceulemans et al. 1999), unless other resources or experimental conditions do not constrain the plant, in particular root growth (Norby et al. 1999; Scarascia-Mugnozza et al. 2001). It has been debated as to whether high CO_2 availability may ameliorate adverse O_3 effects in trees (Karnosky et al. 2001a). However, the available knowledge is rather ambiguous, given the influence of the inter- and intraspecific genetic variability on tree sensitivity (Karnosky et al. 1996; Pääkkönen et al. 1997; Houpis et al. 1999). Ameliorating CO_2 effects have indeed been observed with respect to macroscopic foliar injury (Karnosky et al. 1999) or photosynthesis and growth under O_3 stress (Manes et al. 1998; Volin et al. 1998), but depend on nutrition and water supply or seasonal influences (Lippert et al. 1996; Grams et al. 1999; Lütz et al. 2000). High CO_2 may raise the capacity for detoxification and repair (cf. Grams and Matyssek 1999) or lower stomatal conductance and, by this, O_3 influx. Such stomatal effects may be accompanied, however, by a reduced responsiveness to moisture deficits (Broadmeadow et al. 1999), and there might be bias also by experimental restrictions on root growth that can incite down-regulation of photosynthesis and associated stomatal closure (Saxe et al. 1998). O_3 tolerance appears to be promoted also by the tree's capability to create C sinks that may facilitate compensatory responses such as high rates of leaf or lammas shoot formation, while the creation of such sinks strongly depends on soil nutrition and CO_2 supply (Kolb and Matyssek 2001).

Findings also exist, on the other hand, on enhanced O_3 sensitivity under high CO_2 levels (Polle et al. 1993; Kull et al. 1996; Oksanen et al. 2000) or absence of compensatory CO_2 effects on O_3 impact (Barnes et al. 1995; Kellomäki and Wang 1998a,b). There is evidence that the capacity of O_3 defense is lowered under high CO_2 supply (Schwanz et al. 1996; Niewiadomska et al. 1999), although the regulation that determines amelioration or exacerbation of O_3 stress still demands further research. In addition, ozone and CO_2 interact in changing the plant's C/N ratio (Maurer and Matyssek 1997; Hättenschwieler et al. 1996) and, as a result, the balance between growth and competitiveness versus the capacity in parasite defense (Herms and Mattson 1992). Host–parasite relationships in a changing environment have been discussed extensively; however, the pathways of interaction that may actually become effective largely remain uncertain (Manning and von Tiedemann 1995). Since elevated CO_2 and O_3 levels have the potential to modify the crown and root architecture, relevance emerges for competitiveness (see below, Reekie and Bazzaz 1989; Matyssek et al. 1992, 1993b; Hättenschwieler et al. 1996). A phytotron study indicated, however, minor effects of combined O_3/CO_2 regimes only on the efficiencies in space and carbon sequestration of spruce and beech that grew in pure and mixed plantations (Grams et al. 2002). Given the multiple external and internal influences on the O_3/CO_2 interaction in trees, ameliorating CO_2 effects are rather unlikely to occur throughout a tree's entire lifespan (Karnosky et al. 2001a).

ζ) O_3/Biotic Factors

Many exposure studies with young trees, potted in more or less site-relevant soil substrates, may have suffered from the disturbance or lack of natural mycorrhization. This is problematic regarding conclusions about changing resource allocation between root and shoot under O_3 stress, as mycorrhizae are know to represent a considerable sink for carbon in woody plants (Andersen and Rygiewicz 1991). The reduced C flux to the root can limit the metabolism of the mycorrhizal fungi and the root, when the foliage is impacted by ozone (Scherzer and McClenahen 1989; Andersen et al. 1991; Edwards 1991; Kytöviita et al. 1999). However, it is conceivable also that the sink strength of the fungi may overrule the C demand in the O_3-exposed foliage for stress compensation, turning mycorrhization from a "benefit" into a "burden" in trees under O_3 stress. In fact, it was observed in *Pinus taeda* that mycorrhization prevented the decline in the root/shoot biomass ratio often observed under O_3 exposure (Mahoney et al. 1985). Such "conflicts" between the C demand of mycorrhizae and that of the O_3-exposed foliage appear to have relevance in herbaceous plants (Miller et al. 1997). Remarkably, both respiration and the fungal and bacterial biomass in the rhizosphere were enhanced in seedlings of *Pinus ponderosa* when grown under environmentally relevant O_3 stress (Andersen and Scagel 1997), but this response was found only if the nutrient availability of the soil was low. It remains unclear if the raised metabolic activity of roots and mycorrhizae indicate enhanced nutrient uptake for sustaining the O_3-stressed foliage. O_3-related changes in the mycorrhizal morphology and increases in the fungal and presumably non-mycorrhizal biomass in the rhizosphere are perhaps indications of an enhanced risk of pathogenic root infestation (Duckmanton and Widden 1994). Decline in mycorrhization under O_3

stress (Ericsson et al. 1996) may lead to a lowered metabolic defense capacity (Gehring et al. 1997). The latter can favor infestations by root pathogens and enhance the attraction to phytophagous insects (Langebartels et al. 1997; Bonello et al. 1993), indicating interrelationships between O_3 impact, mycorrhization and parasite attack.

Since at the molecular and biochemical level plant response to ozone cannot be distinguished from defense reactions against pathogenic infestation (Sandermann 1996, Sandermann et al. 1998), ozone may strengthen the tree's resistance to parasite attack (through stress-induced lignin formation perhaps even in a permanent way, Sandermann 2000b). On the other hand, tree resistance may be weakened if chronic O_3 impact, on the long-term scale, "over-strains" the defense capacity. This latter possibility is discussed as one cause of successional changes in the heavily O_3-impacted forests of California (Miller and McBride 1999). Here, ozone is believed to predispose trees to biotic stress that eventually leads to lethal injury, primarily in the less O_3-resistant genotypes. Predisposition seems to act through reduced C gain that eventually causes the resin pressure in the stem to decrease, by this alleviating bark beetle attack (Pronos et al. 1999). Resulting changes in the gene pool of tree populations affect the long-term species composition, food webs and fire ecology of forests (McBride and Miller 1999). A thorough review of the contradictory empirical literature on ozone/biotic disease interaction has been published by Manning and von Tiedemann (1995). Recent progress in the biochemistry and genetics of O_3 effects suggests that the methods of molecular epidemiology should help to make the complex ozone/biotic disease interactions more predictable (Sandermann 2000b).

e) Can Ozone Predispose to Further Stress?

The allocation of resources to the various needs of plant metabolism is conceived to follow trade-offs (Zangerl and Bazzaz 1992; Stitt and Schulze 1994). Such trade-offs may exist between the demands for growth, reserve storage, fructification and defense, altogether characterizing individual plant fitness (Bazzaz 1997). Concerning the chronic O_3 impact, defense is required for coping with this stress. Herms and Mattson (1992) have focused their concept on the trade-off between parasite defense and growth, conceiving the latter as a means for staying competitive with neighboring plants in terms of space sequestration and resource gain (Fig. 2, Grams et al. 2002). Investment into defense (including repair) may sustain O_3 tolerance (Sandermann 2000a), one special recent observation being the ozone-induced emission of volatile organic compounds (Heiden et al. 1999). However, the plant's effort in doing so may occur at the expense of competitiveness (Fig. 2 "plant A", cf. Miller and McBride 1999), as investments into space sequestration

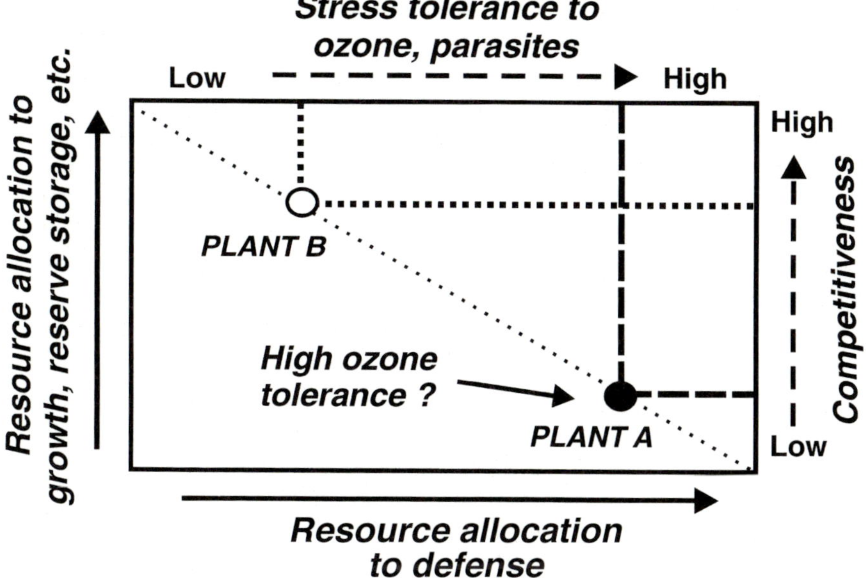

Fig. 2. Conceptual relationships between resource allocation to growth and defense, and resulting consequences for competitiveness and stress tolerance – highlighting implications in the presence of ozone. (Adapted from Herms and Mattson 1992)

gradually become limited. Also, predisposition to constraints by low water and nutrient availability in the soil (via inhibited root growth) and reduced carbon gain (via reduced branching and foliation) might increase. A reduced resource uptake capacity can, in turn, weaken defense – increasing mortality under biotic stress (Pronos et al. 1999). In conclusion, pathways through which ozone may predispose to further stress are well imaginable. On the other hand, vigorous trees of high competitiveness, displaying low efforts in defense, may be susceptible as well to both O_3 impact and parasite attack (Fig. 2 "plant B", Laurence et al. 1994; Herms and Mattson 1992).

f) Can O_3 Responses Be Scaled to Adult Forest Trees?

Since the physiological performance changes during ontogeny (Kozlowski and Pallardy 1997), juvenile trees can hardly serve as surrogates for adult trees, regarding the sensitivity to ozone (Kelly et al. 1995; Kolb et al. 1997; Kolb and Matyssek 2001). Scaling of O_3 sensitivity from young to adult trees requires, as a prerequisite, the clarifications which relate to tree dimension, within-stand O_3 and light regimes, and age-dependent O_3 susceptibility.

α) Tree Dimension

Photosynthesis and stomatal conductance tend to decrease with tree age, probably as a consequence of increasing hydraulic resistances, as the stem and crown extend in size, and gravitational constraints impede water flux with increasing tree height (Sperry et al. 1993; Waring and Silvester 1993; Hubbard et al. 1999; Ryan et al. 1997, 2000; Bond 2000; Kolb and Stone 2000). Hence, stomata of adult trees tend to respond more sensitively to the external and internal water availability and to reduce aperture so that old rather than young trees might inherently be protected from high O_3 uptake (cf. Reich 1987). Also, young trees possess a high proportion of foliage to total biomass (Grulke and Retzlaff 2001), not only enabling high annual productivity, but also representing a large surface area vulnerable to O_3 impact. Moreover, the high metabolic activity may render juvenile trees susceptible to stress (Laurence et al. 1994). On the other hand, the restricted photosynthesis might confine the capacity for defense and repair in adult trees: this limitation may even be exacerbated by increasing respiratory demands in adult trees for maintaining the high proportion of non-green biomass (Waring and Schlesinger 1985). In addition, trees in forest stands typically develop large proportions of shade foliage which may substantially contribute to the carbon gain, but which may be rather sensitive to O_3 impact (see Sect. 2.d). The extent to which ontogenetic stages differ in O_3 tolerance cannot be resolved at present, but – as will be pointed out below – age- and site-dependent differences in O_3 uptake, foliage properties and defense capacity appear to be of paramount importance.

β) O_3 Gradients Within Forest Canopies

O_3 exposure in forests typically differs from that in chamber experiments with young trees in that canopies can be dense and boundary layers high, which does impede the O_3 influx into leaves (Pleijel et al. 1996), although stomata may be open. O_3 levels underneath the canopy of a spruce stand were found in the morning to stay almost as high as above, whereas, during dusk and at night, below-canopy levels were distinctly lowered (Matyssek et al. 1997c). Pleijel et al. (1996) found the O_3 concentration at night to be higher inside than outside a spruce stand, when the wind velocity was low. O_3 levels decrease towards the forest floor, as the canopy is an O_3 sink, leading to the decomposition of this gaseous agent upon its deposition on leaf and branch surfaces (Taylor and Hanson 1992; Coe et al. 1995; Skelly et al. 1996; Samuelson and Kelly 1997). In a beech forest, O_3 levels were, on average, about 33% lower near the forest floor and by about 20% reduced at the lower as compared with the upper edge of the canopy, although canopy closure was high

(Baumgarten et al. 2000). Samuelson and Kelly (1997) reported similar vertical gradients in a dense hardwood forest, whereas small differences in O_3 exposure were observed between canopies of mature black cherry trees, saplings of a forest opening and seedlings in a field near-by (cf. Kolb and Matyssek 2001). Also, NO released by soil microorganisms can lower O_3 levels near the forest floor (Baumbach and Baumann 1989), and volatile hydrocarbons produced by trees can both favor and prevent high O_3 levels within forests (Enders et al. 1989; Stockwell et al. 1997). Apparently, canopy architecture and site location determine the spatial and temporal patterns of O_3 exposure within forests. Stands with unevenly aged and tall trees or species which contrast in crown shape or foliage type demand particular attention given the propagation in forestry of mixed forests (Bode 1997).

γ) Light Regimes in Canopies and O_3 Sensitivity

Stomatal conductance can predispose to O_3 stress (Reich 1987), and in fact, juvenile rather than mature trees have been found to display high g_{H2O} (see above). However, such a generalization is valid only if both tree ages are exposed to comparably high irradiance. Kolb and Matyssek (2001) show in their review that young trees growing underneath the forest canopy typically display a lower stomatal conductance relative to that of sun-lit leaves in mature trees. The low stomatal conductance may be inherently linked, in addition, with the low area-related N levels of shade leaves (Ellsworth and Reich 1993; Samuelson and Kelly 1997), which is in marked contrast to the leaf differentiation of the often light-exposed juvenile trees of controlled chamber experiments. Viewing ecologically meaningful site conditions, adult trees might be prone, therefore, to higher O_3 uptake, the more so as the vertical O_3 gradient across canopies typically declines towards the forest floor. Also, in the absence of light, stomata of black cherry were open wider in mature rather than young trees (Fredericksen et al. 1996b), which can be of significance at sites where O_3 exposure is high at night. Growth in the forest understory can reduce stomatal conductance also due to water limitation, as shallow-rooted seedlings and saplings may suffer from drought rather than deep-rooted adult trees (Dawson and Ehleringer 1993; Dawson 1996).

Nevertheless, shade leaves can show greater foliar O_3 injury and low WUE as compared with sun leaves (Volin et al. 1993; Tjoelker et al. 1995; Fredericksen et al. 1996a,b), although there are exceptions (Kolb and Matyssek 2001). Chronic exposure to the ambient O_3 regime led in shade leaves of seedlings, saplings and adult trees of black cherry to greater injury, while the ratio of O_3 versus photosynthetic CO_2 uptake was higher than in the sun leaves (Fredericksen et al. 1996a,b).

δ) O_3 Sensitivity and Tree Ontogeny

To what extent is O_3 sensitivity similar in young and adult trees, apart from the fact that stomatal conductance and, hence, susceptibility to ozone may vary due to tree dimension or environmental conditions? Indeed, some consistencies have been found, e.g., similar changes in N and carbohydrate levels of seedlings, branches and mature trees of Douglas fir and red oak in response to ozone (Smeulders et al. 1995; Samuelson et al. 1996), or similar antioxidant levels in young and old spruce trees (Polle et al. 1999, Wieser and Havranek 2001). In beech, macroscopic leaf symptoms were incited at a similar cumulative O_3 uptake for seedlings and mature trees of the same forest site (Baumgarten et al. 2000), and Wieser and Havranek (2001) made similar observations regarding the reduction of photosynthesis in spruce. Perhaps some of the metabolic responses to O_3 stress are independent of tree ontogeny.

Sensitivity to ozone may also depend on the phenological pattern of shoot growth, i.e., indeterminate length growth and leaf production which can proceed throughout most of the growing season, or determinate growth which is restricted to spring. These growth patterns are species-specific (Kozlowski and Pallardy 1997), although indeterminate growth can be promoted by high resource availability (Maurer and Matyssek 1997), and determinate growth may gain in importance beyond the juvenile stage (Roloff 2001). In indeterminate-growing birch cuttings, nutrition decided on the 'strategy' of coping with O_3 stress (Matyssek et al. 1997b): continuing formation of new, uninjured leaves at high nutrition, or maintenance of O_3-injured leaves at low nutrition – with the shoot growth becoming more determinate in the latter case, and being accompanied by major shifts in the whole-tree C allocation. In black cherry, young rather than adult trees follow the indeterminate pattern so that many seedling leaves are exposed to ozone for only part of the growing season, resulting in a lower cumulative O_3 uptake per average leaf than in adult trees (Fredericksen et al. 1995). However, seedlings of black cherry required a lower cumulative O_3 uptake to develop leaf injury while stomatal conductance was high, and the extent of injury eventually exceeded that of adult trees (Fredericksen et al. 1996a,b).

Often, ozone has been found to have stronger impact on leaf function in young rather than adult trees when the irradiance was high (Kolb and Matyssek 2001). In this case, the lower O_3 sensitivity of the adult trees appears to relate to their lower stomatal conductance, and perhaps the resulting reduced O_3 influx of the latter tree age aids coping with potential shortages in the C budget for repair and stress compensation (Kolb and Matyssek 2001, see above). In giant sequoia, stomatal conductance was higher in the juveniles, but light limitation may have contributed to their higher O_3 sensitivity relative to the older trees (Grulke and Miller

1994). Using grafts of *Picea rubens*, Rebbeck and Jensen (1993) found a lower O_3 sensitivity and stomatal conductance in scions from mature than from juvenile trees so that the phenomenon of cyclophysis is reflected also in relation to O_3 sensitivity. In addition, leaf respiration can be raised in juvenile rather than adult trees, even though photosynthesis may stay unchanged under O_3 stress (Momen et al. 1996).

However, there is a notable exception in that the photosynthesis of adult red oak trees was reported to be more sensitive to ozone as compared with the juveniles (Hanson et al. 1994; Samuelson 1994). It is difficult to conclude about the underlying mechanisms. Nevertheless, also in red oak, the O_3 sensitivity positively correlated with the stomatal conductance that, in this case, was higher in the adult trees. One may conclude that leaf-level responses to ozone in juvenile and adult trees are inherently linked with stomatal conductance (so that tree age may sometimes be of minor importance), and that this linkage is very similar to that found in response comparisons between species and foliage types (Reich 1987). Such trends suggest consistent principles for scaling O_3 impact across tree ontogeny and site conditions. However, the issue may be more complicated than merely weighing O_3 uptake by tree age (Samuelson and Kelly 2001), as differences in defense capacity, oxidative burst activity and second messenger molecules may not be negligible. The influence of second messenger systems on ozone sensitivity has been well documented for hybrid poplars. Sensitivity was correlated with a lack of defense-gene activation (Riehl-Koch et al. 1998) and with insensitivity to both salicylic acid and jasmonic acid (Riehl-Koch et al. 2000).

3 Analysis of O_3 Impact on Forest Trees Under Site Conditions

Given the genetic variability in plant responsiveness and multi-factorial modifications of O_3 sensitivity, the scaling schemes discussed above need to be validated for the specific forest sites under consideration. One first step is to examine whether leaf symptoms of O_3 injury do occur under field conditions. However, incipient, macroscopic O_3 injury may be rather unspecific and vary between species so that rigorous cause-effect relationships are difficult to establish (Matyssek and Innes 1999). This shortcoming is exacerbated, as a defined "control" of trees that would grow at the same site but under distinctly lowered O_3 levels or O_3-free air is typically unavailable. Attempts to relate tree response to O_3 impact have employed comparisons of plants exposed to different O_3 regimes in outdoor "open-top" chambers, one of the regimes serving as a "control" (Musselman and Hale 1997). Such approaches have been applied to mainly small trees, and in large trees also bags or cuvettes have been used to expose branches to contrasting O_3 regimes. As such approaches require plant enclosure and restrictions on growth conditions, tree response may suffer from artifacts in experimentation (see below). More realistic exposure conditions may be achieved by recently developed "free-air" O_3 fumigation techniques which do not require plant enclosure, allow the ambient air at the site to serve as a "control", and – most importantly – can perform whole-plant O_3 fumigations on adult forest trees (Karnosky et al. 2001b). The following account will examine, first, the extent to which leaf symptoms of O_3 injury have been documented under actual site conditions. The analytical potential will then be highlighted of O_3 exposure techniques for adult trees in the field.

a) Evidence of O_3 Impact on Trees at Field Sites

Numerous studies in the USA have assessed ozone as an agent responsible for symptom development in leaves. Examples are *Pinus ponderosa* and *Pinus jeffreyi* growing in the San Bernardino Mountains (California) that have been known since the 1950s to be constrained by high O_3 regimes (Haagen-Smit et al. 1952; Miller et al. 1963; Miller and Millecan 1971; Miller and McBride 1999). Macroscopic leaf injury in forest trees has been reported also from other parts of the USA (Davis and Skelly 1992; Neufeld et al. 1992; Simini et al. 1992; Miller et al. 1994; Chappelka et al. 1997; Skelly et al. 1997; Chappelka and Samuelson 1998), and symptoms ascribed to ozone were characterized as red, brownish ("bronzing"), purple or black stipple in sun-exposed leaf sides of angiosperm trees, or chlorotic mottling in conifer needles (Brantley et al. 1994; Skelly et al. 1987; Flagler 1998; Innes et al. 2001). One major region of research were the Smokey Mountains in the eastern USA, where leaf symptoms related to ozone were often found in the mid and lower canopy parts, although trees displayed a broad range of sensitivity (Chappelka et al. 1999a). Black cherry is even regarded as a suitable indicator tree of O_3 stress (Chappelka et al. 1999b).

Less evidence of O_3 injury at field sites has been reported for tree species in Europe (Davison and Barnes 1998), although there is an increasing number of reports from the Mediterranean region that relate leaf injury in trees to O_3 impact (Inclan et al. 1999; Gravano et al. 1999; Soda et al. 2000). A number of pine species (e.g., *Pinus halepensis*, *P. cembra*, *P. nigra*) and *Abies cephalonica* appear to be prone to O_3 stress (Gimeno et al. 1992; Velissariou et al. 1992, 1996; Dahlsten et al. 1997), and O_3 injury has been confirmed through experimentation in *P. halepensis* (Anttonen et al. 1998). Since in many parts of the Mediterranean region the highest O_3 levels occur during mid- to late-summer, when drought becomes most severe, O_3 uptake may be low in plants that display drought-induced stomatal closure or leaf loss; however, O_3 uptake may vary, depending on the edaphic water availability (cf. Damesin and Rambal 1995; Radoglou 1996).

The southern edge of the Alps is a region of high precipitation during the summer months, but also high insolation and elevated O_3 levels advected from northern Italy (Wunderli and Gehrig 1990; Staffelbach et al. 1997; Matyssek and Innes 1999). Thus, the O_3 uptake of trees may be high. A survey conducted in Ticino (southern Switzerland) indeed revealed a number of autochthonous and introduced species to exhibit symptoms of leaf injury (Skelly et al. 1999) that had previously been ascribed to O_3 impact (Skelly et al. 1987). Offsprings of symptomatic plants were grown in exposure chambers and submitted to either ambient O_3 regimes or treatments of reduced O_3 levels (filtered air). In offsprings of sensitive plants, the same leaf symptoms were reproduced under the ambient O_3 fumigation regime as found in the open field (Vanderheyden et al. 2001) – or their occurrence was delayed in filtered air with lowered O_3 levels (Skelly et al. 1999). This study is one of the rare cases in Europe that have fulfilled *Koch's postulates* in linking rigorously – through experimentation – leaf symptoms that occur in the field to ozone as the causal agent. In this way, an appreciable number of woody species was demonstrated to be O_3-sensitive at their natural field sites, e.g., *Fraxinus excelsior, Fagus sylvatica, Betula pendula, Sambucus racemosa, Viburnum lantana, Morus nigra, Rhamnus catharticus* or *Salix viminalis.* Amongst the introduced species, black cherry again tended to be sensitive, although the study confirmed genetic constitution as one determinant in tree responsiveness.

O_3 injury of birch leaves was identified in Central Europe when young plants grown under ambient air were compared with individuals exposed to controlled O_3 fumigations in outdoor exposure chambers (Günthardt-Goerg et al. 1998). The same kind of stippling occurred when leaves of both sets of plants had experienced similar O_3 exposure. In Finland, field experimentation with birch ("free-air" O_3 fumigation, see Sect. 3.b), although restricted to young plants, confirmed findings from chamber studies in that elevated O_3 levels reduced the height growth and foliage mass, promoted leaf discoloration and ultrastructural chloroplast decline, and induced partial stomatal closure, but increased the stomatal density in leaves that had differentiated under O_3 impact (Pääkkönen et al. 1993, 1997). As observed elsewhere (Günthardt-Goerg et al. 1993; Matyssek et al. 1998), leaves formed by indeterminate growth in mid-summer were more sensitive to ozone than those which had flushed during spring. "Carry-over" effects were observed during years that succeeded experimental O_3 exposures, resulting in persistent reductions in shoot and root biomass, inhibited bud development, leaf growth and branching as well as decreases in photosynthesis and stomatal conductance (Langebartels et al. 1998; Pääkkönen et al. 1999; Oksanen and Saleem 1999). Birch clones differed in O_3 sensitivity by their region of origin in Finland (Pääkkönen et al. 1997) – again highlighting the role of the genotype in stress tolerance.

b) Experimentation on Adult Trees in the Field

One approach towards risk assessment of O_3 impact on adult trees in the field has been the employment of fumigation bags or cuvettes for exposing individual branches (or branch sections) within tree crowns to defined O_3 regimes (Houpis et al. 1991; Havranek and Wieser 1994; Musselman and Hale 1997). Bags/cuvettes conveying the ambient or reduced O_3 regimes typically served as a "control". Such assessments elucidated O_3 uptake, for example, as depending on air humidity and soil moisture (Wieser and Havranek 1995, 1996), or the biochemical acclimation of trees that grow under the persistently enhanced O_3 levels of high-altitude sites (Polle et al. 1995; Sandroni et al. 1994). As this methodology requires branch enclosure within a pneumatic system, boundary layers around leaves are altered with consequences for O_3 uptake, and unless bags/cuvettes are not climatized, the micro-climatic bias can distinctly change the metabolic responsiveness of leaves (Landolt and Lüthy-Krause 1991). In addition, it remains open to what extent biochemical interaction with non-fumigated branch or crown parts may modify the stress response (Sandermann et al. 1997), although "net carbon fluxes" may reflect carbon autonomy of branches (Matyssek and Schulze 1988; Dickson and Isebrands 1991; Stitt and Schulze 1994). Micro-climatic bias appears to be problematic also if tall trees are experimentally exposed to O_3 regimes in large "open-top" chambers (e.g., Hanson et al. 1994).

Bias can be overcome by using "free-air" fumigation systems that avoid branch enclosure. Such a system that released ozone to leaf clusters on several sun and shade branches was operated in a sugar maple stand (Tjoelker et al. 1993, 1995), confirming an enhanced O_3 sensitivity of photosynthesis in shade leaves under stand conditions. The methodology for exposing entire canopies of adult trees or stands to "free-air" fumigations, and – by this – for inducing whole-tree responses to gaseous agents, has enabled appreciable progress during the past decade (Karnosky et al. 2001b), although the number of such experiments conducted with ozone is scarce. Some of these experiments, conducted in experimental plots, were restricted to juvenile trees that progressed in ageing throughout several years (McLeod et al. 1992; Wulff et al. 1992; Oksanen 2001). Findings resembled, to some extent, those known from young individuals under chamber conditions (birch: Pääkkönen et al. 1993) – or, exposure resulted only in a few significant effects of ozone, as in the case of spruce (Wellburn et al. 1997).

To date, there have been only a few "free-air" fumigations on large ageing trees, some of these studies being restricted to air pollutants other than ozone or clarifications of tree response to elevated CO_2 levels (Hendry et al. 1992, 1999; Lewin et al. 1994; Gielen et al. 2001). In one study, conducted in fast-growing aspen plantations, combinations of variable CO_2 and O_3 regimes are being examined, indicating ameliorating effects of elevated CO_2 levels on O_3 responses to be questionable in the field on the long-term scale (Karnosky et al. 1999, 2001a). In a managed 50- to 60-year-old, mixed beech/spruce forest (*Fagus sylvatica/Picea abies*), a "free-air" O_3 fumigation system is being operated at 17–25 m aboveground, within the foliated canopy (Häberle et al. 1999; Karnosky et al. 2001b). O_3 is released – under feedback control by on-line O_3 monitors – through a tubing system suspended across the canopy (ozonated volume about 1500 m^3, Werner and Fabian 2001), and O_3 exposure is assessed through passive samplers (Werner 1992) at weekly intervals. Ten trees (five spruce and beech individuals each) are exposed to a chronic, twice-ambient O_3 regime, being confined to a maximum of 150 nl O_3 l^{-1}, while a corresponding number of tree individuals in unchanged air serves as a "control". The comparison between the two O_3 regimes enables a quantitative risk assessment of a broad spectrum of molecular, biochemical and ecophysiological tree responses under the given site conditions, allowing the examination of tree processes which are at risk or already reflect incipient injury under the unchanged O_3 regime. Each of the experimental trees is considered as an individual case study, aimed at deriving consistency patterns from synchronously occurring O_3 responses at the cell, organ and whole-tree level. This exposure experiment is planned to continue for several years, and, to date, beech appears to be O_3-sensitive in displaying incipient leaf injury and accelerated leaf loss in autumn under experimentation (Nunn, Reiter, Häberle, Günthardt-Goerg, pers. comm.). "Free-air" fumigations appear to be the "ultimate choice" for approaching a quantitative risk assessment of the chronic O_3 exposure of adult trees and forest stands (Matyssek and Innes 1999; Karnosky et al. 2001b).

c) Hypotheses on the Action of Ozone in Adult Forest Trees

Kolb and Matyssek (2001) have derived seven hypotheses that may be relevant to scaling O_3 responses across tree ontogeny (Table 3) but largely demand for validation under forest conditions. Only hypotheses 3, 4, and 7 (see Table 3) could be evaluated from direct data comparisons between juvenile and mature trees: In the case of no. 3, inconsistencies were apparent between adult and juvenile trees, regarding carbon availability for defense, compensation, and repair responses to O_3 stress, #4 tended to be supported in that reduced stomatal conductance limits O_3 uptake and mesophyll injury in adult rather than juvenile trees, as long as focus was on sun leaves. However, #4 did not substantiate when com-

Table 3. Hypotheses of relevance for scaling chronic O_3 impact from juvenile to adult trees under forest conditions. (Adapted from Kolb and Matyssek 2001)

Hypothesis 1:	Carbon allocation to stem and roots, and sinks associated with defense or tolerance to other stress are reduced by chronic O_3 impact in trees with determinate rather than indeterminate shoot growth.
Hypothesis 2:	In trees with determinate shoot growth, recurrent flushing or formation of lammas shoots increase as a compensatory response to chronic O_3 stress, if the availability of reserve compounds or resource supply (N, C) are high.
Hypothesis 3:	Assimilate availability for defense, compensation and repair responses to O_3 stress are lower in mature than juvenile trees, because of higher respiratory costs for maintaining the living, non-green biomass and a lower photosynthetic capacity of mature trees.
Hypothesis 4:	Stomatal aperture limits O_3 uptake into leaves more for mature than juvenile trees because of greater hydraulic resistance to water transport.
Hypothesis 5:	Scaling O_3 effects from juvenile to mature trees must consider modification of tree competitiveness as mediated through structural and functional changes in resource uptake at the crown and root level.
Hypothesis 6:	Scaling O_3 impacts from juvenile to mature trees must consider influences of O_3 on mycorrhizae, and secondary effects of mycorrhizae on stress tolerance and pest resistance.
Hypothesis 7:	Light limitations on photosynthesis of shaded leaves increases O_3 sensitivity because of decreased carbon availability for defense and repair.

paring shaded leaves with rather low stomatal conductances in young with sun-exposed leaves of adult trees. The high O_3 sensitivity of shade leaves may relate to imbalances between O_3 uptake and assimilates available for repair or defense, which appears to be in support of no. 7. The other four hypotheses (1, 2, 5, 6, Table 3) were difficult to evaluate, given the limited knowledge on the O_3 sensitivity of adult forest trees.

4 Conclusions

The current state of knowledge on the impact of ozone on trees provides consistent response patterns across the levels of cells, organs and the whole plant (cf. Fig. 1). Hence, linkages emerge that form the conceptual basis for scaling O_3 effects within trees. However, this kind of knowledge basically derives from experimentation with juvenile (often potted and single) trees as grown under controlled chamber conditions. Such experiments have nevertheless been valuable in elucidating principles of O_3 action on the metabolism and structural differentiation of trees –

proving ozone as a pollutant of injurious potential to plants - and in yielding information not only about effects of acute, but also chronic exposure regimes. However, databases of this kind have little ecological relevance unless they are validated under "real" stand conditions and for trees beyond the ontogenetic stage of juvenility. This requirement has to be met because of the multi-factorial, biotic and abiotic impacts that modify and mimic O_3 effects in the field, and because of ontogenetic changes in resource allocation patterns that distinctly differ between juvenile and adult trees. Such changes relate to physical (tree dimension) and environmental constraints (varying resource availability due to competition and parasitic/phytophagous impacts) that affect the physiological performance, and to gene expression varying during ontogeny. Age-dependent resource allocation appears to be a major determinant in the responsiveness to stress; however, little is known about the mechanistic basis in the control of allocation. This issue has been recognized as a general and crucial challenge in ecophysiological research (Bazzaz 1997; Lerdau and Gershenzon 1997), and this holds true, in particular, concerning the ecophysiology of adult forest trees (Matyssek 2001b). On these grounds, many questions still remain open regarding the sensitivity of forest trees to chronic O_3 impact under site conditions.

For example, there is agreement on the O_3 flux into leaves to drive both acclimation to and injury from O_3 stress, although it is the burst reaction, rather than the physicochemical impact of ozone, that initiates, via metabolic signalling, the stress response. However, the "unifying theory", as proposed by Reich (1987), for explaining O_3 sensitivity in relationship to foliage type and leaf longevity through basic, biological principles still demands for experimental validation with respect to adult trees at forest stands. Given the O_3 influx, the latter must be viewed in close relation to the metabolic capacities of defense and repair and to memory effects in stress response, in order to thoroughly understand O_3 tolerance. However, biochemical assessment is a difficult task to be achieved in the field, in particular, if findings of seasonal representativeness need to be elaborated with tall trees. Although a number of critical questions can be raised regarding this latter task, answers cannot be provided to an extent that would warrant a comprehensive understanding of the biochemical mechanisms responsible for the performance of adult trees and forest ecosystems (Heath 1999). Still, knowledge about biochemical mechanisms is an indispensable prerequisite for using ecophysiological processes as "scaling units" towards the higher levels of system integration (cf. Figs. 1, 2; Reynolds et al. 1993). In conclusion, predictions about the O_3 sensitivity of adult trees in the field must largely rely still on findings derived from juvenile trees and controlled chamber studies, although the newly emerging field of molecular epidemiology (Sandermann 2000) has the potential to bridge the gap between the laboratory experiments and the field situation.

At present, a number of hypotheses that appear to have major relevance for the long-term development of trees and forest under chronic O_3 impact cannot be answered in a conclusive way (cf. Table 3). One approach for overcoming this deficit may be through modelling, and, in fact, a number of concepts have been pursued (e.g., Reich et al. 1990; Weinstein et al. 1991, 1998; Taylor et al. 1994; Constable and Taylor 1997; Ollinger et al. 1997; Retzlaff et al. 1997; Kärenlampi 1999; Laurence et al. 2001). Nevertheless, such models are mostly based on the performance of young trees, which appears to be critical with respect to long-term predictions of tree and ecosystem development under chronic O_3 stress. Validation of such models is required regarding adult trees and actual stand conditions, and a way for providing adequate reference data are novel "free-air" fumigation techniques to be employed to stand canopies. Even though the use of such methodologies will be restricted to a few locations and tree species because of logistic reasons, the insights provided into the physiology and O_3 sensitivity of adult trees will valuably improve the parametrization of long-term O_3 simulation models. Only the combination of ecologically relevant experimentation and modelling will provide the kind of quantitative risk assessment that is required for environmental policy making, concerning the chronic O_3 impact on trees and forests.

O_3 regimes prevail today in many parts of the world that are known from controlled chamber studies to have the potential to adversely affect tree growth. For chronic O_3 stress, which for most scenarios appears to have higher ecological relevance than acute O_3 impact, it seems that the regulatory capacity of resource allocation rather than productivity may be significant, in the long term, for the individual fitness and survival of trees. The O_3 levels currently encountered at many locations do not seem to pose an acute danger to adult trees and forests and are unlikely to lead to widespread dieback in the short term. Nevertheless, O_3 episodes do occur that are conducive to irreversible O_3 injury – more likely in deciduous rather than coniferous trees. Chronic O_3 stress can eventually lead to losses in species and genetic diversity, and it needs to be examined to what extent forest decline through diseases or changes in competitiveness between trees species may arise from predispositions induced by O_3 exposure (cf. Fig. 2). There has been increasing awareness that genetic disposition may govern O_3 tolerance, and that responsiveness may vary more strongly between genotypes than species. There is growing awareness also that enhanced O_3 regimes, as they act on a long-term scale, are "players" in "global change" scenarios which are typically viewed in relation to elevated CO_2 and N deposition levels (and other associated effects) rather than chronic O_3 stress. Tropospheric O_3 levels are likely to stay high in the future and, thus, may mitigate the carbon sink strength of trees and forests which is expected to increase, to some extent, under the elevated CO_2 supply. Due attention must be given to the interaction between ozone, CO_2 and N availability, as they alone and in combination do pose risks, via induced changes in resource allocation, to the long-term stability of forests, including predisposition to plant diseases. Given the current state of knowledge, scientifically founded arguments do exist already for justifying a precautionary, environmental policy to prevent adverse effects of ozone on trees and forests - despite the shortcomings that currently still exist towards quantitative risk assessment.

Acknowledgements. The assistance of Dr. H. Blaschke in literature research is highly appreciated.

References

Adams MB, Ewards NT, Taylor GE, Skaggs BL (1990) Whole-plant carbon-14 photosynthate allocation in *Pinus taeda*, seasonal patterns at ambient and elevated ozone levels. Can J For Res 20:152–158

Alscher RG, Amundson RG, Cumming JR, Fellows S, Fincher J, Rubin G, Van Leuken P, Weinstein LH (1989) Seasonal changes in pigments, carbohydrates and growth of red spruce as affected by ozone. New Phytol 113:211–223

Andersen CP, Rygiewicz PT (1991) Stress interactions and mycorrhizal plant response: understanding carbon allocation priorities. Environ Pollut 73:217–244

Andersen CP, Scagel CF (1997) Nutrient availability alters belowground respiration of ozone-exposed ponderosa pine. Tree Physiol 17:377–387

Andersen CP, Hogsett WE, Wessling R, Plocher M (1991) Ozone decreases spring root growth and root carbohydrate content in ponderosa pine the year following exposure. Can J For Res 21:1288–1291

Andersen CP, Wilson R, Plocher M, Hogsett WE (1997) Carry-over effects of ozone on root growth and carbohydrate concentrations of ponderosa pine seedlings. Tree Physiol 17:805–811

Anttonen S, Kittilä M, Kärenlampi L (1998) Impacts of ozone on Aleppo pine needles: visible symptoms, starch concentrations and stomatal responses. Chemosphere 36:663–668

Arndt U, Seufert G (1990) Introduction to the Hohenheim long-term experiment. Environ Pollut 68:195–204

Baldocchi DD (1993) Scaling water vapor and carbon dioxide exchange from leaves to a canopy: rules and tools. In: Ehleringer JR, Field CB (eds) Scaling physiological processes, leaf to globe. Academic Press, San Diego, pp 77–114

Barbo DN, Chappelka AH, Somers GL, Miller-Goodman MS, Stolte K (1998) Diversity of an early successional plant community as influenced by ozone. New Phytol 138:653–662

Barnes JD, Eamus D, Brown KA (1990a) The influence of ozone, acid mist and soil nutrient status on Norway spruce (*Picea abies* (L.) Karst). I. Plant-water relations. New Phytol 114:713–720

Barnes JD, Eamus D, Brown KA (1990b) The influence of ozone, acid mist and soil nutrient status on Norway spruce (*Picea abies* (L.) Karst). II. Photosynthesis, dark respiration and soluble carbohydrates of trees during late autumn. New Phytol 115:149–156

Barnes JD, Pfirrmann T, Steiner K, Lütz C, Busch U, Küchenhoff H, Payer H-D (1995) Effects of elevated CO_2, O_3, and K deficiency on Norway spruce (*Picea abies* [L.] Karst.). II. Seasonal changes in photosynthesis and non-structural carbohydrate content. Plant Cell Environ 18:1345–1357

Baumbach G, Baumann K (1989) Ozone in forest stands – examinations to its occurrence and degradation. In: Georgii HW (ed) Mechanisms and effects of pollutant-transfer into forests. Kluwer, Dordrecht, pp 37–44

Baumgarten M, Werner H, Häberle K-H, Emberson LD, Fabian P, Matyssek R (2000) Seasonal ozone response of mature beech trees (*Fagus sylvatica*) at high altitude in the Bavarian Forest (Germany) in comparison with young beech grown in the field and in phytotrons. Environ Pollut 109:431–442

Bazzaz FA (1997) Allocation of resources in plants: state of the science and critical questions. In: Bazzaz FA, Grace J (eds) Plant resource allocation. Academic Press, San Diego, pp 1–38

Bennett JP, Rassat P, Berrang P, Karnosky DF (1992) Relationships between leaf anatomy and ozone sensitivity of *Fraxinus pennsylvanica* Marsh. and Prunus serotina Ehrh. Environ Exp Bot 32:33–41

Berrang P, Karnosky DF, Bennett JP (1989) Natural selection for ozone tolerance in *Populus tremuloides*: field verification. Can J For Res 19:519–522

Berrang P, Karnosky DF, Bennett JP (1991) Natural selection for ozone tolerance in *Populus tremuloides*: an evaluation of nation-wide trends. Can J For Res 21:1091–1097

Beyers JL, Riechers GH, Temple PJ (1992) Effects of long-term ozone exposure and drought on the photosynthetic capacity of ponderosa pine (*Pinus ponderosa* Laws.). New Phytol 122:81–90

Bielenberg DG, Lynch JP, Pell EJ (2001) A decline in nitrogen availability affects plant response to ozone. New Phytol 151:413–425

Black VJ, Black CR, Roberts JA, Stewart CA (2000) Impact of ozone on the reproductive development of plants. New Phytol 147:421–447

Bode W (ed) (1997) Naturnahe Waldwirtschaft. Prozeßschutz oder biologische Nachhaltigkeit? Deukalion, Holm, pp 396

Bond JB (2000) Age-related changes in photosynthesis of woody plants. Trends Plant Sci Rev 5:349–353

Bonello P, Heller W, Sandermann H (1993) Ozone effects on root-disease susceptibility and defence responses in mycorrhizal and non-mycorrhizal seedlings of Scots pine (*Pinus sylvestris* L.). New Phytol 124:653–663

Bortier K, Ceulemans R, Temmermann de L (2000) Effects of ozone exposure on growth and photosynthesis of beech seedlings (*Fagus sylvatica*). New Phytol 146:271–280

Bortier K, Dekelver G, Temmermann de L, Ceulemans R (2001) Stem injection of *Populus nigra* with EDU to study ozone effects under field conditions. Environ Pollut 111:199–208

Brantley EA, Anderson RL, Smith G (1994) How to identify ozone injury on eastern forest bioindicator plants. Protection Report R8-PR 25, US Department of Agriculture Forest Service, Southern Region and Northeastern Area, Asheville, NC, and Durham, NH, p 2

Broadmeadow MSJ, Jackson SB (2000) Growth responses of *Quercus petraea*, *Fraxinus excelsior* and *Pinus sylvestris* to elevated carbon dioxide, ozone and water supply. New Phytol 146:437–451

Broadmeadow MSJ, Heath J, Randle TJ (1999) Environmental limitations to O_3 uptake – some key results from young trees growing at elevated CO_2 concentrations. Water Air Soil Pollut 116:299–310

Cape JN, Fowler D, Eamus D, Murray MB, Sheppard LJ, Leith ID (1990) Effects of acid mist and ozone on frost hardiness of Norway spruce seedlings. In: Payer HD, Pfirrmann T, Mathy P (eds) Environmental research with plants in closed chambers. Air pollution research report 26. CEC DG XII, Brussels, pp 331–335

Ceulemans R, Janssens IA, Jach ME (1999) Effects of CO_2 enrichment on trees and forests: lessons to be learned in view of future ecosystem studies. Ann Bot 84:577–590

Chameideis WL, Fehsenfeld F, Rodgers MO, Cardelino C, Martinez J, Parrish D, Lonneman W, Lawson DR, Rasmussen RA, Zimmerman P, Greenber GJ, Middleton P, Wang T (1992) Ozone precursor relationships in the ambient atmosphere. J Geophys Res 97:6037–6055

Chappelka AH, Chevone BI (1992) Tree response to ozone. In: Lefohn AS (ed) Surface level ozone exposures and their effects on vegetation. Lewis, Chelsea, MI, pp 271–324

Chappelka AH, Samuelson LJ (1998) Ambient ozone effects on forest trees of the eastern United States: a review. New Phytol 139:91–108

Chappelka AH, Kush JS, Meldahl RS, Lockaby BG (1990) An ozone-low temperature interaction in loblolly pine (*Pinus taeda L.*). New Phytol 114:721–726

Chappelka AH, Renfro J, Somers G, Nash B (1997) Evaluation of ozone injury on foliage of black cherry (*Prunus serotina*) and tall milkweed (*Asclepias exaltata*) in Great Smoky Mountains National Park. Environ Pollut 95:13–18

Chappelka A, Somers G, Renfro J (1999a) Visible ozone injury on forest trees in Great Smokey Mountains National Park, USA. Water Air Soil Pollut 116:255–260

Chappelka A, Skelly J, Somers G, Renfro J, Hildebrand E (1999b) Mature black cherry used as a bioindicator of ozone injury. Water Air Soil Pollut 116:261–266

Chiron H, Drouet A, Claudot A-C, Eckerskorn C, Trost M, Heller W, Ernst D, Sandermann H Jr (2000a) Molecular cloning and functional expression of a stress-induced multifunctional *O*-methyltransferase with pinosylvin methyltransferase activity from Scots pine (*Pinus sylvestris* L.). Plant Mol Biol 44:733–745

Chiron H, Drouet A, Lieutier F, Payer H-D, Ernst D, Sandermann H Jr (2000b) Gene induction of stilbene biosynthesis in Scots pine in response to ozone treatment, wounding, and fungal infection. Plant Physiol 124:865–872

Clayton H, Knight MR, Knight H, McAinsh MR, Hetherington AM (1999) Dissection of the ozone-induced calcium signature. Plant J 17:575–579

Coe H, Gallagher MW, Choularton TW, Dore C (1995) Canopy scale measurements of stomatal and cuticular O_3 uptake by Sitka spruce. Atmos Environ 29:1413–1423

Coleman MD, Dickson RE, Isebrands JG, Karnosky DF (1995) Carbon allocation and partitioning in aspen clones varying in sensitivity to tropospheric ozone. Tree Physiol 15:721–726

Constable JVH, Taylor GE Jr (1997) Modeling the effects of elevated tropospheric O_3 on two varieties of *Pinus ponderosa*. Can J For Res 27:527–537

Dahlsten DL, Rowney DL, Kickert RN (1997) Effects of oxidant air pollutants on western pine beetle (*Coleoptera, Scolytidae*) populations in southern California. Environ Pollut 96:415–423

Damesin C, Rambal S (1995) Field study of leaf photosynthetic performance by a Mediterranean deciduous oak tree (*Quercus pubescens*) during a severe summer drought. New Phytol 131:159–167

Davies TD, Schuepbach E (1994) Episodes of high ozone concentrations at the Earth's surface resulting from transport down from the upper troposphere/lower stratosphere: a review and case studies. Atmos Environ 28:53–68

Davis DD, Skelly JM (1992) Foliar sensitivity of eight eastern hardwood tree species to ozone. Water Air Soil Pollut 62:269–277

Davison AW, Barnes JD (1998) Effects of ozone on wild plants. New Phytol 139:135–151

Dawson TE, Ehleringer JR (1993) Gender-specific physiology, carbon isotope discrimination, and habitat distribution in boxelder, *Acer negundo*. Ecology 74:798–815

Dawson TE (1996) Determining water use by trees and forests from isotopic, energy balance and transpiration analyses: the role of tree size and hydraulic lift. Tree Physiol 16:263–272

Dickson RE, Isebrands JG (1991) Leaves as regulators of stress response. In: Mooney HA, Winner WE, Pell EJ (eds) Response of plants to multiple stresses. Academic Press, San Diego, pp 4–34

Dizengremel P (2001) Effects of ozone on the carbon metabolism of forest trees. Plant Physiol Biochem 39:729–742

Drogoudi PD, Ashmore MR (2000) Does elevated ozone have differing effects in flowering and deblossomed strawberry? New Phytol 147:561–569

Duckmanton L, Widden P (1994) Effect of ozone on the development of vesicular-arbuscular mycorrhizae in sugar maple saplings. Mycologia 86:181–186

Edwards GS, Pier PA, Kelly JM (1990) Influence of ozone and soil magnesium status on the cold hardiness of loblolly pine (*Pinus taeda* L.) seedlings. New Phytol 115:157–164

Edwards NT (1991) Root and soil respiration reponses to ozone in *Pinus taeda* L. seedlings. New Phytol 118:315–322

Einig W, Lauxmann U, Hauch B, Hampp R, Landolt W, Maurer S, Matyssek R (1997) Ozone-induced accumulation of carbohydrates changes enzyme activities of carbohydrate metabolism in birch leaves. New Phytol 137:673–680

Ellsworth DS, Reich PB (1993) Canopy structure and vertical patterns of photosynthesis and related leaf traits in a deciduous forest. Oecologia 96:169–178

Emberson LD, Wieser G, Ashmore MR (2000a) Modelling of stomatal conductance and ozone flux of Norway spruce: comparison with field data. Environ Pollut 109:393–402

Emberson LD, Ashmore MR, Cambridge HM, Simpson D, Tuovinen J-P (2000b) Modelling stomatal ozone flux across Europe. Environ Pollut 109:403–413

Emberson LD, Ashmore MR, Murray F, Kuylenstierna JCI, Percy KE, Izuta T, Zheng Y, Shimizu H, Sheu BH, Liu CP, Agrawal M, Wahid A, Abdel-Latif NM, Tienhoven van M, Bauer de LI, Domingos M (2001) Impacts of air pollutants on vegetation in developing countries. Water Air Soil Pollut 130:107–118

Enders G, Teichmann U, Kramm G (1989) Profiles of ozone and surface layer parameters over a mature spruce stand. In: Georgii HW (ed) Mechanisms and effects of pollutant-transfer into forests. Kluwer, Dordrecht, pp 21–35

Ericsson T, Rytter L, Vapaavuori E (1996) Physiology and allocation in trees. Biomass Bioenergy 11:115–127

Farquhar GD, Ehleringer JR, Hubick KT (1989) Carbon isotope discrimination and photosynthesis. Annu Rev Plant Physiol Plant Mol Biol 40:503–537

Fink S (1989) Pathological anatomy of conifer needles subjected to gaseous air pollutants or mineral deficiencies. Aquilo Ser Bot 27:1–6

Flagler RB (1998) Recognition of air pollution injury to vegetation: a pictorial atlas. AandWMA, Pittsburgh

Fowler D, Cape JN, Coyle M, Flechard C, Kuylenstienra J, Hicks K, Derwent D, Johnson C, Stevenson D (1999) The global exposure of forests to air pollutants. Water Air Soil Pollut 116:5–32

Fowler D, Flechard C, Cape JN, Storton-West RL, Coyle M (2001) Measurements of ozone deposition to vegetation quantifying the flux, the stomatal and non-stomatal components. Water Air Soil Pollut 130:63–74

Foyer CH, Lelandais M, Edwards EA, Mullineaux PM (1991) The role of ascorbate in plants, interactions with photosynthesis, and regulatory significance. In: Pell E, Steffen K (eds) Active oxygen/oxidative stress and plant metabolism. American Society of Plant Physiologists, Rockville, pp 131–144

Fredericksen TS, Joyce BJ, Skelly JM, Steiner KC, Kolb TE, Kouterick KB, Savage JE, Snyder KR (1995) Physiology, morphology, and ozone uptake of leaves of black cherry seedlings, saplings, and canopy trees. Environ Pollut 89:273–283

Fredericksen TS, Skelly JM, Steiner KC, Kolb TE, Kouterick KB (1996a) Size-mediated foliar response to ozone in black cherry trees. Environ Pollut 91:53–63

Fredericksen TS, Skelly JM, Snyder KR, Steiner KC, Kolb TE (1996b) Predicting ozone uptake from meteorological and environmental variables. J Air Waste Manage Assoc 46:464–469

Frey B, Scheidegger C, Günthardt-Goerg MS, Matyssek R (1996) The effects of ozone and nutrient supply on stomatal response in birch (*Betula pendula*) leaves as determined by digital image-analysis and X-ray microanalysis. New Phytol 132:135–143

Friend AL, Tomlinson PT (1992) Mild ozone exposure alters 14C dynamics in foliage of *Pinus taeda* L. Tree Physiol 11:214–227

Fuhrer J, Achermann B (1994) Critical levels for ozone. UN-ECE Workshop Report, Liebefeld-Bern

Fuhrer J, Achermann B (1999) Critical levels for ozone – level II. Environmental documentation 115. Swiss Agency for the Environment, Forests and Landscape, Berne, Switzerland

Fuhrer J, Skärby L, Ashmore MR (1997) Critical levels for ozone effects on vegetation in Europe. Environ Pollut 97:91–106

Gehring CA, Cobb NS, Whitman TG (1997) Three way interactions among ectomycorrhizal mutualists, scale insects, and resistant and susceptible pinyon pines. Am Nat 149:824–841

Gerant D, Podor M, Grieu P, Afif D, Cornu S, Morabito D, Banvoy J, Robin C, Dizengremel P (1996) Carbon metabolism, enzyme activities and carbon partitioning in *Pinus halepensis* Mill. to mild drought and ozone. J Plant Physiol 148:142

Gielen B, Calfapietra C, Ceulemans R (2001) Effects of elevated CO_2 on crown structure, leaf area and growth of poplar genotypes in the POPFACE experiment. Proceedings of the 19th International Meeting for Specialists in Air Pollution Effects on Forest Ecosystems, Houghton, USA

Gimeno BS, Velissariou D, Barnes JD, Inclan R, Peña JM, Davison AW (1992) Daños visibles por ozono en acículas de *Pinus halepensis* Mill. en Grecia y España. Ecología 6:131–134

Götz B (1996) Ozon und Trockenstreß, Wirkungen auf den Gaswechsel von Fichte. Libri Botanici Bd. 16. IHW-Verlag, Eching, pp 149

Gorissen I, Joosten SS, Smeulders SM, Vanveen JA (1994) Effects of short-term ozone exposure and soil water availability on the carbon partitioning of juvenile Douglas fir. Tree Physiol 14:647–657

Grams TEE, Matyssek R (1999) Elevated CO_2 counteracts the limitation by chronic ozone exposure on photosynthesis in *Fagus sylvatica* L.: comparison between chorophyll fluorescence and leaf gas exchange. Phyton 39:31–40

Grams TEE, Anegg S, Häberle K-H, Langebartels C, Matyssek R (1999) Interactions of chronic exposure to elevated CO_2 and O_3 levels in the photosynthetic light and dark reactions of European beech (*Fagus sylvatica*). New Phytol 144:95–107

Grams TEE, Kozovits AR, Reiter IM, Winkler JB, Sommerkorn M, Blaschke H, Häberle K-H, Matyssek R (2002) Quantifying competitiveness in woody plants. Plant Biol 4:153–158

Gravano E, Ferretti M, Bussotti F, Grossoni P (1999) Foliar symptoms and growth reduction of *Ailanthus altissima* Dest. in an area with high ozone and acidic deposition in Italy. Water Air Soil Pollut 116:267–272

Greitner CS, Winner WE (1989) Nutrient effects on responses of willow and alder to ozone. In: Olson RK, Lefohn AS (eds) Transaction: effects of air pollution on western forests. Air and Waste Management Association, Anaheim, CA, pp 493–511

Greitner CS, Pell EJ, Winner WE (1994) Analysis of aspen foliage exposed to multiple stresses: ozone, nitrogen deficiency and drought. New Phytol 127:579–589

Grennfelt P, Beck JP (1994) Ozone concentrations in Europe in relation to different concepts of the critical levels. In: Fuhrer J, Achermann B (eds) Critical levels for ozone: a UN-ECE workshop report. Schriftenreihe der FAC Liebefeld, 184–194. Swiss Federal Research Station for Agricultural Chemistry, Liebefeld-Bern

Grünhage L, Krause GHM, Köllner B, Bender J, Weigel H-J, Jäger H-J, Guderian R (2001) A new flux-orientated concept to derive critical levels for ozone to protect vegetation. Environ Pollut 111:355–362

Grulke NE, Balduman L (1999) Deciduous conifers: high N deposition and O_3 exposure effects on growth and biomass allocation in Ponderosa pine. Water Air Soil Pollut 116:235–248

Grulke NE, Miller PR (1994) Changes in gas exchange characteristics during the life span of giant sequoia - implications for response to current and future concentrations of atmospheric ozone. Tree Physiol 14:659–668

Grulke NE, Retzlaff WA (2001) Changes in physiological attributes of ponderosa pine from seedlings to mature trees. Tree Physiol 21:275–286

Guderian R (1985) Air pollution by photochemical oxidants, formation, transport, control, and effects on plants. Ecological studies 52. Springer, Berlin Heidelberg New York, pp 346

Günthardt-Goerg MS, Matyssek R, Scheidegger C, Keller T (1993) Differentiation and structural decline in the leaves and bark of birch (*Betula pendula*) under low ozone concentration. Trees 7:104–114

Günthardt-Goerg MS, McQuattie CJ, Scheidegger C, Rhiner C, Matyssek R (1997) Ozone-induced cytochemical and ultrastructural changes in leaf mesophyll cell walls. Can J For Res 27:453–463

Günthardt-Goerg MS, Maurer S, Frey B, Matyssek R (1998) Birch leaves from trees grown in two fertilization regimes: diurnal and seasonal responses to ozone. In: De Kok LJ, Stulen I (eds) Responses of plant metabolism to air pollution and global change. Backhuys Publishers, Leiden, Netherlands, pp 315–318

Günthardt-Goerg MS, Maurer S, Bolliger J, Clark AJ, Landolt W, Bucher JB (1999) Responses of young trees (five species in a chamber exposure) to near-ambient ozone concentrations. Water Air Soil Pollut 116:323–332

Günthardt-Goerg MS, McQuattie CJ, Maurer S, Frey B (2000) Visible and microscopic injury in leaves of five deciduous tree species related to current critical ozone levels. Environ Pollut 109:489–500

Haagen-Smit AJ, Darley EF, Zaitlin M, Hull H, Nobel WM (1952) Investigation of injury to plants from air pollution in the Los Angeles area. Plant Physiol 27:18–34

Häberle K-H, Werner H, Fabian P, Pretzsch H, Reiter I, Matyssek R (1999) "Free-air" ozone fumigation of mature forest trees: a concept for validating AOT40 under stand conditions. In: Fuhrer J, Achermann B (eds) Critical level for ozone - level II. Swiss Agency for the Environment, Forests and Landscape (SAEFL), Berne, pp 133–137

Hättenschwieler S, Schweingruber FH, Körner C (1996) Tree ring responses to elevated CO_2 and increased N deposition in *Picea abies*. Plant Cell Environ 19:1369–1378

Hagemeier J, Schneider B, Oldham NJ, Hahlbrock K (2001) Accumulation of soluble and wall-bound indolic metabolites in *Arabidopsis thaliana* leaves infected with virulent or avirulent *Pseudomonas syringae* pathovar tomato strains. Proc Natl Acad Sci USA 98:753–758

Hanisch B, Kilz E (1990) Monitoring of forest damage. Ulmer, Stuttgart, 334 pp

Hanson PJ, Samuelson LJ, Wullschleger SD, Tabberer TA, Edwards GS (1994) Seasonal patterns of light-saturated photosynthesis and leaf conductance for mature and seedling *Quercus rubra* L. foliage: differential sensitivity to ozone exposure. Tree Physiol 14:1351–1366

Hartmann G, Nienhaus F, Butin H (1995) Farbatlas Waldschäden, Diagnose von Baumkrankheiten. Ulmer, Stuttgart, pp 288

Havranek WM, Wieser G (1994) Design and testing of twig chambers for ozone fumigation and gas exchange measurements in mature trees. Proc R Soc Edinb Sect B 102:541–546

Heath RL (1980) Initial events in injury to plants by air pollutants. Annu Rev Plant Physiol 31:395–401

Heath RL, Taylor GE (1997) Physiological processes and plant responses to ozone exposure. In: Sandermann H, Wellburn AR, Heath RL (eds) Forest decline and ozone, a comparison of controlled chamber and field experiments. Ecological studies 127. Springer, Berlin Heidelberg New York, pp 317–368

Heggestad HE, Middleton JT (1959) Ozone in high concentrations as cause of tobacco leaf injury. Science 129:208–210

Heiden AC, Hoffmann T, Kahl J, Kley D, Klockow D, Langebartels C, Mehlhorn H, Sandermann H, Schraudner M, Schuh G, Wildt J (1999) Emission of volatile organic compounds from ozone-exposed plants. Ecol Appl 9:1160–1167

Hendrey GR, Lewin KF, Kolb ER Z, Evans LS (1992) Controlled enrichment system for experimental fumigation of plants in the field with sulfur dioxide. J Air Waste Manage Assoc 42:1324–1327

Hendrey GR, Ellsworth DS, Lewin KF, Nagy J (1999) A free-air enrichment system for exposing tall forest vegetation to elevated atmospheric CO_2. Global Change Biol 5:293–309

Herms DA, Mattson WJ (1992) The dilemma of plants: to grow or defend. Q Rev Biol 67:283–335

Houpis JL, Costella MP, Cowles S (1991) A branch exposure chamber for fumigating ponderosa pine to atmospheric pollution. J Environ Qual 20:467–474

Houpis JLJ, Anderson PD, Pushnik JC, Anschel DJ (1999) Among-provenance variability of gas exchange and growth in response to long-term CO_2 exposure. Water Air Soil Pollut 116:403–412

Hubbard RM, Bond BJ, Ryan MG (1999) Evidence that hydraulic conductance limits photosynthesis in old *Pinus ponderosa* trees. Tree Physiol 19:165–172

Inclan R, Ribas A, Penuelas J, Gimeno BS (1999) The relative sensitivity of different Mediterranean plant species to ozone exposure. Water Air Soil Pollut 116:273–277

Innes JL, Ghosh S, Dobbertin M, Rebetez M, Zimmermann S (1997) Kritische Belastungen und die Sanasilva-Inventur. Forum für Wissen 1997. Eidg. Forschungsanstalt Wald, Schnee Landschaft, Birmensdorf, pp 73–83

Innes JL, Skelly JM, Schaub M (2001) Ozone and broadleaved species - a guide to the identification of ozone-induced foliar injury. Flück-Wirth, Teufen, Switzerland, pp 136

Kärenlampi L (1999) Small decrease in assimilation rate can result in considerable loss of yield - tentative long-term model calculations on the impact of ozone on Silver Birch plantation. In: Fuhrer J, Achermann B (eds) Critical level for ozone - level II. Environmental documentation No. 115. SAEFL, Berne, Switzerland, pp 150–156

Kärenlampi L, Skärby L (1996) Critical levels for ozone in Europe: testing and finalizing the concepts. University of Kuopio, Kuopio

Karlsson PE, Medin E-L, Wickström H, Selldén G, Wallin G, Ottoson S, Skärby L (1995) Ozone and drought stress: interactive effects on the growth and physiology of Norway spruce (*Picea abies* L. Karst.). Water Air Soil Pollut 85:1325–1330

Karlsson PE, Pleijel H, Karlsson GP, Medin EL, Skärby L (2000) Simulations of stomatal conductance and ozone uptake to Norway spruce saplings in open-top chambers. Environ Pollut 109:443–451

Karnosky DF (1981) Changes in eastern white pine stands related to air pollution stress. Mitt Forstl Bundes-Versuchsanst Wien 137:41–45

Karnosky DF, Gagnon ZE, Dickson RE, Coleman MD, Lee EH, Isebrands JG (1996) Changes in growth, leaf abscision, and biomass associated with seasonal tropospheric ozone exposure of Populus tremuloides clones and seedlings. Can J For Res 26:23–37

Karnosky DF, Mankovska B, Percy K, Dickson DE, Podila GK, Sober A, Noormets G, Hendrey MD, Coleman M, Kubiske M, Pregitzer KS, Isebrands JG (1999) Effects of tropospheric O_3 on trembling aspen and interaction with CO_2: results from a O_3 gradient and a FACE experiment. Water Air Soil Pollut 116:311–322

Karnosky DE, Oksanen E, Dickson RE, Isebrands JG (2001a) Impacts of interacting greenhouse gases on forest ecosystems. In: Karnosky DF, Scarascia-Mugnozza G, Ceulemans R, Innes JL (eds) The impacts of carbon dioxide and other greenhouse gases on forest ecosystems. CABI Press, Wallingford, UK, pp 253–267

Karnosky DF, Gielen B, Ceulemans R, Schlesinger WH, Norby RJ, Oksanen E, Matyssek R, Hendrey GR (2001b) FACE systems for studying the impacts of greenhouse gases on forest ecosystems. In: Karnosky DF, Scarascia-Mugnozza G, Ceulemans R, Innes JL (eds) The impacts of carbon dioxide and other greenhouse gases on forest ecosystems. CABI Press, Wallingford, UK, pp 297–324

Karpinski S, Karpinska B, Meltzer M, Hällgren J-E, Wingsle G (2001) Signalling and antioxidant defence mechanisms in higher plants. In: Huttunen S, Heikkilä H, Bucher J, Sundberg B, Jarvis P, Matyssek R (eds) Trends in European forest tree physiology research. Kluwer, Dordrecht, pp 93–114

Keller T, Häsler R (1987) The influence of a fall fumigation with ozone on the stomatal behavior of spruce and fir. Oecologia 64:284–286

Kellomäki S, Wang K-Y (1998a) Daily and seasonal CO_2 changes in Scots pine grown under elevated O_3 and CO_2: experiment and stimulation. Plant Ecol 136:229–248

Kellomäki S, Wang K-Y (1998b) Growth, respiration and nitrogen content in needle of Scots pine exposed to elevated ozone and carbon dioxide in the field. Environ Pollut 101:263–274

Kelly JM, Taylor GE Jr, Edwards NT, Adams MB, Edwards GS, Friend AL (1993) Growth, physiology, and nutrition of loblolly pine seedlings stressed by ozone and acid precipitation: a summary of the ROPIS-south project. Water Air Soil Pollut 69:363–391

Kelly JM, Samuelson L, Edwards G, Hanson P, Kelting D, Mays A, Wullschleger S (1995) Are seedlings reasonable surrogates for trees? An analysis of ozone impacts on Quercus rubra. Water Air Soil Poll 85:1317–1324

Kolb TE (2001) Ageing as an influence on tree response to ozone: theory and observations. In: Huttunen S, Häkkela H, Bucher JB, Sundberg B, Jarvis P, Matyssek R (eds) Trends in European forest tree physiology research. Kluwer, Dordrecht, pp 127–155

Kolb TE, Matyssek R (2001) Limitations and perspectives about scaling ozone impact in trees. Limitations and perspectives. Environ Pollut 115:373–393

Kolb TE, Stone JE (2000) Differences in leaf gas exchange and water relations among species and tree sizes in an Arizona pine-oak forest. Tree Physiol 20:1–12

Kolb TE, Fredericksen TS, Steiner KC, Skelly JM (1997) Issues in scaling tree size and age responses to ozone: a review. Environ Pollut 98:195–208

Kozlowski TT, Pallardy SG (1997) Physiology of woody plants. Academic Press, New York, 411 pp

Küppers M (1994) Canopy gaps: competitive light interception and economic space filling – a matter of whole-plant allocation. In: Caldwell MM, Pearcy RW (eds) Exploitation of environmental heterogeneity by plants – ecophysiological processes above- and below-ground. Academic Press, San Diego, pp 111–144

Kull O, Sober A, Coleman MD, Dickson RE, Isebrands JG, Gagnon Z, Karnosky DF (1996) Photosynthetic responses of aspen clones to simultaneous exposures of ozone and CO_2. Can J For Res 26:639–648

Kytöviita M-M, Pell Oux J, Fontaine V, Botton B, Dizengremel P (1999) Elevated CO_2 does not amerliorate effects of ozone on carbon allocation in *Pinus halepensis* and *Betula pendula* in symbiosis with *Paxillus involutus*. Physiol Planta 106:370–377

Laisk A, Kull O, Moldau H (1989) Ozone concentration in leaf intercellular air spaces is close to zero. Plant Physiol 90:1163–1167

Landolt W, Lüthy-Krause B (1991) Wirkungen umweltrelevanter Ozon-Konzentrationen auf verschiedene Pflanzen. In: Stark L (ed) Luftschadstoffe und Wald – Lufthaushalt, Luftverschmutzung und Waldschäden in der Schweiz, vol 5. Verlag der Fachvereine Zürich, Zürich, pp 127–134

Landolt W, Pfenninger I, Lüthy-Krause B (1989) The effect of ozone and season on the pool sizes of cyclitols in Scots pine (*Pinus sylvestris*). Trees 3:85–88

Landolt W, Günthardt-Goerg MS, Pfenninger I, Einig W, Hampp R, Maurer S, Matyssek R (1997) Effect of fertilization on ozone-induced changes in the metabolism of birch leaves (*Betula pendula*). New Phytol 137:389–397

Lange BM, Trost M, Heller W, Langebartels C, Sandermann H Jr (1994) Elicitor-induced formation of free and cell-wall-bound stilbenes in cell-suspension cultures of Scots pine (*Pinus sylvestris* L.) Planta 194:143–148

Langebartels C, Heller W, Kerner K, Leonardi S, Rosemann D, Schraudner M, Trost M, Sandermann H (1990) Ozone-induced defense reactions in plants. In: Payer HD, Pfirrmann T, Mathy P (eds) Environmental research with plants in closed chambers. Air pollution research report 26. CEC DG XII, Brussels, pp 358–368

Langebartels C, Ernst D, Heller W, Lütz C, Payer H-D, Sandermann H Jr (1997) Ozone responses of trees: results from controlled chamber exposures at the GSF phytotron. In: Sandermann H Jr, Wellburn AR, Heath RL (eds) Forest decline and ozone. Springer, Berlin Heidelberg New York, pp 163–200

Langebartels C, Heller W, Führer G, Lippert M, Simons S, Sandermann H Jr (1998) Memory effects in the action of ozone on conifers. Ecotox Environ Safety 41:62–72

Langebartels C, Schraudner M, Heller W, Ernst D, Sandermann H (2001) Oxidative stress and defense reactions in plants exposed to air pollutants and UV-B radiation. In: Inzé D, Van Montagu M (eds) Oxidative stress in plants. Harwood Acad Publ, Amsterdam, pp 105–135

Laurence JA, Amundson RG, Friend AL, Pell EJ, Temple PJ (1994) Allocation of carbon in plants under stress: an analysis of the ROPIS experiment. J Environ Qual 23:412–417

Laurence JA, Retzlaff WA, Kern JS, Lee EH, Hogsett WE, Weinstein DA (2001) Predicting the regional impact of ozone and precipitation on the growth of loblolly pine and yellow-poplar using linked TREGRO and ZELIG models. For Ecol Manage 146:247–263

Laurila T, Tuovinen J-P (1996) Monitored data in relation to exceedances of AOT40. In: Fuhrer J, Achermann B (eds) Critical levels for ozone: a UN-ECE workshop report. Schriftenreihe der FAC Liebefeld, 115–124. Swiss Federal Research Station for Agricultural Chemistry, Liebefeld-Bern

Lefohn AS (1992) Surface level ozone exposure and their effects on vegetation. Lewis Publishers, Chelsea, USA, pp 366

Lefohn AS, Jones CK (1986) The characterization of ozone and sulfur dioxide air quality data for assessing possible vegetation effects. J Air Pollut Control Assoc 36:1123–1129

Leipner J, Oxborough K, Baker NR (2001) Primary sites of ozone-induced perturbations of photosynthesis in leaves: identification and characterization in *Phaseolus vulgaris* using high resolution chlorophyll fluorescence imaging. J Exp Bot 52:1689–1696

Lerdau M, Gershenzon J (1997) Allocation theory and chemical defense. In: Bazzaz FA, Grace J (eds) Plant resource allocation. Academic Press, San Diego, pp 265–277

Lewin KF, Hendrey GR, Nagy J, McMorte RL (1994) Design and application of a free-air carbon dioxide enrichment facility. Agric For Meteorol 70:15–29

Lippert M, Steiner K, Payer H-D, Simons S, Langebartels C, Sandermann H Jr (1996) Assessing the impact of ozone on photosynthesis of European beech (*Fagus sylvatica* L.) in environmental chambers. Trees 10:268–275

Lock J, Price AH (1994) Evidence that disruption of cytosolic calcium is critically important in oxidative plant stress. Proc R Soc Edinb Sect B Biol Sci 102:261–264

Lu C, Zhang J (1998) Changes in PSII function during senescence of wheat leaves. Physiol Plant 104:239–247

Lucas PW, Wolfenden J (1996) The role of plant hormones as modifiers of sensitivity to air pollutants. Phyton 36:51–56

Lüthy-Krause B, Pfenninger I, Landolt W (1990) Effects of ozone on organic acids in needles of Norway spruce and Scots pine. Trees 4:198–204

Lütz C, Anegg S, Gerant D, Alaoui-Sosse B, Gerard J, Dizengremel P (2000) Beech trees exposed to high CO_2 and to simulated summer ozone levels: effects on photosynthesis, chloroplast components and leaf enzyme activity. Physiol Planta 109:252–259

Mahoney MJ, Chevone BI, Skelly JM, Moore LD (1985) Influence of mycorrhizae on the growth of loblolly pine *Pinus taeda* seedlings exposed to ozone and sulphur dioxide. Phytopathology 75:679–682

Maier-Maercker U (1998) Image analysis of the stomatal cell walls of *Picea abies* (L.) Karst. in pure and ozone-enriched air. Trees 12:181–185

Manes F, Vitale M, Donato E, Paoletti E (1998) O_3 and O_3+CO_2 effects on a Mediterranean evergreen broadleaf tree, Holm Oak (*Quercus ilex* L.). Chemosphere 36:801–806

Manninen S, Le Thiec D, Rose C, Nourrisson G, Radnai F, Garrec JP, Huttunen S (1999) Pigment concentrations and ratios of Aleppo pine seedlings exposed to ozone. Water Air Soil Pollut 116:333–338

Manning WJ, von Tiedemann A (1995) Climate change: potential effects of increased atmospheric carbon dioxide (CO_2), ozone (O_3), and ultraviolet-B (UV-B) radiation on plant diseases. Environ Pollut 88:219–225

Matyssek R (1986) Carbon, water and nitrogen relations in evergreen and deciduous conifers. Tree Physiol 2:177–187

Matyssek R (1998) Ozon – ein Risikofaktor für Bäume und Wälder? BiologieUnserer Zeit 28:348–361

Matyssek R (2001a) How sensitive is birch to ozone? Responses in structure and function. J For Sci 47:8–20

Matyssek R (2001b) Trends in forest tree physiological research: biotic and abiotic interactions. In: Huttunen S, Heikkilä H, Bucher J-B, Sundberg B, Jarvis PG, Matyssek R (eds) Trends in European forest tree physiological research. Kluwer, Dordrecht, pp 241–246

Matyssek R, Innes JL (1999) Ozone – a risk factor for trees and forests in Europe? Water Air Soil Pollut 116:199–226

Matyssek R, Schulze E-D (1988) Carbon uptake and respiration in above-ground parts of a *Larix decidua* x *leptolepis* tree. Trees 2:233–241

Matyssek R, Maruyama S, Boyer JS (1988) Rapid wall relaxation in elongating tissues. Plant Physiol 86:1163–1167

Matyssek R, Günthardt-Goerg MS, Keller T, Schneidegger C (1991) Impairment of gas exchange and structure in birch leaves (*Betula pendula*) caused by low ozone concentrations. Trees 5:5–13

Matyssek R, Günthardt-Goerg MS, Saurer M, Keller T (1992) Seasonal growth, $\delta^{13}C$ in leaves and stem, and phloem structure of birch (*Betula pendula*) under low ozone concentrations. Trees 6:69–76

Matyssek R, Günthardt-Goerg MS, Landolt W, Keller T (1993a) Whole-plant growth and leaf formation in ozonated hybrid poplar (*Populus* x *euramericana*). Environ Pollut 81:207–212

Matyssek R, Keller T, Koike T (1993b) Branch growth and leaf gas exchange of *Populus tremula* exposed to low ozone concentrations throughout two growing seasons. Environ Pollut 79:1–7

Matyssek R, Reich PB, Oren R, Winner WE (1995a) Response mechanisms of conifers to air pollutants. In: Smith WK, Hinckley TH (eds) Physiological ecology of coniferous forests. Physiological ecology series. Academic Press, New York, pp 255–308

Matyssek R, Günthardt-Goerg MS, Maurer S, Keller T (1995b) Nighttime exposure to ozone reduces whole-plant production in *Betula pendula*. Tree Physiol 15:159–165

Matyssek R, Havranek WM, Wieser G, Innes JL (1997a) Ozone and the forests in Austria and Switzerland. In: Sandermann H Jr, Wellburn AR, Heath RL (eds) Forest decline and ozone: a comparison of controlled chamber and field experiments. Ecological studies 127. Springer, Berlin Heidelberg New York, pp 95–134

Matyssek R, Maurer S, Günthardt-Goerg MS, Landolt W, Saurer M, Polle A (1997b) Nutrition determines the 'strategy' of *Betula pendula* for coping with ozone stress. Phyton 37:157–167

Matyssek R, Maurer S, Fabian P, Pretzsch H (1997c) 'Scaling' von Ozonwirkungen in Holzpflanzen bis zur Bestandsebene: Ausgangsbasis, Erfordernisse und Perspektiven. EcoSys Kiel 20:49–58

Matyssek R, Günthardt-Goerg MS, Schmutz P, Saurer M, Landolt W, Bucher JB (1998) Response mechanisms of birch and poplar to air pollutants. J Sustainable For 6:3–22

Maurer S (1995) Einfluß der Nährstoffversorgung auf die Ozon-Empfindlichkeit der Birke (*Betula pendula*). Doctoral Thesis, University of Basel, Basel, Switzerland

Maurer S, Matyssek R (1997) Nutrition and the ozone sensitivity of birch (*Betula pendula*). II. Carbon balance, water-use efficiency and nutritional status of the whole plant. Trees 12:11–20

Maurer S, Matyssek R, Günthardt-Goerg MS, Landolt W, Einig W (1997) Nutrition and the ozone sensitivity of birch (*Betula pendula*). I. Responses at the leaf level. Trees 12:1–10

McBride JR, Miller PR (1999) Implications of chronic air pollution in the San Bernadino Mountains for forest management and future research. In: Miller PR, McBride JM (eds) Oxidant air pollution impacts in the montane forest of southern California. Ecological studies 134. Springer, Berlin Heidelberg New York, pp 405–415

McLeod AR, Shaw PJA, Holland MR (1992) The Liphook forest fumigation project: studies of sulphur dioxide and ozone effects on coniferous trees. For Ecol Manage 51:121–127

Menser HA (1964) Response of plants to air pollutants. III. A relation between ascorbic acid levels and ozone susceptibility of light-preconditioned tobacco plants. Plant Physiol 39:564–567

Middleton JT, Kendrick JB Jr, Schwalm HW (1950) Injury to herbaceous plants by smog or air pollution. Plant Dis Rep 34:245–252

Mikkelsen TN, Ro-Poulsen H (1995) Exposure of Norway spruce to ozone increases the sensitivity of current-year needles to photoinhibitions and desiccation. New Phytol 128:153–163

Mikkelsen TN, Ro-Poulsen H, Hovmand MF, Hummelshoj P, Jensen NO (1996) Carbon and water balance for a mixed forest stand in relation to ozone uptake. In: Kärenlampi L, Skärby L (eds) Critical levels for ozone in Europe: testing and finalizing the concept. UN-ECE Workshop Report, Kuopio, Finland, pp 269–274

Mikkelsen TN, Ro-Poulsen H, Pilegaard K, Hovmand MF, Jensen NO, Christensen CS, Hummelshoej P (2000) Ozone uptake by an evergreen forest canopy: temporal variation and possible mechanisms. Environ Pollut 109:423–429

Millán MM, Salvador R, Mantilla E, Artíñano B (1996) Meteorology and photochemical air pollution in southern Europe: experimental results from EC research projects. Atmos Environ 30:1909–1924

Millán MM, Salvador R, Mantilla E (1997) Photooxidant dynamics in the Mediterranean basin in summer: results from European research projects. J Geophys Res 102:8811–8823

Miller JD, Arteca RN, Pell EJ (1999) Senescence-associated gene expression during ozone-induced leaf senescence in *Arabidopsis*. Plant Physiol 120:1015–1023

Miller PR, McBride JM (1999) Oxidant air pollution impacts in the montane forest of southern California. Ecological studies 134. Springer, Berlin Heidelberg New York, pp 317–336

Miller PR, Millecan AA (1971) Extent of air pollution damage to some pines and other conifers in California. Plant Dis Rep 55:555–559

Miller PR, Parmeter JR, Taylor OC, Cardiff EA (1963) Ozone injury to the foliage of *Pinus ponderosa*. Phytopathology 53:1072–1076

Miller PR, Grulke NE, Stolte KW (1994) Effects of air pollution on giant sequoia ecosystems. In: Aune PS (Technical Coordinator) Proceedings of the Symposium on Giant Sequoias: their place in the ecosystem and society. General Technical Report PSW-151. USDA Forest Service, Pacific Southwest Research Station, Albany, CA, pp 90–98

Miller PR, Arbaugh MJ, Temple PJ (1997) Ozone and its known effects on forests in western United States. In: Sandermann H Jr, Wellburn AR, Heath RL (eds) Forest decline and ozone. Springer, Berlin Heidelberg New York, pp 39–67

Momen B, Helms JA, Criddle RS (1996) Foliar metabolic heat rate of seedlings and mature trees of *Pinus ponderosa* exposed to simulated acid rain and elevated ozone. Plant Cell Environ 19:747–753

Mooney HA, Winner WE (1991) Partitioning response of plants to stress. In: Mooney HA, Winner WE, Pell EJ (eds) (1991) Response of plants to multiple stresses. Academic Press, San Diego, pp 129–141

Mooney HA, Winner WE, Pell EJ (eds) (1991) Response of plants to multiple stresses. Academic Press, San Diego, pp 422

Musselman RC, Hale BA (1997) Methods for controlled and field ozone exposures of forest tree species in North America. In: Sandermann H Jr, Wellburn AR, Heath RL (eds) Forest decline and ozone: a comparison of controlled chamber and field experiments. Ecological studies 127. Springer, Berlin Heidelberg New York, pp 277–315

Musselman RC, McCool PM, Lefohn AS (1994) Ozone descriptors for an air quality standard to protect vegetation. J Air Waste Manage Assoc 44:1383–1290

Neals TF, McLeod AL (1992) Do leaves contribute to the abscisic acid present in the xylem of 'droughted' sunflower plants? Plant Cell Environ 14:979–986

Nebel B, Fuhrer J (1995) Inter- and intraspecific differences in ozone sensitivity in semi-natural plant communities. Angew Bot 68:116–121

Neufeld HS, Renfro JR, Hacker WD, Silsbee D (1992) Ozone in Great Smoky Mountains National Park: dynamics and effects on plants. In: Berglund RL (ed) Transactions: tropospheric ozone and the environment II. Air and Waste Management Assoc, Pittsburgh, PA, USA, pp 594–617

Niewiadomska E, Gaucher-Veilleux C, Chevrier N, Maufette Y, Dizengremel P (1999) Elevated CO_2 does not provide protection against ozone considering the activity of several antioxidant enzymes in the leaves of sugar maple. J Plant Physiol 155:70–77

Norby RJ, Wullschleger SD, Gunderson CA, Johnson DW, Ceulemans R (1999) Tree response to rising CO_2 in experiments field: implications for the future forests. Plant Cell Environ 22:683–714

Nowak DJ, Civerolo KL, Trivikrama R, Sistla G, Luley CJ, Crane DE (2000) A modelling study of the impact of urban trees on ozone. Atmos Environ 34:1601–1613

Oksanen EJ (2001) Increasing tropospheric ozone level reduced birch (*Betula pendula*) dry mass within a five year period. Water Air Soil Pollut 130:947–952

Oksanen E, Saleem A (1999) Ozone exposure results in various carry-over effects and prolonged reduction in biomass in birch (*Betula pendula* Roth). Plant Cell Environ 22:1401–1411

Oksanen EJ, Wustman BA, Podilla GK, Isebrands JG, Karnosky DF (2000) CO_2/ozone interactions in trees. Proceedings of the 19th meeting for specialists in air pollution effects on forest ecosystems. Houghton, Michigan, p 61 (abstract)

Ollinger SV, Aber JD, Reich PB (1997) Simulating ozone effects on forest productivity: interactions among leaf, canopy and stand level processes. Ecol Applic 7:1237–1251

Overmyer K, Tuominen H, Kettunen R, Betz C, Langebartels C, Sandermann H Jr, Kangasjärvi J (2000) Ozone-sensitive *Arabidopsis rcd1* mutant reveals opposite roles for ethylene and jasmonate signaling pathways in regulating superoxide-dependent cell death. Plant Cell 12:1849–1862

Pääkkönen E, Holopainen T (1995) Influence of nitrogen supply on the response of clones of birch (Betula pendula Roth.) to ozone. New Phytol 129:595–603

Pääkkönen E, Paasisalo S, Holopainen T, Kärenlampi L (1993) Growth and stomatal responses of birch (*Betula pendula* Roth.) clones to ozone in open-air and chamber fumigations. New Phytol 125:615–623

Pääkkönen E, Holopainen T, Kärenlampi L (1996) Relationships between open-field ozone exposures and growth and senescence of birch (*Betula pendula* and *Betula pubescens*). In: Skärby L, Pleijel H (eds) Critical levels for ozone – experiments with crops, wild plants and forest tree species in the Nordic countries. TemaNord, vol 582. Nordic Council and Council of Ministers, Copenhagen, Denmark, pp 39–48

Pääkkönen E, Holopainen T, Kärenlampi L (1997) Variation of ozone sensitivity among clones of *Betula pendula* and *Betula pubescens*. Environ Pollut 95:37–44

Pääkkönen E, Vahala J, Pohjolai M, Holopainen T, Kärenlampi L (1998a) Physiological, stomatal and ultrastructural ozone responses in birch (*Betula pendula* Roth.) are modified by water stress. Plant Cell Environ 21:671–684

Pääkkönen E, Seppänen S, Holopainen T, Kokko H, Kärenlampi S, Kärenlampi L, Kangasjärvi J (1998b) Induction of genes for the stress proteins PR-10 and PAL in relation to growth, visible injuries and stomatal conductance in birch (*Betula pendula*) clones exposed to ozone and/or drought. New Phytol 138:295–305

Pääkkönen E, Holopainen T, Kärenlampi L (1999) Ozone impact remains in birch (*Betula pendula* Roth) one and two seasons after exposure. In: Fuhrer J, Achermann B (eds) Critical level for ozone – level II. Environmental documentation No. 115. SAEFL, Berne, Switzerland, pp 139–143

Panek JA, Goldstein AH (2001) Response of stomatal conductance to drought in ponderosa pine: implications for carbon and ozone uptake. Tree Physiol 21:337–344

Pearson M, Mansfield TA (1993) Interacting effects of ozone and water stress on the stomatal resistance of beech (*Fagus sylvatica* L.) New Phytol 123:351–358

Pearson M, Mansfield TA (1994) Effects of exposure to ozone and water stress on the following season's growth of beech (*Fagus sylvatica L.*). New Phytol 126:511–515

Pell EJ, Dann MS (1991) Multiple stress-induced foliar senescence and implications for whole-plant longevity. In: Mooney HA, Winner WE, Pell EJ (eds) Response of plants to multiple stresses. Academic Press, San Diego, pp 189–204

Pell EJ, Temple PJ, Friend AL, Mooney HA, Winner WE (1994) Compensation as a plant response to ozone and associated stresses: an analysis of ROPIS experiments. J Environ Qual 23:429–436

Pell EJ, Schlagnhaufer CD, Arteca RN (1997) Ozone-induced oxidative stress: mechanisms of action and reaction. Physiol Plant 100:264–273

Pell EJ, Sinn JP, Brendley BW, Samuelson L, Vinten-Johansen C, Tien M, Skillmann J (1999) Differential response of four tree species to ozone-induced acceleration of foliar senescence. Plant Cell Environ 22:779–790

Pellinen R, Palva T, Kangasjärvi J (1999) Subcellular localization of ozone-induced hydrogen peroxide production in birch (*Betula pendula*) leaf cells. Plant J 20:349–356

Pleijel H, Wallin G, Karlsson PE, Skärby L (1996) Ozone gradients in a spruce forest stand in relation to wind speed and time of the day. Atmos Environ 30:4077–4084

Polle A (1998) Photochemical oxidants: uptake and detoxification mechanisms. In: DeKok LJ, Stulen I (eds) Responses of plant metabolism to air pollution. Backhuys Publishers, Leiden, pp 95–116

Polle A, PfiirrmanNT, Chakrabarti S, Rennenberg H (1993) The effects of enhanced ozone and enhanced carbon dioxide concentrations on biomass, pigments, and antioxidative enzymes in spruce needles (*Picea abies* L.) . Plant Cell Environ 16:311–316

Polle A, Wieser G, Havranek WM (1995) Quantification of ozone influx and apoplastic ascorbate content in needles of Norway spruce trees (*Picea abies* L., Karst.) at high altitude. Plant Cell Environ 18:681

Polle A, Baumbach LO, Oschinski C, Eiblmeier M, Kuhlenkamp V, Vollrath B, Schol F, Rennenberg H (1999) Growth and protection against oxidative stress in young clones and mature trees (*Picea abies* L.) at high altitudes. Oecologia 121:149–156

Polle A, Matyssek R, Günthardt-Goerg MS, Maurer S (2000) Defense strategies against ozone in trees: the role of nutrition. In: Agrawal SB, Agrawal M (eds) Environmental pollution and plant responses. Lewis Publishers, New York, pp 223–245

Pronos J, Merril L, Dahsten D (1999) Insects and pathogens in a pollution-stressed forest. In: Miller PR, McBride JM (eds) Oxidant air pollution impacts in the montane forest of southern California. Ecological studies 134. Springer, Berlin Heidelberg New York, pp 317–336

Proyou AG, Toupance G, Perros PE (1991) A 2-year study of ozone behavior at rural and forested sites in eastern France. Atmos Environ 25A:2145–2153

Pye JM (1988) Impact of ozone on the growth and yield of trees: a review. J Environ Qual 17:347–360

Quirino BF, Noh Y-S, Himelblau E, Amasino RM (2000) Molecular aspects of leaf senescence. Trends Plant Sci 5:278–282

Radoglou K (1996) Environmental control of CO_2 assimilation rates and stomatal conductance in five oak species growing under field conditions in Greece. Ann Sci For 53:269–278

Rao MV, Davis KR (2001) The physiology of ozone induced cell death. Planta 213:682–690

Rebbeck J, Jensen KF (1993) Ozone effects on grafted mature and juvenile red spruce: photosynthesis, stomatal conductance, and chlorophyll concentration. Can J For Res 23:450–456

Reekie EG, Bazzaz FA (1989) Competition and patterns of resource use among seedlings of five tropical trees grown at ambient and elevated CO_2. Oecologia 79:212–222

Reich PB (1983) Effects of low concentrations of O_3 on net photosynthesis, dark respiration, and chlorophyll contents in aging hybrid poplar leaves. Plant Physiol 73:291–296

Reich PB (1987) Quantifying plant response to ozone: a unifying theory. Tree Physiol 3:63–91

Reich PB, Ellsworth DS, Kloeppel BD, Fownes JH, Gower ST (1990) Vertical variation in canopy structure and CO_2 exchange of oak-maple forests: influences of ozone, nitrogen, and other factors on simulated canopy carbon gain. Tree Physiol 7:329–345

Rennenberg H, Herschbach C, Polle A (1996) Consequences of air pollution on shoot-root interaction. J Plant Physiol 148:296–301

Retzlaff WA, Weinstein DA, Laurence JA, Gollands B (1997) Simulating the growth of a 160-year-old sugar maple (*Acer saccharum*) tree with and without ozone exposure using the TREGROW model. Can J For Res 27:783–789

Retzlaff WA, Arthur MA, Grulke NE, Weinstein DA, Gollands B (2000) Use of a single-tree simulation model to predict effects of ozone and drought on growth of a white fir tree. Tree Physiol 20:195–202

Reynolds JF, Hilbert DW, Kemp PR (1993) Scaling ecophysiology from the plant to the ecosystem: a conceptual framework. In: Ehleringer JR, Field CB (eds) Scaling physiological processes, leaf to globe. Academic Press, San Diego, pp 127–140

Riehl Koch J, Scherzer AJ, Eshita SM, Davis KR (1998) Ozone sensivity in hybrid poplar is correlated with a lack of defense-gene activation. Plant Physiol 118:1243–1252

Riehl Koch J, Creelman RA, Eshita SM, Seskar M, Mullet JE, Davis KR (2000) Ozone sensivity in hybrid poplar correlates with insensivity to both salicylic acid and jasmonic acid. The role of programmed cell death in lesion formation. Plant Physiol 123:487–496

Roloff A (2001) Baumkronen – Verständnis und praktische Bedeutung eines komplexen Naturphänomens. Ulmer, Stuttgart, pp 162

Rosemann D, Heller W, Sandermann H (1991) Biochemical plant responses to ozone. II. Induction of stilbene biosynthesis in scots pine (*Pinus sylvestris* L.). Plant Physiol 97:1280–1286

Ryan MG, Binkley D, Fownes JH (1997) Age-related decline in forest productivity: pattern and process. Adv Ecol Res 27:213–262

Ryan MG, Bond BJ, Law BE, Hubbard RM, Woodruff D, Cienciala E, Kucera J (2000) Transpiration and whole-tree conductance in ponderosa pines trees of different heights. Oecologia 124:553–560

Samuel MA, Miles GP, Ellis BE (2000) Ozone treatment rapidly activates MAP kinase signalling in plants. Plant J 22:367–376

Samuelson LJ (1994) The role of micro-climate in determining the sensitivity of *Quercus rubra* L. to ozone. New Phytol 128:235–241

Samuelson LJ, Kelly JM (1997) Ozone uptake in *Prunus serotina, Acer rubrum* and *Quercus rubra* forest trees of different sizes. New Phytol 136:255–264

Samuelson L, Kelly JM (2001) Scaling ozone from seedlings to forest trees. New Phytol 149:21–41

Samuelson LJ, Kelly JM, Mays PA, Edwards GS (1996) Growth and nutrition of *Quercus rubra* seedlings and mature trees after three seasons of ozone exposure. Environ Pollut 91:317–320

Sandermann H Jr (1996) Ozone and plant health. Annu Rev Phytopathol 34:347–366

Sandermann H Jr (2000a) Active oxygen species as mediators of plant immunity: three case studies. Biol Chem 381:649–653

Sandermann H Jr (2000b) Ozone/biotic disease interactions: molecular biomarkers as a new experimental tool. Environ Pollut 108:327–332

Sandermann H Jr, Langebartels C, Heller W (1990) Ozonstreß bei Pflanzen. Frühe und "Memory"-Effekte von Ozon bei Nadelbäumen. UWSF-Z Umweltchem Ökotox 2:14–15

Sandermann H Jr, Wellburn AR, Heath RL (1997) Forest decline and ozone: synopsis. In: Sandermann H Jr, Wellburn AR, Heath RL (eds) Forest decline and ozone: a comparison of controlled chamber and field experiments. Ecological studies 127. Springer, Berlin Heidelberg New York, pp 369–377

Sandermann H Jr, Ernst D, Heller W, Langebartels C (1998) Ozone: an abiotic elicitor of plant defence reactions. Trends Plant Sci 3:47–50

Sandroni S, Bacci P, Botta G, Pellegrini U, Ventura A (1994) Tropospheric ozone in the pre-alpine and alpine regions. Sci Total Environ 156:169–182

Sanz MJ, Millán MM (1998) The dynamics of aged airmasses and ozone in the western Mediterranean: relevance to forest ecosystems. Chemosphere 36:1089–1094

Sasek TW, Richardson CJ (1989) Effects of chronic doses of ozone on loblolly pine: photosynthetic characteristics in the third growing season. For Sci 35:745–755

Saurer M, Maurer S, Matyssek R, Landolt W, Günthardt-Georg MS, Siegenthaler U (1995) The influence of ozone and nutrition on $\delta^{13}C$ in *Betula pendula*. Oecologia 103:397–406

Saxe H, Ellsworth DS, Heath J (1998) Tree and forest functioning in an enriched CO_2 atmosphere. New Phytol 139:395–436

Scarascia-Mugnozza GE, Karnosky DF, Ceulemans R, Innes JL (2001) The impact of CO_2 and other greenhouse gases on forest ecosystems: an introduction. In: Karnosky DF, Scarascia-Mugnozza G, Ceulemans R, Innes JL (eds) The impacts of carbon dioxide and other greenhouse gases on forest ecosystems. CABI Press, Wallingford, UK, pp 1–16

Scherzer AJ, McClenahen JR (1989) Effects of ozone or sulphur dioxide on pitch pine seedlings. J Environ Qual 18:57–61

Schier GA, McQuattie CJ, Jensen KF (1990) Effects of ozone and aluminum on pitch pine (*Pinus rigida*) seedlings growth and nutrient relations. Can J For Res 20:1714–1719

Schraudner M, Möder W, Wiese C, van Camp W, Inze D, Langebartels C, Sandermann H (1998) Ozone-induced oxidative burst in the ozone biomonitor plant. Tobacco Bel W3. Plant J 16:235–245

Schubert R, Fischer R, Hain R, Schreier PH, Bahnweg G, Ernst D, Sandermann H Jr (1997) An ozone-responsive region of the grapevine resveratrol synthase promoter differs from the basal pathogen-responsive sequence. Plant Mol Biol 34:417–426

Schulze E-D (1994) The regulation of plant transpiration: interactions of feedforward, feedback, and futile cycles. In: Schulze E-D (ed) Flux control in biological systems. Academic Press, New York, pp 203–235

Schulze E-D, Hall AE (1982) Stomatal responses, water loss, and nutrient relations in contrasting environments. In: Lange OL, Nobel PS, Osmond CB, Ziegler H (eds) Encyclopedia of plant ecology 12B, physiological plant ecology II. Springer, Berlin Heidelberg New York, pp 182–230

Schupp R, Rennenberg H (1988) Diurnal changes in the glutathione content of spruce needles (*Picea abies* L.). Plant Sci 57:113–117

Schwanz P, Häberle K-H, Polle A (1996) Interactive effects of elevated CO_2, ozone and drought stress on the activities of antioxidative enzymes in needles of Norway spruce trees (*Picea abies* L.) grown with luxurious N supply. J Plant Physiol 148:351–355

Schweizer B, Arndt U (1990) CO_2/H_2O gas exchange parameters of one- and two-year-old needles of spruce and fir. Environ Pollut 68:275–292

Shao M, Zhao M, Zhang Y, Peng L, LI J (2000) Biogenic VOCs emissions and its impacts on ozone formation in major cities of China. J Environ Sci Health A35:1941–1950

Shavnin S, Maurer S, Matyssek R, Bilger W, Scheidegger C (1999) The impact of ozone fumigation and fertilization on chlorophyll fluorescence of birch leaves (*Betula pendula*). Trees 14:10–16

Sheppard LJ (1994) Causal mechanisms by which sulphate, nitrate and acidity influence frost hardiness in red spruce. Review and hypothesis. New Phytol 127:69–82

Simini M, Skelly JM, Davis DD, Savage JE (1992) Sensitivity of four hardwood species to ambient ozone in north central Pennsylvania. Can J For Res 22:1789–1799

Skärby L (1994) Critical levels for ozone to protect forest trees. In: Fuhrer J, Achermann B (eds) Critical levels for ozone: a UN-ECE workshop report. Schriftenreihe der FAC Liebefeld, 74–87. Swiss Federal Research Station for Agricultural Chemistry, Liebefeld-Bern

Skärby L, Karlsson PE (1996) Critical levels for ozone to protect forest trees – best available knowledge from the Nordic countries and the rest of Europe. In: Fuhrer J, Achermann B (eds) Critical levels for ozone: a UN-ECE workshop report. Schriftenreihe der FAC Liebefeld, Swiss Federal Research Station for Agricultural Chemistry, Liebefeld-Bern, pp 72–85

Skärby L, Troeng E, Boström C (1987) Ozone uptake and effects on transpiration, net photosynthesis and dark respiration in Scots pine. For Sci 33:801–808

Skärby L, Ro-Poulsen H, Wellburn FAM, Sheppard LJ (1998) Impacts of ozone on forests: a European perspective. New Phytol 139:109–122

Skelly JM, Davis DD, Merrill W, Cameron EA, Brown, HD, Drummond DB, Dochinger LS (eds) (1987) Diagnosing injury to eastern forest trees: a manual for identifying damage caused by air pollution, pathogens, insects, and abiotic stresses. National Acidic Precipitation Program, Forest Response Program, Vegetation Survey Research Cooperative. University Park, PA, Agricultural Information Services, College of Agriculture, Department of Plant Pathology, Pennsylvania State University, 122 pp

Skelly JM, Fredericksen TS, Savage JE, Snyder KR (1996) Vertical gradients of ozone and carbon dioxide within a deciduous forest in central Pennsylvania. Environ Pollut 94:235–240.

Skelly JM, Chappelka AH, Laurence JA, Fredericksen TS (1997) Ozone and its known and potential effects on forests in eastern United States. In: Sandermann H, Wellburn AR, Heath RL (eds) Forest decline and ozone, a comparison of controlled chamber and field experiments. Ecological studies 127. Springer, Berlin Heidelberg New York, pp 400

Skelly JM, Innes JL, Savage JE, Snyder KR, Vanderheyden D, Zhang J, Sanz MJ (1999) Observation and conservation of foliar ozone symptoms of native plant species of Switzerland and southern Spain. Water Air Soil Pollut 116:227–234

Smeekens S (2000) Sugar-induced signal transduction in plants. Annu Rev Plant Physiol 51:49–81

Smeulders SM, Gorissen A, Joosten NN, Vanveen JA (1995) Effects of short-term ozone exposure on the carbon economy of mature and juvenile Douglas firs [*Pseudotsuga menziesii* (Mirb) Franco]. New Phytol 129:45–53

Soda C, Busotti F, Grossoni P, Barnes J, Mori B, Tani C (2000) Impacts of urban levels of ozone on *Pinus halepensis* foliage. Environ Exp Bot 44:69–82

Somers GL, Chappelka AH, Rosseau P, Renfro JR (1998) Empirical evidence of growth decline related to visible ozone injury. For Ecol Manage 104:129–137

Spence DR, Rykiel EJ, Sharpe PJH (1990) Ozone alters carbon allocation in loblolly pine assessment with carbon-11 labeling. Environ Pollut 64:93–106

Sperry JS, Alder NN, Eastlack SE (1993) The effect of reduced hydraulic conductance on stomatal conductance and xylem cavitation. J Exp Bot 44:1075–1082

Spiecker H, Mielikäinen K, Köhl M, Skovegaard J (1996) Growth trends in European forests. Springer, Berlin Heidelberg New York, pp 372

Staehelin J, Schmid W (1991) Trend analysis of tropospheric ozone concentrations utilizing the 20-year data set of balloon soundings over Payerne (Switzerland). Atmos Environ 25A:1739–1749

Staehelin J, Thudium J, Buehler R, Volz-Thomas A, Graber W (1994) Trends in surface ozone concentrations at Arosa (Switzerland). Atmos Environ 28:75–88

Staffelbach T, Neftel A, Blattner A, Gut A, Fahrni M, Stähelin J, Prévôt A, Hering A, Lehning M, Neininger B, Bäumle M, Kok GL, Dommen J, Hutterli M, Anklin M (1997) Photochemical oxidant formation over southern Switzerland. 1. Results from summer 1994. J Geophys Res 102:23345–23362

Stitt M, Schulze ED (1994) Plant growth, storage and resource allocation: from flux control in metabolic chain to the whole-plant level. In: Schultze ED (ed) Flux control in biological systems: from enzymes to populations and ecosystems. Academic Press, San Diego, pp 57–118

Stockwell WR, Kramm G, Scheel H-E, Mohnen VA, Seiler W (1997) Ozone formation, destruction and exposure in Europe and the United States. In: Sandermann H, Wellburn AR, Heath RL (eds) Forest decline and ozone, a comparison of controlled chamber and field experiments. Ecological studies 127. Springer, Berlin Heidelberg New York, pp 400

Takemoto BK, Bytnerowicz A, Fenn ME (2001) Current and future effects of ozone and atmospheric nitrogen deposition on California's mixed conifer forests. For Ecol Manage 144:159–173

Taylor GE Jr, Hanson PJ (1992) Forest trees and tropospheric ozone: role of canopy deposition and leaf uptake in developing exposure-response relationships. Agric Ecosyst Environ 42:255–273

Taylor GE Jr, Johnson DW, Andersen CP (1994) Air pollution and forest ecosystems: a regional to global perspective. Ecol Appl 4:662–689

Temple PJ, Riechers GH (1995) Nitrogen allocation in ponderosa pine seedlings exposed to interacting ozone and drought stresses. New Phytol 130:97–104

Thomas HT, Stoddart JT (1980) Leaf senescence. Annu Rev Plant Physiol 31:83–111

Tingey DT, Wilhour RG, Standley C (1976) The effect of chronic ozone exposures on the metabolite content of Ponderosa pine seedlings. For Sci 22:234–241

Tjoelker MG, Luxmoore RJ (1991) Soil nitrogen and chronic ozone stress influence physiology, growth and nutrient status of *Pinus taeda* L. and *Liriodendron tulipifera* L. seedlings. New Phytol 119:69–81

Tjoelker MG, Volin JC, Oleksyn J, Reich PB (1993) Light environment alters response to ozone stress in seedlings of *Acer saccharum* Marsh. and hybrid *Populus* L. I. In situ net photosynthesis, dark respiration and growth. New Phytol 124:627–636

Tjoelker MG, Volin JC, Oleksyn J, Reich PB (1995) Interaction of ozone pollution and light effects on photosynthesis in a forest canopy experiment. Plant Cell Environ 18:895–905

Tobiessen P (1982) Dark opening of stomata in successional trees. Oecologia (Berl) 52:356–359

Tremmel DC, Bazzaz FA (1995) Plant architecture and allocation in different neighborhoods – implications for competitive success. Ecology 76:262–271

Urbach W, Schmidt W, Kolbowski J, Rümmele S, Reisber GE, Steigner W, Schreiber U (1989) Wirkungen von Umweltschadstoffen auf Photosynthese und Zellmembranen von Pflanzen. In: Reuther M, Kirchner M (eds) 1. Statusseminar der PBWU zum Forschungsschwerpunkt Waldschäden. GSF, München, pp 195–206

Utriainen M, Kokko H, Auriola S, Sarrazin O, Kärenlampi S (1998) PR-10 protein is induced by copper stress in roots and leaves of a Cu/Zn tolerant clone of birch, *Betula pendula*. Plant Cell Environ 21:821–828

Vanderheyden D, Skelly J, Innes J, Hug C, Zhang J, Landolt W, Bleuler P (2001) Ozone exposure thresholds and foliar injury on forest plants in Switzerland. Environ Pollut 111:321–331

Velissariou D, Davison AW, Barnes JD, Pfirrmann T, MaClean DC, Holevas CD (1992) Effects of air pollution on *Pinus halepensis* (Mill.): pollution levels in Attica, Greece. Atmos Environ 26:363–380

Velissariou D, Gimeno BS, BadianI M, Fumagalli I, Davison AW (1996) Records of O_3 visible injury in the ECE Mediterranean region. In: Kärenlampi L, Skärby L (eds) Critical levels for ozone in Europe: testing and finalizing the concepts. University of Kuopio, Kuopio, pp 343–350

Volin JC, Tjoelker MG, Oleksyn J, Reich PB (1993) Light environment alters response to ozone stress in *Acer saccharum* Marsh. and hybrid *Populus* L. seedlings. II. Diagnostic gas exchange and leaf chemistry. New Phytol 124:637–646

Volin JC, Reich PB, Givnish T (1998) Elevated carbon dioxide ameliorates the effects of ozone on photosynthesis and growth: species respond similarly regardless of photosynthetic pathway or plant functional group. New Phytol 138:315–325

Walters MB, Kruger EL, Reich PB (1993) Relative growth rate in relation to physiological and morphological traits for northern hardwood tree seedlings: species, light environment and ontogenetic considerations. Oecologia 96:219–231

Waring RH (1993) How ecophysiologists can help scale from leaves to landscapes. In: Ehleringer JR, Field CB (eds) Scaling physiological processes, leaf to globe. Academic Press, San Diego, pp 159–166

Waring RH, Schlesinger WH (1985) Forest ecosystems, concepts and management. Academic Press, Orlando, pp 340

Waring RH, Silvester WB (1993) Variation in foliar $\delta^{13}C$ values within the crowns of *Pinus radiata* trees. Tree Physiol 14:1203–1213

Weinstein DA, Beloin RM, Yanai RD (1991) Modeling changes in red spruce carbon balance and allocation in response to interacting ozone and nutrient stresses. Tree Physiol 9:127–146

Weinstein DA, Samuelson LJ, Arthur MA (1998) Comparison of the response of red oak (*Quercus rubra*) seedlings and mature trees to ozone exposure using simulation modeling. Environ Pollut 102:307–320

Wellburn FAM, Lau K-K, Milling MK, Wellburn AR (1996) Drought and air pollution affect nitrogen cycling and free-radical scavenging in *Pinus halepensis* Mill. J Exp Bot 47:1361–1367

Wellburn AR, Barnes JD, Lucas PW, McLeod AR, Mansfield TA (1997) Controlled O_3 exposures and field observations of O_3 effects in the UK. In: Sandermann H, Wellburn AR, Heath RL (eds) Forest decline and ozone: a comparison of controlled chamber and field experiments. Ecological studies 127. Springer, Berlin Heidelberg New York, pp 201–236

Werner H (1992) Das Indigopapier, sensitives Element zum Aufbau von Passivsammlern zur Messung von Ozonimmissionen. Forstl Forschungsberichte München, vol 122, pp 1–147

Werner H, Fabian P (2001) Free-air fumigation on mature trees: a novel system for controlled ozone enrichment in grown-up beech and spruce canopies. Environ Sci Pollut Res 9:117–121

Wieser G, Havranek WM (1993) Ozone uptake in the sun and shade crown of spruce: quantifying the physiological effects of ozone exposure. Trees 7:227–232

Wieser G, Havranek WM (1994) Exposure of mature Norway spruce to ozone in twig-chambers: effects on gas exchange. Proc R Soc Edinb Sect B 102:119–125

Wieser G, Havranek WM (1995) Environmental control of ozone uptake in *Larix decidua* Mill.: a comparison between different altitudes. Tree Physiol 15:253–258

Wieser G, Havranek WM (1996) Evaluation of ozone impact on mature spruce and larch in the field. J Plant Physiol 148:189–194

Wieser G, Havranek MW (2001) Effects of ozone on conifers in the timberline ecotone. In: Huttunen S, Bucher JB, Sundberg B, Jarvis P, Matyssek R (eds) Trends in European forest tree physiology research. Kluwer, Dordrecht, pp 115–125

Wieser G, Tegischer K, Tausz M, Häberle K-H, Grams TEE, Matyssek R (2002) Approach for comparing susceptibility to ozone uptake in Norway spruce [(*Picea abies* (L.) Karst.] across tree age: linking stress avoidance with defense. Tree Physiol (in press)

Wiskich JT, Dry IB (1985) The tricarboxylic acid cycle in plant mitochondria: its operation and regulation. In: Douce R, Day DA (eds) Higher plant cell respiration, encyclopaedia of plant physiology. New series, vol 18. Springer, Berlin Heidelberg New York, pp 281–313

Wolfenden J, Mansfield TA (1991) Physiological disturbances in plants caused by air pollutants. Proc R Soc Edinb 97B:117–138

Wulff A, Hänninen O, Tuomainen A, Kärenlampi L (1992) A method for open-air exposure of plants to ozone. Ann Bot Fennici 29:253-262

Wunderli S, Gehrig R (1990) Surface ozone in rural, urban and alpine regions of Switzerland. Atmos Environ 24A:2641–2646

Zangerl AB, Bazzaz FA (1992) Theory and pattern in plant defense allocation. In: Fritz RS, Simms EL (eds) Plant resistance to herbicides and pathogens. The University of Chicago Press, Chicago, pp 363–391

Zinser C, Ernst D, Sandermann H Jr (1998) Induction of stilbene synthase and cinnamyl alcohol dehydrogenase mRNAs in Scots pine (*Pinus sylvestris* L.) seedlings. Planta 204:169–176

Zinser C, Jungblut T, Heller W, Seidlitz HK, Schnitzler J-P, Ernst D, Sandermann H Jr (2000) The effect of ozone in Scots pine (*Pinus sylvestris* L.): gene expression, biochemical changes and interactions with UV-B radiation. Plant Cell Environ 23:975–982

Rainer Matyssek
Lehrstuhl für Ökophysiologie der Pflanzen
Technische Universität München
Am Hochanger 13
85354 Freising/Weihenstephan, Germany

e-mail: Matyssek@bot.forst.tu-muenchen.de

Heinrich Sandermann, Jr
Institut für Biochemische Pflanzenpathologie
GSF – Forschungszentrum für Umwelt und Gesundheit
Ingolstädter Landstrasse 1
85764 Neuherberg, Germany

e-mail: sandermann@gsf.de

Fine Root Biomass of Temperate Forests in Relation to Soil Acidity and Fertility, Climate, Age and Species

Christoph Leuschner and Dietrich Hertel

1 Introduction

Root growth represents an important component of ecosystem carbon cycling because, in a global perspective, belowground carbon storage is more important than aboveground storage (Schlesinger 1997). Together with above-ground litter fall, root production provides the primary input of organic carbon to soils. Due to methodological problems, only slow progress in our understanding of ecosystem belowground processes has occurred and, as a consequence, carbon storage in, and carbon flow through the root system is only poorly represented in most models on plant and ecosystem carbon turnover. However, prediction of the effects of global warming, nitrogen deposition or soil acidification on plant growth and carbon sequestration remains questionable if root processes are not adequately covered.

In forest ecosystems, roots <2 mm in diameter conventionally are termed fine roots, whereas roots >2 mm form the coarse root fraction (Böhm 1979). Because of their role in water and nutrient absorption, fine roots represent the functionally most important component of tree root systems. Although this root diameter class may contribute less than 2% of tree biomass in mature forests, fine root growth can consume up to 75% of the carbon fixed by the canopy (Fogel and Hunt 1983). Tree fine roots are organs with a rapid turnover not only of water and nutrients, but also of carbon because their life expectancy ranges from weeks to a few years, depending on species and environmental conditions (Hendrick and Pregitzer 1993; Bloomfield et al. 1996).

Fine root growth and root mortality are, in part, controlled by soil temperature, moisture and nutrient availability (Nadelhoffer et al. 1985; Pregitzer et al. 1993; Burton et al. 2000). Drought and elevated concentrations of aluminium, hydrogen ions or other potentially harmful elements in the soil solution may act as stressors that can increase fine root mortality of trees (e.g., Marshall 1986; Murach and Ulrich 1988).

There is increased interest in estimating the fine root biomass of forests because of (1) its prominent role in the ecosystem carbon budget,

Progress in Botany, Vol. 64

(2) its importance in the uptake of water and the cycling of nutrients, and (3) a possible indicator function of fine root vitality for the physical and chemical state of the soil and its change over time (Polomski and Kuhn 1998).

In this review we synthesise fine root biomass data of temperate forests and relate them to soil chemical parameters (pH, soil fertility), temperature, rainfall, elevation and stand age. The data were analysed with the aim (1) to compare fine root biomass of temperate broad-leaved and coniferous forests, (2) to detect differences among important temperate tree genera (*Fagus, Quercus, Acer, Pinus, Picea, Pseudotsuga*), and (3) to investigate relationships between root biomass and the six abiotic and biotic factors mentioned above.

This analysis of a large set of forest root mass data differs from earlier reviews in four ways. (1) Only one biome (the forest ecosystems of the cool temperate zone) with two major tree life forms (broad-leaved cold-deciduous and needle-leaved evergreen) was considered which contrasts with other studies that covered a multitude of vegetation types in a worldwide perspective (e.g., Vogt et al. 1996; Cairns et al. 1997). (2) Compared with earlier reviews the selection criteria for root data were more rigid: we rejected data of root diameters >3 mm and carefully attempted to include data on live fine root mass only. (3) Systematic errors due to highly variable sampling depths in the studies were reduced by calculating biomass data standardised to a profile depth of 70 cm. (4) Compared with earlier reviews with a North American focus, this data basis for temperate forests was substantially enlarged by including numerous Central European studies. With this comprehensive and sufficiently precise data basis we were able to analyse relationships between fine root biomass and influencing variables not only for the two foliage types (n=34–67), but also for a number of tree species/genera that have been characterised by a sufficiently high number of studies (n=10–24 per species/genus).

2 The Database

We assembled a database of about 70 published studies (including 4 unpublished data sets) that analysed fine root biomass in 133 stands and met the following criteria:

1. The studied forest stands are located in the cool temperate zone of Eurasia, North America, southeastern Australia, New Zealand, or southern South America. We included a small number of studies from the hemiboreal zone (e.g., southernmost Finland and southern Canada) if both broad-leaved and coniferous forests were studied in close vicinity to each other.

2. Sampling was carried out with corers or by the monolith technique yielding volume-related fine root densities. Root biomass data from trenching studies were excluded because this technique normally counts branch root endings; this type of information cannot be transformed to volume-related biomass data. Most studies did sampling on at least two occasions per year; the maximal number of sampling dates was eight. A few studies extracted soil cores only once. We selected seasonal averages of fine root biomass if available.
3. Only those studies were selected that sampled in at least the uppermost 15 cm of the mineral soil and the forest floor. Most studies covered 30–60 cm of the profile with certain investigations reaching 100 cm depth. Root data that refer to the forest floor only were excluded.
4. Only those studies were considered that included the separation of live and dead fine roots under a microscope.
5. All studies with data on fine root biomass <2 mm or <3 mm were considered. A few investigations which refer to the <1-mm fraction were included. In contrast to the data sets of Vogt et al. (1996) and Cairns et al. (1997), we excluded all studies that present data on the <5-mm fraction only because 2–5 mm roots have only negligible function in nutrient or water absorption. Moreover, this fraction will markedly increase the data heterogeneity. For example, about 25% of the root biomass <5 mm was contributed by the 2–5 mm fraction in *Pinus sylvestris* stands (Vanninen and Mäkelä 1999) but the proportion was 43–50% for *Fagus sylvatica* and 83–90% for *Quercus petraea* in a mixed stand (Leuschner et al. 2001b).
6. In most sources, data refer to studies of forest stands that had a closed canopy and were >20 years old. We also included five investigations with stands of 10–12-year-old pioneer trees (*Populus* spp., *Pinus radiata, P. sylvestris*) because these species grow rapidly and close their canopy after a decade or so. The data of Fredericksen and Zedakar (1995; Nos. 59–61 in Table 1), which refer to 3-year-old trees, have only been considered in the context of biomass–stand age relationships.
7. If the authors distinguished between tree and herbaceous roots in their samples we considered the data on tree root biomass only.
8. All stands in which soil manipulation (e.g., liming, fertilisation) had recently been conducted were omitted.

An inspection of the database assembled in Table 1 shows that the maximum soil depth sampled in the studies was highly variable ranging from 15–100 cm in the profile. This has the consequence that comparison among the studies is difficult and may lead to the wrong conclusions. Earlier reviews (e.g., Vogt et al. 1996; Cairns et al. 1997) simply

Table 1. Fine root biomass and necromass in 133 temperate forest stands and related site parameters. *BEL* Belgium; *CHI* China; *FRA* France; *GER* Germany; *NEZ* New Zealand; *NDL* Netherlands; *POL* Poland; *SLO* Slowakia; *SWE* Sweden; *TCH* Czech Republic. Necromass data: " + " -- precise analysis under a microscope; pH data: * pH measured in KCl or $CaCl_2$. For soil fertility index, see text

Cit. No	Species	Age (years)	Elevation (m a.s.l.)	Precipitation (mm yr^{-1})	Temperature (°C)	Fine root biomass (g d.m. m^{-2})	Fine root necromass (g d.m. m^{-2})	Fine root diameter (mm)	Topsoil pH (H_2O)	Soil fertility index
1	Abies amabilis	23	1150	2730	5.4	764	–	2.0	–	2
2	Abies amabilis	180	1150	2730	5.4	894	–	2.0	–	2
3	Abies balsamea, Picea glauca	–	–	823	0.6	707	–	2.0	5.4	3
4	Abies balsamea, Picea glauca	–	–	823	0.6	868	–	2.0	4.8	1
5	Abies balsamea, Thuja occidentalis	232	300	823	0.6	525	–	2.0	–	2
6	Abies grandis	25	450	750	8.0	488	446	2.0	4.3*	2
7	Picea abies	88	440	636	–	297	–	2.0	–	1
8	Picea abies	113	235	780	7.5	380	149	2.0	3.6	2
9	Picea abies	140	1720	1800	3.0	568	–	2.0	4.8*	3
10	Picea abies	28	110	–	–	273	728	1.0	3.9	1
11	Picea abies	140	1720	1800	3.0	470	97	2.0	5.0	3
12	Picea abies	88	945	1200	4.5	118	–	2.0	–	1
13	Picea abies	88	945	1200	4.5	69	–	2.0	–	1
14	Picea abies	140	780	1300	8.0	542	–	2.0	3.5*	2
15	Picea abies	35	–	953	–	333	–	3.0	–	2
16	Picea abies	44	120	585	–	220	270	2.0	3.8	1
17	Picea abies	35	550	1450	6.1	134	391	1.0	–	2
18	Picea abies	130	1435	1650	4.8	640	–	2.0	–	3
19	Picea abies	–	–	–	–	367	–	2.0	–	2
20	Picea abies	88	945	1200	4.5	148	–	2.0	–	1
21	Picea abies	90	550	911	6.4	240	100	2.0	3.5	2
22	Picea abies	88	945	1200	4.5	114	–	2.0	–	1
23	Picea abies	100	500	1050	6.8	350	170	2.0	3.5	2
24	Picea abies	100	500	1048	6.8	250	275	2.0	3.7	2
25	Picea abies	50	770	1000	5.0	202	55	2.0	3.2*	1
26	Picea abies	90	250	680	7.8	258	–	2.0	3.5*	2
27	Picea abies	50	670	1000	5.0	222	79	2.0	2.8*	1
28	Picea abies	104	500	1088	6.5	415	–	2.0	3.2*	2

Table 1 (continued)

Cit. No	Species	Age	Elevation	Precipitation	Temperature	Fine root biomass	Fine root necromass	Fine root diameter	Topsoil pH (H_2O)	Soil fertility index
29	Picea abies	88	945	1200	4.5	138	–	2.0	–	1
30	Picea abies	140	780	1300	8.0	437	101	2.0	4.0	3
31	Picea abies	104	500	1088	6.5	360	–	2.0	2.9*	2
32	Picea engelmanii, Abies lasiocarpa	350	3500	1000	1.5	193	677	2.0	–	2
33	Picea engelmannii, Abies lasiocarpa	200, 500	3500	1000	1.5	900	–	2.0	–	2
34	Pinus koraiensis	150	800	671	2.2	508	–	2.0	–	2
35	Pinus radiata	20	630	791	–	260	–	2.0	–	1
36	Pinus radiata	12	590	1491	–	204	547	2.0	–	2
37	Pinus resinosa	35	–	–	–	620	–	0.5	5.1	2
38	Pinus resinosa	100	–	–	–	402	–	2.5	–	2
39	Pinus resinosa	35	274	953	6.9	441	–	2.5	–	2
40	Pinus resinosa	53	–	1070	–	450	370	3.0	–	1
41	Pinus strobus	100	–	–	–	289	–	2.5	–	1
42	Pinus strobus	35	274	953	6.9	372	–	2.5	–	2
43	Pinus sylvestris	131	40	700	8.9	725	263	2.0	3.4	1
44	Pinus sylvestris	127	145	630	8.9	348	67	2.0	4.3	2
45	Pinus sylvestris	97	68	700	8.9	347	93	2.0	3.5	2
46	Pinus sylvestris	101	180	610	8.6	275	41	2.0	6.7	3
47	Pinus sylvestris	102	80	799	8.4	237	–	2.0	3.3*	1
48	Pinus sylvestris	12	–	526	7.7	310	–	2.0	5.8	1
49	Pinus sylvestris	38	–	558	4.0	265	55	2.0	3.8	1
50	Pinus sylvestris	101	85	740	8.4	510	263	2.0	3.4	1
51	Pinus sylvestris	107	70	740	8.4	600	291	2.0	3.5	1
52	Pinus sylvestris	12	–	526	7.7	270	–	2.0	5.8	2
53	Pinus sylvestris	95	55	600	8.3	441	88	2.0	3.8	2
54	Pinus sylvestris	–	–	–	–	300	–	2.0	3.8	1
55	Pinus sylvestris	98	78	705	8.3	602	311	2.0	3.6	1
56	Pinus sylvestris, Betula pendula	30	95	700	8.5	446	–	2.0	–	1

Table 1 (continued)

Cit. No	Species	Age	Elevation	Precipitation	Temperature	Fine root biomass	Fine root necromass	Fine root diameter	Topsoil pH (H_2O)	Soil fertility index
57	Pinus sylvestris, Quercus petraea	80	55	594	8.8	303	354	2.0	4.3	2
58	Pinus taeda	3	450	1240	–	63.9	9.1	2.0	–	–
59	Pinus taeda, Acer rubrum	3	450	1240	–	41.6	8.3	2.0	–	–
60	Pinus taeda, Robinia pseudoacacia	3	450	1240	–	67.1	7.9	2.0	–	1
61	Pseudotsuga menziesii	36	–	–	–	311	–	2.0	3.5*	1
62	Pseudotsuga menziesii	60	–	–	–	780	–	2.0	–	1
63	Pseudotsuga menziesii	40	320	100	–	550	–	2.0	–	1
64	Pseudotsuga menziesii	28	–	–	–	231	–	2.0	3.8*	1
65	Pseudotsuga menziesii	120	520	2290	–	315	406	1.0	–	3
66	Pseudotsuga menziesii	170	790	200	–	350	816	1.0	–	2
67	Pseudotsuga menziesii	40	320	100	–	270	–	2.0	–	3
68	Pseudotsuga menziesii	40	–	–	–	1500	–	2.0	–	1
69	Pseudotsuga menziesii	40	–	100	–	135	–	2.0	–	2
70	Pseudotsuga menziesii	40	–	100	–	562	–	2.0	–	2
71	Pseudotsuga menziesii	70	610	1800	8.0	253	1076	1.0	–	2
72	Acer rubrum	3	450	1240	–	18.6	1.1	2.0	–	2
73	Acer saccharum	100	–	–	–	323	–	2.5	–	2
74	Acer saccharum	35	274	953	6.9	428	–	2.0	–	2
75	Acer saccharum mixed	200	165	–	–	398	370	2.0	4.9	2
76	Acer saccharum mixed	135	440	–	–	582	698	2.0	3.7	1
77	Acer saccharum (mixed)	74	–	810	5.8	743	–	2.0	4.7	2
78	Acer saccharum (mixed)	78	–	850	7.6	872	–	2.0	4.7	2
79	Acer saccharum, Fagus grandifolia	76	600	1395	5.6	452	225	2.0	–	2
80	Acer saccharum, Prunus serotina	–	–	–	–	293	–	1.0	–	2
81	Acer saccharum, Fagus grandifolia	80	530	1060	–	241	307	3.0	–	2
82	Betula spec.	35	274	953	–	318	–	2.0	–	2
83	Fagus sylvatica	142	605	1100	6.0	556	–	2.0	3.2*	1

84	Fagus sylvatica	90	450	1450	6.1	118	194$^+$	1.0	–	2
85	Fagus sylvatica	100	320	750	7.5	225	750$^+$	2.0	4.1	2
86	Fagus sylvatica	140	350	600	7.5	229	–	2.0	5.1	3
87	Fagus sylvatica	64	–	–	–	960	–	2.0	–	1
88	Fagus sylvatica	130	142	690	8.5	319	607$^+$	2.0	4.1	2
89	Fagus sylvatica	30	300	820	9.2	800	–	2.0	–	2
90	Fagus sylvatica	38	–	–	–	720	–	2.0	–	2
91	Fagus sylvatica	127	510	800	7.0	549	190	2.0	3.2*	2
92	Fagus sylvatica	150	510	1050	6.8	308	1632$^+$	2.0	3.4	2
93	Fagus sylvatica	151	470	650	7.0	290	–	2.0	4.8*	3
94	Fagus sylvatica	150	510	1031	6.9	397	1021$^+$	2.0	3.7	2
95	Fagus sylvatica	140	510	1050	6.8	617	258	2.0	3.7	2
96	Fagus sylvatica	100	115	801	8.1	448	2955$^+$	2.0	3.3	1
97	Fagus sylvatica	110	450	780	7.5	724	–	2.0	4.1	2
98	Fagus sylvatica	120	280	462	8.6	171	768$^+$	2.0	4.7	2
99	Fagus sylvatica	120	420	647	8.7	328	327$^+$	2.0	7.0	3
100	Fagus sylvatica	–	–	–	–	283	212	2.0	–	2
101	Fagus sylvatica	108	400	1050	7.5	570	202	2.0	3.4*	2
102	Fagus sylvatica	120	275	700	8.5	155	183$^+$	2.0	5.9	3
103	Fagus sylvatica, Pseudotsuga menziesii	76	–	–	–	550	–	2.0	–	1
104	Fagus sylvatica, Pseudotsuga menziesii	43	–	–	–	950	–	2.0	–	1
105	Fagus sylvatica, Quercus petraea	100, 190	115	801	8.1	554	–	2.0	3.3	1
106	Northern mixed Hardwoods	110	620	1395	5.6	452	–	2.0	–	1
107	Populus tremuloides	pioneer	–	823	0.6	707	–	2.0	5.4	3
108	Populus tremuloides	pioneer	–	823	0.6	777	–	2.0	4.7	1
109	Populus tremuloides	10	–	800	–	1013	–	3.0	4.5	2
110	Populus tremuloides	32	–	800	–	909	–	3.0	4.5	2
111	Populus tremuloides	20	–	800	–	741	–	3.0	4.5	2

Table 1 (continued)

Cit. No	Species	Age	Elevation	Precipitation	Temperature	Fine root biomass	Fine root necromass	Fine root diameter	Topsoil pH (H_2O)	Soil fertility index
112	Populus tremuloides, Betula papyrifera	48	300	823	0.6	710	–	2.0	–	2
113	Populus tremuloides, Betula papyrifera	122	300	823	0.6	445	–	2.0	–	2
114	Populus tristis x balsamifera	11	–	–	–	356	–	0.5	5.7	2
115	Quercus alba	35	274	953	6.9	341	–	2.5	–	2
116	Quercus alba	100	–	–	–	515	–	2.5	–	1
117	Quercus alba	105	240	939	13.0	298	100	2.0	–	2
118	Quercus petraea	20	–	–	–	297	–	2.0	3.6	2
119	Quercus petraea	131	150	650	8.0	163	–	2.0	4.4	2
120	Quercus petraea	99	80	750	8.0	415	–	2.0	4.3	2
121	Quercus petraea	107	110	739	8.4	320	–	2.0	3.2*	2
122	Quercus petraea	99	80	750	8.0	305	–	2.0	4.3	2
123	Quercus petraea	50	480	1200	8.0	346	–	2.0	3.6	2
124	Quercus petraea	131	150	650	8.0	367	–	2.0	4.3	2
125	Quercus robur	180	20	750	7.5	260	–	2.0	3.9	2
126	Quercus robur, Carpinus betulus	–	–	–	–	1017	–	2.0	–	2
127	Quercus robur, Tilia cordata	160	60	644	7.5	172	–	2.0	4.7	2
128	Quercus rubra	35	274	953	6.9	270	–	2.0	–	2
129	Quercus rubra	mature	–	792	6.9	640	–	2.0	4.7	2
130	Quercus rubra	100	–	–	–	389	–	2.5	–	2
131	Quercus rubra, Acer rubrum	80	–	1070	–	590	430	3.0	–	2
132	Quercus velutina	35	274	953	6.9	270	–	2.5	–	2
133	Robinia pseudoacacia	3	450	1240	–	28	5	2.0	–	–

Table 1 (continued)

Cit. No	Location	Country	Longitude	Latitude	Soil type	Sampling depth in soil (cm)	Authors
1	Washington	USA	121°35' W	47°19' N	Andisol (Spodosol)	40	Grier et al. 1981; Vogt et al. 1982
2	Washington	USA	121°35' W	47°19' N	Andisol (Spodosol)	40	Grier et al. 1981; Vogt et al. 1982
3	Quebec	CAN	79°20' W	48°30' N	Grey Luvisol	20	Bauhus & Messier 1999
4	Quebec	CAN	79°20' W	48°30' N	Humo-ferric Podzol (till)	20	Bauhus & Messier 1999
5	Quebec	CAN	79°20' W	48°30' N	Grey Luvisol (clayey lake deposits)	30	Finér et al. 1997
6	Rhineland-Palatinate	GER	07° E	49°50' N	Gleyic Acrisol	45	Xu et al. 1997
7	Zelivka	TCH	–	–	–	30	Vins & Sika 1977
8	Hils, Lower Saxony	GER	–	–	Dystric Cambisol	60	Wiedey 1991
9	Garmisch-Partenkirchen, Bavaria	GER	–	–	Rendzina	30	Sandhage 1991
10	Halmstad	SWE	13°13' E	56°33' N	Haplic Podzol (loamy sand)	100	Persson et al. 1995
11	Garmisch-Partenkirchen, Bavaria	GER	–	–	Terrafusca-Rendzina	30	Sandhage-Hoffmann & Zech 1993
12	Serlich	TCH	–	–	Humo-ferric Podzol	30	Vins & Sika 1977
13	Serlich	TCH	–	–	Humo-ferric Podzol	30	Vins & Sika 1977
14	Garmisch-Partenkirchen, Bavaria	GER	–	–	Rendzina	30	Sandhage 1991
15	Wisconsin	USA	89°24' W	43°02' N	Alfisol (glac.) (silt, loam)	–	Nadelhoffer et al. 1985
16	Heinola	FIN	26°03' E	61°10' N	Spodo-dystric Cambisol	30	Helmisaari & Hallbäcken 1999
17	Ardennes	BEL	06° 06' E	50°32' N	Inceptisol (acid brown forest soil)	70	van Praag et al. 1988
18	Garmisch-Partenkirchen, Bavaria	GER	–	–	Rendzina	30	Sandhage 1991
19	Schwarzwald, Baden-Wuerttemberg	GER	–	–	–	–	Raspe et al. 1989
20	Serlich	TCH	–	–	Humo-ferric Podzol	30	Vins & Sika 1977
21	Witzenhausen, Hesse	GER	–	–	Spodo-dystric Cambisol (loamy)	32	Eichhorn 1987
22	Serlich	TCH	–	–	Humo-ferric Podzol	30	Vins & Sika 1977
23	Solling, Lower Saxony	GER	09°35' E	51°46' N	Spodo-dystric Cambisol (loamy)	30	Murach 1984
24	Solling, Lower Saxony	GER	09°35' E	51°46' N	Dystric Cambisol	40	Murach 1991
25	Fichtelgebirge, Bavaria	GER	–	–	Podzol	30	Schneider 1990
26	Spanbeck, Lower Saxony	GER	–	–	Spodo-dystric Cambisol	40	Murach & Wiedemann 1988
27	Fichtelgebirge, Bavaria	GER	–	–	Spodo-dystric Cambisol	30	Schneider 1990
28	Solling, Lower Saxony	GER	09°35' E	51°46' N	Spodo-dystric Cambisol	40	Murach & Wiedemann 1988

Table 1 (continued)

Cit. No	Location	Country	Longitude	Latitude	Soil type	Sampling depth in soil (cm)	Authors
29	Serlich	TCH	–	–	Humo-ferric Podzol	30	Vins & Sika 1977
30	Garmisch-Partenkirchen, Bavaria	GER	–	–	Luvisol/Chromic Cambisol	30	Sandhage-Hoffmann & Zech 1993
31	Solling, Lower Saxony	GER	09°35' E	51°46' N	Spodo-dystric Cambisol	40	Murach & Wiedemann 1988
32	Colorado	USA	–	–	Cryoboralfs	30	Arthur & Fahey 1992
33	Colorado	USA	–	–	Cryoboralfs	–	Arthur & Fahey 1992
34	–	CHI	128°60' E	42°25' N	Dark brown volcanic ash	–	Jianping et al. 1993
35	Canberra	AUS	148°56' E	35°21' S	Yellow podzolic typic Albaqualf	20	Ryan et al. 1996
36	Puruki	NEZ	176°14' E	38°25' S	Yellow brown pumice (silty sandy loam)	80	Santantonio & Grace 1987, Santantonio & Santantonio 1987
37	Wisconsin	USA	89°25' W	45°38' N	Alfic Haplorthods (glacial sandy loam)	70	Coleman et al. 2000
38	Wisconsin	USA	89°45' W	43°40' N	Colluvial Entisol	15	Aber et al. 1985
39	Wisconsin	USA	89°24' W	43°02' N	Alfisol (glac.)	20	Aber et al. 1985
40	Massachusetts	USA	–	42° N	Spodosol, Entic Haplorthod (stony, glacial)	45	Aber et al. 1985, McClaugherty et al. 1982
41	Wisconsin	USA	89°45' W	43°40' N	Spodosol (Entic Haplorthod)	15	Aber et al. 1985
42	Wisconsin	USA	89°24' W	43°02' N	Alfisol (glac.)	20	Aber et al. 1985
43	Lower Saxony	GER	–	–	Humo-ferric Podzol (sandy)	100	Scherfose 1990
44	Lower Saxony	GER	–	–	Gley-Cambisol (loamy, sandy)	100	Scherfose 1990
45	Lower Saxony	GER	–	–	Dystric Cambisol (sandy)	100	Scherfose 1990
46	Niedersachsen, Lower Saxony	GER	–	–	Terra fusca, eutrophic Cambisol	100	Scherfose 1990
47	Lüneburger Heide, Lower Saxony	GER	–	–	Spodo-dystric Cambisol	40	Murach & Wiedemann 1988
48	Kornik	POL	17°04' E	52°15' N	–	45	Oleksyn et al. 1999
49	–	FIN	22°09' E	61°19' N	Ferric Podzol	30	Helmisaari et al. 1999
50	Lower Saxony	GER	–	–	Spodo-dystric Cambisol (sandy)	100	Scherfose 1990
51	Lower Saxony	GER	–	–	Syrosem (sandy)	100	Scherfose 1990
52	Kornik	POL	17°04' E	52°15' N	–	45	Oleksyn et al. 1999
53	Lower Saxony	GER	–	–	Dystric Luvisol (sandy, loamy)	100	Scherfose 1990
54	–	RUS	–	–	–	–	Saurina + Kameneckaja 1969 (in Santantonio et al. 1977)
55	Lower Saxony	GER	–	–	Oligotrophic fen (peat soil)	100	Scherfose 1990
56	Lüneburger Heide, Lower Saxony	GER	10°30' E	52°45' N	Dystric Cambisol	45	Becker 1997

57	Berlin	GER	13°14' E	52°28' N	Dystric Cambisol	30	Kalhoff 2000
58	Virginia	USA	80° W	37° N	Typic Hupludults (clayey)	30	Fredericksen & Zedakar 1995
59	Virginia	USA	80° W	37° N	Typic Hupludults (clayey)	30	Fredericksen & Zedakar 1995
60	Virginia	USA	80° W	37° N	Typic Hupludults (clayey)	30	Fredericksen & Zedakar 1995
61	–	NDL	05°46' E	52°11' N	Leptic Podzol	80	Olsthoorn 1991
62	Veluwe	NDL	05°41' E	52°16' N	Siliceous, mesic Entic Haplorthod	90	Hendriks & Bianchi 1995
63	Washington	USA	122° W	46° N	Typic Haplorthod (sandy loam)	45	Keyes & Grier 1981
64	–	NDL	05°41' E	52°15' N	Orthic Podzol, Luvisol	80	Olsthoorn 1991
65	Oregon	USA	122°13' W	44°14' N	Typic Haplohumult	75	Santantonio & Hermann 1985
66	Oregon	USA	122°13' W	44°14' N	Entic Haplubrept (loamy clay)	75	Santantonio & Hermann 1985
67	Washington	USA	122° W	46° N	Colluvial soil with lake sedim.	45	Keyes & Grier 1981
68	Veluwe	NDL	05°41' E	52°16' N	Siliceous, mesic Entic Haplorthod	90	Hendriks & Bianchi 1995
69	Washington	USA	122° W	47° N	Typic Haplorthod (sandy loam)	–	Vogt et al. 1990
70	Washington	USA	122° W	47° N	Typic Haplorthod (sandy loam)	–	Cole & Gessel 1968; Vogt et al. 1990
71	Oregon	USA	122°13' W	44°14' N	Typic Dystrochrept (glacial loam)	75	Santantonio & Hermann 1985
72	Virginia	USA	80° W	37° N	Typic Hupludults (clayey)	30	Fredericksen & Zedakar 1995
73	Wisconsin	USA	89°45' W	43°40' N	Alfisol (sandy clayey loam)	15	Aber et al. 1985
74	Wisconsin	USA	89°24' W	43°02' N	Alfisol (glac.)	20	Aber et al. 1985, Nadelhoffer et al. 1985
75	Vermont	USA	72°59' W	44°48' N	Fine-textured Haplorthod (stone, silt, loam)	40	Liu & Tyree 1997
76	Vermont	USA	72°53' W	44°31' N	Coarse-textured Haplorthod (coarse loam)	40	Liu & Tyree 1997
77	Michigan	USA	85°50' W	44°23' N	Alfic + typic Haplorthods (glac.)	30	Hendrick & Pregitzer 1993
78	Michigan	USA	86°09' W	43°40' N	Entic + typic Haplorthods (glac.)	30	Hendrick & Pregitzer 1993
79	New Hampshire	USA	71°45' W	43°56' N	Haplorthods, Spodosols (sandy, loamy)	45	Fahey & Hughes 1994
80	–	USA	–	–	–	–	Wilczywski & Pickett 1993
81	New York	USA	74°13' W	44°00' N	Typic Haplorthod (fine sandy loam)	28	Burke & Raynal 1994
82	Wisconsin	USA	89°24' W	43°02' N	Alfisol (glac.)	–	Nadelhoffer et al. 1985
83	Harz, Lower Saxony	GER	–	–	Spodo-dystric Cambisol	40	Wiedemann 1991

Table 1 (continued)

Cit. No	Location	Country	Longitude	Latitude	Soil type	Sampling depth in soil (cm)	Authors
84	Ardennes	BEL	06°06' E	50°35' N	Inceptisol (loamy, brown, acid)	70	van Praag et al. 1988
85	Hameln, Lower Saxony	GER	–	–	Spodo-dystric Cambisol	20	Hertel (unpublished)
86	Hainich, Thuringia	GER	–	–	Terrafusca-Rendzina	20	Koch 2002
87	Veluwe	NDL	05°41' E	52°16' N	Siliceous, mesic Entic Haplorthod	90	Hendriks & Bianchi 1995
88	Bad Münder, Lower Saxony	GER	–	–	Spodo-dystric Cambisol	20	Hertel (unpublished)
89	Nancy	FRA	07°05' E	48°40' N	Cambisol (loamy, sandy)	–	Epron et al. 1999
90	Veluwe	NDL	05°41' E	52°16' N	Siliceous, mesic Entic Haplorthod	90	Hendriks & Bianchi 1995
91	Kaufunger Wald, Hesse	GER	–	–	Spodo-dystric Cambisol	40	Wiedemann 1991
92	Solling, Lower Saxony	GER	09°35' E	51°46' N	Spodo-dystric Cambisol	40	Hertel 1999
93	Zierenberg, Hesse	GER	–	–	Eutrophic Cambisol	60	Paar 1994
94	Solling, Lower Saxony	GER	09°35' E	51°46' N	Spodo-dystric Cambisol	20	Bauhus & Bartsch 1996
95	Solling, Lower Saxony	GER	09°35' E	51°46' N	Spodo-dystric Cambisol	40	Rapp 1991
96	Lüneburger Heide, Lower Saxony	GER	10°30' E	52°45' N	Spodo-dystric Cambisol	40	Hertel 1999
97	Hils, Lower Saxony	GER	–	–	Spodo-dystric Cambisol	70	Raben 1988
98	Ziegelroda Forest, Thuringia	GER	11°26' E	51°23' N	Spodo-dystric Cambisol	40	Hertel 1999
99	Göttinger Wald, Lower Saxony	GER	10°03' E	51°31' N	Rendzina	40	Hertel 1999
100	Göttingen, Lower Saxony	GER	–	–	–	40	Cassens-Sasse 1987
101	Egge, Lower Saxony	GER	–	–	Spodo-dystric Cambisol	40	Wiedemann 1991
102	Göttingen, Lower Saxony	GER	–	–	Cambisol	20	Hertel (unpublished)
103	Veluwe	NDL	05°41' E	52°16' N	Siliceous, mesic Entic Haplorthod	90	Hendriks & Bianchi 1995
104	Veluwe	NDL	05°41' E	52°16' N	Siliceous, mesic Entic Haplorthod	90	Hendriks & Bianchi 1995
105	Lüneburger Heide, Lower Saxony	GER	10°30' E	52°45' N	Spodo-dystric Cambisol	60	Büttner & Leuschner 1994
106	New Hampshire	USA	71°30' W	44°00' N	Podzolic Haplorthod	–	Cole & Rapp 1981, Covington 1981
107	Quebec	CAN	79°20' W	48°30' N	Grey Luvisol (clay)	20	Bauhus & Messier 1999
108	Quebec	CAN	79°20' W	48°30' N	Humo-ferric Podzol (till)	20	Bauhus & Messier 1999
109	Wisconsin	USA	89°45' W	45°45' N	Entic Haplorthod (glac.)	30	Ruark & Bockheim 1987
110	Wisconsin	USA	89°45' W	45°45' N	Entic Haplorthod (glac.)	30	Ruark & Bockheim 1987
111	Wisconsin	USA	89°45' W	45°45' N	Entic Haplorthod (glac.)	30	Ruark & Bockheim 1987

112	Quebec	CAN	79°20' W	48°30' N	Grey Luvisol (clayey lake deposits)	30	Finer et al. 1997
113	Quebec	CAN	79°20' W	48°30' N	Grey Luvisol (clayey lake deposits)	30	Finer et al. 1997
114	Wisconsin	USA	89°25' W	45°38' N	Alfic Haplorthod (glac.) sandy-loamy	70	Coleman et al. 2000
115	Wisconsin	USA	89°24' W	43°02' N	Alfisol (glac.)	20	Aber et al. 1985
116	Wisconsin	USA	89°45' W	43°40' N	Alfisol (sandy clay, loam)	15	Aber et al. 1985
117	Missouri	USA	92°12' W	38°40' N	Aquic Hapludalf (silt, loam)	22	Joslin & Henderson 1987
118	Ardennes	FRA	–	–	Spodo-dystric Cambisol (loamy)	55	Bakker 1999
119	Lappwald, Lower Saxony	GER	10°80' E	52°20' N	Cambi-stagnic Gleysol (loamy, sandy)	70	Thomas & Hartmann 1998
120	Sprakensehl, Lower Saxony	GER	10°50' E	52°70' N	Dystric Cambisol (sandy)	70	Thomas & Hartmann 1998
121	Lüneburger Heide, Lower Saxony	GER	–	–	Spodo-dystric Cambisol	50	Murach & Wiedemann 1988
122	Sprakensehl, Lower Saxony	GER	10°50' E	52°70' N	Dystric Cambisol (sandy)	70	Thomas & Hartmann 1998
123	Ardennes	FRA	–	–	Dystric Cambisol	55	Bakker 1999
124	Lappwald, Lower Saxony	GER	10°80' E	52°20' N	Cambi-stagnic Gleyosol (loamy, sandy)	70	Thomas & Hartmann 1998
125	Neuenburg, Lower Saxony	GER	08°40' E	53°20' N	Stagnic Cambisol (loamy)	70	Thomas & Hartmann 1998
126	–	SLO	–	–	–	60	Simonovic 1988
127	Linnebjer	SWE	13°18' E	55°44' N	Gleyic Cambisol	60	Andersson 1970
128	Wisconsin	USA	89°24' W	43°02' N	Alfisol (glac.)	20	Aber et al. 1985, Nadelhoffer et al. 1985
129	Wisconsin	USA	91°20' W	44°10' N	Typic Hapludalf	30	Yin et al. 1989
130	Wisconsin	USA	89°45' W	43°40' N	Alfisol (sandy clay, loam)	15	Aber et al. 1985
131	Massachusetts	USA	72° W	42° N	Spodosol, Entic Haplorthods (glac.)	45	McClaugherty et al. 1982, Vitousek et al. 1982
132	Wisconsin	USA	89°24' W	43°02' N	Alfisol (glac.)	20	Aber et al. 1985
133	Virginia	USA	80° W	37° N	Typic Hupludults (clayey)	30	Fredericksen & Zedakar 1995

ignored this bias in their root database. In an attempt to reduce root biomass variability among the sites resulting from different sampling depths, we chose a standard profile of 0–70 cm (plus forest floor) and re-calculated the original cumulative fine root biomass data for this standard profile. This was done in all those stands that were sampled to smaller or larger soil depths than 70 cm. We assumed that fine root density decreases exponentially with soil depth and used the exponential equations derived by Jackson et al. (1996) from nine temperate coniferous forests and eight temperate broad-leaved forests:

$$y = 1 - \beta^d \tag{1}$$

where y is the cumulative root fraction (a proportion between 0 and 1) from the soil surface to depth d (cm), and β is the fitted extinction coefficient. For broad-leaved and coniferous forests, β values of 0.966 (r^2=0.97) and 0.976 (r^2=0.93), respectively, were found by these authors. While this standardisation procedure is likely to reduce the depth-related error it can well introduce additional errors because no account is made for factors that could influence vertical root distribution (e.g., soil fertility, species influences). Therefore, we conducted all analyses with both data sets, the original fine root biomass numbers given in the source and the data that were standardised for 70 cm soil depth.

Because precise data on soil nutrient content were lacking in the majority of studies, we estimated soil fertility from information on soil type and texture and classed the sites as being either of very low fertility (No. 1), low to medium fertility (No. 2), or high fertility (No. 3). Sandy soils from glacial or eolian deposits were classed as 1, sandy-loamy or loamy soils outside alluvial regions as 2, and alluvial, carbonatic or heavy clayey soils as 3. Two-factor analyses of variance (ANOVAs) were used to examine the effect of soil fertility class on fine root biomass (original data) in the two foliage types.

3 Fine Root Biomass in Relation to Environmental Factors

a) Fine Root Biomass in Broad-Leaved and Coniferous Forests

Soil-coring studies in more than 130 forests of the temperate zone showed a significantly larger mean fine root biomass <2 or 3 mm in diameter (B_{cum}) in profiles under broad-leaved than under coniferous stands (difference of means: 75 and 93 g m^{-2} in the original and standardised data sets, respectively). Analysis of the original data on living fine root biomass in profiles of variable depths (15–100 cm) yielded 482 g dry matter m^{-2} as a mean for 60 broad-leaved forests and 407 g m^{-2} for 69 coniferous forests (all temperate forests: 442 g m^{-2}; Table 2). A re-

Table 2. Summary table of fine root biomass (B_{cum}), fine root necromass (N_{cum}) and biomass/necromass ratio (B_{cum}/N_{cum}) in various temperate forest types (mean, standard error and range of n stands per forest type; according to various sources cited in Table 1). The data are cumulative values over mineral soil profiles that were sampled to variable depths of 15–100 cm (the average sampled profile depth is indicated); the root mass of the forest floor is added. For fine root biomass, both the original (non-extrapolated) data and values standardised for a profile depth of 70 cm are given (see text). Different small Latin letters indicate significant ($p<0.05$) differences between tree species (genera), capitals mark differences between pioneer and late-successional trees, and Greek letters differences between the broad-leaved and coniferous forest data sets

	Bcum (original data)			Soil depth (cm)	Ncum (original data)		Bcum/Ncum	Bcum (standard. data)	
	(g m^{-2}) Mean ± SE	Range	*n*	$\bar{x}$	(g m^{-2}) Mean ± SE	*n*	Mean ± SE	(g m^{-2}) Mean ± SE	*n*
All temperate forests	442 ± 21	69–1500	129	47	426 ± 72	47	1.86 ± 0.24	588 ± 32	117
Broad-leaved forests	$482\pm31^{\alpha}$	69–1017	60	43	$602\pm157^{\alpha}$	19	$1.20\pm0.22^{\alpha}$	$637\pm44^{\alpha}$	56
Coniferous forests	$407\pm29^{\beta}$	114–1500	69	50	$306\pm49^{\alpha}$	28	$2.30\pm0.35^{\beta}$	$544\pm44^{\beta}$	61
Pioneer trees	498 ± 46^{A}	163–1013	24	56	166 ± 39^{A}	11	3.18 ± 0.62^{A}	625 ± 85^{A}	23
Late-successional trees	428 ± 26^{A}	69–1500	94	45	493 ± 93^{B}	35	1.45 ± 0.21^{B}	563 ± 36^{A}	87
Fagus sylvatica	470 ± 52^{a}	118–960	23	49	715 ± 222^{a}	13	1.05 ± 0.28^{ad}	538 ± 47^{af}	22
Acer spp.	492 ± 64^{a}	241–872	10	33	406 ± 81^{ab}	5	1.22 ± 0.22^{ae}	726 ± 91^{bef}	9
Quercus spp.	396 ± 46^{ab}	163–1017	19	46	177 ± 130^{ab}	3	1.45 ± 0.86^{abd}	532 ± 66^{acf}	19
Picea abies	302 ± 30^{b}	69–640	25	38	220 ± 60^{bd}	11	2.28 ± 0.47^{bde}	434 ± 53^{a}	23
Pinus sylvestris	399 ± 38^{ab}	237–725	15	74	183 ± 39^{bc}	10	3.50 ± 0.59^{bc}	422 ± 31^{a}	14
Pseudotsuga menziesii	478 ± 116^{ab}	135–1500	11	73	766 ± 195^{a}	3	0.48 ± 0.16^{d}	502 ± 125^{ad}	9
Other *Pinus* spp.	394 ± 39^{ab}	204–620	10	36	459 ± 88^{ad}	2	0.79 ± 0.42^{ad}	687 ± 96^{cde}	8

analysis of this data set for the 70-cm standard profile using the root biomass–depth relationships of Jackson et al. (1996) resulted in B_{cum} values that were by about 25% higher than the original data. The means were 637 and 544 g m^{-2} for broad-leaved and coniferous forests, respectively (all temperate forests: 588 g m^{-2}).

Earlier reviews reported contrasting results with respect to systematic differences in fine root biomass between temperate broad-leaved and coniferous forests. For example, Jackson et al. (1996) found means of 440 and 500 g m^{-2} for a sample of 10 to 14 broad-leaved and coniferous forests, respectively, whereas Vogt et al. (1996), who included several studies with roots <5 mm, give higher means (655 and 525 g m^{-2}; n=18–25) and reported a reversed order of the two foliage types which is supported by our much more comprehensive database. The larger fine root biomass of broad-leaved stands is astonishing because this foliage type typically has a smaller leaf mass and area than temperate coniferous forests (O'Neill and DeAngelis 1981). Thus, our data indicate significantly smaller fine root/leaf (needle) mass ratios in coniferous than broad-leaved forests. If nutrient uptake were linked to fine root biomass in trees, these results should indicate a smaller nutrient uptake per leaf mass in conifers compared with broad-leaved trees that would match the higher nutrient economy observed in evergreen conifers (Aerts 1995).

b) Fine Root Biomass of Different Tree Species and Genera

Pioneer tree stands contained a mean fine root biomass that was not significantly different from that of late-successional stands, although the means differed by approx. 60 g m^{-2} (original and standardised data) in favour of the pioneers (Table 2). This result is only weak support for the hypothesis of Gale and Grigal (1987) who postulated a deeper root penetration and the occupation of a larger soil volume by pioneer trees compared to late-successional trees. According to this hypothesis, stands of the latter species would be characterised by a more shallow root system with an adaptation to soil surface-directed nutrient cycling. However, the hypothesis of Gale and Grigal (1987) was formulated for vertical root distribution patterns only, and does not refer to cumulative fine root biomass. Whether a larger root space in pioneer trees corresponds to higher B_{cum} values requires further testing by root mass data.

The pioneer tree group included the genera *Betula*, *Populus*, and *Robinia*, and *Pinus sylvestris*, the late-successional group the genera *Fagus*, *Acer*, *Quercus*, *Abies* and *Pseudotsuga*, and *Picea abies*. If these genera are analysed separately, comparably large mean B_{cum} values were found for certain North American *Acer* and *Pinus* species, and the European species *Fagus sylvatica*; particularly low ones occurred in *Pinus sylvestris* and *Picea abies* forests from central Europe (Table 2). However, the difference between these taxa in mean fine root biomass of the 70-cm standard profile was surprisingly small: the *Acer* stands exceeded the *Pinus sylvestris* stands by not more than 40%. According to Table 2, differences among the taxa in B_{cum}, which are visible in the original data, have to be interpreted with caution because the mean sampling depths in these species differ substantially. For example, comparably high B_{cum} values in *Pseudotsuga* stands must in part be seen as a consequence of greater average sampling depths (73 cm) in these forests.

Species that build stands with a high fine root mass could differ from those with fewer roots in functional terms, for example, with respect to water and/or nutrient demand, stand productivity, nutrient economy of carbon gain, belowground competitive ability, or root litter production and associated soil carbon accumulation. Moreover, species-specific differences in fine root biomass and turnover are likely to have a profound influence on soil properties. The link between fine root biomass and related soil processes in forests is largely unexplored and needs to be pursued further.

c) Fine Root Biomass and Climatic Factors

Based on a global data set, Vogt et al. (1996) identified both climatic variables and the soil nutrient pool as being important controlling fac-

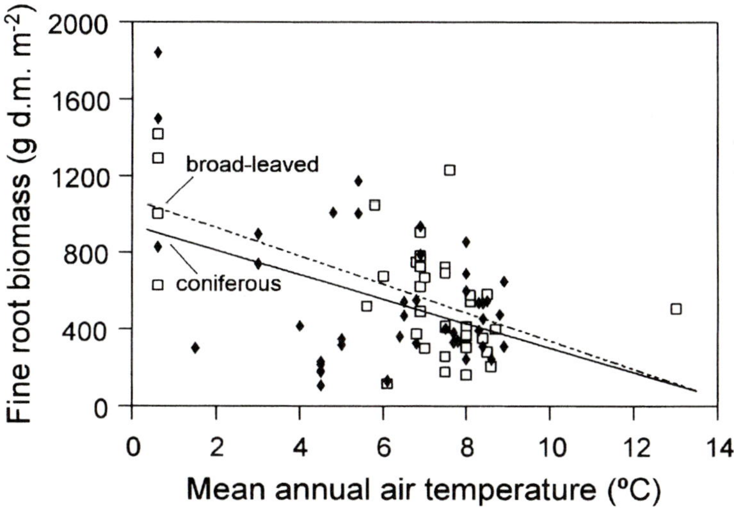

Fig. 1. Relationship between cumulative tree fine root biomass and mean annual air temperature for temperate broad-leaved (n=38) and coniferous forests (n=44). Linear regression based on depth-standardised data (broad-leaved forests: y=1077–74x; coniferous forests: y=939–64x). For results of a regression analysis with the original data, see Table 3. *Broken line* and *unfilled squares* broad-leaved forests; *full line* and *filled rhombs* coniferous forests

Table 3. Summary of relationships between original (non-extrapolated) and standardised fine root biomass data, and abiotic and biotic variables [topsoil pH(H_2O), mean annual temperature, mean annual precipitation, elevation, stand age]. Original data refer to the cumulative fine root biomass in profiles of variable sampling depths (15–100 cm), standardised data were extrapolated to a 70-cm profile depth using Eq. (1), see text. Relationships with $p < 0.05$ are printed in bold. All correlation coefficients refer to linear equations (for parameters a and b, see figure legends) except for three non-linear equations that are marked with an asterisk

	pH(H_2O)			Temperature			Precipitation			Elevation			Age		
	r	p	*n*	r	p	*n*	r	p	*n*	r	p	*n*	r	p	*n*
Original data															
All temperate forests	−0.02	0.415	70	**−0.28**	**0.005**	**86**	+0.12	0.105	104	+0.10	0.182	86	+0.08	0.203	124
Broad-leaved forests	−0.08	0.316	36	**−0.40**	**0.005**	**40**	+0.03	0.427	46	+0.26	0.055	38	−0.20	0.064	57
Coniferous forests	+0.02	0.463	34	−0.20	0.086	46	+0.20	0.060	58	+0.09	0.277	48	**+0.24**	**0.023**	**67**
Fagus sylvatica	**−0.46**	**0.035**	**16**	-0.10	0.350	17	+0.51	0.017	17	+0.14	0.309	16	**−0.66**	**0.001**	**21**
Quercus spp.	+0.18	0.316	10	-0.20	0.274	12	+0.31	0.142	14	+0.21	0.256	12	+0.02	0.48	17
Picea abies	**+0.59**	**0.010**	**15**	+0.14	0.284	20	+0.42	0.022	23	**+0.36**	**0.047**	**23**	**+0.65**	**0.001**	**24**
Pinus sylvestris	**−0.60**	**0.015**	**13**	+0.32	0.131	14	+0.47	0.044	14	−0.48	0.067	11	**+0.47**	**0.045**	**14**

Standardised data															
All temperate forests	+0.10	0.203	70	**–0.48**	**0.001**	**82**	+0.13	0.101	96	**+0.27**	**0.006**	**82**	+0.04	0.334	116
Broad-leaved forests	+0.28*	0.103*	36	**–0.58**	**0.001**	**38**	+0.03	0.423	43	**+0.30**	**0.037**	**35**	–0.20	0.070	54
Coniferous forests	**+0.35***	**0.046***	**34**	**–0.42**	**0.002**	**44**	**+0.27***	**0.047***	**53**	**+0.30**	**0.018**	**47**	+0.20	0.054	62
Fagus sylvatica	**–0.54**	**0.015**	**16**	**–0.47**	**0.032**	**16**	**+0.69**	**0.002**	**16**	+0.24	0.201	15	**–0.44**	**0.026**	**20**
Quercus spp.	+0.26	0.244	10	**–0.63**	**0.013**	**12**	**+0.46**	**0.049**	**14**	**+0.53**	**0.037**	**12**	–0.11	0.335	17
Picea abies	**+0.63**	**0.006**	**15**	+0.04	0.432	20	**+0.48**	**0.011**	**22**	**+0.48**	**0.010**	**23**	**+0.75**	**0.001**	**23**
Pinus sylvestris	**–0.61**	**0.013**	**13**	+0.06	0.422	14	+0.30	0.145	14	**–0.63**	**0.018**	**11**	+0.11	0.355	14

tors for the amount of tree root biomass in forests. However, different forest types appeared to be influenced by different variables. For example, maximum monthly temperature explained 65% of the variation in fine root biomass for needle-leaved forests, but not for broad-leaved forests (Vogt et al. 1986). In our data set, coniferous and broad-leaved forests differed in their response of fine root biomass to mean annual temperature, too. In 40 broad-leaved forests, a significant increase in B_{cum} with decreasing temperature was visible (original data: p=0.005; standardised data: p<0.001); for 46 coniferous forests, a corresponding relation was detected only in the standardised data set (p=0.002; Fig. 1; Table 3). On the species (genus) level, only *Fagus* and *Quercus* showed a significant increase of fine root biomass with a decrease in temperature (standardised data). In contrast, B_{cum} of the two conifers *Picea abies* and *Pinus sylvestris* was not significantly correlated with temperature.

There was a close positive correlation between fine root biomass and elevation for both the coniferous and the broad-leaved forests (standardised data: p=0.018 and 0.037, respectively; Fig. 2). In the original data, no significant relation between B_{cum} and elevation appeared (Table 3). In the case of the broad-leaved forests, the general validity of the observed relationship is biased due to the lack of data from sites

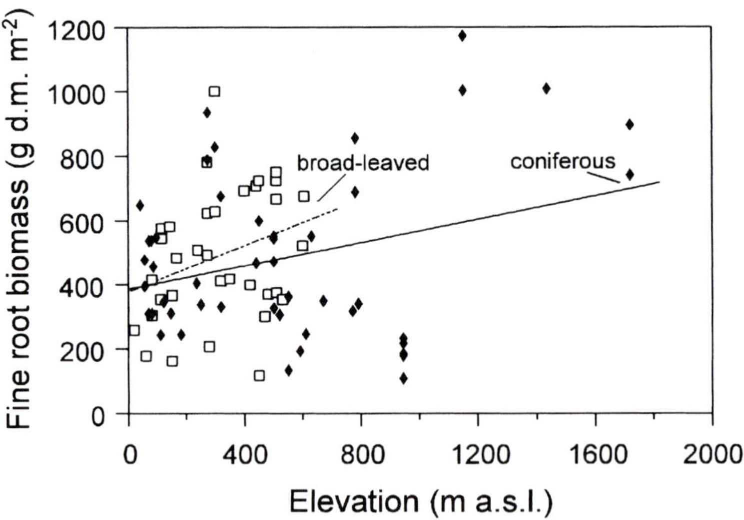

Fig. 2. Relationship between cumulative tree fine root biomass and elevation for temperate broad-leaved (n=35) and coniferous forests (n=47). Linear regression based on depth-standardised data (broad-leaved forests: y=379+0.358x; coniferous forests: y=387+0.181x). For results of a regression analysis with the original data, see Table 3. *Broken line* and *open squares* broad-leaved forests; *full line* and *filled rhombs* coniferous forests

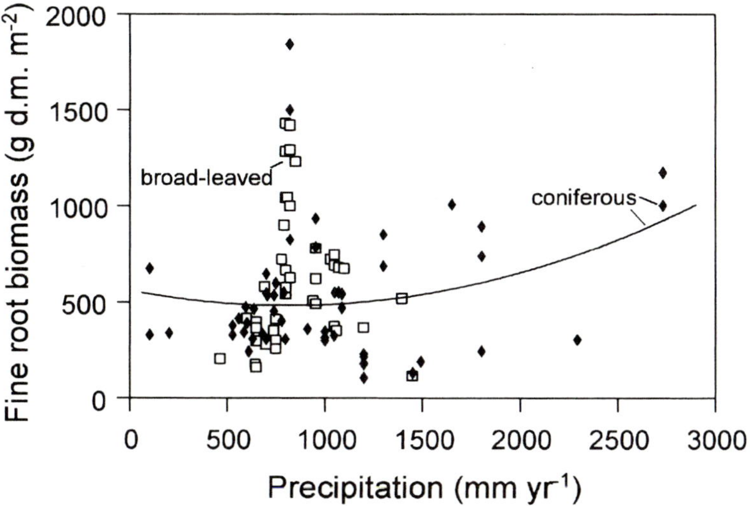

Fig. 3. Relationship between cumulative tree fine root biomass and mean annual precipitation for temperate broad-leaved (n=43) and coniferous forests (n=53). Coniferous forests: polynomial regression based on depth-standardised data ($y=563-0.193x+0.0001x^2$). For results of a regression analysis with the original data, see Table 3. *Open squares* broad-leaved forests; *full line* and *filled rhombs* coniferous forests

>700-m elevation. The closest positive relation to elevation was found for the root biomass of *Picea abies* stands with p=0.010 (standardised data).

Analysis of the fine root biomass–precipitation relationship shows a slight non-linear increase of B_{cum} with rainfall for a data set of 53 coniferous forests (p=0.047; Fig. 3; Table 3). For the broad-leaved stands (n=43), in contrast, no correlation with mean annual precipitation appeared (p=0.423). The contrasting sensitivity to a rainfall influence in the two foliage types may be the consequence of (1) the restriction of temperate broad-leaved forests to lower elevations (<ca. 600 m) with lower rainfall as compared to the coniferous forests where montane stands are included in the data set, and (2) possible differences in drought tolerance of the respective tree species. Indeed, a more detailed investigation of the tree species level revealed a highly significant positive relationship between B_{cum} and precipitation for stands of *Fagus sylvatica* and *Picea abies* (Fig. 4). In contrast, for *Pinus sylvestris*, a positive correlation appeared only in the original but not the standardised data, and for the oak species, significance was visible only in the standardised data (Table 3).

The results of these correlation analyses between fine root biomass and precipitation match well with published data on drought sensitivity of leaf physiological processes and growth of *Fagus sylvatica*, *Picea*

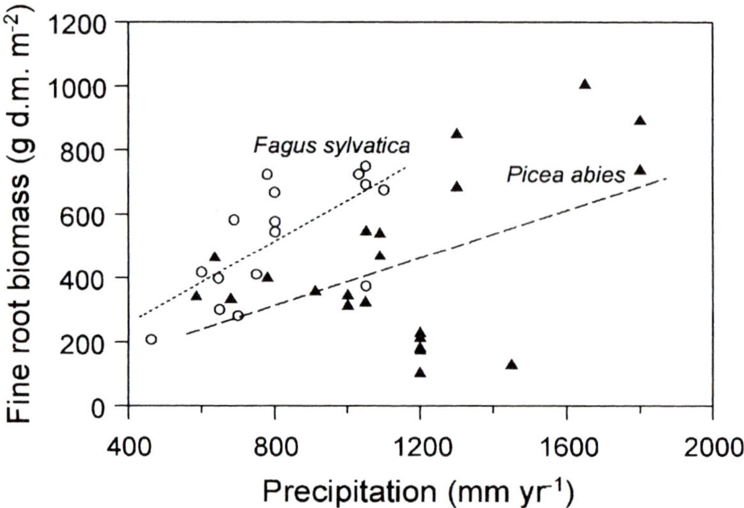

Fig. 4. Relationship between cumulative tree fine root biomass and mean annual precipitation for 16 stands of *Fagus sylvatica* and 22 stands of *Picea abies*. Linear regression based on depth-standardised data (*Fagus*: y=3.85+0.638x; *Picea*: y=16.74+0.372x). For results of a regression analysis with the original data, see Table 3. *Dashed line* and *filled triangles P. abies*; *dotted line* and *open circles F. sylvatica*

abies, *Pinus sylvestris* and temperate *Quercus* species. *Fagus* and *Picea* have been found to be drought-sensitive species (Hellkvist et al. 1980; Backes and Leuschner 2000), a fact that could explain the root biomass decrease with decreasing rainfall in our data set. In contrast, *Pinus sylvestris* and many temperate *Quercus* species are known to be rather drought insensitive (Abrams 1990; Leuschner et al. 2001a), which is in line with a weak or lacking positive relation of fine root biomass to precipitation height as observed in these species.

d) Fine Root Biomass and Soil Chemical Factors

How tree root systems respond to elevated soil acidity depends on two factors, (1) the sensitivity of metabolism and growth of individual fine roots and associated mycorrhizal fungi to high concentrations of toxic ions or unfavourable ion ratios (Kochian 1995), and (2) the disposition of a tree species to favour a build-up of thick organic layers on the forest floor by producing slowly decaying litter on poor soil (Berendse et al. 1989). In a global comparison, Vogt et al. (1996) found that particularly high fine root biomasses existed in forests on soils with a dominance of the soil chemistry by Al and Fe, and a high accumulation of soil organic matter as occurs in Ultisols, Spodosols or Inceptisols. These authors

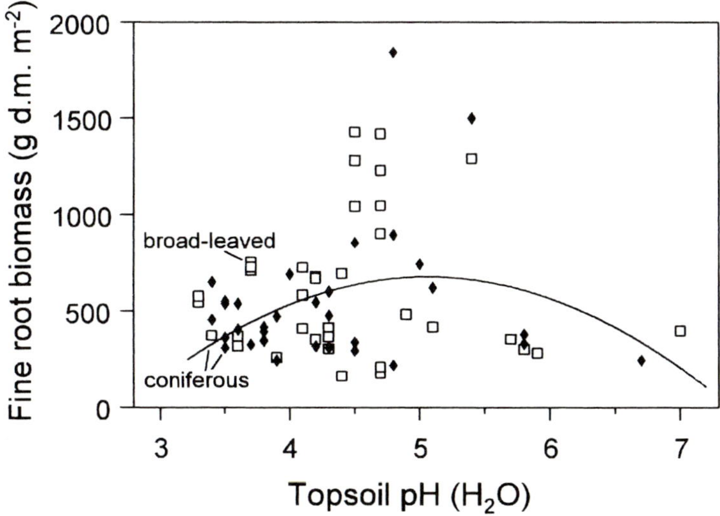

Fig. 5. Relationship between cumulative tree fine root biomass and pH (measured in water) of the mineral topsoil for temperate broad-leaved (n=37) and coniferous forests (n=34). Polynomial regression based on depth-standardised data (broad-leaved forests: $y=-1842+1041x-106x^2$; coniferous forests: $y=-2522+1263x-125x^2$). For results of a regression analysis with the original data, see Table 3. *Open squares* broad-leaved forests; *full line* and *filled rhombs* coniferous forests

identified coniferous forests of the temperate zone, where Spodosols are widespread, as one of the forest types with highest fine root biomasses around the world. Apparently, only those tree species are successful on acidic soils that are able to exploit the organic material on the forest floor for nutrients and water by expansive fine root growth in this medium, and thus to avoid unfavourable conditions in the mineral soil.

For the broad-leaved forest data set, fine root biomass showed no significant relationship to the hydrogen ion concentration [pH(H_2O)] in the solution of the mineral topsoil (original data: p=0.316; standardised data: p=0.103; Fig. 5; Table 3), although a biomass maximum seems to exist at moderate soil acidity (pH 4.0–5.5). In contrast, B_{cum} of the coniferous forests apparently decreases with increasing soil acidity in the pH range 4.5–3.3. This relationship is only significant in the standardised data set (p=0.046) but not in the original data (p=0.463).

In a second step, we analysed the relationship between cumulative fine root biomass (standardised data) and topsoil acidity separately for the four intensively studied tree species or genera (*Fagus sylvatica*, *Quercus* spp., *Pinus sylvestris* and *Picea abies* with n=10–16 stands per species/genus). These analyses indicate that different tree species or genera seem to respond differently to variation in soil acidity. Fine root biomass showed no correlation with topsoil pH for the *Quercus* stands,

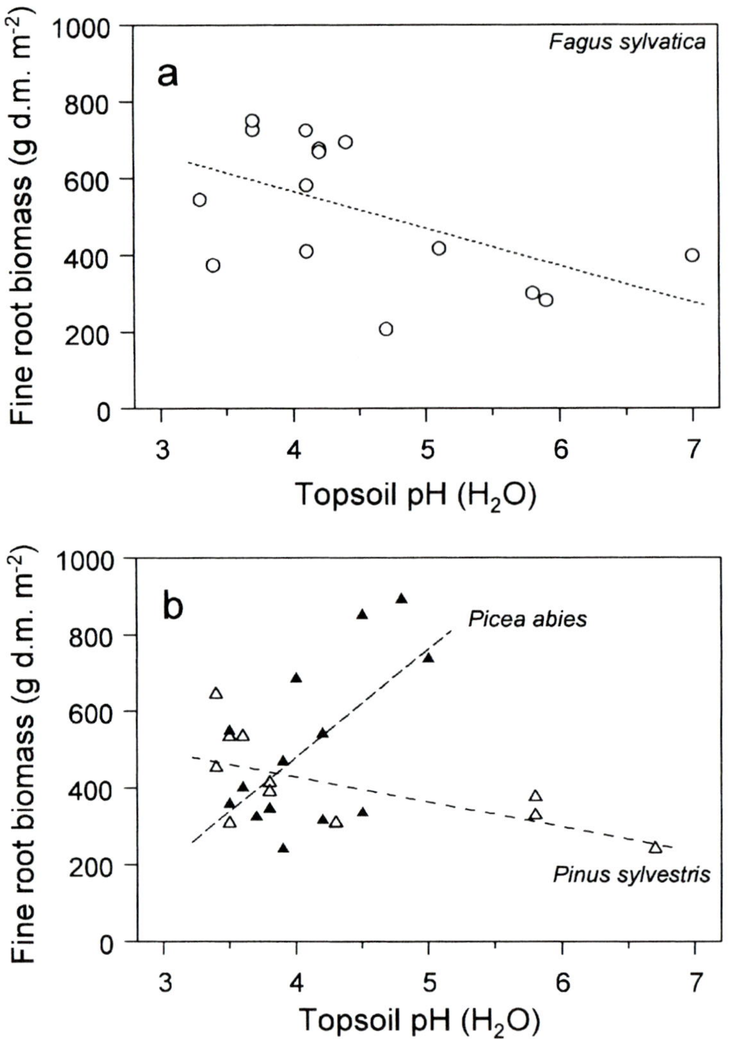

Fig. 6. Relationship between cumulative tree fine root biomass and pH (measured in water) of the mineral topsoil for (**a**) 16 stands of *Fagus sylvatica* and (**b**) 15 and 13 stands of *Picea abies* and *Pinus sylvestris*, respectively. Linear regression based on depth-standardised data (*Fagus*: y=930–92x; *Picea*: y=–650+282x; *Pinus*: y=695–65x). For results of a regression analysis with the original data, see Table 3. *Dotted line* and *open circles F. sylvatica*; *dashed line* and *filled triangles P. abies*; *dashed line* and *open triangles P sylvestris*

but revealed a significant positive correlation for the *Picea abies* stands (p=0.006; Table 3), and a negative correlation for the *Pinus sylvestris* (p=0.013) and *Fagus sylvatica* forests (p=0.015; Fig. 6). These results may reflect differences in both ecology and geographic range of the three species.

Fagus sylvatica and *Pinus sylvestris* seem to represent species that are more successful in exploring thick organic layers than is *Picea abies*. The fine root system of *Fagus* apparently is not negatively affected by high soil acidity, since B_{cum} was found to increase significantly with decreasing pH. Similarly, *Pinus sylvestris* stands showed highest fine root biomasses at highest soil acidity. High root biomass totals at acidic nutrient-poor sites have been interpreted as a response to high nitrogen and phosphorus residence times in biologically inactive soils (Vogt et al. 1996). According to this hypothesis, a larger root mass would ensure sufficient nutrient absorption even in nutrient-poor and highly patchy soils.

A different picture was found in the *Picea abies* stands with a significant and striking decrease of root biomass with decreasing pH that may indicate a sensitivity of spruce fine roots to hydrogen or aluminium ions. Indeed, *Picea* root systems have been observed to retreat from soil horizons with high Al concentrations or elevated Al/Ca ratios in the soil solution, a response that could eventually lead to reduced fine root biomass totals at the stand level.

The number of studies that have reported fine root necromass data (n=13 and 21 for broad-leaved and coniferous forests, respectively) is much smaller than for biomass data. Conclusions on root necromass patterns are additionally biased by the fact that data on profile totals of

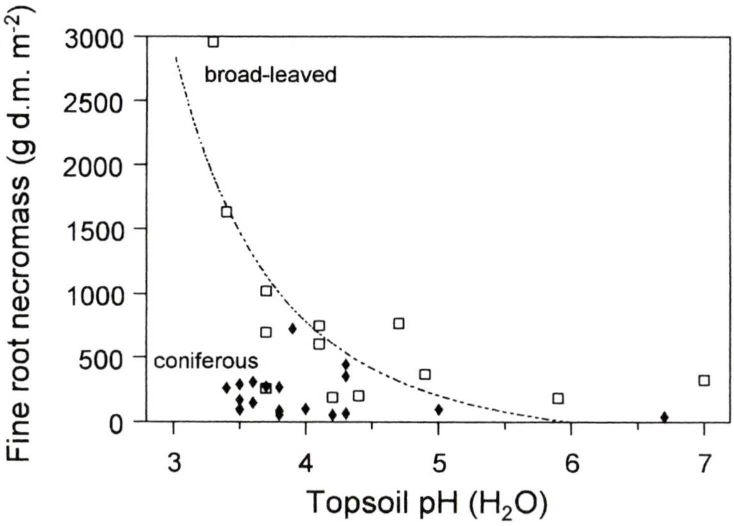

Fig. 7. Relationship between cumulative tree fine root necromass and pH (measured in water) of the mineral topsoil for temperate broad-leaved (n=13) and coniferous forests (n=21). Broad-leaved forests: exponential regression based on the original data ($y = -195+250355x^{-4}$; $r=0.79$, $p<0.001$). *Broken line* and *unfilled squares* broad-leaved forests; *filled rhombs* coniferous forests

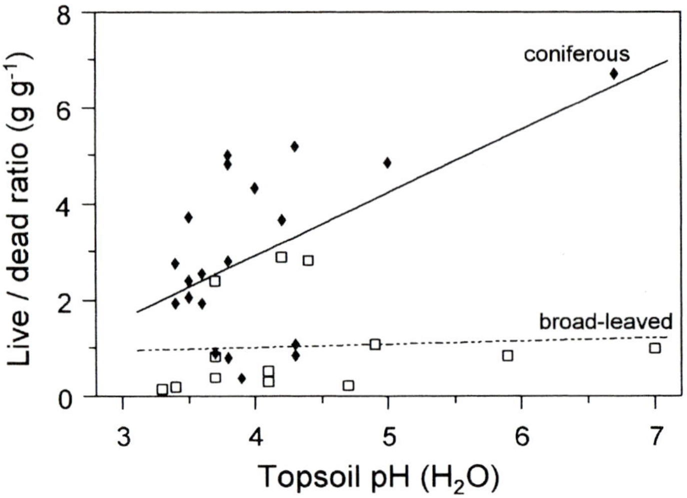

Fig. 8. Relationship between the live/dead ratio of fine root mass and pH (measured in water) of the mineral topsoil for temperate broad-leaved (n=13) and coniferous forests (n=21). Coniferous forests: linear regression based on the original data (y= –2.313+1.309x; r=+0.55, p=0.045). *Broken line* and *unfilled squares* broad-leaved forests; *full line* and *filled rhombs* coniferous forests

necromass (N_{cum}) should be less reliable than biomass totals because the amount of necromass extracted from a soil sample is likely to depend on the laboratory involved and the technique used. A precise extraction of small dead root fragments under a microscope can yield up to ten times higher necromass values compared with simple washing techniques (Hertel 1999). Nevertheless, the recent data set of 13 broad-leaved forests revealed a highly significant and striking decrease in necromass from highly acidic to neutral and basic soils (p=0.001; Fig. 7). Coniferous forests, in contrast, showed no such relationship (p=0.308) and had markedly smaller necromass totals on acidic sites than were found in broad-leaved forests. According to this data set, comparably small N_{cum} values corresponded to higher live/dead ratios of fine root mass (typically 1–3 g g^{-1}) in the coniferous stands as compared with the broad-leaved forests (0.3–3; Fig. 8). Thus, at least in broad-leaved stands (including *Fagus*), a high standing crop of fine root biomass in acidic soils seems to be linked to elevated root mortalities and turnover rates which indicates that acidic soils represent a stressful environment not only for coniferous, but also for broad-leaved tree roots. This raises the question whether trees of *Fagus* or other broad-leaved species invest particularly large amounts of carbon in fine roots in order to compensate for a rapid root death in a stressful environment, and whether they differ in this respect from *Picea* trees (Hertel 1999).

Table 4. Results of two-factor ANOVAs for the effects of soil fertility (relative values estimated from soil texture and geology) and tree species/genus on cumulative fine root biomass (B_{cum}, original data) showing the F values and associated significance. The analysis was conducted separately for broad-leaved stands (n=60) and coniferous stands (n=69); values of p in bold are <0.05

Source	SS	df	F	p
		Coniferous forests		
Model	**2,214,386**	8	**2.79**	**0.012**
Tree species/genus	**1,159,974**	2	**5.85**	**0.005**
Soil fertility	283,624	2	1.43	0.25
Species × fertility	770,788	4	1.95	0.12
Error	4,953,193	50		
		Broad-leaved forests		
Model	**2,600,283**	7	**5.35**	**<0.001**
Tree species/genus	**1,598,154**	2	**11.52**	**<0.0001**
Soil fertility	**520,031**	2	**3.75**	**0.031**
Species × fertility	482,098	3	2.32	0.088
Error	3,260,954	47		

Soil nutrient content or nutrient pools are a second soil chemical property that has been found to influence tree fine root biomass (Vogt et al. 1996) and fine root longevity (Keyes and Grier 1981; Aber et al. 1985; Nadelhoffer et al. 1985; Burton et al. 2000). The results of two-factor ANOVAs showed no effect of soil fertility, as expressed by relative numbers between 1 and 3, on fine root biomass of the coniferous forests in our data set (Table 4). For the broad-leaved forests, a significant effect of fertility existed (p=0.031). However, in both foliage types, the effect of tree species (genus) on B_{cum} was much more important (p<0.0001) than the fertility influence.

Our analysis can give only vague hints on the role of the fertility factor because of the lack of precise data in the majority of studies. Other authors (e.g., Keyes and Grier 1981; Vogt et al. 1983, 1987, 1995; Vitousek and Sanford 1986) have reported inverse relationships between soil fertility and either root biomass or R/S ratio particularly in very nutrient-poor soils: the higher the nitrogen availability, the lower the fine root biomass in these ecosystems. The positive relationship between soil acidity and B_{cum} in *Fagus sylvatica* and *Pinus sylvestris* as observed in our data set also points to a stimulation of fine root growth in nutrient-poor soils, thus supporting an inverse fertility/root biomass relationship in these species.

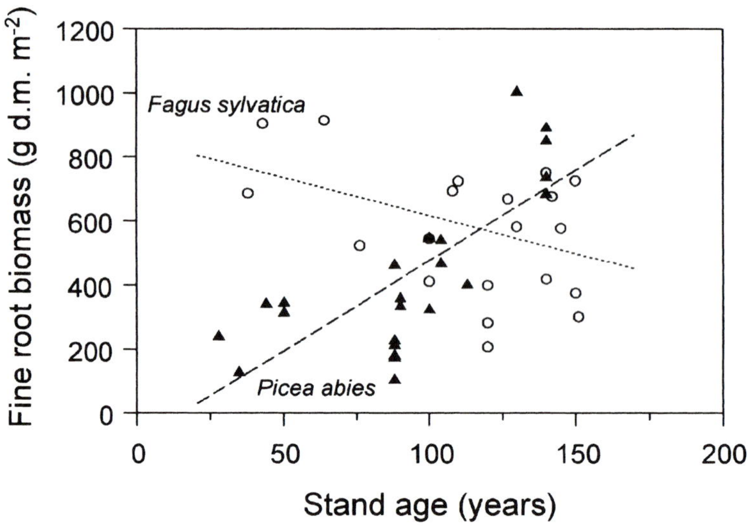

Fig. 9. Relationship between cumulative tree fine root biomass and stand age for stands of *Fagus sylvatica* (n=20) and *Picea abies* (n=23). Linear regression based on depth-standardised data (*Fagus*: y=857–2.54x; *Picea*: y=–86+5.62x). For results of a regression analysis with the original data, see Table 3. *Dashed line* and *filled triangles P. abies*; *dotted line* and *open circles F. sylvatica*

e) Fine Root Biomass and Stand Age

Fine root biomass correlated more closely with stand age in the coniferous stands than in the broad-leaved forests (Table 3). B_{cum} increased with age in the conifer stands between 15 and >200 years (original data: p=0.023; standardised data: p=0.054), whereas for the broad-leaved stands, a (non-significant) negative relationship rather than a positive one appeared from the data (original data: p=0.064; standardised data: p=0.070). *Picea abies* showed a highly significant increase of fine root biomass with increase in age between 25 and 140 years. In contrast, B_{cum} decreased with an increase in age for the *Fagus sylvatica* stands (Fig. 9). The reason for this contradictory behaviour is unclear. One possible explanation would be that the old *Picea* and *Fagus* stands, which were included in this review, existed under contrasting temperature and soil acidity conditions, thus masking a relationship between B_{cum} and age.

Earlier attempts to relate tree fine root biomass to stand age also produced contradictory evidence. Several authors working in temperate coniferous forests found increases of B_{cum} in young stands until canopy closure occurred and a subsequent leveling-off or decrease in fine root biomass (e.g., Vogt et al. 1987; Vanninen and Mäkelä 1999). Contrasting patterns can, in part, be explained by differences in soil fertility. However, it is still unclear whether conifers and broad-leaved trees, in general, or different tree species have a similar root biomass–age relationship or not.

4 Conclusions

Correlation analyses between the fine root biomass in temperate broad-leaved and coniferous forests and four important tree species (genera) and certain abiotic and biotic variables revealed highly different responses among these groups. The temperate conifer sample showed an increase in root biomass with decreasing temperature, increasing elevation or increasing precipitation (in the range 1000–2700 m a.s.l.). Broad-leaved forests also had higher root biomasses at higher elevation or lower temperature (two factors that are negatively correlated), but were not significantly influenced by precipitation. In general, the fine root biomass of *Pinus sylvestris* and the *Quercus* species was influenced by fewer variables (temperature, elevation and soil pH) than that of *Fagus sylvatica* and *Picea abies* (temperature, precipitation, elevation, pH, age; Table 3). This result would support Bazzaz's (1979) hypothesis on the ecology of early- and late-successional trees. He postulated that pioneer species principally are less sensitive to environmental fluctuation than late-successional species, a prediction that has never been tested for tree root systems. Systematic studies on forest fine root biomass are needed that apply standardised sampling techniques and also investigate relevant soil chemical and physical variables including nutrient pools and nutrient fluxes. A particular emphasis should be laid on tree species differences in the root biomass–environment relationship because tree fine roots may serve as sensible indicators of soil and climate conditions.

References

Aber JD, Melillo JM, Nadelhoffer KJ, McClaugherty CA, Pastor J (1985) Fine root turnover in forest ecosystems in relation to quantity and form of nitrogen availability: a comparison of two methods. Oecologia 66:317–321

Abrams MD (1990) Adaptations and responses to drought in *Quercus* species of North America. Tree Physiol 7:227–238

Aerts R (1995) The advantages of being evergreen. Trends Ecol Evol 10:402–407

Andersson F (1970) Ecological studies in a Scandinavian woodland and meadow area, southern Sweden. II. Plant biomass, primary production and turnover of organic matter. Bot Not 123:8–51

Arthur MA, Fahey TJ (1992) Biomass and nutrients in an Engelmann spruce-subalpine fir forest in north central Colorado: pools, annual production, and internal cycling. Can J For Res 22:315–325

Backes K, Leuschner C (2000) Leaf water relations of competitive *Fagus sylvatica* L. and *Quercus petraea* (Matt.) Liebl. trees during four years differing in soil drought. For Ecol Manage 84:219–229

Bakker MR (1999) The effect of lime and gypsum applications on a sessile oak (*Quercus petraea* (M.) Liebl.) stand at La Croix-Scaille (French Ardennes). II. Fine root dynamics. Plant Soil 206:109–121

Bauhus J, Bartsch N (1996) Fine-root growth in beech (*Fagus sylvatica*) forest gaps. Can J For Res 26:2153–2159

Bauhus J, Messier C (1999) Soil exploitation strategies of fine roots in different tree species of the south boreal forest of eastern Canada. Can J For Res 29:260–273

Bazzaz FA (1979) The physiological ecology of plant succession. Annu Rev Ecol Syst 10:351–371

Becker C (1997) Das Feinwurzelsystem von Birke, Kiefer und Drahtschmiele in Konkurrenzsituation in einem Mischbestand. Diploma thesis, University of Göttingen, Göttingen, pp 1–58

Berendse F, Bobbink R, Rouwenhorst G (1989) A comparative study on nutrient mineralization. Oecologia 78:338–348

Bloomfield J, Vogt K, Wargo PM (1996) Tree root turnover and senescence. In: Waisel Y, Eshel A, Kafkafi U (eds) Plant roots. The hidden half, 2nd edn. Marcel Dekker, New York, pp 363–381

Böhm W (1979) Methods for studying root systems. Springer, Berlin Heidelberg New York

Büttner V, Leuschner C (1994) Spatial and temporal patterns of fine root abundance in a mixed oak-beech forest. For Ecol Manage 70:11–21

Burke MK, Raynal DJ (1994) Fine root growth phenology, production, and turnover in a northern hardwood forest ecosystem. Plant Soil 162:135–146

Burton AJ, Pregitzer KS, Hendrick RL (2000) Relationships between fine root dynamics and nitrogen availability in Michigan northern hardwood forests. Oecologia 125:389–399

Cairns MA, Brown S, Helmer EH, Baumgardner GA (1997) Root biomass allocation in the world's upland forests. Oecologia 111:1–11

Cassens-Sasse E (1987) Witterungsbedingte saisonale Versauerungsschübe im Boden zweier Waldökosysteme. Ber Forsch Zentr Waldökosysteme Univ Gött A 30:1–287

Cole DW, Rapp M (1981) Elemental cycling in forests. In: Reichle DE (ed) Dynamic properties of forest ecosystems. Cambridge University Press, London, pp 341–409

Coleman MD, Dickson RE, Isebrands JG (2000) Contrasting fine-root production, survival and soil CO_2 efflux in pine and poplar plantations. Plant Soil 225:129–139

Eichhorn J (1987) Vergleichende Untersuchungen von Feinwurzelsystemen bei unterschiedlich geschädigten Altfichten (*Picea abies* Karst.). Forschungsber Hessischen Forstl Versuchsanst 3:1–179

Epron D, Farque L, Lucot E, Badot P-M (1999) Soil CO_2 efflux in a beech forest: the contribution of root respiration. Ann Sci For 56:289–295

Fahey TJ, Hughes JW (1994) Fine root dynamics in a northern hardwood forest ecosystem, Hubbard Brook Experimental Forest, NH. J Ecol 82:533–548

Finér L, Messier C, De Grandpré L (1997) Fine-root dynamics in mixed boreal conifer-broad-leafed forest stands at different successional stages after fire. Can J For Res 27:304–314

Fogel R, Hunt G (1983) Contribution of mycorrhizae and soil fungi to nutrient cycling in a Douglas-fir ecosystem. Can J For Res 13:219–232

Fredericksen TS, Zedaker SM (1995) Fine root biomass, distribution, and production in young pine-hardwood stands. New For 10:99–110
Gale MR, Grigal DF (1987) Vertical root distributions of northern tree species in relation to successional status. Can J For Res 17:829–834
Grier CC, Vogt KA, Keyes MR, Edmonds RL (1981) Biomass distribution and above- and below-ground production in young and mature *Abies amabilis* zone ecosystems of the Washington Cascades. Can J For Res 11:155–167
Hellkvist J, Hillerdal-Hagströmer K, Mattson-Djos E (1980) Field studies of water relations and photosynthesis in Scots pine using manual techniques. In: Persson T (ed) Structure and function of northern coniferous forests – an ecosystem study. Ecol Bull 32:183–204
Helmisaari H-S, Hallbäcken L (1999) Fine-root biomass and necromass in limed and fertilized Norway spruce (*Picea abies* (L.) Karst.) stands. For Ecol Manage 119:99–110
Helmisaari H-S, Makkonen K, Olsson M, Viksna A, Mälkönen E (1999) Fine-root growth, mortality and heavy metal concentrations in limed and fertilized *Pinus sylvestris* (L.) stands in the vicinity of a Cu-Ni smelter in SW Finland. Plant Soil 209:193–200
Hendrick RL, Pregitzer KS (1993) The dynamics of fine root length, biomass, and nitrogen content in two northern hardwood ecosystems. Can J For Res 23:2507–2520
Hendriks CMA, Bianchi FJJA (1995) Root density and root biomass in pure and mixed forest stands of Douglas-fir and beech. Neth J Agric Sci 43:321–331
Hertel D (1999) Das Feinwurzelsystem von Rein- und Mischbeständen der Rotbuche Struktur, Dynamik und interspezifische Konkurrenz. Dissertationes botanicae 317. Cramer, Berlin, pp 1–190
Jackson RB, Canadell J, Ehleringer JR, Mooney HA, Sala OE, Schulze E-D (1996) A global analysis of root distributions for terrestrial biomes. Oecologia 108:389–411
Jianping S, Dali T, Sidong Z, Miao W (1993) Fine root dynamics of broadleaved Korean pine forest in Changbai Montains, China. Chin J Appl Ecol 4:241–245
Joslin JD, Henderson GS (1987) Organic matter and nutrients associated with fine root turnover in a white oak stand. For Sci 33:330–346
Kalhoff M (2000) Das Feinwurzelsystem in einem Kiefern-Eichen-Mischbestand. Dissertationes botanicae 332. Cramer, Berlin, pp 1–199
Keyes MR, Grier CC (1981) Above- and belowground net production in 40-year-old Douglas fir stand on low and high productivity sites. Can J For Res 11:599–605
Koch O (2002) Wasserumsatz eines Buchenbestandes im Hainich. Diploma thesis, University of Göttingen, Göttingen
Kochian L (1995) Cellular mechanisms of aluminium toxicity and resistance in plants. Annu Rev Plant Physiol Plant Mol Biol 46:237–260
Leuschner C, Backes K, Hertel D, Schipka F, Schmitt U, Terborg O, Runge M (2001a) Drought responses at leaf, stem and fine root levels of competitive *Fagus sylvatica* L. and *Quercus petraea* (Matt.) Liebl. Trees in dry and wet years. For Ecol Manage 149:33–46
Leuschner C, Hertel D, Coners H, Büttner V (2001b) Root competition between beech and oak: a hypothesis. Oecologia 126:276–284
Liu X, Tyree MT (1997) Root carbohydrate reserves, mineral nutrient concentrations and biomass in a healthy and a declining sugar maple (*Acer saccharum*) stand. Tree Physiol 17:179–185
Marshall JD (1986) Drought and shade interact to cause fine-root mortality in Douglas-fir seedlings. Plant Soil 91:51–60
McClaugherty CA, Aber JD, Melillo JM (1982) The role of fine roots in the organic matter and nitrogen budgets of two forested ecosystems. Ecology 63:1481–1490
Murach D (1984) Die Reaktion der Feinwurzeln von Fichte (*Picea abies* Karst.) auf zunehmende Bodenversauerung. Gött Bodenkundl Ber 7:1–128

Murach D (1991) Feinwurzelumsätze auf bodensauren Fichtenstandorten. Forstarchiv 62:12–17

Murach D, Ulrich B (1988) Destabilization of forest ecosystems by acid deposition. Geo J 17 2:253–260

Murach D, Wiedemann H (1988) Dynamik und chemische Zusammensetzung der Feinwurzeln von Waldbäumen als Maß für die Gefährdung von Waldökosystemen durch toxische Luftverunreinigungen. Ber Forsch Zentr Waldökosysteme Univ Gött B10:1–287

Nadelhoffer KH, Aber JD, Melillo JM (1985) Fine roots, net primary production, and soil nitrogen availability: a new hypothesis. Ecology 66:1377–1390

Oleksyn J, Reich PB, Chalupka W, Tjoelker MG (1999) Differential above- and below-ground biomass accumulation of European *Pinus sylvestris* populations in a 12-year-old provenance experiment. Scand J For Res 14:7–17

Olsthoorn AFM (1991) Fine root density and root biomass of two Douglas-fir stands on sandy soils in the Netherlands. I. Root biomass in early summer. Neth J Agric Sci 39:49-60

O'Neill RV, DeAngelis DL (1981) Comparative productivity and biomass relations of forest ecosystems. In: Reichle DE (ed) Dynamic properties of forest ecosystems. Cambridge University Press, Cambridge, pp 411–450

Paar U (1994) Untersuchungen zum Einfluß von Ammonium und Nitrat auf wurzelphysiologische Reaktionsmuster der Buche. Ber Forsch Zentr Waldökosysteme Univ Gött A115:1-124

Persson H, Von Fircks Y, Majdi H, Nilsson LO (1995) Root distribution in a Norway spruce (*Picea abies* (L.) Karst.) stand subjected to drought and ammonium sulphate application. Plant Soil 168/169:161–165

Polomski J, Kuhn N (1998) Wurzelsysteme. P Haupt Verlag, Bern

Pregitzer KS, Hendrick RL, Fogel R (1993) The demography of fine roots in response to patches of water and nitrogen. New Phytol 125:575–580

Raben GH (1988) Untersuchungen zur raumzeitlichen Entwicklung boden- und wurzelchemischer Stressparameter und dessen Einfluß auf die Feinwurzelentwicklung in bodensauren Waldgesellschaften des Hils. Ber Forsch Zentr Waldökosysteme Univ Gött A38:1–253

Rapp C (1991) Untersuchungen zum Einfluß von Kalkung und Ammoniumsulfat-Düngung auf Feinwurzeln und Ektomykorrhizen eines Buchenaltbestandes im Solling. Ber Forsch Zentr Waldökosysteme Univ Gött A72:1–293

Raspe S, Feger KH, Zöttl HW (1989) Erfassung der Elementvorräte in der Wurzelbiomasse eines 100jährigen Fichtenbestandes (*Picea abies* Karst.) im Schwarzwald. J Appl Bot 63:145–163

Ruark GA, Bockheim JG (1987) Below-ground biomass of 10-, 20-, and 32-year-old *Populus tremuloides* in Wisconsin. Pedobiologia 30:207–217

Ryan MG, Hubbard RM, Pongracic S, Raison RJ, McMurtrie RE (1996) Foliage, fine-root, woody-tissue and stand respiration in *Pinus radiata* in relation to nitrogen status. Tree Physiol 16:333–343

Sandhage A (1991) Dynamik und Elementgehalte von Fichtenfeinwurzeln. Untersuchungen von Beständen in verschiedener Höhenlage auf Böden aus Kalkstein. Mitt Dtsch Bodenkundl Ges 66:709–712

Sandhage-Hofmann A, Zech W (1993) Dynamik und Elementgehalte von Fichtenwurzeln in Kalkgesteinböden am Wank (Bayerische Kalkalpen). J Plant Nutr Soil Sci 156:181–190

Santantonio D, Grace JC (1987) Estimation of fine-root production and turnover from biomass and decomposition data: a compartment-flow model. Can J For Res 17:900–908

Santantonio D, Hermann RK (1985) Standing crop, production, and turnover of fine roots on dry, moderate, and wet sites of mature Douglas fir in western Oregon. Ann For Sci 42:113–142

Santantonio D, Santantonio E (1987) Effects of thinning on production and mortality of fine roots in a *Pinus radiata* plantation on a fertile site in New Zealand. Can J For Res 17:919–928

Santantonio D, Hermann RK, Overton WS (1977) Root biomass studies in forest ecosystems. Pedobiologia 17:1–31

Saurina NE, Kameneckaja IV (1969) Bull Mosk Obsc Ispyt Prir 74:96 (cit. in Santantonio et al. 1977)

Scherfose V (1990) Feinwurzelverteilung und Mykorrhizatypen von *Pinus sylvestris* in verschiedenen Bodentypen. Ber Forsch Zentr Waldökosysteme Univ Gött A 62:1–204

Schlesinger WH (1997) Biogeochemistry. An analysis of global change, 2nd edn. Academic Press, San Diego

Schneider BU (1990) Wachstum und Ernährung von Feinwurzeln unterschiedlich immissionsbelasteter Fichtenbestände des Fichtelgebirges. PhD thesis, University of Bayreuth, Bayreuth, pp 1–154

Simonovic V (1988) Biomass of roots in a natural oak-hornbeam ecosystem. IInd Int ISSR-Symp: plant roots and their environment. Abstracts, vol 2, p 32

Thomas FM, Hartmann G (1998) Tree rooting patterns and soil water relations of healthy and damaged stand of mature oak (*Quercus robur* L. and *Quercus petraea* (Matt.) Liebl.). Plant Soil 203:145–158

Vanninen P, Mäkelä A (1999) Fine root biomass of Scots pine stands differing in age and soil fertility in southern Finland. Tree Physiol 19:823–830

Van Praag HJ, Sougnez-Remy S, Weissen F, Carletti G (1988) Root turnover in a beech stand of the Belgian Ardennes. Plant Soil 105:87–103

Vinš B, Šika A (1977) Biomass of the above-ground and underground parts of sample trees of Norway spruce (in Czech). Pr Vyzk Ustavu Lesn Hospod Myslivosti 51:125–150

Vitousek P, Sanford RL (1986) Nutrient cycling in moist tropical forests. Annu Rev Ecol Syst 17:137–167

Vogt KA, Grier CC, Meier CE, Edmonds RL (1982) Mycorrhizal role in net primary production and nutrient cycling in *Abies amabilis* ecosystems in western Washington. Ecology 63:370–380

Vogt KA, Moore EE, Vogt DJ, Redlin MJ, Edmonds RL (1983) Conifer fine root and mycorrhizal root biomass within the forest floors of Douglas-fir stands of different ages and site productivities. Can J For Res 13:429–437

Vogt KA, Grier CC, Vogt DJ (1986) Production, turnover, and nutrient dynamics of above- and belowground detritus of world forests. Adv Ecol Res 15:303–377

Vogt KA, Vogt DJ, Moore EE, Fatuga BA, Redlin MR, Edmonds RL (1987) Conifer and angiosperm fine-root biomass in relation to stand age and site productivity in Douglas-fir forests. J Ecol 75:857–870

Vogt KA, Vogt DJ, Gower ST, Grier CC (1990) Carbon and nitrogen interactions of forest ecosystems. In: Persson H (ed) Above- and belowground interactions in forest trees in acidified soils. Air pollution report 32. Commission of the European Communities. Directorate-General for Science, Research and Development. Environment Research Programme, Brussels, Belgium

Vogt KA, Vogt DJ, Asbjornsen H, Dahlgren RA (1995) Roots, nutrients and their relationship to spatial patterns. Plant Soil 168/169:113–123

Vogt KA, Vogt DJ, Palmiotto PA, O'Hara J, Asbjornsen H (1996) Review of root dynamics in forest ecosystems grouped by climate, climatic forest type and species. Plant Soil 187:159–219

Wiedemann H (1991) Feinwurzeluntersuchungen in Buchenwaldökosystemen in Abhängigkeit vom Bodenchemismus. Ber Forsch Zentr Waldökosysteme Univ Gött A76:1–289

Wiedey G-A (1991) Ökosystemare Untersuchungen in zwei unterschiedlich exponierten Fichtenaltbeständen und in einem Kalkungs- und Düngungsversuch im Hils. Ber Forsch Zentr Waldökosysteme Univ Gött A 63:1–204

Wilczynski CJ, Pickett STA (1993) Fine root biomass within experimental canopy gaps: evidence for a below-ground gap. J Veg Sci 4:571–574

Xu Y, Röhrig E, Fölster H (1997) Reaction of root systems of grand fir (*Abies grandis* Lindl.) and Norway spruce (*Picea abies* Karst.) to seasonal waterlogging. For Ecol Manage 93:9–19

Yin XW, Perry JA, Dixon RK (1989) Fine-root dynamics and biomass distribution in a *Quercus* ecosystem following harvesting. For Ecol Manage 27:159–177

Prof. Dr. Christoph Leuschner
Dr. Dietrich Hertel
Georg-August-Universität Göttingen
Albrecht-von-Haller-Institut für Pflanzenwissenschaften
Abt. Ökologie und Ökosystemforschung
Untere Karspüle 2
37073 Göttingen, Germany

e-mail: cleusch@gwdg.de
e-mail: dhertel@gwdg.de

Light Heterogeneity and Plants: from Ecophysiology to Species Coexistence and Biodiversity

Fernando Valladares

1 Introduction

Light is produced by changes in the energy level of electrons (when an electron changes from a high-energy, or excited, state to a low-energy state, its atom will emit a photon), but its dual nature makes it scientifically puzzling: it moves through space as a wave, but when it encounters matter it behaves like a particle, the quantum (Achenbach 2001). In practice "light" is used for the portion of the electromagnetic spectrum in the vicinity of visible light. In plant biology and ecology, photosynthetically active radiation (PAR), which essentially coincides with visible light, is probably the most relevant measure of light. The PAR region is where energy is most abundant (it represents on average 43% of the solar irradiance), and it is strong enough to drive electron transport in photosynthesis, yet weak enough to avoid excessive damage to biological molecules. However, there exists no worldwide network for PAR measurements like the network of actinometric stations, where global, direct, diffuse and reflected solar radiation are measured using unified instruments and methodology and are metrologically based on the world radiometric reference (Ross and Sulev 2000). The dual nature of light, i.e., particle versus wave, affects the way light in general and PAR in particular is measured. One group of sensors measures energy (e.g., $W\ m^{-2}$), the other group quanta (e.g., $\mu mol\ m^{-2}\ s^{-1}$). General conversion factors are 1.814, 1.758, 2.127, and 0.462 μmol PAR W^{-1} for global, direct, diffuse, and reflected radiation, respectively (Ross and Sulev 2000), but these factors should be used with care since they may be different under different environmental conditions.

Light provides the energy used in photosynthesis and the signals used in photoregulation of plant growth and development, and is, among the factors affecting plants, perhaps the most spatially and temporally heterogeneous (Pearcy 1999). PAR in the understory ranges from 50–80% of full sunlight under leafless deciduous trees, to 10–15% in even-aged pine stands, 2.5% in closed spruce canopies, 0.2–0.4% in dense beech forests, and even less than 0.1% in certain tropical rainforests (Barnes et al. 1998). In addition to this quantitative variability, light is also qualita-

Progress in Botany, Vol. 64

tively heterogeneous. A total of five basic light environments can be found in terrestrial ecosystems according to the color of light: (1) forest shade (greenish or yellow-green light due to selective absorption of red and blue by vegetation), (2) woodland shade (bluish or bluish-grey light due to the dominance of the radiation from the sky), (3) small gaps (yellowish-reddish light due to direct sunlight), (4) large gaps, open, or any habitat under cloudy conditions (whitish light due to combination of sun and sky light, or because of the dominance of the white light radiating from clouds), and (5) any habitat early and late in the day, when sun is below 10° from the horizon (purplish light; Endler 1993; Kiltie 1993). Once the vegetation is established, it becomes the main cause of the remarkable spatial heterogeneity of light in most natural habitats (Fig. 1). Forest overstory canopies never close completely and there is a continuous gradient in gap size from those between 1 cm^2 and 1 m^2 due to foliage clumping or wind-induced abrasion of adjacent crowns, to those between 50 and 600 m^2 due to tree falls. Changes in the light environment associated with successional changes are relatively slow and predictable, and allow individual plants to anticipate and respond; the same applies to both seasonal and diurnal variations. Sun flecks, which often contribute a substantial fraction of the total light available in the understory (Pearcy 1983; Chazdon 1988), cause the most rapid

Fig. 1. Influence of light heterogeneity on the structure and dynamics of plant populations and communities can be explored via mechanistic approaches (*left panels*) or correlation studies (*right panels*). In the understory the low, diffuse, background light is punctuated by sun flecks of various durations and intensities, which can be seen in diurnal cycles estimated from hemispherical photographs of the canopy (*central photo* in *left panels*). This temporally heterogeneous light can be exploited by plant species of different crown architectures, i.e., with different light capture efficiencies and allocation patterns. Light capture efficiency by the whole crown can be simulated by realistic 3-D reconstructions using computer models (plant images in *left panels* were obtained with Y-plant; Pearcy and Yang 1996). Plant species also differ in their photosynthetic capacity to use this variable and dynamic light (Valladares et al. 1997). Frequently, many of these differences are genetically fixed (Arntz and Delph 2001). These different plants coexist in the understory either by being functionally equivalent (e.g., two contrasting architectures might have the same light capture efficiency, two different photosynthetic physiologies might lead to the same daily total carbon gain), by converging by phenotypic plasticity, or by the fact that other traits (tolerance to herbivory or water stress, efficient reproduction, dispersal or recruitment) counteract suboptimal morphologies and physiologies. Functional convergence in response to shade could explain the relatively high levels of biodiversity in tropical understories where light is the limiting factor (see, e.g., Valladares et al. 2002b). However, light in most natural ecosystems is heterogeneous not only in time but also in space and many forest ecosystems exhibit a complex spatial distribution of mean daily photosynthetically active radiation (PAR). Correlation studies considering spatial patterns explicitly (e.g., by means of geostatistics) are enhancing our understanding on where, when and to which extent light heterogeneity contributes to species coexistence and promotes biodiversity (see, e.g., Nicotra et al. 1999; Bascompte and Rodriguez 2001)

scale of temporal heterogeneity: from seconds to minutes. The spatial scale of sun flecks typically vary from 0.1 to 1 m, so that often only part of the crown of an understory plant will be influenced by a given sun fleck (Baldocchi and Collineau 1994). Spatial autocorrelation explored with arrays of photosensors revealed very fine grain heterogeneity (autocorrelation of 0.4 within 0.2 m and almost none for 0.5 m) due to the interplay of the overstory canopy and the self-shading within the crown of the understory plant (Chazdon et al. 1988). The spatial heterogeneity

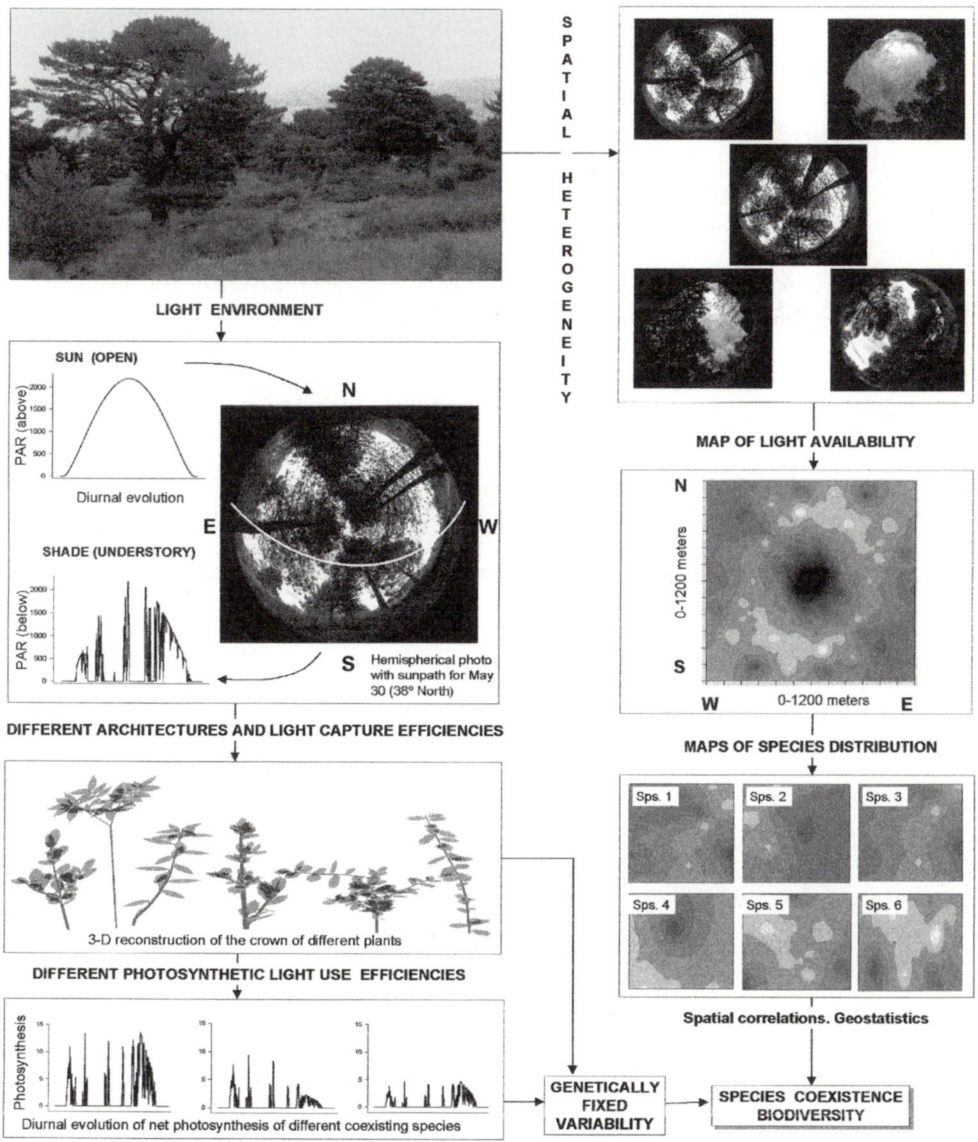

observed for a day is noticeably decreased when monthly or annual means are considered due to seasonal changes in the solar elevation angle. For this reason, spatial analysis carried out with hemispherical photographs (Fig. 1) generally reveals heterogeneity of coarser grain: autocorrelation was still significant at 2.5 m in a tropical forest (Becker and Smith 1990). Autocorrelation was significant at even longer distances (>10 m) when only the spatial heterogeneity of diffuse light was considered in a study of different tropical forests (Nicotra et al. 1997). The distinction between spatial and temporal heterogeneity of light is convenient but also rather artificial (Pearcy 1999). The spatial and temporal scales are highly correlated and, in terms of analysis, they are frequently interchangeable (Baldocchi and Collineau 1994).

Light quantity and quality affect ecosystem properties in general and plant performance in particular, but the effects are in many cases difficult to unveil due to correlations (e.g., high light–high temperature–high water stress) and complex interactions with other environmental factors (e.g., water stress reduces the photosynthetic capacity to use light, ultraviolet light affects plants directly but it also alters herbivore and microbial activity which in turn affects plants both directly and indirectly via changes in nutrient cycling). The purpose of this chapter is to review a selection of the abundant literature on this rapidly developing field to illuminate some of the most promising areas for future research. The task of integrating the information has been challenging due not only to the large number of publications, but also to the variety of interrelated topics, techniques and approaches that have appeared since the earlier and now classical attempts of exploring light as an ecological factor (e.g., Shirley 1929; Bainbrige et al. 1966; Horn 1971).

2 Light Capture and Photosynthesis: Scaling Up to the Plant

a) Crown Architecture, a Compromise of Multiple Functions

Leaves are the ultimate sink for light and their optical properties can significantly affect light capture by the whole plant, while their photosynthetic characteristics dramatically influence whole plant photosynthetic performance. Leaf morphological and physiological adaptations to high and low irradiances received intense attention during the 1970s and 1980s, which led to a thorough description of the so-called sun and shade types of leaf (Björkman 1981; Evans et al. 1988; Larcher 1995). The last decades of the twentieth century have witnessed a change in the research priorities of most plant ecologists, with the whole plant and not the leaf as the main target. This in turn has led to a renovated, more functional interest in plant architecture (Valladares 1999; Valladares and Pearcy 2000).

A striking feature of understory vegetation in neotropical rainforests is the large variation in plant appearance and leaf morphology (Bongers and Popma 1990; Leigh 1998; Turner et al. 2000). Are these contrasting habits and morphologies rendering a functionally equivalent architecture in terms of light capture efficiency? Branching pattern and leaf arrangement in the crown of a plant have a direct impact on the efficiency of leaf display and consequently on light capture and photosynthesis (Valladares 1999). However, crown architecture influences numerous aspects of whole-plant function and not only light capture efficiency (Pearcy and Valladares 1999). Thus, real plants in low-light environments may deviate considerably from the predicted, optimal pattern for light capture as a result of other constraints on crown form (e.g., water transport, mechanical stability). For example, optimal display of foliage for light capture may be prohibited by requirement for high resources for supporting tissues (King 1991; Pearcy and Yang 1998; Poorter and Werger 1999). Or, selection in trees for a fast increase in height to attain a position in the canopy can be at the cost of a narrow crown far from optimal in terms of light capture during the understory phase (King 1990; Clark and Clark 2001). Crown architecture has a static function (i.e., efficient leaf display at the current light environment) and, since light is predictably more available at increasing heights, also a dynamic function (i.e., increasing in height at the lowest construction cost), which translates into a trade-off between height growth and leaf area extension (Kohyama and Hotta 1990). All these constraints and trade-offs could explain why among all the plants in a comparative study of 24 co-occurring understory plants none had a light absorption efficiency greater than 0.75 (Valladares et al. 2002b). When natural selection acts on two or more traits simultaneously, evolutionary responses may be limited by genetic correlations that reflect genetic, physiological and/or developmental constraints (Ackerly 2000). Diversity in architecture may reflect different ways of using light resources and thus may enable species coexistence (Aiba and Kohyama 1997; Fig. 1). As argued by Kohyama (1987), the architectural and allometric diversity found in the forest understory can be related to the alternative of "choosing" any of these combinations of traits.

b) Plant Movements Influenced by Light

Crown architecture is rather dynamic. Changes in its geometry and size can be due to external factors (e.g., mechanical damage) and also to internal responses to environmental clues (Valladares 1999). Leaves can be arranged either to increase or to decrease light capture, and, as in the case of chloroplasts, leaf movements can be triggered by light itself (Koller 1990). Leaves of many terrestrial plants perform "sleep movements", assuming a compactly folded configuration during the night and an unfolded one in daytime, increasing interception of light by the lamina (Koller 2000). Photonastic movements (such as sleep movements) take place in a determined direction but are independent of the direction of the light, while phototropic movements take place according to the direction of light and are more prone to modifications by environmental stresses (Ehleringer and Werk 1986; Koller 2000). Heliotropic movements track the daily solar transit and after sunset they reverse direction; they can be diaheliotropic, when leaves are kept perpendicular to the sunrays, or paraheliotropic when they are kept parallel to the sunrays, which minimizes light and heat loads and can be advantageous under water stress (Ehleringer and Werk 1986). Leaf movements are effective in regulating light capture by the whole crown only when mutual shading by leaves is reduced, i.e., when individuals are widely spaced and leaf area indices are lower than 1.5 (Ehleringer and Forseth 1989). Leaf movements require a high direct-to-diffuse light ratio, so they are restricted to the upper layer of a canopy and are scant in the understory, except certain downwards movements in the presence of high light that can prevent photoinhibition of deep shade plants during sun flecks (Powles and Björkman 1981). Not only leaves, but also apical parts of the shoot may exhibit heliotropic movements, which have been described not only in the domestic sunflower and its wild relatives, but also in *Crozophora tinctoria*, *Xanthium strumarium*, and in a number of Arctic and alpine plants (Kevan 1975; Stanton and Galen 1993; Koller 2000).

c) Morphological and Structural Photoprotection

Certain features of the shape and architecture of the crown, such as self-shading and steep leaf elevation angles, can prevent excessive light from reaching the photosystems together with leaf attributes such as pubescence or thick cuticular or epidermal layers (Valladares 1999; Valladares and Pugnaire 1999). This morphological or structural protection can be efficient not only against excessive PAR, avoiding photoinhibition and overheating, but also against ultraviolet radiation, avoiding DNA mutations and growth alterations (Day 1993; Skaltsa et al. 1994; Kyparissis et al. 1995; Grammatikopoulos et al. 1998). In fact, certain plant life forms have been shown to be more effective than others in screening out UV-B radiation (Day et al. 1992). Structural and physiological protection against excessive light interact and complement each other. Carotenoids, primarily photoprotective (Young 1991), exhibited higher concentrations in leaves than in stems in a comparison of six leguminous, green-stem shrubs (Valladares et al. submitted), which was in agreement with the different exposure to high light and risk of photoinhibition associated with the different mean elevation angle of leaves and stems (Ehleringer and Cooper 1992; Valladares and Pearcy 1999). This finding agrees with previous studies where the particular orientation and daily light interception of each individual leaf correlated with the concentration of the photoprotective carotenoids of the xanthophyll cycle (Lovelock and Clough 1992).

d) Geometry of the Crown, Curiosity or Function?

In most plants, leaf primordia at the apex appear as far as possible from each other (Hofmeister's rule; Jean 1984), which in plants with helical leaf arrangement commonly leads to a divergence angle between consecutive leaves near the "golden angle" of 137.5° at which there is no complete overlap between any two leaves on a vertical stem (Bell 1993). For plants in shaded habitats, minimum leaf overlap is a predicted characteristic. For that reason, phyllotaxis (the geometry of leaf arrangement on stems) and the mathematically intriguing properties of the Fibonacci series in divergence angles have been historically dealt with in terms of selection pressures favoring light interception (Sekimura 1995; Adler et al. 1997; Jean and Barabé 1998). Unusual patterns such as the monostichous phyllotaxis of *Costus*, with a divergence angle as low as 30–40°, has attracted and puzzled many botanists. However, after more than 120 years of study, the mystery of costoid phyllotaxy still remains (see Kirchoff and Rutishauser 1990 and references cited therein). The small divergence angles of *Costus* are correlated in certain species with helical twining of the stem to give the shoot the appearance of a spiral staircase (in this case the phyllotaxis is referred as spiromonostichous; Bell 1993). Is this staircase-like crown compensating for the small divergence angle between consecutive leaves? The answer after a comparative study of 24 species cooccurring in a tropical understory is yes (Valladares et al. 2002b). Distichy, characteristically orthodistichy, is the typical kind of phyllotaxy among monocotyledons (Wilder 1992 and references cited therein); this implies the existence of two straight rows of leaves and thus, a potentially large mutual shading among leaves in vertical stems. This potentially inefficient leaf display was overridden by either an arced stem, by plagiotropic shoots, or by long petioles in the same comparative study of understory plants (Valladares et al. 2002b). The extent to which phyllotaxis influences light capture of real plants still remains poorly understood, but the fact that a suite of morphological features can compensate for suboptimal phyllotactic arrangements (Niklas 1988; Brites and Valladares, submitted) suggests that phyllotaxis involves more geometric and mathematic curiosity than functional implications. Other geometric characteristics of the foliage, such as the elevation angle of the leaf and the internodal length, have been confirmed as important determinants of light capture in a number of quantitative, functional studies of plant crowns (McMillen and McClendon 1979; Smith and Ullberg 1989; Herbert 1996; Valladares and Pearcy 1998).

e) Photosynthetic Stems

It is well known that plant parts other than leaves can be green and photosynthetically active, and, in the case of green stems, they can have remarkable implications for the overall carbon gain of the plant (Pfanz and Aschan 2001). Net photosynthetic capacity of green stems contributes to an average of 50% of whole plant carbon gain in habitats like deserts, tropical dry or thorn woodlands and Mediterranean type ecosystems (Nilsen 1992a,b). Stem photosynthesis has a number of possible benefits: extension of period of carbon gain in environments with periodic droughts (Smith and Osmond 1987; Nilsen 1992a), heat (DePuit and Caldwell 1975; Smith and Osmond 1987) or excessive irradiance (Valladares 1999; Valladares and Pugnaire 1999), and tolerance to herbivory (Bossard and Rejmanek 1992). Even though green stems are photosynthetically not efficient in a number of woody species (Comstock and Ehleringer 1988; Pfanz and Aschan 2001), photosynthetic activity of the cortical tissues of leguminous shrubs like *Retama*, *Spartium* or *Cytisus* is as high and efficient as that reported for leaves (Bossard and Rejmanek 1992; Nilsen 1992b; Haase et al. 1999). Leguminous shrubs with green stems differ in the relative importance of leaves for whole plant carbon gain, from species such as *Retama sphaerocarpa*, practically leafless, to species like *Cytisus scoparius* and *Spartium junceum*, where leaves contribute 50–70% of whole plant carbon gain (Nilsen 1992b; Nilsen et al. 1993). A recent comparative study of six leguminous species has shown that the relative amount of leaves determines the short-term survival in deep shade (Valladares et al., submitted). Thus, species with green stems and a low number of leaves seem to be morphologically restricted to well-lit habitats. In fact, the poor light harvesting efficiency of vertical stems becomes a potential advantage in high-light habitats since it provides protection against excessive radiation, enhancing survival and performance in arid environments (Valladares and Pugnaire 1999).

f) Leaf Phenology and Light Capture

Leaf phenology, or the arrangement of leaves in time, is also an important aspect of the light-harvesting strategy of plants (Kikuzawa 1989, 1991). Interrelationships among the three components of leaf phenology, i.e., leaf longevity, leaf habit (evergreenness or deciduousness) and leaf emergence pattern (simultaneous or successive), together with shoot architecture and photosynthetic capacity, affect plant productivity (Kikuzawa 1995; Kikuzawa et al. 1996). Kikuzawa (1995) presented a synthesis between resource-based and climate-based models of leaf phenology. In his model, plants with successive leaf emergence and short leaf longevity (he assumes that the first, unshaded leaf is only replaced by a second leaf when its photosynthetic ability declines) have straight, multilayered shoots and attain high production in resource-rich environments (e.g., early successional stages), while plants with a simultaneous leaf emergence increase the longevity of individual leaves and avoid self-shading by inclining their apical shoot (monolayer canopies) to attain high production in resource-poor environments. The connections between light environment, seasonal variability, canopy architecture, photosynthetic capacity and leaf phenology are appealing but the model oversimplifies the reality. Some of the assumptions (e.g., self-shading is not a serious problem for plants with successive leaf emergence, the replacement takes place only when the photosynthetic ability of the leaf declines) are not universally valid (Valladares and Pearcy 1998, 1999), and the selective pressure for maximizing light capture does not hold in many habitats since self-shading in high-light environments where plant growth is limited by water or extreme temperatures might not be a constraint but an advantage (Valladares 1999; Valladares and Pugnaire 1999).

g) Advantages and Limits of Computer Models

Computer models can handle and integrate complex and interrelated variables and thus are powerful tools to simulate the role of light for net carbon assimilation and plant growth. They have been used in a range of applications, from leaf to canopy photosynthesis (Thornley 1998; Beyschlag and Ryel 1999; Farquhar et al. 2001), and from crown architecture (Pearcy and Yang 1996; Fig. 1) to plant growth (Kramer 1994). The fascinating geometric properties of plant architecture in particular have received considerable attention from modellers (Prusinkiewicz and Lindenmayer 1996; Prusinkiewicz 1999). Empirically based models and mechanistic approaches demand detailed knowledge in a frequently too extensive list of parameters, while goal-seeking methods based on optimality arguments are usually simpler. However, when only a single variable is chosen for optimization, not only realism but also accuracy in the predictions is significantly reduced (Wirtz 2000). Despite the realistic appearance of certain computer simulations, they frequently make broad and sometimes too general assumptions on the physiology underlying the models. The lack of congruence between model predictions and experimental data for the number of primary branches in a 3-D modeling of photomorphogenesis of *Trifolium repens* in neutral versus green shading and control light environments (Gautier et al. 2000) reveals that there is still a long way to go before we fully understand and can accurately simulate the integrated response of plants to different light signals. Many processes underlying leaf trait dynamics are quantitatively or even structurally unknown, which confines the number of variables used in mechanistic models such that certain traits that are dynamic have to be treated as constant (Wirtz 2000). For more general reviews and references on models of plant growth that incorporate light capture and photosynthesis, see Hari et al. (1991), Norman and Arkebauer (1991), Room et al. (1996), De Reffye and Houllier (1997), Prusinkiewicz (1998), and Valladares (1999).

3 Acclimation to Light, Tolerance of the Extremes and Phenotypic Plasticity

a) Photomorphogenesis and Whole-Plant Response to Light

Terrestrial plants are developmentally versatile since as sedentary photoreceptors they must accommodate diurnal, seasonal, and long-term changes in light environments (Bradshaw 1965; Schlichting 1986; Niklas 1997). Plants have been shown to adjust their morphology and physiology to available light at different hierarchical levels: crown allometry (e.g., height to diameter ratio, supporting to non-supporting tissues weight ratio), crown architecture (branching pattern, foliage arrangement, leaf area ratio), leaf phenology (longevity, habit, and emergence pattern), specific leaf area (area to weight ratio of leaves), leaf optical properties (absorbance, transmittance), and finally leaf gas exchange properties (photosynthesis, respiration). Responses to variations in the light environment are expressed over a variety of time scales, from the rapid movements of antenna proteins and pigments between the two photosystems (considered part of the acclimation response), to the relatively slow changes in plant architecture (considered part of the plastic response).

The spectral change experienced by the light transmitted through a canopy is sensed by the plants as a low-red/far-red ratio via the phytochrome system, and is capable of modulating a suite of architectural responses often described as the shade-avoidance syndrome (Smith 1982; Ballare 1999). The most important feature of the shade-avoidance response is the elongation of vertical stems and petioles, which results in leaf blades located higher in the canopy profile (Aphalo et al. 1999). Shade avoidance seems to represent an alternative to shade tolerance (Henry and Aarssen 1997). In fact, tropical shade-tolerant species have not been found to respond to changes in light quality (Kwesiga and Grace 1986; Riddoch et al. 1991).

b) Phenotypic Plasticity

Plants exhibit a remarkable phenotypic plasticity, and a large effort has gone into the investigation of their plastic response to light (Hirose 1987; Strauss-Debenedetti and Bazzaz 1991; Chazdon 1992; Sultan and Bazzaz 1993; Agyeman et al. 1999; Valladares 2000; Valladares et al. 2000b). Phenotypic plasticity is the expression of environmentally induced variability among individuals of identical genotype (Bradshaw 1965; Pigliucci 2001). Since identical genotypes are often difficult to identify and compare, phenotypic plasticity is frequently taken as the variability among individuals of similar genotype (same species and population) that can be unambiguously attributed to environmental variability (Valladares et al. 2000a,b). Even though the degree of expression of plasticity is under genetic control (Hoffmann and Parsons 1991), plasticity is not necessarily adaptive (Pigliucci 2001). Phenotypic plasticity acts as a buffer against spatial or temporal variability in habitat conditions, and serves three main functions: maintenance of homeostasis, foraging for resources, and defense (Grime et al. 1986). Despite the recurring idea that plasticity and developmental instability might be correlated (Tarasjev 1995), there is little evidence for this correlation, and, on biological grounds, the relationship is not to be expected if the misconception that plasticity is a disturbance of the developmental trajectory is discarded (Pigliucci 2001). With the exception of sun- and shade-obligate species, phenotypic plasticity can extend the ranges of most plant species along light gradients, since patterns of individual response to environment are a major element in the realized ecological breadth of species (Sultan et al. 1998 and references therein). Consequently, species with a high degree of phenotypic plasticity should express less clinal and/or ecotypic differentiation (Saxe et al. 2001). Plasticity has been traditionally viewed as an alternative to specialization (Sultan 1992), but new approaches and studies have revealed that plasticity itself can be the result of specialization: plasticity of specialized genotypes is enhanced when specialization is associated with favorable ends of an environmental gradient (Lortie and Aarssen 1996; Valladares et al. 2000b; Balaguer et al. 2001). In comparisons of invasive and native species of Hawaii, it was found that the invaders were more plastic in their photosynthetic and dark respiration responses to light than the native species (Pattison et al. 1998; Baruch et al. 2000; Durand and Goldstein 2001). However, even though plasticity is expected to enhance growth under varying environmental conditions, and thus might be an adaptive feature of invasive species, plasticity does not necessarily lead to increased growth rates (Walters and Reich 1999).

In the case of green-stem species, a plastic response to light can be mediated by responses of either leaves, green stems, or both (Pfanz and Aschan 2001). An inverse relationship between the plasticity of the stems and that of the leaves was found in a comparison of six leguminous shrubs differing in their relative amount of leaves: while mean plasticity of the stem increased with the relative amount of leaves, the mean plasticity of leaves decreased (Valladares et al., submitted). The relative amount of leaves was also a plastic feature, and the plasticity for this trait exhibited significant differences among the species. While morphological adaptations and responses to light tend to enhance shade survival (e.g., enhancing light capture efficiency, as discussed in Valladares et al. 2002b), physiological plasticity seems to enhance high light tolerance (Valladares et al. 2002a). The relationship between plasticities of a series of traits and reproductive fitness is very complex due to phenotypic and plasticity integration (Schlichting 1989). It was found that adaptive constancy of phenotypic traits central to fitness (phenotypic stability) is achieved by plasticity in developmentally related traits (Sultan 1995).

c) Evolution of Phenotypic Plasticity

Phenotypic plasticity can evolve by natural selection as a trait in its own right, but determining how plasticity evolves is not easy (Scheiner and Callahan 1999; Pigliucci 2001). Adaptive phenotypic plasticity can evolve as a strategy to cope with environmental heterogeneity, but it might not be the only possible outcome. Important empirical information can be obtained in comparisons of the plasticity of closely related species, but comparative studies of plasticity are not common, and very few include an explicitly phylogenetic-comparative method approach (Pollard et al. 2001). Examples of interspecific comparisons of plastic responses show that plasticity can evolve rapidly, since it can be very different in closely related species (Valladares et al. 2000b; Pigliucci 2001). Despite the still numerous uncertainties, there is increasing support to the notion that phenotypic plasticity is central to evolution and is not a minor phenomenon secondary to genetic adaptation (Sultan 1987, 1992; West-Eberhard 1989;Scheiner 1993; Soares et al. 1999). Plasticity can play a role in macroevolutionary phenomena by allowing persistence of populations under stressful or novel environments or by generating phenotypic novelties, and the complete spectrum of these evolutionary mechanisms must be incorporated into our interpretation of phenotypic evolution (Pigliucci 2001).

d) To Respond or Not to Respond? Costs of Plasticity

Resources are heterogeneous and plasticity enhances their harvesting; however, there is no plant species perfectly able to adapt by plasticity to any and all environmental conditions. Thus, the evolution of phenotypic plasticity must have some limits. In fact, a conservative resource use strategy involving a low responsiveness seems to be adaptive under stressful or generally adverse conditions (Valladares et al. 2000a,b; Balaguer et al. 2001). In temporally varying environments, the production of a highly differentiated phenotype under favorable conditions might compromise survival when conditions return to normal levels (Waller 1991). Combining the revisions of De Witt et al. (1998) and Pigliucci (2001), the limits and costs of plasticity can be distinguished and summarized as follows:

Limits of plasticity:

1. Information reliability of environmental cues.
2. Deficient sensory capabilities to perceive environmental change.
3. Time lag between the environmental change and the response.
4. Developmental range of possible phenotypes.
5. Functional integration resulting in trade-offs between traits and their plasticities.
6. Lack of genetic variation for plasticity, which can be due to reduced heritability or to strong correlation between plasticity and other traits.

Costs of plasticity:

1. Maintenance of sensory and regulatory mechanisms.
2. Production costs of producing structures plastically versus through fixed genetic responses.
3. Energy expenditures for information acquisition.
4. Developmental instability associated with reduced canalization and developmental "imprecision".
5. Genetic costs due to deleterious effects of plasticity genes through linkage, pleiotropy, or epistasis with other genes.

Once we have a detailed description of costs and limits, how do we actually measure them? As acknowledged by De Witt et al. (1998), it may be very difficult to experimentally disentangle these categories. The most immediate exploration of the costs of plastic responses is by regressions of plasticity and the trait mean within one environment against a measure of fitness (van Tienderen 1991); however, the approach is noisy and frequently fails to reveal costs unless a very large number of genotypes are used (van Klausen et al. 2000). Other approaches involve the experimental usage of hormones to induce different phenotypes and the quantification of the fitness of these phenotypes under different environmental conditions. This approach was used in the exploration of costs on stem elongation in the shade-avoidance response of the common bean (Cipollini and Schultz 1999). However, costs of plasticity must not be confused with the fitness decrement associated with the expression of a wrong phenotype in a given environment.

4 Population Plasticity by the Coexistence of Genotypes

Environmental variation is important for the maintenance of genetic variation in phenotypic plasticity (De Jong and Gavrilets 2000). Genetic variation in any trait, including phenotypic plasticity, allows for a plastic response to environmental heterogeneity at the population level, which, added to the phenotypic plasticity of the individuals, enhances the chances of species survival in changing or highly heterogeneous environments. In a comparison of three populations of Kermes oak (*Quercus coccifera*), individuals from the most favorable and at the same time most heterogeneous site had both the greatest genetic variability and the highest phenotypic plasticity in response to light (Balaguer et al. 2001), which agrees with the proposal that selection in heterogeneous environments leads to the coexistence of genotypes with greater phenotypic plasticity (Sultan 1987). Environmental heterogeneity of a given habitat interacts with its mean degree of adversity, making generalizations on the effects of heterogeneity on genetic variation and plasticity at the individual and population levels less clear. Favorable environments are frequently more heterogeneous than adverse environments (e.g., tropical forests versus arid shrublands), and they tend to allow for the coexistence of a larger number of genotypes, but whether this is due to heterogeneity, benignity or both is unclear.

The theory of metapopulations has increased our understanding of the consequences of demographic dynamics for patchily distributed plant species (Hanski 1997; Hanski and Gilpin 1997; Porembski and Barthlott 2000). Research on metapopulation dynamics is expected to answer questions about minimum size of populations and the minimum amount of suitable habitat required for long-term metapopulation persistence. Spatially realistic metapopulation models with a limited number of parameters may represent the right compromise between realism and practical prediction capacity (Hanski 1997).

5 Species Replacement Along Light Gradients: the Tropical and Temperate Models

a) Competition and Succession: Negative Correlations Between Adaptations to Sun and Adaptations to Shade

Most of our knowledge on the role of light in vegetation dynamics has been obtained from temperate and tropical forest ecosystems. In fact, light is thought to be an important, and in many cases the most important, determinant of plant establishment, growth and survival in tropical rainforests (Osunkoya et al. 1992; Nicotra et al. 1999). Over time, a typical successional pattern is expressed: plants of greater shade tolerance than that of the overstory species tend to establish and grow-up in the shaded understory (Bazzaz 1996; Barnes et al. 1998). According to this observation, woody species are classified as early or late successional. The common pattern of species replacement in tropical and temperate models of succession is largely a consequence of the negative correlation across species between adaptation to high light, which involves maximization of carbon fixation per unit leaf protein, and adaptation to low light, which involves maximization of light capture (Bazzaz 1979, 1991; Givnish 1988). However, scaling up from species to community responses is complicated by the fact that physiological responses do not remain constant (Bazzaz and Stinson 1999), and important mechanisms enabling coexistence of species differ throughout the developing stages of their life history (Tokeshi 1999; Nakashizuka 2001). Community assembly rules are determined by the response of plant life-history traits to community profiles (horizontal and vertical heterogeneity, disturbances and biotic interactions), but the relative contribution of chance factors is still far from well established. Chance factors seem to be less important in temperate tree communities than in tropical ones (Nakashizuka 2001).

b) Tolerance or Avoidance: Two Alternatives to the Shade?

The sun versus shade adaptation model is too simplistic, since adaptations of plants to cope with shade actually involve two rather opposing strategies: tolerance (the classical view of shade adaptation) and avoidance. However, shade avoidance might not be expressed consistently throughout the life of plants and can be modulated by resource availability (Monaco and Briske 2000), so shade tolerance and shade avoidance may not always be mutually exclusive within one species. In fact, some species display an intermediate combination of both strategies. Henry and Aarssen (1997) presented a model for the trade-off or negative correlation between shade tolerance and shade avoidance, linking successional stage with shade strategy and phenotypic plasticity. In this model, early successional species have both weak shade tolerance and weak shade avoidance, and shade avoidance should be most common in the early to intermediate stages of succession, where the probability of encountering high irradiance through vertical growth is still high. Plasticity may not only extend the ranges of species along light gradients, but also make the negative correlation between shade avoidance and shade tolerance more difficult to detect.

6 Species Replacement: Particularities and Uncertainties in Dry Environments

a) Competition and Succession in Light Gradients with Water Constraints

Radiation in arid environments is usually excessive due to the photosynthetic limitations imposed by water stress. However, the shade in Mediterranean-type ecosystems is also dry, and plant allocation to the aerial part for enhanced light capture is in conflict with allocation to the underground part for enhanced water uptake, according to the light-moisture trade-off hypothesis (Tilman 1988). While interactions between high light and drought have been the subject of many studies (e.g., Valladares and Pearcy 1997; Faria et al. 1998), interactions between shade and drought have been far less explored (Abrams and Mostoller 1995; Holmgren 2000; Sack et al. 2002; Valladares and Pearcy 2002). Due to these two facts, the interaction of light with drought and the lack of knowledge on plant responses to shade plus drought, succession is less understood in woody Mediterranean-type formations than in temperate and tropical forests, where species can be ranked according to their shade tolerance, which reflects their successional status (Vila and Sardans 1999; Zavala et al. 2000). Plant response to shade is nutrient and moisture dependent and canopy structure is also expected to vary depending on soil moisture and nutrients (Abrams et al. 1992; Abrams and Mostoller 1995; Barnes et al. 1998; Battaglia et al. 1998; Valladares et al. 2000a). Thus, edaphic factors can influence both the light environment experienced by understory plants by affecting the forest canopy, and the ability of understory plants to survive and respond to this light environment. Consequently, in agreement with Henry and Aarssen (1997), both the relationship between sun-shade adaptation and succession, and the relationship between shade tolerance and shade avoidance, must be examined within the context of variable soil moisture and nutrients. This conclusion is especially pertinent in the case of arid environments.

The long-term historical impact of humans on forest composition and dynamics in Mediterranean-type formations superimposed on already heterogeneous environments has increased biodiversity and complicated the identification of trends in species replacement and coexistence (Zavala et al. 2000). The frequency and importance of disturbances in these formations also introduce complexity in this search for generalizations. Differences in root-to-shoot ratios seem to explain oak species position along soil-moisture gradients (Matsuda et al. 1989), but expectations from the hypothesis of light-moisture allocation trade-off are not found in a number of Mediterranean species. For instance, certain species are relatively drought sensitive despite their large allocation to roots in comparisons of obligate seeders and resprouters (Retana et al. 1999). This is due to the fact that root-to-shoot ratio is only an approximate index of allocation pattern for acquisition of water. Roots serve as storage organs of carbohydrates, nutrients and meristematic tissue in many resprouter species, well adapted to cope with the common disturbances of fire, herbivory, droughts and coppicing (Canadell and Roda 1991).

Under conditions of high light, ultraviolet light can play an important role at both the individual and the community level especially in high latitudes or elevations. Recent studies in Tierra del Fuego suggest that the increase in UV-B radiation associated with the erosion of the ozone layer might be affecting the functioning of this ecosystem by differentially inhibiting the growth of some plant species and by altering plant–insect interactions (Rousseaux et al. 2001).

b) Facilitation and Positive Plant Interactions

While many ecological studies focus on competition and its implications for species replacement and community organization, less attention has been paid to the roles of cooperative relationships among organisms in promoting species coexistence and biodiversity, but such a view needs to be substantially changed due to ubiquity and importance of cooperation in nature (Wilson and Agnew 1992; Callaway and Pugnaire 1999; Tokeshi 1999). The strong irradiance of most arid environments is likely to cause photoinhibition and overheating of the photosynthetic surfaces, compromising plant survival. For this reason, shrubs, trees or any objects that cast shade can facilitate the establishment of plants in these harsh environments (Pugnaire et al. 1996). It has been argued that competition in these shaded, more favorable microhabitats can be larger than in the open, but it does not seem to be the case in desert or very arid sites (Tielborger and Kadmon 2000a). The balance between competition or interference and facilitation in arid sites is affected by environmental gradients (Tielborger and Kadmon 2000b; Pugnaire and Luque 2001). A recent comparison of the effect of shrubs on their understory plant communities in a Spanish semi-arid location has pointed out that not all the different shades created by different shrub species are equally good shelters due to interactions with the soil properties affected by each shrub species (Pugnaire et al., submitted). The effect of the shrub on the soil properties can vary from enhanced moisture and nutrient content, to enhanced accumulation of inhibitory compounds, which exerts from positive to negative effects on the productivity of the understory plant communities.

7 Species Coexistence and Biodiversity

a) Unveiling Patterns and Integrating Processes

Many attempts have been made to link plant biodiversity and patterns in species richness with abiotic parameters (Tokeshi 1999; Porembski and Barthlott 2000). For instance, O'Brien et al (1998) reported that most of the variation in richness of woody plants in southern Africa was accounted for by climate, with richness being a linear function of liquid water and a parabolic function of energy. However, the fact that similar environments in different geographic regions deviate largely in species richness challenges the view that abiotic patterns are the main determinants of diversity patterns (Latham and Ricklefs 1993), and emphasizes the importance of evolutionary and historical aspects (Herrera 1992; Porembski and Barthlott 2000). Coexistence of species violates the competitive exclusion principle and more than 100 mechanisms have been proposed that delay or prevent competitive exclusion (Tokeshi 1999; Wright 2002). One of the conditions required to realize competitive exclusion is that the environment is temporally constant and has no spatial variation, which is not the case in many natural environments, as shown earlier for the case of light. It is increasingly evident that local levels of habitat diversity significantly promote biodiversity. Pieces of evidence have been provided by studies where habitat diversity was enhanced by either large island size (Kohn and Walsh 1994), small-scale disturbances (Phillips et al. 1994), or fire (Braithwaite 1996). Biologically generated spatial patterns, such as the spatial heterogeneity of light (Fig. 1), have also been shown to allow for the coexistence of competing species (Levin and Pacala 1997; Pacala and Levin 1997; Bascompte and Rodriguez 2001), but spatial variation alone (e.g., without temporal variation) is unlikely to explain high levels of plant biodiversity such as those observed in tropical forests (Wright 2002). The trade-off between survivorship in the shade and growth rate (Welden et al. 1991) must be coupled with spatiotemporal resource variation to lead to species coexistence, as shown in models of temperate forest dynamics (Pacala et al. 1996). Nakashizuka (2001) has recently reviewed the mechanisms enabling coexistence in temperate forests.

Another condition required by the competitive exclusion principle is that species have the opportunity to compete. As discussed by Wright (2002), this condition is violated in the forest shade, where suppressed understory plants rarely come into competition with one another. Thus, every species able to tolerate the understory could potentially coexist. The implications of this possible lack of competition are profound, especially in habitats where shade and understory conditions are temporally and spatially important, as in the tropical rainforests (Wright 2002). Kohyama (1994) has suggested another theoretical framework for how trees in different forest strata can coexist without horizontal spatial heterogeneity, by which large, maximum-size trees with low per capita recruitment affect small, maximum-size trees with high recruitment rates. Larger trees suppress growth and recruitment of smaller trees, but the latter do not affect the former (one-sided competition; Kohyama 1994; Nakashizuka 2001). Survival and coexistence in the understory can be mediated by functional convergence of taxonomically distant and morphologically contrasting species. The efficiency of foliage display and light capture was very similar among 24 contrasting species coexisting in the dark understory of a tropical rainforest (Valladares et al. 2002b). This kind of convergence was found in other cases where differences in crown architecture between plants growing in different environments (sun-shade) were more significant than those between species within the same light environment (Poorter and Werger 1999). Therefore, light environment can play two roles in the maintenance of species diversity by violating two conditions of the competitive exclusion principle: (1) environment is not constant in time and space, which allows coexistence due to tradeoffs in sun-shade adaptations, and (2) persistent, extensive low light conditions in the understory allow coexistence due to either lack of competition or one-sided competition associated with a trade-off between maximum size and per capita recruitment. The relative importance of each role seems to be very habitat specific, since factors like water and nutrient availability or intensity of the suppression by herbivores and canopy plants can profoundly modify the importance of competition. As argued in the case of the coexistence of genotypes within plant populations, habitat heterogeneity and mean degree of adversity interact so trends observed in the field must be combined with manipulative experiments and studies under controlled conditions to disentangle the differential effects of environmental variation versus mean environmental benignity. Both empirical and theoretical approaches are still required to achieve significant advances in this field.

The spatial heterogeneity of light in a forest can have implications on plant populations and communities via its effect on pollinators. The general view is that the mobility of pollinators (and also seed dispersal agents) precludes spatial heterogeneity at scales relevant to alpha diversity. However, an irradiance mosaic in a Mediterranean forest allowed for a wide range of diurnal temperatures in the flowers, and since different insect pollinators differ in their thermal requirements, this heterogeneity generated within- and between-habitat variation in the composition and size structure of pollinator assemblages, with clear implications for plant reproduction (Herrera 1997).

b) Management of Forests for the Conservation of Biodiversity

Forest management alters forest structure and influences a number of processes, some purposely, as with productivity, some inadvertently, as with environmental heterogeneity, population dynamics and biodiversity. Intensive forestry according to traditional methods has little ecological similarity to natural disturbances, which generate a more variable patch mosaic than does intensive forestry (Sharitz et al. 1992). The most vulnerable habitats in an intensive forestry regime are associated with forests older than harvest age, which are unique and biologically rich (Perry 1998). The standard approach to conserving forest biodiversity has been to establish reserves, but in most cases they do not adequately protect regional diversity (Perry 1998). The new forestry is aimed at mimicking nature, acknowledging that biodiversity is a complex concept that includes many different aspects of ecosystems. For instance, one of the three main objectives of leaving a certain number of large, green trees at harvest is to enhance the vertical and horizontal heterogeneity (Kohm and Franklin 1997). Mediterranean-type ecosystems are an example of how evolution can accommodate biodiversity under different human activities, and human intervention in this case is not only compatible with biodiversity, but also a source of habitat heterogeneity (Zavala and Oria 1995). The true role of light spatial and temporal heterogeneity for species coexistence is not fully understood yet, and it is clear that it varies significantly among habitats. Sustainable management of forests requires advances in this knowledge since forest management is to a large extent the management of light. Even though we cannot manage what we do not understand, adaptive management can tell us whether we are doing it right. Research programs driven solely by the immediate needs of management and not focused on a detailed understanding of ecosystem processes risk overlooking new insights (Christensen 1997). Scientists are playing a role of increasing importance in forest management, and the scientific challenges of the new, multiple-use forestry are those confronting ecology: understanding the relationships among structure, function, and spatiotemporal dynamics of complex systems interconnected at many scales (Zavala and Oria 1995; Hanski 1997; Perry 1998).

8 Concluding Remarks: the Challenge of Scaling and Integrating Processes

As presented in this paper, a large research effort has gone into the characterization of plant responses to sun and shade, from molecular and leaf biochemical features to the whole plant, but comprehensive scaling exercises are scant (Table 1). What is the relative contribution to plant performance and success of each scale of the response to light? Responses at each scale are likely to vary in both their global costs and their benefits for the plant, and responses at one scale might be obscured by responses (or the lack of adequate responses) at another scale. For instance, all the very precise structural and physiological responses of leaves to the low but highly dynamic light of the forest understory (efficient chloroplast distribution, enhanced light harvesting by pigments and antenna complexes, enhanced photosynthetic utilization of sunflecks) may have no implications for whole plant light capture and carbon gain if the mutual shading among leaves is not minimized by an efficient crown architecture. Thus, what is the hierarchical order of importance in the series of adaptations and acclimations leading to an efficient light capture and utilization by the plant? It is clear that the hierarchical order of importance differs depending on whether light is limiting or excessive, but is it constant across habitats or under the influence of other factors (drought, heat, low temperatures, wind)? Despite the significant scientific progress in the last decades in different complementary and interrelated areas within plant molecular biology, functional morphology, physiology, ecology and evolution, the specific role of light in plant ecology and evolution and the accurate description of both the heterogeneity of light in many natural environments and the integrated responses of plants to it still deserve more attention (Table 1). The cost-benefit analysis of fascinating aspects of plant response to light, such as those affecting the optical properties of the leaf, light capture by the whole crown and the general stability of the phenotype by phenotypic integration, is far from well explored. More knowledge (technical and scientific) is required for realistic scaling exercises, both in computer modeling of light utilization by crowns and canopies, and in remote sensing of ecosystem processes analyzing the radiation reflected on the vegetation. Research is also needed, and expected to be challenging, on the connection between individual responses to light and the ecological and evolutionary responses to future environmental conditions, where light interacts with many factors and constraints in a global change scenario.

Table 1. Fifteen areas of research where there have been relevant progress or reviews in the last decade, and 15 areas where current knowledge is still insufficient and more research is required. Areas are arranged according to the structure of this paper. Key references are provided as either examples or starting points for further discussion and research

	Relevant progress	Insufficient knowledge
1.	Techniques for quantitative description of the light environment: hemispherical photography, miniaturized data-loggers and light sensors (Pontailler and Genty 1996; Englund et al. 2000)	Precise descriptions of the light environment (quantity and quality of light available for plants, temporal dynamics, spatial heterogeneity) in natural habitats (Canham et al. 1990)
2.	Photosynthetic pigments (Nishio 2000)	Plant photoreceptors: blue-light receptors and interactions among phytochromes and cryptochromes (Lin 2000; Mazzella and Casal 2001)
3.	Leaf anatomy and optical properties (Smith et al. 1997)	Plant response and survival in shade combined with drought (Sack et al. 2001; Valladares and Pearcy 2002)
4.	Remote analysis of plant biomass and physiological status (Filella and Peñuelas 1998; Gamon and Qiu 1999)	Costs of photoinhibition under natural conditions (Werner et al. 2001)
5.	Global trends in leaf form and function (Press 1999)	Relative importance and general trends in structural versus physiological protection against excessive light (Valladares 1999; Werner et al. 1999)
6.	Whole plant general responses to shade (Henry and Aarssen 1997)	Precise information on the influence of light on plant morphogenesis for parameterization of realistic 3-D modeling of plant growth (Gautier et al. 2000)
7.	Quantum yield and photosynthesis models (Farquhar et al. 2001; Singsaas et al. 2001)	Whole plant photosynthetic photon use efficiency under real conditions (Pearcy and Sims 1994)
8.	Models of photoinhibition (Han et al. 2000; Marshall et al. 2000)	Phenotypic correlation and functional integration of traits (Schlichting 1989; Shipley and Lechowicz 2000)
9.	Photoprotective pigments (Bungard et al. 1999; Demmig-Adams and Adams 2000)	Costs of phenotypic plasticity (DeWitt et al. 1998; Cipollini and Schultz 1999)

Table 1 (continued)

	Relevant progress	Insufficient knowledge
10.	Effects of ultraviolet light and plant response to UV-B (Mackerness 2000; Paul 2001; Rousseaux et al. 2001)	Trends among species and habitats in phenotypic plasticity in response to light (Valladares et al. 2000a,b)
11.	Photosynthesis under fluctuating light (Pearcy et al. 1995)	Implications of morphology and physiology for fitness and population and community dynamics (Bell 1984; Arntz et al. 2000)
12.	Canopy modeling for light capture and carbon gain (Beyschlag and Ryel 1999)	Genetic basis of phenotypic variability in functional traits (Arntz and Delph 2001)
13.	Incorporation of the general effects of light in 3-D models of plant growth (Prusinkiewicz 1999)	Links between ecophysiology and evolution (Ackerly et al. 2000)
14.	Functional and adaptive meaning of phenotypic plasticity in response to light (Dudley and Schmitt 1996; Sultan 2000)	Precise knowledge on how light, specially its spatial and temporal heterogeneity, influence coexistence and community composition (Nicotra et al. 1999; Schnitzer and Carson 2001)
15.	Models for the coexistence of competing species with emphasis on the role of light (Pacala and Rees 1998; Deutschman et al. 1999)	Light and plant communities of arid environments: facilitation, competition or neutral effects? (Zavala et al. 2000; Pugnaire et al., submitted)

Acknowledgements. Fruitful discussions with Robert W. Pearcy, Luis Balaguer, Francisco Pugnaire, Miguel Angel Zavala and Teodoro Marañón have made this review possible. Thanks to Lucía Ramirez and Alba Valladares for understanding and patience. Financial support was provided by the Spanish CICYT-Ministry of Science and Technology, grant REN2000-0163-P4-05 (ECOFIARB).

References

Abrams MD, Mostoller SA (1995) Gas exchange, leaf structure and nitrogen in contrasting successional tree species growing in open and understory sites during a drought. Tree Physiol 15:361–370

Abrams MD, Kloeppel BD, Kubiske ME (1992) Ecophysiological and morphological responses to shade and drought in two contrasting ecotypes of *Prunus serotina*. Tree Physiol 10:343–355

Achenbach J (2001) The power of light. Natl Geogr 200:2–31

Ackerly DD (2000) Taxon sampling, correlated evolution, and independent contrasts. Evolution 54:1480–1492

Ackerly DD, Dudley SA, Sultan SE, Schmitt J, Coleman JS, Randal Linder C, Sandquist DR, Geber MA, Evans AS, Dawson TE, Lechowicz MJ (2000) The evolution of plant ecophysiological traits: recent advances and future directions. Bioscience 50:979–995

Adler I, Barabe D, Jean V (1997) A history of the study of phyllotaxis. Ann Bot 80:231–244

Agyeman VK, Swaine MD, Thompson J (1999) Responses of tropical forest tree seedlings to irradiance and the derivation of a light response index. J Ecol 87:815–827

Aiba S, Kohyama T (1997) Crown architecture and life-history traits of 14 tree species in a warm-temperate rain forest: significance of spatial heterogeneity. J Ecol 85:611–624

Aphalo PJ, Ballare CL, Scopel AL (1999) Plant-plant signalling, the shade-avoidance response and competition. J Exp Bot 50:1629–1634

Arntz AM, Delph LF (2001) Patterns and process: evidence for the evolution of photosynthetic traits in natural populations. Oecologia 127:455–467

Arntz AM, DeLucia EH, Jordan N (2000) From fluorescence to fitness: variation in photosynthetic rate affects fecundity and survivorship. Ecology 81:2567–2576

Bainbrige R, Evans GC, Rackham O (1966) Light as an ecological factor. Blackwell Scientific, London

Balaguer L, Martínez-Ferri E, Valladares F, Pérez-Corona ME, Baquedano FJ, Castillo FJ, Manrique E (2001) Population divergence in the plasticity of the response of *Quercus coccifera* to the light environment. Funct Ecol 15:124–135

Baldocchi D, Collineau S (1994) The physical nature of solar radiation in heterogeneous canopies: spatial and temporal attributes. In: Caldwell MM, Pearcy RW (eds) Exploitation of environmental heterogeneity by plants. Academic Press, San Diego, pp 21–71

Ballare CL (1999) Keeping up with the neighbours: phytochrome sensing and other signalling mechanisms. Trends Plant Sci 4:97–102

Barnes BB, Zak DR, Denton SR, Spurr SH (1998) Forest ecology. Wiley, New York

Baruch Z, Pattison RR, Goldstein G (2000) Responses to light and water availability of four invasive Melastomataceae in the Hawaiian Islands. Int J Plant Sci 161:107–118

Bascompte J, Rodriguez MA (2001) Habitat patchiness and plant species richness. Ecol Lett 4:1–4

Battaglia M, Cherry ML, Beadle CL, Sands PJ, Hingston A (1998) Prediction of leaf area index in eucalypt plantations: effects of water stress and temperature. Tree Physiol 18:521–528

Bazzaz FA (1979) The physiological ecology of plant succession. Annu Rev Ecol Syst 10:351–371

Bazzaz FA (1991) Habitat selection in plants. Am Nat 137: S116–S130

Bazzaz FA (1996) Plants in changing environments: linking physiological, population, and community ecology. Cambridge University Press, Cambridge

Bazzaz FA, Stinson KA (1999) Genetic vs. environmental control of ecophysiological processes: some challenges for predicting community responses to global change. In: Press MC, Scholes JD, Barker MG (eds) Physiological plant ecology. Blackwell Science, Cornwall, pp 283–296

Becker P, Smith AP (1990) Spatial autocorrelation of solar radiation in a tropical moist forest understory. Agric For Meteorol 52:373–379

Bell AD (1984) Dynamic morphology: a contribution to plant population ecology. In: Dirzo R, Sarukhán J (eds) Perspectives on plant population ecology. Sinauer, Sunderland, MA, pp 48–65

Bell AD (1993) Plant form. Oxford University Press, New York

Beyschlag W, Ryel RJ (1999) Canopy photosynthesis modeling. In: Pugnaire FI, Valladares F (eds) Handbook of functional plant ecology. Marcel Dekker, New York, pp 771–804

Björkman O (1981) Responses to different quantum flux densities. In: Lange OL, Nobel PS, Osmond CB, Ziegler H (eds) Physiological plant ecology. Springer, Berlin Heidelberg New York, pp 57–107

Bongers F, Popma J (1990) Leaf characteristics of the tropical rain forest flora of Los-Tuxtlas, Mexico. Bot Gaz 151:354–365

Bossard CC, Rejmanek M (1992) Why have green stems? Funct Ecol 6:197–205
Bradshaw AD (1965) Evolutionary significance of phenotypic plasticity in plants. Adv Genet 13:115–155
Braithwaite R (1996) Biodiversity and fire in the savanna landscape. In: Solbrig DT, Medina E, Silva JF (eds) Biodiversity and savanna ecosystem processes. Springer, Berlin Heidelberg New York, pp 121–140
Bungard RA, Ruban AV, Hibberd JM, Press MC, Horton P, Scholes JD (1999) Unusual carotenoid composition and a new type of xanthophyll cycle in plants. Proc Natl Acad Sci USA 96:1135–1139
Callaway RM, Pugnaire FI (1999) Facilitation in plant communities. In: Pugnaire FI, Valladares F (eds) Handbook of functional plant ecology. Marcel Dekker, New York, pp 623–648
Canadell J, Roda F (1991) Root biomass of *Quercus ilex* in a montane Mediterranean forest. Can J For Res 21:1771–1778
Canham CD, Denslow JS, Platt WJ, Runkle JR, Spies TA, White PS (1990) Light regimes beneath closed canopies and tree-fall gaps in temperate and tropical forests. Can J For Res 20:620–631
Chazdon RL (1988) Sunflecks and their importance to forest understory plants. Adv Ecol Res 18:1–63
Chazdon RL (1992) Photosynthetic plasticity of two rain forest shrubs across natural gap transects. Oecologia 92:586–595
Chazdon RL, Williams K, Field CB (1988) Interactions between crown structure and light environment in five rain forest *Piper* species. Am J Bot 75:1459–1471
Christensen NL (1997) Managing for heterogeneity and complexity on dynamic landscapes. In: Pickett STA, Ostfeld RS, Shachak M, Likens GE (eds) The ecological basis of conservation heterogeneity, ecosystems and biodiversity. Chapman and Hall, New York, pp 167–186
Cipollini DF, Schultz JC (1999) Exploring costs constraints on stem elongation in plants using phenotypic manipulation. Am Nat 153:236–242
Clark DA, Clark DB (2001) Getting to the canopy: tree height growth in a neotropical rain forest. Ecology 82:1460–1472
Comstock JP, Ehleringer JR (1988) Contrasting photosynthetic behavior in leaves and twigs of *Hymenoclea salsola*, a green-twigged warm desert shrub. Am J Bot 75:1360–1370
Day TA (1993) Relating Uv-B radiation screening effectiveness of foliage to absorbing-compound concentration and anatomical characteristics in a diverse group of plants. Oecologia 95:542–550
Day TA, Vogelmann TC, Delucia EH (1992) Are some plant life forms more effective than others in screening out ultraviolet-B radiation? Oecologia 92:513–519
De Jong G, Gavrilets S (2000) Maintenance of genetic variation in phenotypic plasticity: the role of environmental variation. Genet Res 76:295–304
Demmig-Adams B, Adams WW (2000) Photosynthesis – harvesting sunlight safely. Nature 403:371–374
DePuit EJ, Caldwell MM (1975) Stem and leaf gas exchange of two arid land shrubs. Am J Bot 62:954–961
De Reffye P, Houllier F (1997) Modelling plant growth and architecture: some recent advances and applications to agronomy and forestry. Curr Sci (Bangalore) 73:984–992
Deutschman DH, Levin SA, Pacala SW (1999) Error propagation in a forest succession model: the role of fine-scale heterogeneity in light. Ecology 80:1927–1943
DeWitt TJ, Sih A, Wilson DS (1998) Costs and limits of phenotypic plasticity. Trends Ecol Evol 13:77–81
Dudley SA, Schmitt J (1996) Testing the adaptive plasticity hypothesis: density-dependent selection on manipulated stem length in *Impatiens capensis*. Am Nat 147:445–465

Durand LZ, Goldstein G (2001) Photosynthesis, photoinhibition, and nitrogen use efficiency in native and invasive tree ferns in Hawaii. Oecologia 126:345–354

Ehleringer JR, Cooper TA (1992) On the role of orientation in reducing photoinhibitory damage in photosynthetic-twig desert shrubs. Plant Cell Environ 15:301–306

Ehleringer JR, Forseth IN (1989) Diurnal leaf movements and productivity in canopies. In: Russell G, Marshall B, Jarvis PG (eds) Plant canopies: their growth, form and function. Cambridge University Press, Cambridge, pp 129–142

Ehleringer JR, Werk KS (1986) Modifications of solar-radiation absorption patterns and implications for carbon gain at the leaf level. In: Givnish TJ (eds) On the economy of plant form and function. Cambridge University Press, Cambridge, pp 564–571

Endler JA (1993) The color of light in forests and its implications. Ecol Monogr 61:1–27

Englund SR, O'Brien JJ, Clark DB (2000) Evaluation of digital and film hemispherical photography and spherical densiometry for measuring forest light environments. Can J For Res 30:1999–2005

Evans JR, von Caemmerer S, Adams WW III (1988) Ecology of photosynthesis in sun and shade. CSIRO, Melbourne

Faria T, Silvério D, Breia E, Cabral R, Abadía A, Abadía J, Pereira JS, Chaves MM (1998) Differences in the response of carbon assimilation to summer stress (water deficits, high light and temperature) in four Mediterranean tree species. Physiol Planta 102:419–428

Farquhar GD, von Caemmerer S, Berry JA (2001) Models of photosynthesis. Plant Physiol 125:42–45

Filella I, Peñuelas J (1998) Visible and near-infrared reflectance techniques for diagnosing plant physiological status. Trends Plant Sci 3:151–156

Gamon JA, Qiu H (1999) Ecological applications of remote sensing at multiple scales. In: Pugnaire FI, Valladares F (eds) Handbook of functional plant ecology. Marcel Dekker, New York, pp 805–846

Gautier H, Mech R, Prusinkiewicz P, Varlet-Grancher C (2000) 3D architectural modelling of aerial photomorphogenesis in white clover (*Trifolium repens* L.) using L-systems. Ann Bot 85:359–370

Givnish TJ (1988) Adaptation to sun and shade: a whole-plant perspective. Aust J Plant Physiol 15:63–92

Grammatikopoulos G, Kyparissis A, Drilias P, Petropoulou Y, Manetas Y (1998) Effects of UV-B radiation on cuticle thickness and nutritional value of leaves in two Mediterranean evergreen sclerophylls. J Plant Physiol 153:506–512

Grime JP, Crick JC, Rincon E (1986) The ecological significance of plasticity. In: Jennings DH, Trewavas AJ (eds) Plasticity in plants. Company of Biologists, Cambridge, pp 5–29

Haase P, Pugnaire FI, Clark SC, Incoll LD (1999) Diurnal and seasonal changes in cladode photosynthetic rate in relation to canopy age structure in the leguminous shrub *Retama sphaerocarpa*. Funct Ecol 13:640–649

Han BP, Virtanen M, Koponen J, Straskraba M (2000) Effect of photoinhibition on algal photosynthesis: a dynamic model. J Plankton Res 22:865–885

Hanski I (1997) Habitat destruction and metapopulation dynamics. In: Pickett STA, Ostfeld RS, Shachak M, Likens GE (eds) The ecological basis of conservation heterogeneity, ecosystems and biodiversity. Chapman and Hall, New York, pp 217–227

Hanski I, Gilpin ME (1997) Metapopulation biology: ecology, genetics, and evolution. Academic Press, San Diego

Hari P, Nikinmaa E, Korpilahti E (1991) Modeling – canopy, photosynthesis, and growth. Physiol Trees 419–444

Henry HAL, Aarssen LW (1997) On the relationship between shade tolerance and shade avoidance strategies in woodland plants. Oikos 80:575–582

Herbert T (1996) On the relationship of plant geometry to photosynthetic response. In: Mulkey SS, Chazdon RL, Smith AP (eds) Tropical forest plant ecophysiology. Chapman and Hall, New York, pp 139–161

Herrera CM (1992) Historical effects and sorting processes as explanations for contemporary ecological patterns: character syndromes in Mediterranean woody plants. Am Nat 140:421–446

Herrera CM (1997) Thermal biology and foraging responses of insect pollinators to the forest floor irradiance mosaic. Oikos 78:601–611

Hirose T (1987) A vegetative plant growth model: adaptive significance of phenotypic plasticity in dry matter partitioning. Funct Ecol 1:195–202

Hoffmann AA, Parsons PA (1991) Evolutionary genetics and environmental stress. Oxford Science Publications, Oxford

Holmgren M (2000) Combined effects of shade and drought on tulip poplar seedlings: trade-off in tolerance or facilitation? Oikos 90:67–78

Horn HS (1971) The adaptive geometry of trees. Princeton University Press, Princeton, NJ

Jean RV (1984) Mathematical approach to pattern and form in plant growth. Wiley, New York

Jean RV, Barabé D (1998) Symmetry in plants. World Scientific Publishing Co Pte Ltd, Singapore

Kevan PG (1975) Sun-tracking solar furnaces in high Arctic flowers, significance for pollination and insects. Science 189:723–726

Kikuzawa K (1989) Ecology and evolution of phenological pattern, leaf longevity and leaf habit. Evol Trends Plants 3:105–110

Kikuzawa K (1991) A cost benefit analysis of leaf habit and leaf longevity of trees and their geographical pattern. Am Nat 138:1250–1263

Kikuzawa K (1995) Leaf phenology as an optimal strategy for carbon gain in plants. Can J Bot 73:158–163

Kikuzawa K, Koyama H, Umeki K, Lechowicz MJ (1996) Some evidence for an adaptive linkage between leaf phenology and shoot architecture in sapling trees. Funct Ecol 10:252–257

Kiltie RA (1993) New light on forest shade. Trends Ecol Evol 8:39–40

King DA (1990) Allometry of saplings and understory trees of a Panamanian forest. Funct Ecol 4:27–32

King DA (1991) Correlations between biomass allocation, relative growth rate and light environment in tropical forest saplings. Funct Ecol 5:485–492

Kirchoff BK, Rutishauser R (1990) The phyllotaxy of Costus (Costaceae). Bot Gaz 151:88–105

Kohm KA, Franklin JF (1997) Creating a forestry for the 21st century. Island Press, Washington, DC

Kohn DD, Walsh DM (1994) Plant species richness – the effect of island size and habitat diversity. J Ecol 82:367–377

Kohyama T (1987) Significance of architecture and allometry in saplings. Funct Ecol 1:399–404

Kohyama T (1994) Size-structure-based models of forest dynamics to interpret population- and community-level mechanisms. J Plant Res 107:107–116

Kohyama T, Hotta M (1990) Significance of allometry in tropical saplings. Funct Ecol 4:515–521

Koller D (1990) Light-driven leaf movements. Plant Cell Environ 13:615–632

Koller D (2000) Plants in search of sunlight. Adv Bot Res 33:35–131

Kramer K (1994) Selecting a model to predict the onset of growth of *Fagus sylvatica*. J Appl Ecol 31:172–181

Kwesiga FR, Grace J (1986) The role of red/far red ratio in the response of tropical tree seedlings to shade. Ann Bot 57:283–290

Kyparissis A, Petropoulou Y, Manetas Y (1995) Summer survival of leaves in a soft-leaved shrub (*Phlomis furticosa* L. Labiatae) under Mediterranean field conditions: avoidance of photoinhibitory damage through decreased chlorophyll contents. J Exp Bot 46:1825–1831

Larcher W (1995) Physiological plant ecology. Ecophysiology and stress physiology of functional groups. Springer, Berlin Heidelberg New York

Latham RE, Ricklefs RE (1993) Global patterns of tree species richness in moist forests: energy-diversity theory does not account for variation in species richness. Oikos 67:325–333

Leigh EG (1998) Tropical forest ecology: a view from Barro Colorado Island. Oxford University Press, New York

Levin SA, Pacala SW (1997) Theories of simplification and scaling of spatially distributed processes. In: Tilman D, Kareiva P (eds) Spatial ecology: the role of space in population dynamics and interspecific interactions. Princeton University Press, Princeton, NJ, pp 271–295

Lin CT (2000) Plant blue-light receptors. Trends Plant Sci 5:337–342

Lortie CJ, Aarssen LW (1996) The specialization hypothesis for phenotypic plasticity in plants. Int J Plant Sci 157:484–487

Lovelock CE, Clough BF (1992) Influence of solar radiation and leaf angle on leaf xanthophyll concentrations in mangroves. Oecologia 91:518–525

Mackerness SAH (2000) Plant responses to ultraviolet-B (UV-B: 280–320 nm) stress: what are the key regulators? Plant Growth Regul 32:27–39

Marshall HL, Geider RJ, Flynn KJ (2000) A mechanistic model of photoinhibition. New Phytol 145:347–359

Matsuda K, Bride JR, Kimura M (1989) Seedling growth form of oaks. Ann Bot 64:439–446

Mazzella MA, Casal JJ (2001) Interactive signalling by phytochromes and cryptochromes generates de-etiolation homeostasis in *Arabidopsis thaliana*. Plant Cell Environ 24:155–161

McMillen GG, McClendon JH (1979) Leaf angle: an adaptive feature of sun and shade leaves. Bot Gaz 140:437–442

Monaco TA, Briske DD (2000) Does resource availability modulate shade avoidance responses to the ratio of red to far-red irradiation? An assessment of radiation quantity and soil volume. New Phytol 146:37–46

Nakashizuka T (2001) Species coexistence in temperate, mixed deciduous forests. Trends Ecol Evol 16:205–210

Nicotra AB, Chazdon RL, Schlichting CD (1997) Patterns of genotypic variation and phenotypic plasticity of light response in two tropical Piper (Piperaceae) species. Am J Bot 84:1542–1552

Nicotra AB, Chazdon RL, Iriarte SVB (1999) Spatial heterogeneity of light and woody seedling regeneration in tropical wet forests. Ecology 80:1908–1926

Niklas KJ (1988) The role of phyllotactic pattern as a "developmental constraint" on the interception of light by leaf surfaces. Evolution 42:1–16

Niklas KJ (1997) The evolutionary biology of plants. University of Chicago Press, Chicago

Nilsen ET (1992a) The influence of water stress on leaf and stem photosynthesis in *Spartium junceum* L. Plant Cell Environ 15:455–461

Nilsen ET (1992b) Partitioning growth and *photosynthesis between leaves and stems during nitrogen limitation in Spartium* junceum. Am J Bot 79:1217–1223

Nilsen ET, Karpa D, Mooney HA, Field C (1993) Patterns of stem photosynthesis in two invasive legumes (*Spartium junceum*, *Cytisus scoparius*) of the California coastal region. Am J Bot 80:1126–1136

Nishio JN (2000) Why are higher plants green? Evolution of the higher plant photosynthetic pigment complement. Plant Cell Environ 23:539–548

Norman JM, Arkebauer TJ (1991) Predicting canopy light-use efficiency from leaf characteristics. Modeling Plant Soil Systems 31:125–143

O'Brien EM, Whittaker RJ, Field R (1998) Climate and woody plant diversity in southern Africa: relationships at species, genus and family levels. Ecography 21:495–509

Osunkoya OO, Ash JE, Hopkins MS, Graham AW (1992) Factors affecting survival of tree seedlings in North Queensland rainforests. Oecologia 91:569–578

Pacala SW, Levin SA (1997) Biologically generated spatial pattern and the coexistence of competing species. In: Tilman D, Kareiva P (eds) Spatial ecology: the role of space in population dynamics and interspecific interactions. Princeton University Press, Princeton, NJ, pp 204–232

Pacala SW, Rees M (1998) Models suggesting field experiments to test two hypotheses explaining successional diversity. Am Nat 152:729–737

Pattison RR, Goldstein G, Ares A (1998) Growth, biomass allocation and photosynthesis of invasive and native Hawaiian rainforest species. Oecologia 117:449–459

Pacala SW, Canham CD, Saponara J, Silander JA, Kobe RK, Ribbens E (1996) Forest models defined by field measurements: estimation, error analysis and dynamics. Ecol Monogr 66:1–43

Paul N (2001) Plant responses to UV-B: time to look beyond stratospheric ozone depletion? New Phytol 150:5–8

Pearcy RW (1983) The light environment and growth of C3 and C4 tree species in the understory of a Hawaiian forest. Oecologia 58:19–25

Pearcy RW (1999) Responses of plants to heterogeneous light environments. In: Pugnaire FI, Valladares F (eds) Handbook of functional plant ecology. Marcel Dekker, New York, pp 269–314

Pearcy RW, Sims DA (1994) Photosynthetic acclimation to changing light environments: scaling from the leaf to the whole plant. In: Caldwell MM, Pearcy RW (eds) Exploitation of environmental heterogeneity by plants: ecophysiological processes above and below ground. Academic Press, San Diego, pp 145–174

Pearcy RW, Valladares F (1999) Resource acquisition by plants: the role of crown architecture. In: Press M, Scholes JD, Barker MG (eds) Physiological plant ecology. Blackwell Scientific, London, pp 45–66

Pearcy RW, Yang W (1996) A three-dimensional shoot architecture model for assessment of light capture and carbon gain by understory plants. Oecologia 108:1–12

Pearcy RW, Yang W (1998) The functional morphology of light capture and carbon gain in the redwood-forest understory plant, *Adenocaulon bicolor* Hook. Funct Ecol 12:543–552

Pearcy RW, Sassenrath-Cole GF, Krall JP (1995) Photosynthesis in fluctuating light environments. In: Baker NR (ed) Environmental stress and photosynthesis. Kluwer Academic, The Hague, pp 85–93

Perry DA (1998) The scientific basis of forestry. Annu Rev Ecol Syst 29:435–466

Pfanz H, Aschan G (2001) The existence of bark and stem photosynthesis in woody plants and its significance for the overall carbon gain. An eco-physiological and ecological approach. Progress in Botany 62. Springer, Berlin Heidelberg New York, pp 477–510

Phillips OL, Hall P, Gentry AH, Sawyer SA, Vasquez R (1994) Dynamics and species richness of tropical rain forests. Proc Natl Acad Sci USA 91:2805–2809

Pigliucci M (2001) Phenotypic plasticity: beyond nature and nurture. John Hopkins University Press, Baltimore

Pollard H, Cruzan M, Pigliucci M (2001) Comparative studies of reaction norms in Arabidopsis. I. Evolution of response to daylength. Evol Ecol Res 3:129–155

Pontailler JY, Genty B (1996) A simple red-far-red sensor using gallium arsenide phosphide detectors. Funct Ecol 10:535–540

Poorter L, Werger MJA (1999) Light environment, sapling architecture, and leaf display in six rain forest tree species. Am J Bot 86:1464–1473

Porembski S, Barthlott W (2000) Biodiversity research in botany. Progress in Botany 61. Springer, Berlin Heidelberg New York, pp 335–362

Powles SB, Björkman O (1981) Leaf movement in the shade species Oxalis oregana. II. Role in protection against injury by intense light. Carnegie Inst Wash 80:63–66

Press MC (1999) The functional significance of leaf structure: a search for generalizations. New Phytol 143:213–219

Prusinkiewicz P (1998) Modeling of spatial structure and development of plants: a review. Sci Hortic 74:113–149

Prusinkiewicz P (1999) A look at the visual modeling of plants using L-systems. Agronomie 19:211–224

Prusinkiewicz P, Lindenmayer A (1996) The algorithmic beauty of plants. Springer, Berlin Heidelberg New York

Pugnaire FI, Luque MT (2001) Changes in plant interactions along a gradient of environmental stress. Oikos 93:42–49

Pugnaire FI, Haase P, Puigdefábregas J, Cueto M, Clark SC, Incoll LD (1996) Facilitation and succession under the canopy of a leguminous shrub, *Retama sphaerocarpa*, in a semi-arid environment in south-east Spain. Oikos 76:455–464

Retana J, Espelta JM, Gracia M, Riba M (1999) Seedling recruitment. In: Roda F, Retana J, Gracia C, Bellot J (eds) Ecology of Mediterranean evergreen oak forests. Springer, Berlin Heidelberg New York, pp 89–103

Riddoch I, Lehto T, Grace J (1991) Photosynthesis of tropical tree seedlings in relation to light and nutrient supply. New Phytol 119:137–147

Room P, Hanan J, Prusinkiewicz P (1996) Virtual plants: new perspectives for ecologists, pathologists and agricultural scientists. Trends Plant Sci 1:33–38

Ross J, Sulev M (2000) Sources of errors in measurements of PAR. Agric For Meteorol 100:103

Rousseaux MC, Scopel AL, Searles PS, Caldwell MM, Sala OE, Ballare CL (2001) Responses to solar ultraviolet-B radiation in a shrub-dominated natural ecosystem of Tierra del Fuego (southern Argentina). Global Change Biol 7:467–478

Sack L, Grubb PJ, Marañon T (2002) The functional morphology of seedlings tolerant of deep shade plus drought in three Mediterranean-climate forests of southern Spain. Plant Ecol (in press)

Saxe H, Cannell MGR, Johnsen O, Ryan MG, Vourlitis G (2001) Tree and forest functioning in response to global warming. New Phytol 149:369–400

Scheiner SM (1993) Genetics and evolution of phenotypic plasticity. Annu Rev Ecol Syst 24:35–68

Scheiner SM, Callahan HS (1999) Measuring natural selection on phenotypic plasticity. Evolution 53:1704–1713

Schlichting CD (1986) The evolution of phenotypic plasticity in plants. Annu Rev Ecol Syst 17:667–693

Schlichting CD (1989) Phenotypic integration and environmental change. What are the consequences of differential phenotypic plasticity of traits? BioScience 39:460–464

Schnitzer SA, Carson WP (2001) Treefall gaps and the maintenance of species diversity in a tropical forest. Ecology 82:913–919

Sekimura T (1995) The diversity in shoot morphology of herbaceous plants in relation to solar radiation captured by leaves. J Theor Biol 177:289–297

Sharitz RR, Boring LR, Van Lear DH, Pinder JE (1992) Integrating ecological concepts with natural resource management of southern forests. Ecol Appl 2:226–237

Shipley B, Lechowicz M (2000) The functional co-ordination of leaf morphology, nitrogen concentration, and gas exchange in 40 wetland species. Ecoscience 7:183–194

Shirley HL (1929) The influence of light intensity and light quality upon the growth of plants. Am J Bot 16:354

Singsaas EL, Ort DR, DeLucia EV (2001) Variation in measured values of photosynthetic quantum yield in ecophysiological studies. Oecologia 128:15–23

Skaltsa H, Verykokidou E, Harvala C, Karabourniotis G, Manetas Y (1994) UV-B protective potential and flavonoid content of leaf hairs of *Quercus ilex*. Phytochemistry 37:987–990

Smith H (1982) Light quality, photoperception and plant strategy. Annu Rev Plant Physiol 33:481–518

Smith M, Ullberg D (1989) Effect of leaf angle and orientation on photosynthesis and water relations in *Silphium terebinthinaceum*. Am J Bot 76:1714–1719

Smith SD, Osmond CB (1987) Stem photosynthesis in a desert ephemeral, *Eriogonum inflatum*. Morphology, stomatal conductance and water use efficiency in field conditions. Oecologia 72:533–541

Smith WK, Vogelmann TC, DeLucia EH, Bell DT, Shepherd KA (1997) Leaf form and photosynthesis. Bioscience 47:785–793

Soares AG, Scapini F, Brown AC, McLachlan A (1999) Phenotypic plasticity, genetic similarity and evolutionary inertia in changing environments. J Mollus Stud 65:136–139

Stanton ML, Galen C (1993) Blue light controls solar tracking by flowers of an alpine plant. Plant Cell Environ 16:983–989

Strauss-Debenedetti S, Bazzaz FA (1991) Plasticity and acclimation to light in tropical moraceae of different successional positions. Oecologia 87:377–387

Sultan SE (1987) Evolutionary implications of phenotypic plasticity in plants. Evol Biol 21:127–176

Sultan SE (1992) What has survived of Darwin's theory? Phenotypic plasticity and the Neo-Darwinian legacy. Evol Trends Plants 6:61–71

Sultan SE (1995) Phenotypic plasticity and plant adaptation. Acta Bot Neer 44:363–383

Sultan SE (2000) Phenotypic plasticity for plant development, function and life history. Trends Plant Sci 5:537–542

Sultan SE, Bazzaz FA (1993) Phenotypic plasticity in *Polygonum persicaria*.1. Diversity and uniformity in genotypic norms of reaction to light. Evolution 47:1009–1031

Sultan SE, Wilczek AM, Bell DL, Hand G (1998) Physiological response to complex environments in annual *Polygonum* species of contrasting ecological breadth. Oecologia 115:564–578

Tarasjev A (1995) Relationship between phenotypic plasticity and developmental instability in *Iris pumila* L. Russ J Genet 31:1409–1416

Thornley JHM (1998) Dynamic model of leaf photosynthesis with acclimation to light and nitrogen. Ann Bot 81:421–430

Tielborger K, Kadmon R (2000a) Indirect effects in a desert plant community: is competition among annuals more intense under shrub canopies? Plant Ecol 150:53–63

Tielborger K, Kadmon R (2000b) Temporal environmental variation tips the balance between facilitation and interference in desert plants. Ecology 81:1544–1553

Tilman D (1988) Plant strategies and the dynamics and structure of plant communities. Princeton University Press, Princeton, NJ

Tokeshi M (1999) Species coexistence. Ecological and evolutionary perspectives. Blackwell Science, Cambridge

Turner IM, Lucas PW, Becker P, Wong SC, Yong JWH, Choong MF, Tyree MT (2000) Tree leaf form in Brunei: a heath forest and a mixed dipterocarp forest compared. Biotropica 32:53–61

Valladares F (1999) Architecture, ecology and evolution of plant crowns. In: Pugnaire FI, Valladares F (eds) Handbook of functional plant ecology. Marcel Dekker, New York, pp 121–194

Valladares F (2000) Light and plant evolution: adaptation to the extremes versus phenotypic plasticity. In: Greppin H (ed) Advanced studies in plant biology. University of Geneva, Geneve, pp 341–355

Valladares F, Pearcy RW (1997) Interactions between water stress, sun-shade acclimation, heat tolerance and photoinhibition in the sclerophyll *Heteromeles arbutifolia*. Plant Cell Environ 20:25–36

Valladares F, Pearcy RW (1998) The functional ecology of shoot architecture in sun and shade plants of *Heteromeles arbutifolia* M. Roem., a Californian chaparral shrub. Oecologia 114:1–10

Valladares F, Pearcy RW (1999) The geometry of light interception by shoots of *Heteromeles arbutifolia*: morphological and physiological consequences for individual leaves. Oecologia 121:171–182

Valladares F, Pearcy RW (2000) The role of crown architecture for light harvesting and carbon gain under extreme light conditions assessed with a realistic 3-D model. An Jardin Bot Madr 58:3–16

Valladares F, Pearcy RW (2002) Drought can be more critical in the shade than in the sun: a field study of carbon gain and photoinhibition in a Californian shrub during a dry El Niño year. Plant Cell Environ 25:749–759

Valladares F, Pugnaire FI (1999) Tradeoffs between irradiance capture and avoidance in semiarid environments simulated with a crown architecture model. Ann Bot 83:459–470

Valladares F, Allen MT, Pearcy RW (1997) Photosynthetic response to dynamic light under field conditions in six tropical rainforest shrubs occurring along a light gradient. Oecologia 111:505–514

Valladares F, Martinez-Ferri E, Balaguer L, Perez-Corona E, Manrique E (2000a) Low leaf-level response to light and nutrients in Mediterranean evergreen oaks: a conservative resource-use strategy? New Phytol 148:79–91

Valladares F, Wright SJ, Lasso E, Kitajima K, Pearcy RW (2000b) Plastic phenotypic response to light of 16 congeneric shrubs from a Panamanian rainforest. Ecology 81:1925–1936

Valladares F, Chico JM, Aranda I, Balaguer L, Dizengremel P, Manrique E, Dreyer E (2002) Greater high light tolerance of seedlings of *Quercus robur* over *Fagus sylvatica* is linked to a greater physiological plasticity. Trees Struct Funct (in press)

Valladares F, Skillman J, Pearcy RW (2002b) Convergence in light capture efficiencies among tropical forest understory plants with contrasting crown architectures: a case of morphological compensation. Am J Bot 89:1275–1284

van Klausen M, Fischer M, Schmid B (2000) Costs of plasticity in foraging characteristics of the clonal plant *Ranunculus reptans*. Evolution 54:1947–1955

van Tienderen PH (1991) Evolution of generalists and specialists in spatially heterogeneous environments. Evolution 45:1317–1331

Vila M, Sardans J (1999) Plant competition in Mediterranean-type vegetation. J Veg Sci 10:281–294

Waller DM (1991) The dynamics of growth and form. In: Crawley MJ (eds) Plant ecology. Blackwell Scientific, Wiltshire, pp 291–320

Walters MB, Reich PB (1999) Low-light carbon balance and shade tolerance in the seedlings of woody plants: do winter deciduous and broad-leaved evergreen species differ? New Phytol 143:143–154

Welden CW, Hewett SW, Hubbell JG, Foster RB (1991) Sapling survival, growth, and recruitment: relationship to canopy height in a neotropical forest. Ecology 72:35–50

Werner C, Correia O, Beyschlag W (1999) Two different strategies of Mediterranean macchia plants to avoid photoinhibitory damage by excessive radiation levels during summer drought. Acta Oecol 20:15–23

Werner C, Ryel RJ, Correia O, Beyschlag W (2001) Effects of photoinhibition on whole-plant carbon gain assessed with a photosynthesis model. Plant Cell Environ 24:27–40

West-Eberhard MJ (1989) Phenotypic plasticity and the origins of diversity. Annu Rev Ecol Syst 20:249–278

Wilder GJ (1992) Orthodistichous phyllotaxy and dorsiventral symmetry on adult shoots of *Cyclanthus bipartitus* (Cyclanthaceae, Monocotyledoneae). Can J Bot 70:1388–1400
Wilson JB, Agnew ODQ (1992) Positive feedback switches in plant communities. Adv Ecol Res 20:265–336
Wirtz KW (2000) Simulating the dynamics of leaf physiology and morphology with an extended optimality approach. Ann Bot 86:753–764
Wright SJ (2002) Plant diversity in tropical forests: a review of mechanisms of species coexistence. Oecologia 130:1–14
Young AJ (1991) The photoprotective role of carotenoids in higher plants. Physiol Plant 83:702–708
Zavala MA, Oria JA (1995) Preserving biological diversity in managed forests: a meeting point for ecology and forestry. Landscape Urban Planning 31:363–378
Zavala MA, Espelta JM, Retana J (2000) Constraints and trade-offs in Mediterranean plant communities: the case of Holm oak-Aleppo pine forests. Bot Rev 66:119–149

Fernando Valladares
Centro de Ciencias Medioambientales
C.S.I.C., Serrano 115 dpdo
28006 Madrid, Spain

e-mail: valladares@ccma.csic.es

Applications of Stable Isotopes in Plant Ecology

Cristina Máguas and Howard Griffiths

1 Introduction

The use of stable isotopes at natural abundance levels has brought a new dimension to our understanding of plant ecology. In the past two decades there have been remarkable advances in the theoretical understanding of discrimination processes, as well as technical developments in mass spectrometry, leading to an exponential growth in applications for natural systems. Stable isotopes provide plant ecologists with a quantitative and integrative framework for chemical, biological and ecological transformations. The range of stable isotopes currently routinely analysed for ecological applications and their natural abundances are shown in Table 1. Analyses of the relative natural abundances of stable isotopes of carbon ($^{13}C/^{12}C$), oxygen ($^{18}O/^{16}O$), nitrogen ($^{15}N/^{14}N$) and deuterium (D/H) have been used across a wide range of scales, from cellular to community and ecosystem level, contributing much to our understanding of the interactions between biosphere, pedosphere and atmosphere.

Initial work by geochemists on the natural variations in carbon isotope composition (Craig 1954) led to the correlation with C_3 and C_4 photosynthetic pathways (Smith and Epstein 1971), and subsequently to a robust quantitative framework describing fractionation processes

Table 1. Average terrestrial abundance of the stable isotopes of major elements of interest in ecological studies

Element	Isotope	Natural abundance (‰)
Hydrogen	^{1}H	99.985
	^{2}H	0.015
Carbon	^{12}C	1.110
	^{13}C	99.63
Oxygen	^{16}O	99.759
	^{17}O	0.037
	^{18}O	0.204

Progress in Botany, Vol. 64

(Vogel 1980; Farquhar et al. 1982, 1989; O'Leary 1981; Griffiths 1996). In general terms, processes such as diffusion and enzymatic incorporation favour the lighter isotope and lead to depletion of the heavier isotope as compared to source material. This may result in contrasting effects on the isotope content of primary producers: thus ^{13}C is depleted during photosynthesis, but leaf water is enriched in D or ^{18}O during transpiration. Secondary discrimination against ^{13}C, D and ^{18}O may subsequently occur during synthesis of organic material. Such kinetic fractionation differs from equilibrium fractionation, which may lead to an enrichment in the heavy isotope, for instance, during the dissolution of CO_2. Thus, the extent that C_3 and C_4 plants fractionate against ^{13}C is tempered by both processes (see Fig. 1), partly because PEP_C uses HCO_3^- as substrate and shows low fractionation, but also because some leakage from the bundle sheath allows the inherently high fractionation by Rubisco to be expressed.

There has been a wealth of material published recently, which has been reviewed extensively (Griffiths et al. 1999; Brugnoli and Farquhar 2000; Adams and Grierson 2001; Evans 2001; Robinson 2001), with the scope of isotope applications ranging from metabolic trafficking at the cellular level (Gleixner et al. 1993; Meier-Augenstein 1999), to scaling of gaseous exchanges across the planetary boundary level to provide a footprint for contrasting ecosystems (e.g. Lloyd and Farquhar 1994; Ciais et al. 1995). Our aim is to focus on specific areas of ecological relevance for our understanding of natural vegetation at the leaf and community level. We will summarise the background isotope effects and fractionation processes, and show how these translate into observed patterns of carbon isotopes for organic material in lower and higher plants, with particular reference to the balance between photosynthesis and respiratory processes. This will lead to a consideration of the concept of "water use efficiency", often derived as a correlate of carbon isotope discrimination, and how precise determinants of carbon isotope composition in leaves relate to internal leaf/thallus structures and external environmental conditions. Finally, we will briefly consider more general applications using the suite of isotopes mentioned above, and see how such approaches may be integrated to characterise functional groups of plants at an ecological level.

2 Background to Isotope Effects

Mass spectrometry: traditional approaches using dual-inlet mass spectrometers were revolutionised with the development of continuous flow methodologies. The coupling of an elemental analyser and a gas chromatograph to a mass spectrometer operating in single-inlet mode allowed

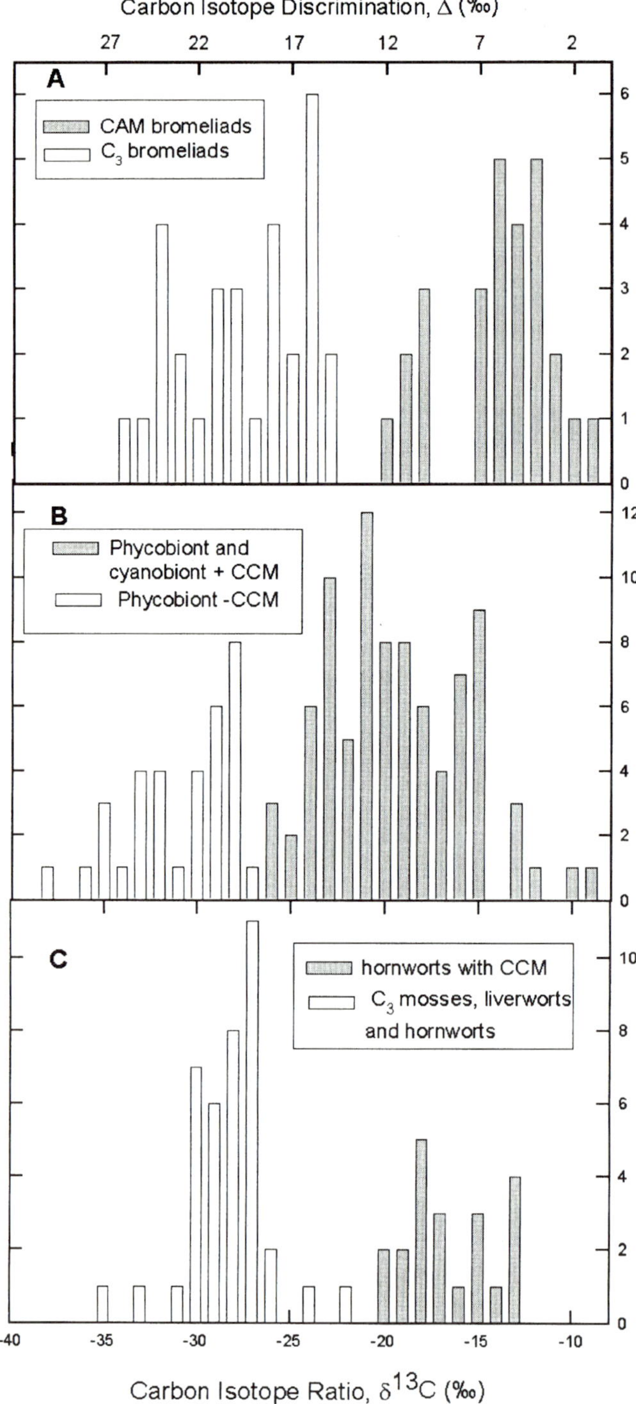

Fig. 1. Distribution of photosynthetic pathways in bryophytes, lichens and CAM members of the Bromeliaceae in Trinidad with the corresponding carbon isotope composition ($\delta^{13}C$) and carbon isotope discrimination values (Δ). **A** C_3 and CAM members of the Bromeliaceae in Trinidad. **B** C_3 and CCM in lichens, distinguished by the occurrence of chloroplast pyrenoid in the phycobiont. **C** C_3 and CCM in bryophytes, with CCM in members of the Anthocerotae with pyrenoid. (Griffiths et al. 1999)

high sample throughput and a progressive reduction in sample size to as little as 10 ng. The latest generation of mass spectrometers interface with an array of combustion/pyrolysis systems for conversion of organic and aqueous samples to gases. Additionally, with the progressive reduction in sample size required for analysis, conventional "flask" techniques vital for maintaining the global network of atmospheric CO_2 sampling programmes (e.g. Ehleringer and Cook 1998) require a much smaller air sample and provide the opportunity for increased replication.

a) Fractionation of Carbon

Mass spectrometric analysis gives high precision measurements of the molar abundance ratio (R) for particular isotopes (e.g. $^{13}C/^{12}C$, measured as mass ratio of 45/44 for carbon), and defined as $R{=}^{13}CO_2/^{12}CO_2$. Whilst the sample is measured as a differential ratio against a defined standard, compared as ($R_{sample}/R_{standard}/-1$), such $\delta^{13}C$ values tend to be negative since most organic material is depleted in ^{13}C with respect to the carbonate standard. However, for biological systems, it is more useful to think in terms of an isotope effect, denoted by α (or fractionation factor), which has been defined as the carbon isotope composition R in the reactant divided by that in the product, for a given reaction (Farquhar et al. 1989):

$$\alpha = R\ \text{reactant}/R\ \text{product} \qquad (1)$$

or

$$\alpha = (^{13}C/^{12}C)\ \text{reactant}/\ (^{13}C/^{12}C)\ \text{product} \qquad (2)$$

A detailed description of the fractionation factors associated with carbon isotopes can be found in Farquhar et al. (1989), with the overall isotope effect, defined as discrimination, Δ (such that $\Delta{=}1{+}\alpha$). In essence, because the heavier isotope ($^{13}CO_2$) is slower to diffuse and react, the cumulative effects of fractionation during CO_2 diffusion through boundary layers, stomata, and then subsequent dissolution and transport of CO_2, and then finally carboxylation, can be calculated for each step during the uptake of CO_2, with fractionation factors associated with each step.

When thinking in terms of the isotope effects between reactant and product for plant material, it is often useful to correct for the contribution that source CO_2 may make, so as to derive the magnitude of biological discrimination as a positive value (Farquhar and Richards 1984; Farquhar et al. 1989). Thus, from the measured isotope ratios ($\delta^{13}C$) of source air (δ_a) and plant material (δ_p), it is possible to calculate a value of discrimination in organic material as:

$$\Delta = (\delta_a - \delta_p)/\ (1 + \delta_p) \qquad (3)$$

Of course, the CO_2 signal in the atmosphere has changed in the global atmosphere due to fossil fuel combustion, declining from –5.7 to –8.0‰ in the last 40 years, although in the northern hemisphere this is superimposed on the seasonal cycle of summer photosynthetic drawdown (enrichment in ^{13}C) and winter respiratory release (depleted in ^{13}C). We will discuss the implications for respiratory CO_2 on overall carbon isotope signals below.

b) Oxygen Isotope Effects

During evaporation, enrichment of ^{18}O occurs at the sites of evaporation in terms of kinetic fractionations and the ratio of vapour pressure from air (e_a) to inside of the leaf (e_i). When the leaf water evaporation rate reaches a steady state, the isotope composition of water entering the leaf must equal that evaporating the leaf, although leaves may vary in their approach to isotopic steady state under natural conditions (Harwood et al. 1999). As evaporation occurs, the back-diffusion of the residual heavy isotope is controlled to some extent by the overall flux of water evaporating (as defined by the Peclet effect: Farquhar and Lloyd 1993). Because the ^{18}O signal is also adopted by CO_2 exchanging with the leaf, there has been some debate as to where within the leaf, from chloroplast to evaporating site, the CO_2 signal represents (Farquhar and Lloyd 1993; Farquhar et al. 1998; Yakir and Sternberg 2000). An additional fractionation of 27‰ occurs when carbonyl oxygen and water exchange during sucrose and cellulose synthesis, and account must also be taken (as a "dampening factor") of leaf water heterogeneity during cellulose synthesis (Saurer et al. 1997).

3 Carbon Isotope Discrimination in Organic Material

a) C3, C4 and CAM

In practice, the primary determinants of the carbon isotope composition of C3 organic material are the balance between the fractionation during diffusion through stomata (fractionation factor α=1.0044 or discrimination, a=4.4‰), and that expressed by Rubisco (α=1.029, or b=29‰). This approach was originally derived by Farquhar et al. (1982), and reviewed by Farquhar et al. (1989) and Brugnoli and Farquhar (2000). In the simplest formulation, if CO_2 in the substomatal cavity is considered to be that partial pressure presented to Rubisco, then:

$$\Delta = a + (b - a)\frac{p_i}{p_a} \qquad (4)$$

where a is discrimination during diffusion in still air; b is net fractionation during carboxylation in C_3 plants, i.e. mostly Rubisco, but also PEPc carboxylation. This derivation provides the link between isotope composition and water use, since both transpiration and discrimination are independently regulated via stomata in higher plants (for a discussion of the correlation for water use efficiency, see Sect. 4). However, more detailed formulation also accounts for the fractionation that occurs during respiration and photorespiration, in terms of the effect on the residual organic material, and the actual CO_2 partial pressure presented to Rubisco, P_c, hence:

$$\Delta = a_b \frac{p_a - p_s}{p_a} + a \frac{p_s - p_i}{p_a} + (e_s + a_1) \frac{p_i - p_c}{p_a} + b \frac{p_c}{p_a} - \frac{e \frac{R_d}{k} + f\Gamma^*}{p_a} \quad (5)$$

where a_b (=2.9‰) is fractionation during diffusion in the boundary layer; a (=4.4‰) is fractionation during binary diffusion in air; e_s (=1.1‰) is discrimination during CO_2 dissolution; a_l (=0.7‰) is fractionation during diffusion in the liquid phase; b (≈28‰) is fractionation during carboxylation in C_3 plants; p_a is atmospheric CO_2 partial pressure; p_s is CO_2 partial pressure in the boundary layer; p_i is intercellular CO_2 partial pressure; p_c is chloroplastic CO_2 partial pressure; e is discrimination during photorespiration; f is discrimination during photorespiration; k is carboxylation efficiency; and Γ^* is the compensation point in the absence of dark respiration.

Direct measurements of the two (photo)respiratory terms (e, f) have been difficult to produce, and the model works well for organic material when respiration is negligible (von Caemmerer and Evans 1991). However, there may be significant impacts for direct measurements of Δ, online during photosynthesis and for gaseous fluxes above vegetation at high temperatures, or for organisms where respiration makes up a high proportion of carbon turnover and refixation, as we address below.

As shown in Fig. 1a, discrimination results in a bimodal distribution of carbon isotope composition, with ^{13}C more depleted in C_3 plants and distinct from that in CAM or C_4 plants. This demonstrates, in part, the large isotope effect of Rubisco, as compared to phosphoenolpyruvate carboxylase (PEPc), the initial carboxylase in C_4 and CAM plants. The two enzymes have distinct kinetic effects: PEPc does not utilise CO_2, but rather the hydrated form of inorganic carbon (HCO_3^-), which has a different isotopic composition at equilibrium (O'Leary 1981, 1988). Together with a low fractionation by the carboxylase (2‰), the stable products of PEPc (the 4-carbon malate) are relatively enriched in ^{13}C by some 6‰ in comparison to source air, determining in part the lower fractionation observed in C_4 and CAM plants. These values are tempered in part by environmental conditions (C_4, Madhavan et al. 1991) or by the extent of the C_3 metabolic pathway in CAM plants (Roberts et al. 1997;

Griffiths et al. 1999). Subsequently, during decarboxylation and refixation in the bundle sheath cells, some CO_2 leaks out allowing Rubisco discrimination to be expressed, and shifts the δ_p of organic material from −12 to −18‰ (equivalent to a Δ of 4–12‰).

In C_4 plants, the main source of fractionation is the so-called leakiness of the bundle sheath. The simplified expression for discrimination is:

$$\Delta = a + (b_4 + b_3\Phi - a)\,(p_i / p_a) \qquad (6)$$

where b_4 is the net fractionation for CO_2 dissolving, hydrating to HCO_3^- and then being fixed by PEPc (−5.7‰), b_3 is the fractionation caused by Rubisco carboxylation, and Φ is the proportion of CO_2 fixed by PEPc carboxylation which subsequently leaks out of the bundle sheath (Farquhar et al. 1982).

A similar model to that of C_4 plants can describe daily carbon gain for CAM plants (Farquhar et al. 1989), with the extent of PEPc fractionation visualised directly by on-line isotope discrimination techniques (Griffiths et al. 1990; Roberts et al. 1997). Whilst the true PEPc discrimination is masked in C_4 plants by daytime Rubisco activity, in CAM plants, when p_i/p_a is high at night, associated Δ values approach the theoretical minimum (Δ=-6‰); under high VPD, as stomata limit gas exchange, values shift towards +4.4‰, reflecting diffusion limitation (Roberts et al. 1997). Of course, optimal growth of CAM plants often occurs when net CO_2 uptake by day is maximised (phase IV net CO_2 uptake in the late afternoon), which tends to contribute a greater proportion of C_3 organic material. Indeed, the bimodal range of Δ values found for C_3 and CAM populations (Fig. 1a) perhaps encompasses many plants in the C_3 range which are not overtly CAM, with recent estimates that a range of CAM plants with up to 30% of their carbon gain at night will show C_3-like Δ values (Winter and Holtum 2002) with similar findings for bromeliads (S. Pierce and H. Griffiths, unpubl. data).

Discrimination in whole plant dry matter (structural carbon and stored carbohydrates) represents long-term integration of fixed carbon, but instantaneous or on-line Δ measurements, where the isotopic composition of the air-CO_2 is analysed before and after passing over a leaf in a cuvette, provides a time scale of minutes. An intermediate measure, on a daily basis, is provided by newly fixed carbon, such as in leaf soluble sugars and starch (Deléens-Provent and Schwebel-Gdugué 1987; Brugnoli et al. 1988; Gleixner et al. 1993; Borland et al. 1994; Brugnoli and Farquhar 2000). Other relevant differences in Δ among metabolites have been found, with the most striking differences in lipids (5–10‰) in ^{13}C bulk material, mainly due to secondary fractionation associated with decarboxylation of pyruvic acid (Schmidt and Gleixner 1998). Others, such as amino acids, showed a wide range of Δ values, and secondary plant metabolites such as lignin presented a strong depletion in ^{13}C

when compared with cellulose (for reviews, see Schmidt and Gleixner 1998; Brugnoli and Farquhar 2000).

b) Poikilohydric Plants

The lack of stomata, as well as the absence of a transport system in bryophytes (mosses and liverworts) and lichens, makes an interesting contrast for control of $\Delta^{13}C$. As poikilohydric organisms, water status is completely dependent on their environment, in such a way that the organisms reach an equilibrium with atmospheric humidity (Green and Lange 1994). Thus, the CO_2 uptake may be transient, a function of the water thallus content, and regulated by atmospheric humidity, dewfall and precipitation. Without the possibility of regulating water content through stomata, they must operate in a delicate equilibrium between sufficient water to reactivate photosynthesis and CO_2 uptake limited by diffusion in a water-saturated tissue.

Variation in Δ values in mosses has been mainly attributed to the effect of water films, and consequently Δ is inversely related to plant water content, given the very high resistance to CO_2 diffusion (Rundel et al. 1979; Teeri 1981; Proctor et al. 1992; Williams and Flanagan 1996). In *Sphagnum*, during progressive dehydration, photosynthesis increases to a maximum when diffusion rates and photosynthetic CO_2 fixation are optimised, with Δ increasing linearly, reflecting the Rubisco fractionation; as dehydration continues, photosynthetic rates decrease and Δ continues to increase, with fractionation reflecting discrimination in the chloroplast and not diffusional processes (Rice and Giles 1996; Williams and Flanagan 1996, 1998). Recently, the correlation between Δ from field samples and RGR (relative growth rate) has shown that Δ is not only influenced by water availability, but also by other environmental constraints such as light and nutrient availability (Rice 2000).

Carbon isotope discrimination in lichen dry matter has been shown to be dependent on discrimination expressed by the photosynthetic partner (Fig. 1b), whether phycobiont (green algae) or cyanobiont. Phycobiont lichen associations are able to reactivate photosynthesis from water vapour in air, while the cyanobiont ones require rewetting with liquid water (i.e., Lange et al. 1988). Initially, the observed differences in Δ were attributed to differences in CO_2 transfer resistance related to the type of photobiont present in the lichen association (Lange and Ziegler 1986; Lange et al. 1988), particularly for cyanobiont lichens, where liquid water is required to reactivate net CO_2 fixation. For both types, Δ ranges between 12 to 16‰, since a biophysical CO_2 concentrating mechanism (CCM) is associated with, respectively, a chloroplast pyrenoid or cyanobacterial carboxysome; when the phycobiont lacks a pyrenoid (often in a

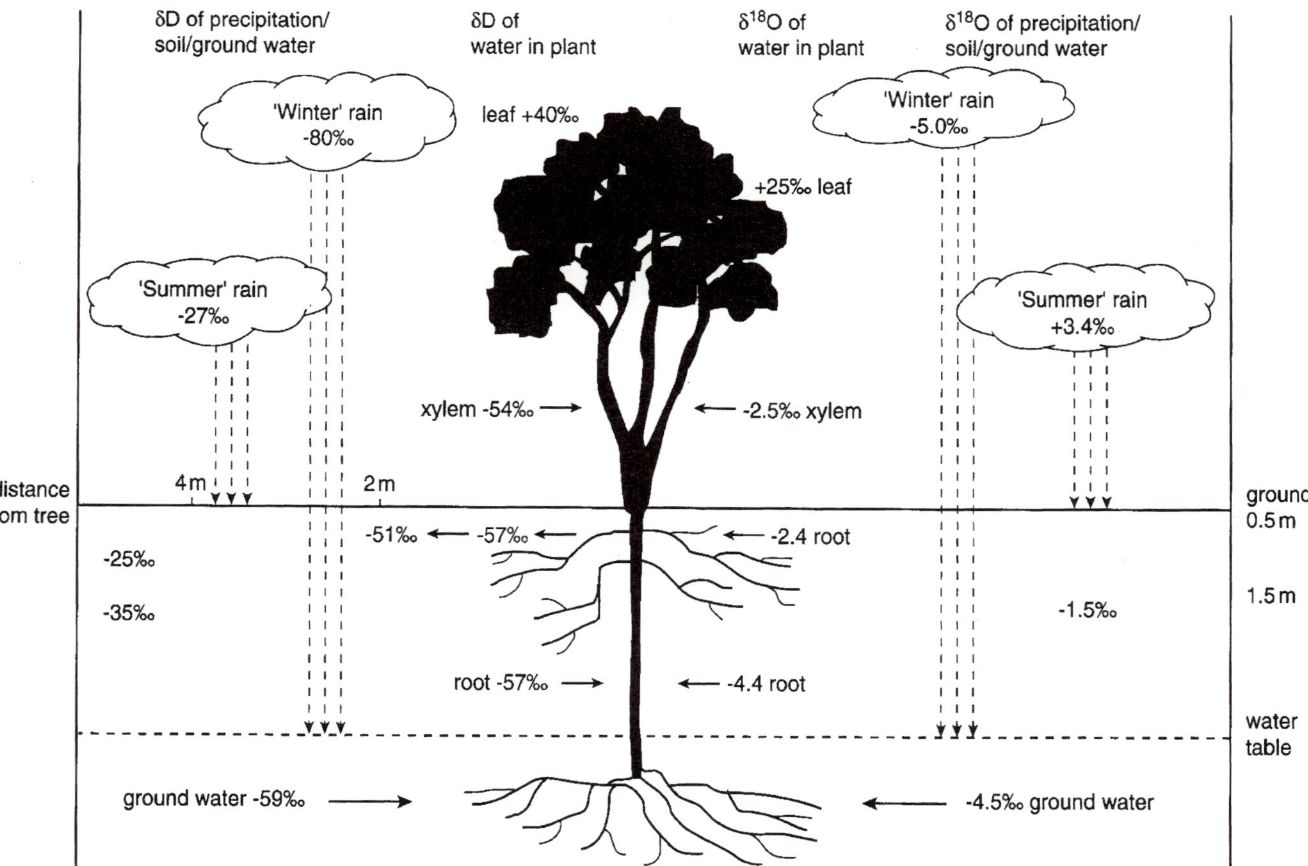

Fig. 2. Schematic diagram of $\delta^{18}C$ and δD as tracers of water sources. The isotopic values shown are average values for mid-latitude/semi-arid environments, and represent the soil-plant-atmosphere continuum. (Griffiths et al. 1999)

specialised tripartite associations with cyanobacteria in cephalodia), Δ ranges from 24–28‰ (Fig. 1b; Máguas et al. 1993, 1995; Palmqvist et al. 1994; Smith and Griffiths 1996). These ranges are matched by instantaneous, on-line Δ values and CO_2 compensation points (Fig. 2; Máguas et al. 1995; Smith and Griffiths 1996).

The effect of a CCM induction has been primarily shown to influence instantaneous Δ in free-living algae, with Δ declining from 28‰ to 4‰ following transfer from elevated CO_2 to air-equilibrated supply (Sharkey and Berry 1985). The CCM may in fact confer a particular advantage to lichens, since the resistances to CO_2 diffusion in the hydrated fungal matrix may be high (Lange et al. 1988; Cowan et al. 1992; Lange and Green 1996). The activity of a CCM would temper the effect of the large CO_2 diffusion resistances within the thallus, since it increases the CO_2 concentration at the carboxylation sites (Farquhar et al., 1989). One of the implications of such CCM activity is a reduction in Δ, which has been reported for C_4 plants (Farquhar et al. 1982, 1989), CAM plants (Griffiths et al. 1990) and algae possessing the CCM (Beardall et al. 1982; Sharkey and Berry 1985).

Additionally, thallus structure and morphology play a fundamental role in determining CO_2 diffusion resistance across the thallus to the carboxylation sites, which will determine an important component of Δ in lichens (Máguas et al. 1995, 1997; Máguas and Brugnoli 1996) or even along a single thallus when marginal and central parts were compared (Máguas and Brugnoli 1996; Máguas et al. 1997). In Fig. 1c we also show that the occurrence of a CCM in members of the Anthocerotae possessing a pyrenoid is associated with low Δ values (Griffiths and Smith 1996).

c) Refixation and Respiration

As discussed above, the impact of fractionation during respiration and photorespiration on residual $\Delta^{13}C$ organic matter is taken into consideration in the more detailed equation [Eq. (5)] derived by Farquhar and coworkers (Farquhar et al. 1989; see also Griffiths et al. 1999; Brugnoli and Farquhar 2000). Discrimination expressed during dark respiration at night represents a net loss of fixed carbon, whereas during respiration and photorespiration in the light the effects will be dependent on the extent or respiratory refixation (Máguas et al. 1993; Griffiths 1996; Griffiths et al. 1999). The extent of fractionation during dark respiration seems to vary between cells in vitro, where *e* may be zero (Lin and Ehleringer 1997), to intact leaves, where Duranceau et al. (2001) showed that respired CO_2 was enriched in ^{13}C by about 6‰. The likelihood of variations in *e* will depend on which substrates are being decarboxylated, particularly given the non-uniform ^{13}C distribution within the hexose molecules (Rossman et al. 1991; Schmidt and Gleixner 1998), and isotope effects during the pyruvate dehydrogenase reaction (i.e., Melzer and Schmidt 1987). More recently, respired CO_2 was ^{13}C-enriched compared to all leaf metabolites in *Nicotina sylvestris* and *Helianthus annuus* (Ghashghaie et al. 2001).

In lichens, where we have a substantial contribution of respired CO_2 during light (mycobiont respiration), the photobiont uses a mixture of different sources: CO_2 supplied by air, respired CO_2 and any photorespired CO_2 (which will be dependent on CCM activity). This results in differences between measurements of instantaneous Δ during photosynthesis and organic material. For example, in phycobiont lichens with or without a CCM, instantaneous Δ values were higher than organic Δ by 3–5‰ when measured under "tank" CO_2 (a source depleted in ^{13}C at −40‰). Cyanobiont lichens, in contrast, with lowest instantaneous Δ, showed a large negative offset between instantaneous and organic Δ, with the instantaneous values 4–5‰ lower. It is tempting to speculate that in the "gelatinous" cyanobacterial lichens there is an additional resistance to diffusion which shifts instantaneous Δ. Such a difference would be analogous to the determination of mesophyll conductance from on-line Δ (Evans et al. 1986) and the use of tank CO_2 and altered O_2 concentrations used to distinguish these effects in higher plants (Gillon and Griffiths 1997; Gillon et al. 1998). In studies on samples from the field, phycobiont lichens showed higher CCM activity in shade as opposed to an exposed habitat, whereas cyanobiont lichens showed highest CCM activity in exposed conditions, although there was little consistent change in Δ of organic material (Smith and Griffiths 1998).

Instantaneous measurements of discrimination during gas exchange (Δ_{obs}) can be compared with Δ_i predicted from Eq. (3), when derived from gas exchange measurements of p_i/p_a. The difference between measured and predicted Δ (Δ_i–Δ_{obs}) represents the additional drawdown of CO_2 from substomatal cavity (p_i) to sites of carboxylation (p_c) and is therefore a function of g_i, the mesophyll conductance (Evans et al. 1986; Evans and von Caemmerer 1996). Initial evidence that instantaneous Δ was closely related to temperature, independent of p_i/p_a (Gillon et al. 1998; Harwood et al. 1998), suggested that Δ was allied to high rates of photorespiration. By manipulating the CO_2 source signal and oxygen concentration, it has been possible to derive a fractionation factor associated with photorespiration, f (Gillon and Griffiths 1997; Gillon et al. 1998) which, having been corrected for refixation of photrespiratory CO_2, leads to a consistent value of f=8‰ for wheat, bean and oak (Griffiths et al. 1999). More recent measurements (G. Lanigan, N.R. Betson, H. Griffiths, unpubl. data) have revised this figure to 10‰. Whilst other authors suggest a wide range of values of f for in vitro systems, our values are closer to those originally predicted by Rooney (1988).

Such effects of respiration and photorespiration may be insignificant for fast growing crops under mesic conditions, but in many areas of natural vegetation the gas-exchange footprint may be markedly affected by the extent of (photo)respiration and refixation processes, and should be taken into account when daytime stable isotope measurements are coupled to eddy covariance (Ciais et al. 1995; Miranda et al., 1997).

4 Carbon Isotope Discrimination as an Ecological Concept

In this section, we address the mechanisms which have related carbon isotope discrimination to water use so effectively for crop plants, which is highly appropriate in the year that the first wheat variety will be brought to market following selection primarily via carbon isotopes (G.D. Farquhar, pers. comm.). However, we then go on to consider the relevance of water use efficiency (WUE) in ecological terms.

a) "Water Use Efficiency" and Discrimination

The quantitative model relating Δ with p_i/p_a, the ratio of internal/external CO_2 partial pressure, via stomata conductance, has provided a remarkably robust relationship (Farquhar et al. 1989; Brugnoli and Farquhar 2000). On the basis of this relationship, Δ is frequently linked to water use efficiency (WUE), the molar ratio of photosynthesis to transpiration, as shown below:

$$WUE = \frac{A}{E} = \frac{p_a - p_i}{1{,}6v} \tag{7}$$

where A is the CO_2 assimilation rate in a leaf, E is the transpiration rate, p_a and p_i are the internal and external partial pressures to CO_2, respectively, and v is the water vapor pressure deficit between the intercellular spaces and the atmosphere. The term 1.6 corrects for the relative diffusivities of CO_2 and H_2O in air.

When considering the whole plant, a fraction of the fixed carbon will be lost through respiration of the tissues (ϕ_c) and a fraction of water (ϕ_w) will also be lost during the night if stomata do not completely close:

$$WUE = \frac{p_a\left(1 - \frac{p_i}{p_a}\right)(1 - \phi_c)}{1.6v(1 + \phi_w)} \tag{8}$$

By rearranging the above equations, a negative linear relationship between WUE and Δ is obtained:

$$W = \frac{p_a\left(\frac{b-d-\Delta}{b-a}\right)(1-\phi_c)}{1{,}6v(1+\phi_w)} \tag{9}$$

From Eq. (9) it is possible to estimate WUE through the isotopic analysis, which is far more practical and rapid than gas-exchange studies. This relationship has been tested in a range of plant species, in particular, agronomical species, under controlled and field conditions (i.e. Farquhar and Richards 1984; Hubick et al. 1986; Ehleringer et al. 1992). On the basis of this relationship between Δ and p_i/p_a, carbon isotope discrimination has been proposed as a selection criterion in C_3 species for improving WUE, particularly under water-limiting conditions.

However, several studies have shown that Δ may be directly related to biomass production or grain yield (i.e., Condon et al. 1987; Johnson et al. 1990; Acevedo 1993; Araus et al. 1993; for a review, see Brugnoli and Farquhar 2000). When stomatal conductance is the major determinant of changes in Δ, genotypes with a high yield will show a positive correlation between high conductances and high assimilation rates, and hence consequently high Δ and yield (Brugnoli and Farquhar 2000). Under well-watered conditions, high WUE will not necessarily be a requirement for high yield. However, if the constraint to Δ is the photosynthetic capacity (which can affect internal CO_2 drawdown and hence p_i/p_a directly), WUE will improve the crop production under both well-watered and dry conditions.

b) Relevance of WUE for Natural Vegetation

Active gas exchange and reduction of water loss through evaporation are contradictory problems for photosynthetic organisms. This trade-off is commonly estimated in vascular plants by the photosynthetic water use efficiency (WUE). Often, WUE is measured simply as the ratio of assimilation/evaporation during gas exchange or, alternatively, in the long term, as the ratio of dry matter carbon gain to water use. Here, we argue that WUE is relevant, from an ecological perspective, mainly for two highly specific examples: CAM plants, and poikilohydric plants such as lichens, where the combination of stomatal resistance and/or boundary layer resistances are important in controlling gas exchange. For most vascular plants, the strategy of having a higher WUE and saving water for the future is not successful if water saved in the soil evaporates directly or is used by other neighbouring plants.

However, for CAM plants, which rely on regular and predictable inputs, water taken up can be stored and used more slowly over time, allowing a plant with higher WUE to gain carbon for a longer time period.

A similar argument might be true for lichens: variations in morphology and/or interior-to-exterior structural differences lead to differences in lichen water use efficiency (LWUE) and to the length of time a lichen can remain photosynthetically active, albeit with a time scale of minutes or hours rather than days-to-weeks for CAM plants. Thallus density, in particular, seems to play an important role in controlling water retention capacity, as shown for 12 *Umbilicaria* species (Valladares et al. 1998) and as found for *Lobaria scrobiculata* (Máguas et al. 1997).

In addition, WUE is a concept primarily used by plant breeders in agronomic studies, where the emphasis has been to improve the productivity of annuals in terms of crop production and yield. There is considerable variation in Δ within and between natural plant populations, as we discuss below. Whilst crop plants for intensive agriculture are highly selected and homogenous in their response, natural vegetation is analogous to low input, sustainable agronomic systems, where perhaps it is appropriate to sow seeds with a range of attributes capable of producing some yield in an unpredictable environment. From an ecological perspective, the relationship between Δ and performance may be particularly revealing, as long is it is not simply taken a measure of WUE (Griffiths et al. 1999). Specifically, it may be more appropriate that Δ integrates overall strategies for growth and fitness, as exemplified in the work of Ehleringer and co-workers (e.g. Ehleringer et al. 1992). For natural vegetation, the trade-off between low productivity, low Δ and survival versus high Δ, high potential growth but susceptibility to climatic extremes, represents much more than the notion of water use efficiency.

5 Carbon Isotopes: Productivity Versus Survival in Natural Ecosystems

a) Control over Δ in Natural Vegetation

Although there is an extensive literature supporting the link between Δ and WUE via p_i/p_a, it has also become evident that a wide range of factors influence leaf Δ (Table 2). For instance, external constraints include environmental factors and physiological variables driving them, whereas the internal balance between stomata and internal conductance to CO_2 and the relationship with photosynthetic capacity are key determinants of Δ, perhaps independently of WUE. Additionally, leaf carbon isotope composition is made up of three sources of carbon: (1) imported carbon or reserves, perhaps laid down in a previous growing season; (2) carbon fixed during leaf expansion, perhaps earlier in the growing season when environmental conditions differ, and (3) current photosynthate (Griffiths et al. 1999). Thus, interpretation of Δ needs to account for these differing sources and the environmental conditions each may integrate,

Table 2. Carbon isotope discrimination control by external and internal factors

Control of Δ: Long term	Control of Δ: Instantaneous	CO_2 supply and demand	Physiological/ morphological correlates	Environmental conditions
	Δ_i ↑		Lifeform/strategy[2]	VPD
		Diffusive exchange		PPFD
WUE ← A/g_s —	p_i		SLM[3]	Temperature
↕	↓ g_i	P_c: Rubisco capacity[1]	Chlorophyll[4]	Rainfall
organic $\delta^{13}C$ ←	p_c			Soil water deficit
($\delta^{18}O$, VPD-led)	↓	(Photo)respiration refixation and exchange	Leaf N[5]	
^{13}C, ^{18}O in CO_2 ←	$\Delta_{observation}$		Ash content[6]	Soil strength/ nutrients

Source references to confirm the physiological and morphological correlates of D indicated as:1: Rubisco capacity; 2: genotype/lifeform; 3: specific leaf mass; 4: chlorophyll content; 5: total leaf nitrogen; 6: ash content.
From Griffiths et al. 1999

particularly when yield, and indeed productivity, may be considered as secondary priorities under stressed conditions.

Decreases in Δ can arise from either reduced g_s (the conductance of CO_2 into a leaf) or increased photosynthetic capacity (A). While both reflect increased stomatal limitation on A, the two sources of variation in Δ may result in opposite effects on plant performance, particularly when coupled with simultaneous changes in leaf area. Plant tissues with high Δ may be more productive than a low-Δ species only if the low Δ is due to a decrease in g_c, rather than an increase in A. Moreover, the amount of photosynthates produced at a given p_i will weight the subsequent organic Δ value. Periods with enough water, and consequently large A and Δ, may dominate the average Δ signal as opposed to dry periods when there is both a low A and low Δ. This is particularly important when we are considering water-limiting habitats and plants that are well adapted to these environments. Indeed, the whole problem of synchronicity between the timing of organic material deposition and Δ as an integrator of stress has been reviewed by Griffiths et al. (1999).

b) Leaf Function and Physiological Characteristics

Mesophyll conductance (g_i) lowers internal conductance to CO_2 diffusion between the substomatal cavities and the chloroplasts, with a consequent drop in CO_2 partial pressure, and is an important factor for estimation of Δ (Evans et al. 1986; Brugnoli and Bjorkman 1992; Maxwell et al. 1997). From instantaneous, on-line Δ it has been possible to estimate g_i by measuring the deviation between on-line Δ and that calculated (Δ_i) using the Farquhar model (Evans et al. 1996; Griffiths et al. 1999), and show that g_i can be very low, particularly in woody species. However, this low g_i may also have an adaptive significance. Another way to study g_i may be through the deviation between Δ_i and that measured in leaf sucrose (Brugnoli et al. 1998; Lauteri et al. 1997; Scartazza et al. 1998) giving a daily integrator of g_i, which can be used to evaluate responses under field conditions. Moreover, the gradient of the relationship between Δ_i–Δ_{obs} vs A/p_a is inversely proportional to the mesophyll conductance (Evans et al. 1996; Griffiths et al. 1999). Generally, the highest values of g_i are presented by fast-growing herbs, in contrast to sclerophilous woody species (Griffiths et al. 1999). Interestingly, work with leaf expansion leaves from evergreen broad-leaved trees showed g_i increased during full leaf expansion (FLE), and two CO_2 transfer components were identified: gas-phase conductance (intercellular air space) and liquid-phase conductance (cell wall, cell membrane, cytosol, chloroplast envelope and stroma; Miyazawa and Terashima 2001). Through the use of on-line measurements, g_i was estimated, and accordingly, CO_2 transfer through air spaces decreased after FLE due to the increase in mesophyll thickness. One extreme case of very low g_i values is provided by CAM plants (Maxwell et al. 1997; Griffiths et al. 1999) associated with the structure and anatomy of succulent plants.

Whether or not leaf thickness affects g_i is not yet clear. There are several indications that g_i decreases with leaf thickness (Vitousek et al. 1990; Evans et al. 1994; Lauteri et al. 1997), and there is a large drawdown from p_i to the carboxylation site in thick hypostomatous leaves (Hanba et al. 1999). However, data from different evergreen tree species showed that variation in g_i did not cause any effect on Δ dry matter in similar functional groups when at the same altitude (Hanba et al. 1999).

c) Environmental Gradients

Considerable effort has gone into the description of the environmental control of Δ variation, particularly along environmental gradients (e.g., De Lucia and Schlesinger 1991; Lajtha and Getz 1993; Panek and Waring 1995, 1997). A strong and consistent decrease in Δ with altitude has been observed in many species (Körner et al. 1988; Vitousek et al. 1990; Marshall and Zhang 1994; Sparks and Ehleringer 1997; Hultine and Marshall 2000). This decrease in Δ is a result of a complex combination of interrelated climatic variables, such as precipitation, atmospheric pressure (Marshall and Zhang 1994), temperature (Panek and Waring 1995) and edaphic factors (soil age, depth, nutrient availability, water-holding capacity: Warren et al. 2001). Thus, Δ should also reflect a combination of a decrease in CO_2 and O_2 partial pressures (Ehleringer and Cerling 1995), variation in morphology and anatomical leaf characteristics and physiological traits, such as leaf thickness (Vitousek et al. 1990), stomatal density and conductance (Körner 1989; Meinzer et al. 1992). However, when variations in stomatal density and conductance will lead to an increase in the CO_2 supply, no decrease in Δ would be expected (Hultine and Marshall 2000). Additionally, correlations between Δ, nitrogen leaf content and leaf thickness have been reported across altitude gradients (Körner 1989; Körner et al. 1991; Sparks and Ehleringer 1997; Vitousek et al. 1990). There are several explanations, such as increasing nitrogen in plant tissues with altitude, increasing CO_2 demand at the carboxylation sites, with thicker leaves possessing a higher quantity of photosynthetic enzymes. Alternatively, a decrease in g_i would reduce the CO_2 supply and reduce Δ: Hultine and Marshall (2000) found that internal resistances, rather than A_{max}, were the main reason for the observed decrease in Δ for different conifer species.

In the last two decades, many studies have related significant variations in leaf $\delta^{13}C$ along light gradients within closed plant canopies, mainly forests (e.g., Francey et al. 1985; Garten and Taylor 1992; Broadmeadow et al. 1992) and within single tree crowns (Broadmeadow and Griffiths 1993; Waring and Silvester 1994; Prasolova et al. 2000; Le Roux et al. 2001). Refixation of respiratory CO_2, normally depleted in ^{13}C, makes a minor contribution (Broadmeadow and Griffiths 1993), but environmental constraints are the main determinant: air humidity has direct effects on stomatal conductance; leaf irradiance directly affects Δ through the ratio p_i/p_a (Pearcy and Pfitsch 1994; Broadmeadow and Griffiths 1993); long-term photosynthetic acclimation leads to changes in photosynthetic capacity, stomatal conductance and internal conductance with decreasing irradiance (Hanba et al. 1997, 1999); a direct effect on water potential (more negative) and the corresponding decrease in p_i in the upper canopy and for branches with different sizes (Walcroft et al. 1996; Ninemets et al. 1999). More recently, Le Roux and coworkers (2001), in a single crown, found that WUE and Δ varied along light intensity gradients, mainly due to short-term responses to local irradiance.

For individual environmental variables, Δ has been related to irradiance (e.g., Pearcy and Pfitsch 1994; Le Roux et al. 2001) and temperature (e.g., Tans and Monk 1980; Leavitt and Long 1982). In particular, there is range of possible variations in Δ due to water availability, since plants should respond to decreasing moisture, both in time and space, with both phenotypic and genotypic differences revealed within and between populations. This may arise because of a decrease in stomatal conductance relative to photosynthetic capacity although such a response would lead to a decrease in competitive ability by favouring other plants better adapted to limited water (e.g., Miller et al. 2002; Comstock and Ehleringer 1993; Miller et al. 2001). Several strategies may be used by plants in order to maintain water supply under water-limiting conditions. Higher carbon allocation to root length and depth, and less to leaf area and shoot weight, can reduce stomatal limitation in drier areas and consequently, result in higher Δ. Such a pattern of higher Δ in plants from dry environments has been reported by several authors using common garden experiments (e.g. Hubick and Gibson 1993; Lauteri et al. 1997; Zhang and Marshall 1995). Alternatively, the opposite effect, with lower Δ, from dry areas has also been observed, which was related to genotypic selection predominating over the phenotypic stomatal conductance (Comstock and Ehleringer 1993; Anderson et al. 1996; Li et al. 2000). However, Miller and co-workers (2001) did not find a consistent decrease in Δ along a decreasing rainfall gradient, with species distribution more related to changes in leaf area index (LAI). Studies using common garden approaches, as well as compared to natural resource gradients, are extremely important in describing the relationship between water availability and plant performance. However, it is clear from several studies that water conservation in dry habitats may be more important than water use for biomass production, which leads us to redefine the use of WUE for ecological studies and those species well adapted to stress, particularly to water limitation.

d) Phenotypic and Genotypic Effects Within and Between Populations

There have been a number of elegant studies which have shown that Δ is an integrator of different growth strategies, perhaps typified by work with the warm-desert shrub *Encelia farinosa* (Ehleringer et al. 1992, 1999). Thus, the overall range of high Δ values for *Encelia* suggests that genotypically it adopts a strategy of "high gain" as compared to *Larrea tridentata* (Ehleringer 1992). However, there is also genotypic differentiation between contrasting sites or within any given population. Plants with high Δ values have the phenotypic plasticity with potential for high rates of growth when neighbours are removed; low Δ plants do not respond to added water supply – they are genotypically conservative. These characteristics have stabilised in different populations, which is dependent on predictability of rainfall. Plants from a drier site adopt a more conservative strategy of growth, whereas those subject to more regular summer rains are more opportunistic, with rapid allocation of newly fixed carbon to growth, flowering and seedset (Ehleringer et al. 1999).

Matzner et al. (2001) compared intraspecific variation in Δ and WUE for juveniles of a Mediterranean long-lived woody species (*Q. douglasii*) growing at two sites with dramatically different annual precipitation levels. Similar to other studies where positive correlations have been obtained between biomass production and WUE (Hubick et al. 1986), a strong positive relationship between growth and water use was found for *Q. douglasii* (Matzner et al. 2001). However, the relationship between WUE and Δ was not significant, which was partly attributed to the within-population variations in plant size, and also to the associated variation in the proportion of respired CO_2 and local VPD.

An additional source of variation is associated with gender in dioecious species, where it has been suggested that the differences in cost of reproduction should impose very different resource demands on the plants, and, consequently, gender specialisation would involve specific resources demand (e.g., Cox 1981; Dawson and Bliss 1989; Dawson and Ehleringer 1993). Generally, female plants usually develop higher reproductive effort than males to produce fruits, in addition to flowers, and thus they have to allocate more biomass to reproduction (Wallace and Rundel 1979; Correia et al. 1992; Antos and Allen 1994; Obeso 1997). For example, as pointed out by Diaz-Barradas and Correia (1999), the distribution of male *Pistacia lentiscus* was associated with less disturbed areas, with lower resource availability. A few studies have dealt with WUE and Δ variation in dioecious plants (Dawson and Ehleringer 1993; Correia and Diaz Barradas 2000; Retuerto et al. 2000).

6 $\delta^{18}O$ in Organic Material and CO_2

Recently, it has been suggested that the oxygen isotope ratio ($\delta^{18}O$) of bulk leaf material may reflect evaporative conditions and may be used to determine whether variations in $\Delta^{13}C$ result from differences in photosynthetic capacity or in stomatal conductance (Farquhar et al. 1989). While ^{13}C is depleted during photosynthesis, leaf water is enriched in ^{18}O during transpiration, which is passed on to organic molecules due to an isotopic exchange of the water oxygen and the carbonyl group oxygen during biosynthesis. The oxygen isotope composition of organic material is further influenced by the isotopic composition of the source water. The degree of leaf water enrichment depends on the ratio of the vapour pressure differences in the atmosphere and the intercellular spaces (e_a/e_i), whereby low relative humidity causes an increase in leaf water enrichment. In general, bulk leaf water is somewhat less enriched than water at the evaporating surface, due to a gradient within the leaf as a result of the shifting balance between the convective evaporation flux of unfractionated water through the leaf, and the back-diffusion of isotopically enriched water away from the evaporating sites (Yakir and Sternberg 2000). This phenomenon, termed the Péclet effect, has been modelled by Farquhar and Lloyd (1993) predicting an inverse relationship between $\delta^{18}O$ of bulk leaf water and the rate of transpiration.

The relative rates of evaporation and back-diffusion of ^{18}O also have important implications for CO_2 exchange, particularly when it is considered that globally, on an annual basis, 40% of atmospheric CO_2 exchanges with plant leaves. During residence within the leaf, isotopic exchange occurs between leaf water and CO_2, catalysed by carbonic anhydrase, with the signal representing some point in the leaf water continuum from chloroplast to evaporative site (Farquhar et al. 1993; Flanagan 1998; Yakir 1998). Concurrent measurements of instantaneous gas exchange, as well as $\delta^{18}O$ in CO_2 and water vapour, can be used to model theoretical and actual discrimination (Farquhar et al. 1993; Flanagan et al. 1994; Harwood et al. 1998, 1999). This exchange is critical for recent models developed at the global level, since whilst ^{13}C signals do not distinguish at the ecosystem level between fossil fuel and respiratory CO_2 sources, the ^{18}O signal in CO_2, representing the leaf water, does allow the contribution from vegetation to be determined (Ciais and Miejer 1998). At the leaf and plant level there is still much work to be undertaken to validate the relationship between mesophyll conductance and impact on ^{18}O signal in CO_2, in relation to the extent of isotopic steady state during evaporation and site of CO_2 equilibration; at the canopy level, the contribution from ground layer and stem respiration also needs to be determined for contrasting ecosystems (Ciais and Meijer 1998; Buchmann et al. 1998; Griffiths et al. 1999).

7 Tracing Water Sources Within Ecosystems

The stable isotopes of hydrogen (δD) and oxygen ($\delta^{18}O$) are commonly used in regional groundwater studies to identify flow regimes and sources of water used by vegetation (Gat 1996). Generally, groundwater retains a constant stable isotope signature on an annual basis, which permits the identification of groundwater recharge accurately from isotopically distinct and well-characterised water sources, such as precipitation and surface water. However, there may be a temporal variability in isotopic composition, since evaporation directly from soil or surface water bodies normally enriches the residual water in $\delta^{18}O$ and δD. The δD of water available to plants is significantly influenced by: (1) physical processes such as evaporation; (2) the size and elevational extent of the catchment basin that contributes to the total source-water pool; (3) the depth and geological characteristics of the groundwater aquifer; and (4) the dissolution characteristics and velocity of water movement in subsurface strata into the groundwater (e.g., Fontes 1980). Precipitation and groundwater vary markedly across continental habitats, as well as reflecting seasonal variations (White et al. 1985; Dawson and Ehleringer 1991) and fresh water and oceanic water were distinguished in coastal habitats in southern Florida (Sternberg and Swart 1987).

Hence, root water, stem or xylem water will reflect the source or sources of water used by any particular species (Wershaw et al. 1966; White et al. 1985; Dawson 1998). Thus water which has not been subjected to evaporative or metabolic fractionation (Yakir et al. 1989; Flanagan and Ehleringer 1991) will reflect the source or sources of water used by any particular species, whether surface, groundwater or occult precipitation (Fig. 2). Poikilohydric plants, such as lichens, may also rehydrate from atmospheric humidity alone. In this way, the analysis of δD and $\delta^{18}O$ in water allows one to trace where plants are obtaining their water (e.g. Dawson and Ehleringer 1991; Dawson 1993a; Thorburn and Walker 1993). This information may, in turn, allow us to determine how differential utilisation of water sources influences plant distribution, coexistence and competition for water within the plant community in both space and time (e.g., Dawson and Ehleringer 1991; Ehleringer and Dawson 1992; Dawson 1993b; Thorburn and Walker 1993; Gebauer and Ehleringer 2000; Fig. 2).

The degree of competitive plasticity among all the members of a particular plant community, from analyses of all water sources and their proportional use by each species, has led to some fascinating insights. For instance, hydraulic lift is the process of water movement from relatively moist to dry soil layers using plant root systems as a conduit. Water released from roots during periods when transpiration ceases (usually at night) into the upper soil layers is then re-absorbed the next day and transpired, or can be acquired by neighbouring plants (Caldwell et al. 1998). Dawson (1993b) reported hydraulic lift around isolated maple trees, based on diel fluctuations in Ψ_s (soil water potential) and the natural deuterium abundance in xylem water of the trees and neighbouring vegetation. The δD signature indicated that groundwater was primarily used and not summer precipitation. Adjacent herbaceous species with shallow roots took up this hydraulically lifted groundwater, as seen in the δD signal and improved water status close to the host tree. In fact, trees may transfer up to 30% of their daily requirements to the soil surface layer at night (Dawson 1993b). Conversely, "inverse hydraulic lift" may occur when there is a downward transport of water in roots. In this case, water enriched with deuterium water was supplied to the soil surface and was transported downwards to be released in the drier soil, allowing phreatophytes to gradually penetrate towards the permanent groundwater supply.

8 Nitrogen

The ratios of the two stable isotopes, ^{14}N and ^{15}N, are apparently not subject to large fractionation during assimilation, fixation and transport under natural conditions. Early reports of the natural abundance of ^{15}N provided valuable information about N sources used by plants and fluxes of N in the ecosystems (e.g., Kohl et al. 1971; Shearer and Kohl 1986). However, there has been some debate on the interrelationship between nitrogen natural abundance ($^{15}N/^{14}N$) in soils and plants, and the use as a tracer of nitrogen source or indicator of fractionation during uptake, assimilation and transport (Evans 2001, Robinson 2001). Primary products of biological atmospheric N_2 fixation, such as by cyanobacteria in desert cryptogamic crusts (Evans and Belnap 1999; Lange et al. 1988), normally have a low $\delta^{15}N$, similar to source atmospheric nitrogen, where the $\delta^{15}N$ of atmospheric N_2 is defined as the mass spectrometric standard at 0‰. Any other additional nitrogen source (e.g., soil NH_4^+ and NO_3^-) is likely to be enriched or depleted in ^{15}N, depending on the extent of secondary fractionations during the N cycle.

Robinson (2001) developed a mixing model to account for contrasting N sources, and while this technique is useful in agricultural systems with few nitrogen sources, it is complicated by multiple species within natural habitats where it is not possible to identify all of the ^{15}N of the nitrogen sources available to vegetation (Robinson 2001). However, useful insights were obtained with woody perennials, allowing the quantification of biological N fixation (Boddey et al. 2000). In some systems, $\delta^{15}N$ can clearly be used as a tracer, assuming that the observed differences in $\delta^{15}N$ in plant material are dependent not only on the nitrogen source, but also on physiological mechanisms within the plant (nitrogen uptake mechanisms, assimilation pathways, and extent of recycling), and when plant demand exceeds nitrogen supply (Evans 2001). Intra-plant $\delta^{15}N$ variation, for example, may be caused by multiple assimilation events, such as organ-specific loss of nitrogen and resorption and reallocation of the nitrogen (Evans 2001). An important source of variation is caused by mycorrhizally assisted nitrogen uptake, which can shift the $\delta^{15}N$ of the plant from that of the original nitrogen source (Högberg 1997; Hobbie et al. 2000).

A variety of fractionations may occur during processes related to nitrogen transformation in soils (e.g., Högberg 1997; Hopkins et al. 1997) and plants (Robinson 2001), which may complicate source-sink relationships. For example, nitrification discriminates against ^{15}N more than N mineralisation, which makes NH_4^+ isotopically heavier than the organic N from which is derived (Högberg 1997). Additionally, the $\delta^{15}N$ of a particular compound may change, and thus, together with the complexity of the N geochemical cycle, the use of $\delta^{15}N$ should be carefully evaluated when applied to natural ecosystems.

9 Use of Combined ^{13}C and ^{18}O To Integrate Performance of Functional Groups

Despite the developments in the applications of individual isotopes, as discussed above, perhaps the greatest insight for natural systems will come about using combinations of isotopes. The use of ^{13}C and ^{18}O has been modeled both quantitatively (Saurer et al. 1997; Farquhar et al. 1998; Barbour et al. 2000) and semiquantitatively (Scheidegger et al. 2000) for a range of natural and crop systems. The contribution that mesophyll conductance makes to carbon isotope discrimination can be resolved because the ^{18}O signal solely reflects evaporative demand as a function of stomatal conductance. In contrast, p_i/p_a (and hence the extent of ^{13}C discrimination) can also be affected by the extent of drawdown to Rubisco via the combined physical and biochemical components of the mesophyll conductance. The combination of $\delta^{18}O$ and $\delta^{13}C$ allows the differentiation between stomatal regulation and carbon assimilation (see Farquhar et al. 1998): when humidity increases both $\delta^{13}C$ and $\delta^{18}O$ will decrease should the source of variation be stomatal conductance. However, when the source of variation is increasing photosynthetic capacity, which will tend to draw down c_i, then carbon isotope discrimination will diminish. Thus, if stomata do not respond, when $\delta^{13}C$ becomes less negative, $\delta^{18}O$ is unaffected.

Analyses of herbaceous vegetation in two hay meadows required that an array of responses be determined, depending on stomatal responses to humidity and the rate of change of assimilation (Sheidegger et al. 2000). Thus, there might either be a positive, neutral or negative relationship between $\delta^{13}C$ and $\delta^{18}O$, with *Festuca rubra*, for instance, showing a positive response; this has also been typified for studies of trees (Saurer et al. 1997; Barbour et al. 2000), suggesting that generally, as assimilation rate increases, c_i declines, with the slope of the response dependent on stomatal sensitivity to humidity. These approaches are highly important for attempts to refine the interpretation of tree ring isotope signals (Farquhar et al. 1998; Barbour et al. 2000) once account is taken of the extent of fractionation during secondary metabolism and synthesis of cellulose.

There is clearly great potential to elucidate the genotypic and phenotypic constraints to carbon isotope discrimination, as discussed above for shrubs from semi-arid deserts and for characterising different functional groups within natural vegetation using combined analyses of ^{13}C and ^{18}O. To separate the effects of structural and functional adaptations on WUE, three species representing three functional groups were analysed under controlled conditions: (1) a drought-tolerant evergreen sclerophyll (*Quercus coccifera*), that endures water stress by high stomatal regulation of water loss and a deep root system; (2) a drought semi-deciduous shrub (*Cistus albidus*) with highly pubescent leaves, this species is seasonal dimorphic with smaller summer leaves, and avoids summer drought by reduction of the transpirational leaf area through partial leaf abscission; and (3) a fast-growing perennial herb (*Vinca difformis*) with mesophytic leaves and low tolerance to water stress, normally found at more humid sites.

For the evergreen sclerophyll, $\Delta^{13}C$ reflected stomatal conductance during leaf expansion, rather than regulation of gas exchange to drought conditions (C. Werner, C. Máguas, unpubl. observ.). Hence, $\Delta^{13}C$ is not a sensitive parameter to measure WUE under stressed conditions in these species with long-lived leaves and a short growing period. This has important implications for the interpretation of $\Delta^{13}C$ in Mediterranean sclerophyllous species, where leaves are generally produced under favourable environmental conditions in spring, but regulation of WUE may be more important during summer drought. The oxygen isotope composition of bulk leaf material allowed a more detailed analysis, indicating clear differences between functional groups: in evergreen species, an increase in stomatal conductance was balanced by a proportional increase in photosynthetic capacity, without alteration in WUE, whereas in semi-deciduous species an increase in stomatal conductance exceeded any increase in photosynthetic capacity, resulting in a lower WUE (C. Werner, C. Máguas, unpubl. observ.). These results again indicate the advantage of combined use of carbon and oxygen isotopes of leaf dry matter for the interpretation of WUE and drought tolerance strategies, especially in long-lived evergreen species.

10 Conclusions

While we have largely reiterated the fundamental equations, models and approaches used to analyse stable isotope composition in gaseous samples and plant material, this is vital for framing the developments that we have highlighted during this review. We emphasise the importance of resolving the additional fractionation factors during (photo)respiratory metabolism, which are important for several reasons: for natural vegetation under conditions of environmental stress, the extent of these processes will affect overall carbon balance, particularly when the extent of internal refixation of these respiratory sources is taken into consideration. Whilst the impact of these processes can be modelled effectively using the more complicated equations describing the appropriate fractionation factors, it will also need to be taken into account in collection and analysis of CO_2 air samples, which are increasingly being used to define the gas-exchange "footprint" for natural ecosystems and the combined contribution to gaseous fluxes from soils and plants.

We also conclude that the widespread use of discrimination terminology to represent integrated water use efficiency is an oversimplification other than for crop plants, and consideration should be given to a number of factors which constrain Δ as expressed in natural vegetation. Specifically, the environmental conditions during leaf expansion (or deposition of carbohydrate reserves the previous year) will determine the overall organic signal far more than responses to drought later in the growing season, although measurements of leaf water (^{18}O) and carbohydrate composition (^{13}C, ^{18}O) would provide a more precise daily record of real-time changes during the onset of water deficits. For certain groups of plants, viz., CAM plants and lichens, we argue that water use efficiency is an appropriate concept. However, the complicated interactions between contrasting "conservative" and "high gain" strategies often preclude such a simple interpretation simply in terms of WUE.

Thus, life form, growth strategy and phenology should all be considered important determinants of Δ. Even though experimentally a significant decrease in $\Delta^{13}C$ can be induced by drought stress, the environmental conditions during leaf formation, the length of the growing period, and leaf longevity are more likely to be the major determinants of WUE. In a study of Mediterranean Macchia plants, all species were highly responsive to drought stress in most ecophysiological and structural parameters, whereas only a buffered response was found in $\Delta^{13}C$ of bulk leaf material conditions (C. Werner, C. Máguas, unpubl. observ.). When consideration is given to the range of competitive interactions for water and nitrogen resources, often pulsed under natural conditions, it is evident that studies must consider variations in resource availability in space and time, as well as genotypic differences within and between populations of plants.

Ultimately, however, many of these differences will be resolved by use of a combination of isotopes, such as ^{13}C and ^{18}O, or ^{15}N. Given the quantitative frameworks which are now being developed for each of these isotopes, it is likely that the current exponential growth in the use of stable isotopes to evaluate natural vegetation and exchanges with the atmosphere will continue to provide increasing insight for the foreseeable future.

Acknowledgements. We particularly wish to thank to Kate Maxwell and Christiane Werner for their kind support and to Pedro Pinho for his help in the editing of the reference list. Cristina Máguas gratefully acknowledges the Fundação para a Ciência e Tecnologia (FCT) and Instituto de Ciência Aplicada e Tecnologia (ICAT) for financial support.

References

Acevedo E (1993) Potential of carbon isotope discrimination as a selection criterion in barley breeding. In: Ehleringer JR, Hall AE, Farquhar GD (eds) Stable isotopes and plant carbon-water relations. Academic Press, San Diego, pp 399–417

Adams MA, Grierson PF (2001) Stable isotopes and natural abundance in terrestrial plant ecology and ecophysioloy: an update. Plant Biol 3:299–310

Anderson JE, Williams J, Kriedemann PE, Austin MP, Farquhar GD (1996) Correlations between carbon isotopic discrimination and climate of native habitats for diverse eucalypt taxa growing in a common garden. Aust J Plant Physiol 23:311–320

Antos JA, Allen GA (1994) Biomass allocation among reproductive structures in the dioecious shrub *Oemleria cerasiformis* – a functional interpretation. J Ecol 82:21–29

Araus JL, Reynolds MP, Acevedo E (1993) Leaf posture, grain yield growth, leaf structure and carbon isotope discrimination. Crop Sci 33:1273–1279

Barbour MM, Schurr, Henry BK, Wong SC, Farquhar GD (2000) Variation in the oxygen isotope ratio of phloem sap of sucrose from castor bean: evidence in support of the Péclet effect. Plant Physiol 123:671–679

Beardall J, Griffiths H, Raven J (1982) Carbon isotope discrimination and the CO_2 accumulating mechanism in *Chlorella emersonii.* J Exp Bot 33:729–737

Boddey RM, Peoples MB, Palmer B (2000) Use of the ^{15}N natural abundance technique to quantify biological nitrogen fixation by woody perennials. Nutr Cycling Agroecosyst 57:235–270

Borland AM, Griffiths H, Broadmeadow MSJ, Fordham M, Maxwell C (1994) Carbon isotope composition of biochemical fractions and the regulation of carbon balance in leaves of the C_3-CAM intermediate *Clusia minor* L. growing in Trinidad. Plant Physiol 106:493–501

Broadmeadow MSJ, Griffiths H (1993) Carbon isotope discrimination and the coupling of CO_2 fluxes within forest canopies. In: Ehleringer JR, Hall AE, Farquhar GD (eds) Stable isotopes and plant carbon-water relations. Academic Press, San Diego, pp 109–129

Broadmeadow MSJ, Griffiths H, Maxwell, Borland AM (1992) The carbon isotope ratio of plant organic material reflects temporal and spatial variations in CO_2 within tropical forest formations in Trinidad. Oecologia 89:435–441

Brugnoli E, Björkman O (1992) Growth of cotton under continuous salinity stress: influence on allocation pattern, stomatal and non-stomatal components of photosynthesis and dissipation of excess light energy. Planta 187:335–347

Brugnoli E, Farquhar GD (2000) Photosynthetic fractionation of carbon isotopes. In: Leegood RG Sharkey TD, Caenumerer (eds) Photosynthesis physiology and metabolism. Kluwer, Dordrecht, pp 399–434

Brugnoli E, Hubick K, von Caemmerer S, Wong S, Farquhar G (1988) Correlation between the carbon isotope discrimination in leaf starch and sugars of C_3 plants and the ratio of intercellular and atmospheric partial pressure of carbon dioxide. Plant Physiol 88:1418–1424

Brugnoli E, Scartazza A, Lauteri M, Montverdi MC, Máguas C (1998) Carbon isotope discrimination in structural and non-structural carbohydrates in relation to productivity and adaptation to unfavourable conditions. In: Griffiths H (ed) Stable isotopes: integration of biological, ecological and geochemical processes. Bios Scientific, Oxford, pp 133–146

Buchmann N, Brooks R, Flanagan L, Ehleringer J (1998) carbon isotope discrimination of terrestrial ecosystems. In: Griffiths H (ed) Stable isotopes: integration of biological, ecological and geochemical processes. Bios Scientific, Oxford, pp 203–218

Caldwell MM, Dawson TE, Richards JH (1998) Hydraulic lift: consequences for water efflux from the roots of plants. Oecologia 113:151–161

Ciais P, Meijer HAJ (1998) The $^{18}O/^{16}O$ isotope ratio of atmospheric CO_2 and its role in global carbon cycle research. In: Griffiths H (ed) Stable isotopes: integration of biological, ecological and geochemical processes. Bios Scientific, Oxford, pp 409–438

Ciais P, Tans PP, Trolier M, White JWC, Francey RJ (1995) A large northern hemisphere terrestrial CO_2 sink indicated by the $^{13}C/^{12}C$ ratio of atmospheric CO_2. Science 269:1098–1101

Comstock J, Ehleringer J (1993) Stomatal response to humidity in common bean (*Phaseoulus vulgaris*): implications for maximum transpiration rate, water-use efficiency and productivity. Aust J Plant Physiol 20:669–691

Condon AG, Richards RA, Farquhar GD (1987) Carbon isotope discrimination is positively correlated with grain yield and dry matter production in field-grown wheat. Crop Sci 27:996–1001

Correia O, Diaz-Barradas MC (2000) Ecophysiological differences between male and female plants of *Pistacia lentiscus* L. Plant Ecol 149:131–142

Correia OA, Martins AC, Catarino FM (1992) Comparative phenology and seasonal foliar nitrogen variation in Mediterranean species of Portugal. Ecol Mediterr 18:7–18

Cowan IR, Lange O, Green A (1992) Carbon-dioxide exchange in lichens: determination of transport and carboxylation characteristics. Planta 18:282–294

Cox PA (1981) Niche partitioning between sexes of dioecious plants. Am Nat 117:295–307

Craig H (1954) Carbon 13 in plants and the relationship between carbon 13 and carbon 14 variations in nature. J Geol 62:115–149

Dawson TE (1993a) Water sources of plants as determined from xylem-water isotopic composition: perspectives on plant competition, distribution, and water relations. In: Ehleringer J, Hall A, Farquhar G (eds) Stable isotopes and plant carbon-water relations. Academic Press, London, pp 465–496

Dawson TE (1993b) Hydraulic lift and water use by plants: implications for performance, water balance and plant-plant interactions. Oecologia 95:565–574

Dawson TE (1998) Fog in California redwood forest: ecosystem inputs and use by plants. Oecologia 117:476–485

Dawson TE, Bliss LC (1989) Patterns of water use and the tissue water relations in the dioecious shrub *Salix artica*: the physiological basis for habitat partitioning between the sexes. Oecologia 79:332–343

Dawson TE, Ehleringer JR (1991) Streamside trees that do not use stream water. Nature 350:335–337

Dawson TE, Ehleringer JR (1993) Gender-specific physiology, carbon isotope discrimination and habitat distribution in box elder, *Elder negundo*. Ecology 74:798–815

Deléens-Provent E, Schwebel-Gdugué N (1987) Demonstration of autrophic state and establishment of full tropical C_4 pathway in maize seedlings by photosynthate carbon isotope composition. Plant Physiol Biochem 25:567–572

DeLucia EH, Schlesinger WH (1991) Resource-use efficiency and drought tolerance in adjacent great basin and Sierran plants. Ecology 72:51–58

Diaz-Barradas MC, Correia O (1999) Sexual dimorphism, sex ratio and spatial distribution of male and female shrubs in the dioecious species *Pistacia lentiscus* L. Fol Geobot 34:163–174

Duranceau M, Ghashghaie J, Brugnoli E (2001) Carbon isotope discrimination during photosynthesis and dark respiration in intact leaves of *Nicotiana sylvestris*: comparison between wild type and mitochondrial mutant plants. Aust J Plant Physiol 28:65–71

Ehleringer JR, Cerling TE (1995) Atmospheric CO_2 and the ratio of intercellular to ambient CO_2 levels in plants. Tree Physiol 15:562–566

Ehleringer JR, Cook CS (1998) Carbon and oxygen isotopes ratios of ecosystem respiration along an Oregon conifer transect: preliminary observations based on small-flask sampling. Tree Physiol 18:513–519

Ehleringer JR, Dawson TE (1992) Water uptake by plants: perspectives from stable isotopes. Plant Cell Environ 15:1073–1082

Ehleringer JR, Philips SL, Schuster WEF, Sandquist DR (1991) Differential utilisation of summer rains by desert plants, implications for competition and climate change. Oecologia 88:430–434

Ehleringer JR, Philips SL, Comstock JP (1992) Seasonal variation in carbon isotopic composition of desert plants. Funct Ecol 6:396–404

Ehleringer JR, Schwinning S, Gebauer RLE (1999) Water use in arid land ecosystems. In: Press MC, Barker MG (eds) Physiological plant ecology. Blackwell, Oxford

Evans JR, Sharkey TD, Berry JA, Farquhar GD (1986) Carbon isotope discrimination measured concurrently with gas exchange to investigate CO_2 diffusion in leaves of higher plants. Aust J Plant Physiol 13:281–292

Evans JR, von Caemmerer S, Setchell BA, Hudson GS (1994) The relationship between CO_2 transfer conductance and leaf anatomy in transgenic tobacco with reduced content of Rubisco. Aust J Plant Physiol 21:475–495

Evans RD (2001) Physiological mechanisms influencing plant nitrogen isotope composition. Tree 6:121–126

Evans RD, Belnap J (1999) Long term consequences of disturbance on nitrogen dynamics in an arid ecosystem. Ecology 80:150–160

Evans JR, von Caemmerer S (1996) Carbon dioxide diffusion within leaves. Plant Physiol 110:339–346

Evans RD, Bloom AJ, Sukrapanna SS, Ehleringer JR (1996) Nitrogen isotope composition of tomato (*Lycopersicon esculentum* Mill. cv. T-5) grown under ammonium of nitrate nutrition. Plant Cell Environ 19:1317–1323

Farquhar GD, Lloyd J (1993) Carbon and oxygen isotope effects in the exchange of carbon dioxide between terrestrial plants and the atmosphere. In: Ehleringer J, Hall A, Farquhar G (eds) Stable isotopes and plant carbon-water relations. Academic Press, London, pp 47–70

Farquhar GD, Richards RA (1984) Isotopic composition of plant carbon correlates with water-use-efficiency of wheat genotypes. Aust J Plant Physiol 11:359–552

Farquhar GD, O'Leary MH, Berry JA (1982) On the relationship between carbon isotope discrimination and the intercellular carbon dioxide concentration in leaves. Aust J Plant Physiol 9:121–137

Farquhar GD, Ehleringer JR, Hubick KT (1989) Carbon isotope discrimination and photosynthesis. Annu Rev Plant Physiol Plant Mol Biol 40:503–537

Farquhar GD, Lloyd J, Taylor JA, Flanagan LB, Syversten JP, Hubik KT, et al (1993) Vegetation effects on the isotopic composition of oxygen in atmospheric CO_2. Nature 363:439–442

Farquhar GD, Barbour MM, Henry BK (1998) Interpretation of oxygen isotope composition of leaf material. In: Griffths H (ed) Stable isotopes: integration of biological, ecological and geochemical processes. Bios Scientific, Oxford, pp 27–62

Flanagan LB (1998) Oxygen isotope effects during CO_2 exchange: from leaf to ecosystem processes. In: Griffiths H (ed) Stable isotopes: integration of biological, ecological and geochemical processes. Bios Scientific, Oxford, pp 185–201

Flanagan LB, Ehleringer JR (1991) Stable isotope composition of steam and leaf water: applications to the study of plant water use. Funct Ecol 5:270–277

Flanagan LB, Phillips SL, Ehleringer JR, Loyd J, Farquhar GD (1994) Effect of changes in leaf water oxygen isotopic composition on discrimination against $C^{18}O^{16}O$ during photosynthetic gas exchange. Aust J Plant Physiol 21:221–234

Fontes JC (1980) Environmental isotopes in groundwater hydrology. In: Fritz P, Fontes JC (eds) Handbook of environmental isotope biochemistry. Elsevier, Amsterdam, pp 75–140

Francey RJ, Gifford RM, Sharkey TD, Weir B (1985) Physiological influences on carbon isotope discrimination in huon pine (*Lagarostrobos franklinii*). Oecologia 66:211–218

Garten CT, Taylor GE (1992) Foliar $\delta^{13}C$ within a temperate deciduous forest: spatial, temporal and species sources of variation. Oecologia 90:1–7

Gat JR (1996) Oxygen and hydrogen isotopes in the hydrologic-cycle. Annu Rev Earth Planet Sci 24:225–262

Gebauer R, Ehleringer J (2000) Water and nitrogen uptake patterns following moisture pulses in a cold desert community. Ecology 81:1415–1424

Ghashghaie J, Badek F, Tcherkez G, Deléens (2001) Carbon isotope discrimination during dark respiration in C_3 plants. Plant Cell Environ 24:505–515

Gillon JS, Griffiths H (1997) The influence of (photo)respiration on carbon isotope discrimination in plants. Plant Cell Environ 20:1217–1230

Gillon JS, Borland AM, Harwood KG, Roberts A, Broadmeadows MSJ, Griffiths H (1998) Carbon isotope discrimination in terrestrial plants: carboxylations and decarboxylations. In: Griffiths H (ed) Stable isotopes: integration of biological, ecological and geochemical processes. Bios Scientific, Oxford, pp 111–131

Gleixner G, Danier HJ, Werner RA, Schmidt HL (1993) Correlations between the ^{13}C content of primary and secondary plant products in different cell compartments and that in decomposing basidiomycetes. Plant Physiol 102:1287–1290

Green TGA, Lange OL (1994) Photosynthesis in poikilohydric plants: a comparison of lichens and bryophytes. In: Schulze E, Caldwell M (eds) Ecophysiology of photosynthesis. Ecological studies. Springer, Berlin Heidelberg New York, pp 319–342

Griffiths H (1996) Evaluation and integration of environmental stress using stable isotopes. In: Baker NR (ed) Photosynthesis and the environment. Kluwer, Dordrecht, pp 451–468

Griffiths H, Broadmeadow MSJ, Bordland AM, Hetherington CS (1990) Short-term changes in carbon isotope discrimination identify transitions between C_3 and C_4 carboxylation during crassulacean acid metabolism. Planta 181:604–610

Griffiths H, Borland A, Gillon J, Harwood K, Maxwell K, Wilson J (1999) Stable isotopes reveal exchanges between soil, plants and the atmosphere. In: Press MC, Scholes JD, Barker MG (eds) Physiological plant ecology. The 39th Symposium of the British Ecological Society, University of York, 7–9 September. Blackwell Science, Oxford, pp 415–441

Hanba YT, Mori S, Lei TT, Koike T, Wada E (1997) Variations in leaf $\delta^{13}C$ along a vertical profile of irradiance in a temperate Japanese forest. Oecologia 110:253–261

Hanba YT, Miyazawa SI, Terashima I (1999) The influence of leaf thickness on the CO_2 transfer conductance and leaf stable carbon isotope ratio for some evergreen tree species in Japanese warm-temperate forests. Funct Ecol 13:632–639

Harwood KG, Gillon JS, Griffiths H, Broadmeadow MSJ (1998) Diurnal variation of $\Delta^{13}CO_2$, $\Delta C^{18}O^{16}O$ and evaporative site enrichment of $\delta H_2{}^{18}O$ in *Piper aduncum* under field conditions in Trinidad. Plant Cell Environ 21:269–283

Harwood KG, Gillon JS, Roberts A, Griffiths H (1999) Stable isotopes (^{13}C and ^{18}O) in ambient CO_2 and water vapour reflect the relationships between sources and sinks for gas exchange within a temperate *Quercus petraea* canopy. Oecologia 119:109–119

Hobbie EA, Macko SA, Williams M (2000) Correlations between foliar 15N and nitrogen concentrations may indicate plant-mycorrhizal interactions. Oecologia 122:273–283

Högberg P (1997) ^{15}N natural abundance in soil-plant systems. New Phytol 137:179–203

Hubick KT, Gibson A (1993) Diversity in the relationship between carbon isotope discrimination and transpiration efficiency when water is limited. In: Ehleringer JR, Hall AE, Faquhar GD (eds) Stable isotopes and plant carbon-water relations. Academic Press, San Diego, pp 311–325

Hubick KT, Farquhar GD, Gshorter R (1986) Correlation between water-use efficiency and carbon isotope discrimination in diverse peanut (*Arachis*) germplasm. Aust J Plant Physiol 13:803–816

Hultine KR, Marshall JD (2000) Altitude trends in conifer leaf morphology and stable carbon isotope composition. Oecologia 123:32–40

Johnson DA, Asay KH, Tieszen LL, Ehleringer JR, Jefferson PG (1990) Carbon isotope discrimination: potential in screening cool-season grasses for water-limited environments. Crop Sci 30:338–343

Körner CH (1989) The nutritional status of plants from high altitudes: a world-wide comparison. Oecologia 68:379–391

Körner CH, Farquahar GD, Roksandic Z (1988) A global survey of carbon isotope discrimination in plants from high altitude. Oecologia 74:623–632

Körner CH, Farquahar GD, Wong SC (1991) Carbon isotope discrimination by plants follows latitudinal and altitudinal trends. Oecologia 88:30–40

Kohl DH, Shearer GB, Commoner B (1971) Fertilizer nitrogen: contribution to nitrate in surface water in a corn belt watershed. Science 174:1331–1334

Lajha K, Getz J (1993) Photosynthesis and water use efficiency in pinyon-juniper communities along an elevation gradient in northern New Mexico. Oecologia 94:95–101

Lange OL, Green TGA (1996) High thallus water content severely limits photosynthesis carbon gain of central Europe epilithic lichen under natural conditions. Oecologia 198:13–20

Lange OL, Ziegler H (1986) Different limiting processes of photosynthesis in lichens. In: Marcelle R, Clitjers H, van Poucke M (eds) Biological control of photosynthesis. Martinus Nijhoff, Dordrecht, pp 147–161

Lange OL, Green TGA, Ziegler H (1988) Water status related photosynthesis and carbon isotope discrimination in species of the lichen genus *Pseudocyphellaria* with green and blue green photobionts and in photosymbiodermes. Oecologia 75:494–501

Lauteri M, Scartazza A, Guido MC, Brugnoli E (1997) Genetic variation in photosynthetic capacity, carbon isotope discrimination and mesophyll conductance in provenances of *Castanea sativa* adapted to different environments. Funct Ecol 11:675–683

Leavitt SW, Long A (1982) Evidence for$^{13}C/^{12}C$ fractionation between tree leaves and wood. Nature 298:742–744

Le Roux Xl, Bariac T, Sinoquet H, Genty B, Piel C, Maritti A, Girardin C, Richard P (2001) Spatial distribution of leaf-use efficiency and carbon isotope discrimination within an isolated tree crown. Plant Cell Environ 24:1021–1032

Li C, Berninger F, Koskela J, Sonninen E (2000) Drought responses of *Eucalyptus michroteca* provenance dependent seasonality of rainfall in their place of origin. Aust J Plant Physiol 27:231–238

Lin G, Ehleringer JR (1997) Carbon isotopic fractionation does not occur during dark respiration in C_3 and C_4 plants. Plant Physiol 114:391–394

Lloyd J, Farquhar GD (1994) ^{13}C discrimination during CO_2 assimilation by the terrestrial biosphere. Oecologia 99:201–215

Madhavan S, Treichel I, O'Leary MH (1991) Effects of relative humidity on carbon isotope fractionation in plants. Bot Acta 104:292–294

Máguas C, Brugnoli E (1996) Spatial variation in carbon isotope discrimination across the thalli of several lichen species. Plant Cell Environ 19:437–446

Máguas C, Griffiths H, Ehleringer JR, Serodio J (1993) Characterisation of photobiont associates in lichens using carbon isotopic discrimination techniques. In: Ehleringer J, Hall AE, Farquhar GD (eds) Stable isotopes and plant carbon-water relations. Academic Press, Oxford, pp 201–212

Máguas C, Griffiths H, Broadmeadow MSJ (1995) Gas exchange and carbon isotope discrimination in lichen: evidence for interactions between CO_2-concentration mechanism and diffusion limitation. Planta 196:95–102

Máguas C, Valladares F, Brugnoli E (1997) Effects of thallus size on morphology and physiology of foliose lichens: new findings with a new approach. Symbiosis 23:149–164

Marshall JD, Zhang J (1994) Carbon isotope discrimination and water use efficiency in native plants of the northern-central Rockies. Ecology 75:1887–1895

Matzner SL, Rice KJ, Richards JH (2001) Factors affecting the relationship between carbon isotope discrimination and transpiration efficiency in bleu-oak (*Quercus douglasii*). Aust J Plant Physiol 28:49–56

Maxwell K, von Caemmerer S, Evans JR (1997) Is a low internal conductance to CO_2 diffusion a consequence of succulence in plants with crassulacean acid metabolism? Aust J Plant Physiol 24:777–786

Meier-Augenstein W (1999) Applied gas chromatography coupled to isotope mass spectrometry. J Chromatogr 824:351–371

Meinzer FC, Rundel PW, Goldstein G, Sharifi MR (1992) Carbon isotope composition in relation to leaf gas exchange and environmental conditions in Hawaiian *Metrosideros polymorpha* population. Oecologia 91:305–311

Melzer A, Schmidt HL (1987) Carbon isotope effects on the pyruvate dehydrogenase reaction and their importance for relative carbon-13 depletion in lipids. J Biol Chem 20:8159–8164

Miller JM, Williams RJ, Farquar GD (2001) Carbon isotope discrimination by a sequence of *Eucalyptus* species along a subcontinental rainfall gradient in Australia. Funct Ecol 15:222–232

Miranda AC, Miranda HS, Lloyd J, Grace J, Francey RJ, Mcintyre JA, Meir P, Riggan P, Lockwood R, Brass J (1997) Fluxes of carbon, water and energy over Brazilian cerrado: an analysis using eddy covariance and stable isotopes. Plant Cell Environ 20:315–328

Miyazawa SI, Terashima I (2001) Slow development of leaf photosynthesis in an evergreen broad-leaved tree, *Castanopsis sieboldii*: relationships between leaf anatomical characteristics and photosynthetic rate. Plant Cell Environ 24:279–291

Niinemets U, Kull O, Tenhunen JD (1999) Variability in leaf morphology and chemical composition as a function of canopy light environment in coexisting deciduous trees. Int J Plant Sci 160:837–848

Obeso JR (1997) Cost of reproduction in *Ilex aquifolium*: effects on tree, branch and leaf levels. J Ecol 85:159–166

O'Leary MH (1981) Carbon isotope fractionation in plants. Phytochemistry 20:553–567

O'Leary MH (1988) Carbon isotopes in photosynthesis. BioScience 38:325–336

Palmqvist K, Máguas C, Badger MR, Griffiths H (1994) Assimilation, accumulation and carbon isotope discrimination of inorganic carbon in lichens: further evidence for the operation of a CO_2 concentration mechanism in cyanobacterial lichens. Crypt Bot 4:218–226

Panek JA, Waring RH (1995) Carbon isotope variation in Douglas-fir foliage: improving the 13C-climate relationship. Tree Physiol 15:657–663

Panek JA, Waring RH (1997) Stable carbon isotopes as indicators of limitations to forest growth imposed by climate stress. Ecol Appl 7:854–863

Pearcy RW, Pfitsch WA (1994) Influence of sunflecks on the $\delta^{13}C$ of *Adenocaulon bicolor* plants occuring in contrasting understorey microsites. Oecologia 86:457–462

Prasolova NV, Xu ZH, Farquhar GD, Saffigna PG, Dieters MJ (2000) Variations in branchlet $\delta^{13}C$ in relation to branchlet nitrogen concentration and growth in 8-year old hoop pine families (*Auracaria cunninghamii*) in subtropical Australia. Tree Physiol 20:1024–1055

Proctor MCF, Raven JA, Rice SK (1992) Stable carbon isotope discrimination measurements in *Sphagnum* and other bryophytes: physiological and ecological implications. J Bryol 17:193–202

Retuerto R, Lema BF, Roiloa SR, Obeso JR (2000) Gender, light and water effects in carbon isotope discrimination, and growth rates in the dioecious tree *Ilex aquifolium*. Funct Ecol 14:529–537

Rice SK (2000) Variation in carbon isotope discrimination within and among *Sphagnum* species in a temperate wetland. Oecologia 123:1–8

Rice SK, Giles L (1996) The influence of water content and leaf anatomy on carbon isotope discrimination and photosynthesis in *Sphagnum*. Plant Cell Environ 19:118–124

Roberts A, Bordland AM, Griffiths H (1997) Discrimination processes and shift in carboxylation during the phases of crassulacean acid metabolism. Plant Physiol 113:1283–1292

Robinson D (2001) $\delta^{15}N$ as an integrator of the nitrogen cycle. Trees 16:153:161

Rooney MA (1988) Short-term carbon isotope fractionation by plants. PhD Thesis, University of Wisconsin, Wisconsin, USA

Rossman A, Butzenlechner M, Schmidt H (1991) Evidence for nonstatistical carbon isotope distribution in natural glucose. Plant Physiol 96:609–614

Rundel PW, Stichler W, Zander RH, Ziegler H (1979) Carbon and hydrogen isotopes ratios of bryophytes from arid and humid regions. Oecologia 44:91–94

Saurer M, Allen K, Siegwolf RTW (1997) Correlating $\delta^{13}C$ and $\delta^{18}O$ in cellulose of trees. Plant Cell Environ 20:1543–1550

Scartazza A, Lauteri M, Guido MC, Brugnoli E (1998) Carbon isotope discrimination in leaf and stem sugars, water use efficiency and mesophyll conductance during different developmental stages in rice subjected to drought. Aust J Plant Physiol 25:489–498

Schmidt HL, Gleixner G (1998) Carbon isotope effects on key reactions in plant metabolism and 13C-patterns in natural compounds. In: Griffiths H (ed) Stable isotopes: integration of biological, ecological and geochemical processes. Bios Scientific, Oxford, pp 13–25

Sharkey TD, Berry J (1985) Carbon isotope fractionation of algae as influenced by an inducible CO_2 concentrating mechanism. In: Lucas WJ, Berry J (eds) Inorganic carbon uptake by aquatic organisms. American Society of Plant Physiology, Rockville, Maryland, pp 389–402

Shearer G, Kohl DH (1986) Information derived from variation in the natural abundance of ^{15}N in complex biological systems. In: Buncel E, Saunders WH (eds) Heavy atom isotope effects. Elsevier, Amsterdam, pp 133–190

Sheidegger Y, Saurer M, Bahn M, Siegwolf (2000) Linking stable oxygen and carbon isotopes with stomatal conductance and photosynthetic capacity: a conceptual model. Oecologia 125:350–357

Smith BN, Epstein S (1971) Two categories of $^{13}C/^{12}C$ ratios for higher plants. Plant Physiol 47:380–384

Smith EC, Griffiths H (1996) The occurrence of the chloroplast pyrenoide is correlated with the activity of a CO_2-concentration mechanism and carbon isotope discrimination in lichens and bryophytes. Planta 198:6–16

Smith EC, Griffiths H (1998) Intraspecific variation in photosynthetic responses of treboixioid lichens with reference to the activity of a carbon-concentrating mechanism. Oecologia 113:360–369

Sparks JP, Ehleringer JR (1997) Leaf carbon isotope discrimination and nitrogen content for riparian trees along elevational transects. Oecologia 109:362–367

Sternberg L, Swart P (1987) Utilization of fresh water and ocean water by coastal plants of southern Florida. Ecology 68:1898–1905

Tans PP, Monk WG (1980) Past atmospheric CO_2 levels and the $^{13}C/^{12}C$ ratios in tree rings. Tellus 32:268–283

Teeri JA (1981) Stable carbon isotopes analysis of mosses and lichens growing in xeric and moist habitats. Bryologist 84:82–82

Thorburn P, Walker G (1993) The source of water transpired by *Eucalyptus camaldulensis*: soil, groundwater or streams? In: Ehleringer J, Hall AE, Farquhar GD (eds) Stable isotopes and plant carbon-water relations. Academic Press, Oxford, pp 111–128

Valladares F, Ascaso C, Sancho LG (1998) Water storage in the lichen family Umbilicariaceae. Bot Acta 111:99–107

Vitousek PM, Field CB, Matson PA (1990) Variations in foliar $\delta^{13}C$ in Hawaiian *Metrosideros polymorpha*: a case of internal resistance? Oecologia 84:362–370

Vogel JC (1980) Fractionation of the carbon isotopes during photosynthesis. Springer, Berlin Heidelberg New York, 29 pp

Von Caemmerer S, Evans J (1991) Determination of the average partial pressure of CO_2 in chloroplasts from leaves of several C_3 plants. Aust J Plant Physiol 18:287–305

Walcroft AS, Silvester WB, Grace J, Carson S, Warring R (1996). Effects of branch length on carbon isotope discrimination in *Pinus radiata*. Tree Physiol 16:281–286

Wallace CS, Rundel PW (1979) Sexual dimorphism and resource allocation in male and female shrubs of *Simmondsia chinensis*. Oecologia 44:34–39

Waring RH, Silvester W (1994) Variation in foliar $\delta^{13}C$ values within tree crowns of *Pinus radiata*. Tree Physiol 14:1203–1213

Warren CR, McGrath JF, Adams MA (2001) Water availability and carbon isotope discrimination in conifers. Oecologia 127:476–486

Wershaw R, Friedman I, Heller S, Frank P (1966) Hydrogen isotope fractionation of water passing through trees. In: Hobson G, Spears G (eds) Advances in organic geochemistry. Pergamon Press, London, pp 55–67

White J, Cook E, Lawrence J, Broecker W (1985) The D/H ratios of sap in trees: implications of water sources and tree-rings D/H ratios. Geochim Cosmochim Act 49:237–246

Williams TG, Flanagan LB (1996) Effect of changes in water content on photosynthesis, respiration and discrimination against $^{13}CO_2$ and $C^{18}O^{16}O$ in *Pleurozium* and *Sphagnum*. Oecologia 108:38–46

Williams TG, Flanagan LB (1998) Measuring and modelling environmental influences on photosynthetic gas exchange in *Shagnum* and *Pleurozium*. Plant Cell Environ 21:555–564

Winter K, Holtum (2002) How closely do the 13C values of CAM plants reflect the proportion of CO_2 fixed during day and night? Plant Physiol (in Press)

Yakir D (1998) Oxygen-18 in leaf water: a crossroad for plant-associated isotopic signals. In Griffiths H (ed) Stable isotopes, integration of biological and geochemical processes. Bios Scientific, Oxford, pp 147–168

Yakir D, Sternberg LDL (2000) The use of stable isotopes to study ecosystem gas exchange. Oecologia 123:297

Yakir D, DeNiro MJ, Rundel PW (1989) Isotopes in homogeneity of leaf water: evidence and implications for the use of isotopic signals transduced by plants. Geochim Cosmochim 53:2769–2773

Zhang JW, Marshall JD (1995) Variation in carbon isotope discrimination and photosynthetic gas exchange among populations of *Pseudotsuga menziesii* and *Pinus ponderosa* in different environments. Funct Ecol 9:402–412

Cristina Máguas
FCUL, Centro de Ecologia e Biologia Vegetal
Depto de Biologia Vegetal
Bloco C2, Campo Grande
1700 Lisboa, Portugal

Howard Griffiths
Department of Plant Sciences
University of Cambridge, Downing Street
Cambridge CB2 3EA, UK

e-mail: cmaguas@fc.ul.pt

Trends in Plant Diversity Research

Stefan Porembski

1 Introduction

There is now little doubt that anthropogenic impacts will lead to further fundamental changes in the biosphere. Detailed information on the dynamic attributes of the major components of global change is urgently needed for predicting future scenarios of biodiversity in a rapidly changing world. Excellent overviews of the different facets of global change and related biodiversity topics were provided by Walker et al. (1999), WBGU (2000) and a series of articles published in "Nature" (Nature 405, 2000).

Biodiversity is still an in-word that continues to attract considerable scientific, socio-economic and political interest. This interest in biodiversity is also reflected by the number of new scientific journals devoted to the topic and by an increasing number of scientific institutions using the term "biodiversity" in their official name. The importance of different global change-related impacts on biodiversity has also become a focal point of several large scientific research initiatives (e.g., the German programme on biodiversity and global change "BIOLOG" funded by the German Federal Ministry of Education and Research, BMBF) which are mainly aimed at deciphering the complex relationships between the richness and composition of communities and basic ecosystem functions.

In contrast to former generations of biologists, who focused their scientific interests on studying undisturbed ecosystems, it is inevitable that contemporary and future biologists will not have the same opportunity to analyse ecosystems unmodified by human activities (Tilman 1999). Today, no region on earth exists that is absolutely free of anthropogenic influences, which are frequently summarized as effects of "global change". The dominance of man-made or semi-natural ecosystems will therefore have considerable consequences for future directions of research in botany. It can be assumed that the so-called naturalist approach of studying hitherto unknown (regarding their inventories) regions will gradually disappear, with only a few individuals and institutions adhering to it. As usual when certain developments are en vogue,

Progress in Botany, Vol. 64

this does not imply that other fields of research that seem to be outdated have completely lost their right of existence. For instance, it is accepted that our knowledge of seemingly simple facts, such as the number of species living on earth, is still far from complete, and inventory work therefore remains necessary. This is underlined by the fact that even for relatively well-known groups of organisms, such as vascular plants, the number of species has been considerably underestimated, as shown by Prance (2001). Monitoring data obtained from inventory studies are also necessary for a better calculation of species extinction rates on global and regional scales. A more detailed knowledge of the extent of species extinction rates might be important for predicting the consequences of the reduction of plant diversity for a variety of ecosystem functions (e.g., nutrient cycling, carbon storage). As the Earth is apparently going through a new era of human-caused mass extinctions (Chapin et al. 2000), the problem of how many species are necessary to maintain important ecosystem functions remains open.

The present review deals with selected aspects of plant diversity research, and focuses on contributions that have appeared during the last 3 years.

2 Global Change: Measuring and Predicting the Consequences for Plant Diversity

Global change, defined as changes in climate, atmospheric composition and land-use, is predicted to have serious consequences for plant diversity. A number of abiotic parameters (e.g., CO_2 content of the atmosphere, temperature, precipitation, UV radiation) have shown significant changes over the last decades due to human activities, including the burning of fossil combustibles and different land-use practices.

One of the greatest challenges for the future is to understand how human alteration of the global environment influences biodiversity. In recent years, a large number of reviews and books dealing with this topic have appeared; comprehensive overviews have been presented by Chapin et al. (2000) and Sala et al. (2000). According to these sources, human dominance of Earth's resources has triggered the sixth major extinction event in the history of life, and also widespread changes in the global distribution of organisms. Since biodiversity provides key ecosystem services such as climate regulation and soil fertility, the loss of species (globally ca. 10% of all plant species are threatened with extinction) during the course of global change might lead to a reduced reliability of ecosystem services. However, one has to emphasize that most estimates of extinction rates suffer from a severe lack of knowledge and are thus highly speculative. For terrestrial ecosystems, land-use change will probably have the largest effect on the composition and structure of vegetation. Should recent trends in agriculture continue, 10^9 ha of natural ecosystems will be converted to agriculture by 2050 (Tilman et al. 2001). The concomitant eutrophication and habitat destruction are likely to result in unprecedented ecosystem simplification, loss of ecosystem services and species extinctions. Globally, tropical deforestation releases 20–30% of anthropogenic greenhouse gases. The benefits generated from forest conservation significantly outweigh logging and agriculture both locally and globally. Nationally, however, the financial benefits of industrial logging are frequently larger than conservation benefits (Kremen et al. 2000). A number of studies have addressed the question of shifts in diversity in times of environmental change. For example, Brown et al. (2001) examined long-term trends in taxonomic richness and composition in different regions. Their results suggest that, while species composition may be highly variable in response to environmental change, species diversity of ecosystems is often maintained within narrow limits.

Man-made alterations to the global carbon cycle will cause feedback effects of the biosphere that are still not understood in detail. Due to the importance of this topic, a number of international research programmes (e.g., IGBP) are concerned with the responses of the biosphere on different levels of ecosystem complexity. According to Körner (1996), ca. 1500 publications were available on plant responses to elevated CO_2 levels. Since then, the number has been rising steadily, with an increasing percentage of studies devoted to the functional relationship between the species richness of plant communities and their reaction to elevated CO_2 levels. The response of ecosystems to increasing CO_2 has received high research priority, leading to the application of experimental approaches. Several integrated approaches are currently underway for better describing the relationships between the carbon cycle, biogeochemical processes and vegetation responses (Falkowski et al. 2000). Of critical importance is the question as to what extent terrestrial ecosystems provide sinks or sources with regard to carbon exchange between the biosphere and the atmosphere. In the 1990s, the terrestrial biosphere became a net carbon sink, and this can mainly be attributed to northern extratropical areas. The development of the terrestrial carbon sink is mainly due to changes in land use, such as regrowth on abandoned agricultural lands and fire prevention, in addition to climate change (Schimel et al. 2001). Many European forest ecosystems also act as carbon sinks. The carbon balance is, however, a delicate equilibrium between photosynthesis and respiration, a fact that is particularly true for boreal European forests, making them very vulnerable to climatic change (Valentini et al. 2000). Saxe et al. (2001) have presented a review of the responses of trees and forests at boreal and temperate latitudes to global warming and increased greenhouse gas emissions. Due mainly to deforestation, the Brazilian Amazon represented a source of carbon over the period 1989–1998 (Houghton et al. 2000).

Of increasing relevance have been attempts to model the consequences of different aspects of global change, such as increasing temperature and atmospheric CO_2 concentration (e.g., Neilson et al. 1998; Betts et al. 2000; Duckworth et al. 2000; Cramer et al. 2001). The models provide a useful basis for the investigation of potential vegetation change and frequently suggest that changes, once set in motion, will continue to evolve for at least a century (Cramer et al. 2001), even if atmospheric CO_2 concentration levels and climate were to be instantaneously stabilized. The complex interactions between elevated atmospheric CO_2 concentration and plant species diversity have been investigated by numerous studies (e.g., Hättenschwiler and Körner 2000; Oechel et al. 2000; Smith et al. 2000; Niklaus et al. 2001; and see review by McLeod and Long 1999). Our current knowledge is best for European grasslands, where elevated CO_2 can substantially alter the composition of plant communities. However, we know much less about the response of species-rich tropical ecosystems.

Global climate change (in particular the increase in temperature) is frequently considered a major conservation threat. Reviews of the rapidly accumulating documentations of changes in species and ecosystems linked to global climate change have been provided by Hughes (2000) and McCarty (2001). The responses of sub-Arctic dwarf shrub-dominated heathland vegetation to global change effects (enhanced UV-B radiation and increased summer precipitation) was investigated by Phoenix et al. (2001) over 7 years in a field experiment. Both enhanced UV-B radiation and increased summer precipitation had no effect on dwarf shrub abundance, suggesting that this vegetation type may be relatively tolerant to changes in the tested environmental factors. A considerable number of terrestrial vegetation types are already affected by changing temperatures in association with the rise in atmospheric CO_2. The most pronounced effects have been observed in arctic (Sturm et al. 2001) and temperate (Alward et al. 1999) plant communities. The accelerated warming of the Alaskan Arctic over the last three decades has caused a widespread increase in shrub abundance, possibly resulting in changes in the high-latitude carbon budget, as well as contributing to important changes in the exchange of surface energy (Sturm et al. 2001). Even without a warming climate, anthropogenic disturbance regimes, such as petroleum development, are already having a significant effect on the future distribution of arctic plant communities (Forbes et al. 2001). Largely because of a rapid increase in minimum temperatures, average temperatures in the short grass steppe of northeastern Colorado have risen 1.3 °C since 1970 (Alward et al. 1999). Increased minimum temperatures have been correlated with decreased net primary production of the dominant C_4 grass (*Bouteloua gracilis*) and with increased abundance and production by exotic and native forbs. Reductions in *B. gracilis* may make this system more vulnerable to invasion by exotic species and less tolerant of drought and grazing. Based on an experimental approach, Smith et al. (2000) came to the conclusion that elevated atmospheric CO_2 might enhance the long-term success and dominance of exotic annual grasses (e.g., *Bromus tectorum*) in the arid regions of western North America. Smith et al. (2000) showed that new shoot production of a dominant perennial shrub is doubled by a 50% increase in atmospheric CO_2 in a high rainfall year. However, the exotic *B. tectorum* showed significantly higher plant density, biomass and reproductive success at elevated CO_2 than native annuals. Enhanced success of exotic annual grasses would expose deserts to an accelerated fire cycle to which they are not adapted. The consequences of the various aspects of global change with regard to the invasiveness of plant communities have been discussed in detail over the last few years. This topic will be treated together with other aspects in the next section.

3 Plant Diversity and Ecosystem Functioning

Changes in land use, habitat fragmentation, nutrient enrichment, and environmental stress often lead to reduced plant diversity in ecosystems, with respect to both species richness and the number of plant functional types (e.g., Vitousek et al. 1997). However, it remains controversial if and how these changes and the accompanying reductions in diversity will affect processes such as energy flow and nutrient cycling. Moreover, our understanding of the reactions of biomes that are currently declining in plant biodiversity at unprecedented extinction rates towards elevated atmospheric CO_2 or nitrogen deposition is not sufficient to make clear predictions for future scenarios. Of particular importance in this respect is the long-established diversity-stability debate. Reviews on the latest developments in this field have been provided by amongst others, Chapin et al. (1998), Schläpfer and Schmid (1999), Tilman (1999, 2000), Loreau (2000), McCann (2000) and Fridley (2001). According to Tilman (1999), plant diversity and primary productivity are positively correlated as a result of both the sampling and the niche differentiation effect. The combination of both effects causes more complete resource utilization at higher diversity. Lower levels of available resources at higher diversity could lower the invasibility of ecosystems. The majority of investigations (e.g., Hector et al. 1999; Lehman and Tilman 2000; Spehn et al. 2000; Wilsey and Potvin 2000; Engelhardt and Ritchie 2001; Reich et al. 2001) suggest that ecosystem processes are generally less stable at lower diversity, and that primary productivity increases with diversity. According to Tilman (1999), diversity should be added as one of the major determinants of ecosystem structure and dynamics, along with composition, disturbance, nutrients and climate, since it forms a central determinant of ecosystem functioning. However, Schwartz et al. (2000), in assessing the empirical literature in order to answer the question whether recent studies support the assumption that a close link exists between species richness and ecosystem function, state that solid evidence for such a relationship is practically non-existent. According to these authors, conservationists should thus take a cautious view of endorsing this linkage to promote conservation goals. Many researchers who found a positive correlation between species diversity and ecosystem properties studied artificial microcosms and semi-natural grasslands. A substantial amount of criticism (e.g., Huston et al. 2000; Wardle et al. 2000) has been articulated concerning the experimental methodology and general conclusions of such studies.

In the field of invasion ecology, the debate as to whether more diverse communities are better buffered against the invasion of exotic invaders continues. Surveys of the suggested mechanisms that determine why certain species may successfully invade a given ecosystem or why some biomes are more prone to invasions than others have been provided by Lavorel et al. (1999), Levine and D'Antonio (1999), Williamson (1999), Prieur-Richard and Lavorel (2000), Rejmánek (2000), Richardson et al. (2000) and Sakai et al. (2001). Recent studies have found contradictory results with regard to the relationship between diversity and invasibility (e.g., Knops et al. 1999; Lonsdale 1999; van der Putten et al. 2000). According to Dukes (2001), plant functional diversity can reduce the success of invaders by reducing resource availability, a concept that is in accordance with the ideas of Tilman (1999). In this experimental study, the invader *Centaurea solstitialis* suppressed growth of species-poor communities more strongly by affecting community evapotranspiration rates. Dukes' study thus suggests that communities with fewer species may be more likely to decline as a consequence of invasion.

4 Effects of Habitat Fragmentation

The fact that fragmentation of natural habitats due to anthropogenic impacts is one of the greatest threats to biodiversity worldwide has been widely accepted for several decades now. Habitat fragmentation affects species diversity mainly by the reduction of total habitat size within a region, reduction in size of individual habitats and increasing isolation between habitat fragments. Given the importance of habitat fragmentation as a major driving force of global change, the huge amount of published information on this topic is not surprising. According to Debinski and Holt (2000), a large number of observational studies exist on fragmented landscapes, and there is a substantial theoretical literature (with particular reference to metapopulation ecology, see Hanski 1999) on the population and community effects of fragmentation. However, experimental analyses of habitat fragmentation are much rarer. The literature survey of Debinski and Holt (2000) revealed only 20 experiments worldwide that focused mainly on the effects of fragmentation on species richness or on the abundance of particular species. Only a few experimental studies have been devoted to plant populations. Several studies (e.g., Dubbert et al. 1998; Zabel and Tscharntke 1998) investigated the influence of fragmentation on ectophagous and endophagous insect communities on plants. Studying the short-term responses of vascular plants and certain groups of invertebrates to the experimental fragmentation of calcareous grassland in Switzerland, Zschokke et al. (2000) found that different taxonomic groups were affected to different extents. In particular, certain grasses occurred in higher densities in fragments. Interesting insights into plant–herbivore interactions in meadow fragments of different size were obtained by Kruess and Tscharntke (2000). They came to the general conclusion that the size of meadows is the major determinant of species diversity and population abundance of endophagous insects, and that the conservation of large and less isolated habitat remnants enhances species diversity and parasitism of potential pest insects. The importance of herbivory in forest fragments was investigated by Rao et al. (2001) during a study on the effects of leaf-cutter ants on plants occurring on differently sized islands in Venezuela. The data showed that plant densities were significantly reduced on small islands, due to elevated leaf-cutter ants herbivory. This in turn was attributable to release from predation. The fact that fragment size is of overriding importance in determining the richness, composition and structure of plant communities was demonstrated by investigations in both temperate and tropical regions. Abiotic and biotic edge effects were studied by Gehlhausen et al. (2000) in forest fragments in Illinois (USA). In particular, microclimatic variables differed in the degree to and distance over which they showed an edge effect. Within the framework of the Kansas fragmentation study, Yao et al. (1999) showed that woody plant colonization in the course of old-field succession is positively correlated with patch size. Habitat fragmentation can significantly alter the reproductive success of Australian mallee woodland plants due to changing the community of pollinators and natural enemies (Cunningham 2000). In many cases, fragmentation causes the isolation and diminution of plant populations. Kéry et al. (2000) showed that reproduction was depressed in small populations of *Primula veris* and *Gentiana lutea*. Moreover, small populations of *P. veris* suffered from genetic deterioration as a result of habitat fragmentation. The competing demands of nature conservation and other forms of land-use such as agriculture and urbanization in the South African Renosterveld formed the focus of a study by Kemper et al. (2000). The extent of fragmentation, fragment size and numbers were determined using

LANDSAT imagery and GIS-based programs for spatial pattern analysis. The degree of fragmentation varied with topographic factors, with fragments of natural vegetation concentrated on steep slopes that excluded agricultural use. The largest remaining area of intact deciduous tropical forest worldwide (Tierras Bajas, Bolivia) is suffering from deforestation and fragmentation rates that are among the highest in the world (Steininger et al. 2001). The authors express the opinion that the fragmentation effects of deforestation can be minimized only if conservation planning occurs at the earliest stages of forest-use. Logging and road construction frequently create forest fragments, and these are characterized by abrupt forest edges. Bowersox and Brown (2001) provided new statistical approaches for describing edge abruptness that is often confounded by the patchiness of edges. A concise survey of edge effects in tropical forest fragments has been provided by Gascon et al. (2000). According to these authors, conservation strategies should (1) maximize the area/perimeter relationship of conservation areas by protecting large forest remnants rather than small ones; (2) protect the forest edge itself against structural damage, and (3) minimize the harshness of the surrounding matrix. The detrimental impacts of poor logging practices in tropical forests, which enforce, among other things, edge effects, were discussed by, e.g., Putz et al. (2000) and Putz (2001). The necessity of a multidisciplinary approach for understanding boundary phenomena in an increasingly edgy world was underlined by Laurance et al. (2001a). The complex interacting factors that determine savanna boundaries were the focus of a paper by Jeltsch et al. (2000).

In the tropics, Brazilian rainforests form a particular focus of fragmentation studies, a fact that is due to the severity of the problem there and the large number of Brazilian groups working on the topic. A central role in this respect is played by the Biological Dynamics of Forest Fragments Project (BDFFP), located in the Amazon ca. 80 km north of Manaus. In the framework of this project, Benitez-Malvido (1998) investigated the abundance of mature-phase tree seedlings in unfragmented and fragmented forests. Seedling density declined significantly from continuous forest to forest fragments. The lower seedling number in forest fragments is explained by increased tree mortality, reduced seed output and dispersal, high seed predation and lower seedling establishment rates due to edge effects. Laurance et al. (2001b) also assessed the consequences of rainforest fragmentation for the structure of Amazonian liana communities, again in the framework of the BDFFP. Liana abundance increased significantly near forest edges and was positively associated with forest disturbance, and negatively associated with tree biomass. The number of trees infested with lianas was higher on fragment edges than in forest interiors. Lianas create physical stress on trees and compete for light and nutrients. Heavy liana infestation appears partly responsible for elevated rates of tree mortality and damage near fragment edges. Rainforest fragmentation in central Amazonia is having a disproportionately severe effect on large trees (Laurance et al. 2000). In the fragmented rainforest of the Amazon, large trees died at a rate nearly three times faster when they were within 300 m of edges than they did in forest interiors. This will have considerable consequences for the structural complexity of forests and might alter biogeochemical cycles (Laurance et al. 1998). Fire susceptibility of forest fragments in the Amazon is particularly high during El Niño events (Cochrane and Schulze 1999; Nepstad et al. 1999). According to Laurance et al. (2001c), the speed of rainforest fragmentation in the Brazilian Amazon will increase drastically if the infrastructure projects proposed by the Brazilian government in the "Avança Brasil" program are realized. Most of the Atlantic forests of Brazil have been converted into agricultural land, leaving forest fragments as small isolated patches. In a remote sensing approach, Saatchi et al. (2001) describe the application of radar data to distinguish forest fragments from cocoa planted in the shade of natural-forest trees. Several works (e.g., Tabarelli et al. 1999; da Silva and Tabarelli 2000) evaluating the effects of human-induced habitat fragmentation in the Atlantic forests of Brazil conclude that a large percentage of tree species will become extinct on a regional scale due to the disruption of key ecological processes, such as pollination and dispersal.

Whereas most studies on fragmentation effects have dealt with human-caused habitat patches, the number of investigations that have been devoted to an understanding of processes such as dispersal, pollination and vegetative adaptive traits in plant communities occurring in naturally fragmented ecosystems like rock outcrops is relatively small. Schmiedel and Jürgens (1999) described the vegetation structure of island-like quartz-fields in the South African Succulent Karoo. The vegetation of fragmented porphyry outcrops in the lower valley of the River Saale (Germany) differing in size and age was investigated by Partzsch (2001). Whereas young outcrops are dominated by ruderal vegetation, old outcrops support typical xerothermic plant communities that can only be maintained in the future by traditional land-use practices. A positive correlation between the size of vegetation islands which were surrounded by bare rock and species richness was found by Michelangeli (2000) on the summit of the Roraima Tepui (Venezuela). Aspects such as conservation, reproductive success and genetic diversity of serpentine endemic plants in northern California have been studied by Wolf et al. (2000), Wolf (2001) and Wolf and Harrison (2001). According to these studies, naturally patchy plants are not invulnerable to effects of habitat fragmentation, and the protection of multiple local populations might be necessary for the long-term conservation of plant species consisting of isolated populations. A survey of the vegetation of granitic and gneissic inselbergs that form natural habitat islands worldwide summarizes the current knowledge of this unique ecosystem (Porembski and Barthlott 2000).

5 Patterns of Species Richness

An improved understanding of the patterns of plant diversity from the global scale to the community level is still one of the most significant objectives for ecologists, biogeographers and conservationists. A survey of recent trends in this field and the influence of certain determinants of the spatial distribution of plant diversity has been given by Gaston (2000). He emphasized that applied issues such as the spread of alien invasive species and the control of diseases will become important in the context of global environmental change. Detailed knowledge of the spatial distribution of diversity is necessary for identifying "biodiversity hotspots", areas that are characterized by exceptionally high numbers of species and endemics. The importance of mountains as centers of plant species diversity and endemism is dealt with in a multi-authored volume edited by Körner and Spehn (2002). Myers et al. (2000) promote the biodiversity hotspot approach as a means for the prioritization of regions where exceptional concentrations of endemic species are undergoing substantial loss of habitat. In total, 25 hotspots were identified containing the remaining habitats of 133. 149 plant species. Tropical forests dominate the list of hot spots, followed by Mediterranean-type habitats. The leading hotspots (i.e., comprising the highest numbers of endemics) are the tropical Andes, Sundaland, Madagascar, Brazil's Atlantic forest and the Caribbean. In 1995, 25 biodiversity hotspots contained nearly 20% of the world's human population. Cincotta et al. (2000) estimated that population growth rates in the hotspots to be substantially higher than the average world population growth rate, thus underlining the conservation significance of the selected hotspots. However, the biodiversity hotspot approach was criticized by Jepson and Canney (2001), because spatial priorities that could be relevant for public policy were determined by Myers et al. (2000) merely on the basis of simple species counts.

Due to the availability of modern computing hardware and software such as GIS and World-Map, analyses of large-scale patterns of plant species diversity have become possible that allow the precise location of centers of species richness and endemism. Qian and Ricklefs (2000) focused on temperate-zone plant genera that are disjunctive between eastern Asia and eastern North America. These genera have twice as many species in eastern Asia as in eastern North America. This diversity anomaly is probably a result of the high physiographical heterogeneity of eastern Asia, which, in conjunction with climate and sea-level change, has provided opportunities for evolutionary speciation.

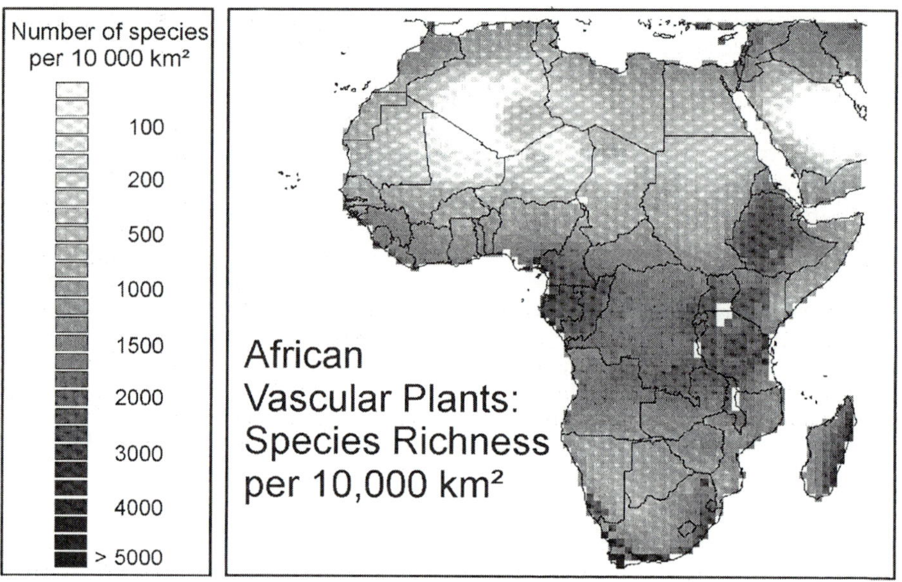

Fig. 1. Geographical distribution of vascular plant species richness in Africa. (After Mutke et al. 2002)

Several studies have been concerned with the determination of patterns of plant species richness and endemism of the African flora (e.g., Kier and Barthlott 2001; Linder 2001; Lovett et al. 2000; Mutke et al. 2002; Fig. 1). Kier and Barthlott (2001) have developed a new methodological approach using an index of "endemism richness" to measure the specific contribution of an area to global biodiversity based on their work on the African flora. According to Linder (2001), African centers of species richness and endemism are the Cape Floristic Region, East Coast, Congo-Zambezi watershed, Kivu, Upper and Lower Guinea. Refugia of endemics north of the equator are relatively small and isolated, whereas, south of the equator, endemics tend to be more generally distributed. Lovett et al. (2000) analysed spatial richness patterns of different plant life-forms in Africa south of the Sahara. Each group showed peaks of diversity at locations different to the other groups. Geophytes were most speciose in the western Cape region and the Mts Crystals (Gabon) formed a center of diversity for mesophytic herbs. Mutke et al. (2002) identified the Capensis, eastern Madagascar, the coastal areas of SE Nigeria, Cameroon, Equatorial Guinea, Gabon, the Drakensberg mountains and the East African Mountains as centers of species richness for vascular plants in Africa. The best correlations of species richness were found with annual sum of NDVI, number of dry months and water balance. The importance of water-energy dynamics as a likely cause of global biodiversity patterns has been demonstrated by several recent studies (e.g., Enquist et al.1999; Whittaker and Field 2000). The additional importance of macro-scale variation in topographic relief was demonstrated by O'Brien et al. (2000). Kleidon and Mooney (2000) asked the question as to how much the global distribution of plant species diversity is a consequence of climatic constraints. The results of this modeling approach suggest that climate explains much of the global distribution of plant species diversity. A theory that aims to explain fundamental patterns of biogeography and biodiversity has been presented by Hubbell (2001). In his stimulating book, he develops a formal mathematical theory that predicts the existence of a universal, dimensionless biodiversity number.

6 Future Research Perspectives

As outlined above, many advances have been made in our understanding of the possible feedback effects of the terrestrial biosphere to the complex man-made changes of the earth. However, in many fields, our knowledge is still too fragmentary to measure "global change", as well as to predict the consequences of global change for plant diversity. For example, the question of species richness and redundancy in relation to ecosystem stability has not been answered, not even for simplified artificial experimental systems. Similarly, the consequences of habitat fragmentation for the maintenance of species-rich communities are not clear in detail. A particular focus of future research should be the linking of landscape dynamics and spatial and temporal patterns of plant species richness with ecosystem processes. Better data from species-rich, relatively untouched ecosystems are urgently needed as this would allow a much better differentiation between the reactions of anthropogenically influenced habitats and primary ecosystems to various parameters of global change. Long-term, interdisciplinary studies (incorporate molecular methods, GIS, for instance) are evidently needed if we are to achieve a better understanding of the patterns of plant species richness in structurally complex and highly diverse ecosystems such as tropical forests. Finally, it is obvious that, in addition to scientific research, the generation of public awareness on the significance of losses of species and possibly ecosystem services related to global change will be of key importance in the future.

Acknowledgements. The critical reading of the manuscript by Dr. Gary Brown and Dr. Dethardt Götze (both Universität Rostock) is gratefully acknowledged. Our own research was supported by the Deutsche Forschungsgemeinschaft and the BMBF.

References

Alward RD, Detling JK, Milchunas DG (1999) Grassland vegetation changes and nocturnal global warming. Science 283:229–231

Benitez-Malvido J (1998) Impact of forest fragmentation on seedling abundance in a tropical rain forest. Conserv Biol 12:380–389

Betts RA, Cox PM, Woodward FI (2000) Simulated responses of potential vegetation to doubled-CO_2 climate change and feedbacks on near-surface temperature. Global Ecol Biogeogr 9:171–180

Bowersox MA, Brown DG (2001) Measuring the abruptness of patchy ecotones. Plant Ecol 156:89–103

Brown JH, Ernest SKM, Parody JM, Haskell JP (2001) Regulation of diversity: maintenance of species richness in changing environments. Oecologia 126:321–332

Chapin FS III, Sala OE, Burke IC et al. (1998) Ecosystem consequences of changing biodiversity. Bioscience 48:45–55

Chapin FS III, Zavaleta ES, Eviner VT et al. (2000) Consequences of changing biodiversity. Nature 405:234–242

Cincotta RP, Wisnewski J, Engelman R (2000) Human population in the biodiversity hotspots. Nature 404:990–992

Cochrane MA, Schulze MD (1999) Fire as a recurrent event in tropical forests of the eastern Amazon: effects of forest structure, biomass and species composition. Biotropica 31:2–16

Cramer W, Bondeau A, Woodward FI et al (2001) Global response of terrestrial ecosystem structure and function to CO_2 and climate change: results from six global vegetation models. Global Change Biol 7:357–373

Cunningham SA (2000) Effects of habitat fragmentation on the reproductive ecology of four plant species in mallee woodland. Conserv Biol 14:758–768

Da Silva JMC, Tabarelli M (2000) Tree species impoverishment and the future flora of the Atlantic forest of northeast Brazil. Nature 404:72–74

Debinski DM, Holt RD (2000) A survey and overview of habitat fragmentation experiments. Conserv Biol 14:342–355

Dubbert M, Tscharntke T, Vidal S (1998) Stem-boring insects of fragmented *Calamagrostis* habitats: herbivore-parasitoid community structure and the unpredictability of grass shoot abundance. Ecol Entomol 23:271–280

Duckworth JC, Bunce RGH, Malloch AJC (2000) Modelling the potential effects of climate change on calcareous grasslands in Atlantic Europe. J Biogeogr 27:347–358

Dukes JS (2001) Biodiversity and invasibility in grassland microcosms. Oecologia 126:563–568

Engelhardt KAM, Ritchie ME (2001) Effects of macrophyte species richness on wetland ecosystem functioning and services. Nature 411:687-689

Enquist BJ, West GB, Charnov EL, Brown JH (1999) Allometric scaling of production and life-history variation in vascular plants. Nature 401:907–911

Falkowski P, Scholes RJ, Boyle E et al. (2000) The global carbon cycle: a test of our knowledge of Earth as a system. Science 290:291–296

Forbes BC, Ebersole JJ, Strandberg B (2001) Anthropogenic disturbance and patch dynamics in circumpolar arctic ecosystems. Conserv Biol 15:954–969

Fridley JD (2001) The influence of species diversity on ecosystem productivity: how, where, and why? Oikos 93:514–526

Gascon C, Williamson GB, Fonseca de GAB (2000) Receding forest edges and vanishing reserves. Science 288:1356–1358

Gaston KJ (2000) Global patterns in biodiversity. Nature 405:220–227

Gehlhausen SM, Schwartz MW, Augspurger CK (2000) Vegetation and microclimatic edge effects in two mixed-mesophytic forest fragments. Plant Ecol 147:21–35

Hättenschwiler S, Körner C (2000) Tree seedling responses to in situ CO_2-enrichment differ among species and depend on understory light availability. Global Change Biol 6:213–226

Hanski I (1999) Metapopulation ecology. Oxford University Press, Oxford

Hector A, Schmid B, Beierkuhnlein C et al. (1999) Plant diversity and productivity experiments in European grasslands. Science 286:1123–1127

Houghton RA, Skole DL, Nobre CA, Hackler JL, Lawrence KT, Chomentowski WH (2000) Annual fluxes of carbon from deforestation and regrowth in the Brazilian Amazon. Nature 403:301–304

Hubbell SP (2001) The unified neutral theory of biodiversity and biogeography. Princeton University Press, Princeton

Hughes L (2000) Biological consequences of global warming: is the signal already apparent? Trends Ecol Evol 15:56–61

Huston MA, Aarssen LW, Austin MP et al. (2000) No consistent effect of plant diversity on productivity. Science 289:1255

Jeltsch F, Weber GE, Grimm V (2000) Ecological buffering mechanisms in savannas: a unifying theory of long-term tree-grass coexistence. Plant Ecol 161:161–171

Jepson P, Canney S (2001) Biodiversity hotspots: hot for what? Global Ecol Biogeogr 10:225–227

Kemper J, Cowling RM, Richardson DM, Forsyth GG, McKelly DH (2000) Landscape fragmentation in south coast Renosterveld, South Africa, in relation to rainfall and topography. Aust Ecol 25:179–186

Kéry M, Matthies D, Spillmann H-H (2000) Reduced fecundity and offspring performance in small populations of the declining grassland plants *Primula veris* and *Gentiana lutea*. J Ecol 88:17–30

Kier G, Barthlott W (2001) Measuring and mapping endemism and species richness: a new methodological approach and its application on the flora of Africa. Biodiv Conserv 10:1513–1529

Kleidon A, Mooney HA (2000) Global distribution of biodiversity inferred from climatic constraints: results from a process-based modelling study. Global Change Biol 6:507–523

Knops JMH, Tilman D, Haddad NM et al. (1999) Effects of plant species richness on invasion dynamics, disease outbreaks, insect abundances and diversity. Ecol Lett 2:286–293

Körner C (1996) The response of complex multispecies systems to elevated CO_2. In: Walker B, Steffen W (eds) Global change and terrestrial ecosystems. Cambridge University Press, Cambridge, pp 21–42

Körner C, Spehn EM (2002) Mountain biodiversity: a global assessment. Parthenon Publishing, London (in press)

Kremen C, Niles JO, Dalton MG, Daily GC, Ehrlich PR, Fay JP, Grewal D, Guillery RP (2000) Economic incentives for rain forest conservation across scales. Science 288:1828–1832

Kruess A, Tscharntke T (2000) Species richness and parasitism in a fragmented landscape: experiments and field studies with insects on *Vicia sepium*. Oecologia 122:129–137

Laurance WF, Laurance SG, Delamônica P (1998) Tropical forest fragmentation and greenhouse gas emissions. For Ecol Manage 110:173–180

Laurance WF, Delamônica P, Laurance SG, Vasconcelos HL, Lovejoy TE (2000) Rainforest fragmentation kills big trees. Nature 404:836

Laurance WF, Didham RK, Power ME (2001a) Ecological boundaries: a search for synthesis. Trends Ecol Evol 16:70–71

Laurance WF, Pérez-Salicrup D, Delamônica P, Fearnside PM, D'Angelo S, Jerozolinski A, Pohl L, Lovejoy TE (2001b) Rain forest fragmentation and the structure of Amazonian liana communities. Ecology 82:105–116

Laurance WF, Cochrane MA, Bergen S, Fearnside PM, Delamônica P, Barber C, D'Angelo S, Fernandes T (2001c) The future of the Brazilian Amazon. Science 291:438–439
Lavorel S, Prieur-Richard A-H, Grigulis K (1999) Invasibility and diversity of plant communities: from patterns to processes. Div Distrib 5:41–49
Lehman CL, Tilman D (2000) Biodiversity, stability, and productivity in competitive communities. Am Nat 156:534–552
Levine JM, D'Antonio CM (1999) Elton revisited: a review of evidence linking diversity and invasibility. Oikos 87:15–26
Linder HP (2001) Plant diversity and endemism in sub-Saharan tropical Africa. J Biogeogr 28:169–182
Lonsdale WM (1999) Global patterns of plant invasions and the concept of invasibility. Ecology 80:1522–1536
Loreau M (2000) Biodiversity and ecosystem functioning: recent theoretical advances. Oikos 91:3–17
Lovett JC, Rudd S, Taplin J, Frimodt-Møller C (2000) Patterns of plant diversity in Africa south of the Sahara and their implications for conservation management. Biodiv Conserv 9:37–46
McCann KS (2000) The diversity-stability debate. Nature 405:228–233
McCarty JP (2001) Ecological consequences of recent climate change. Conserv Biol 15:320–331
McLeod AR, Long SP (1999) Free-air carbon dioxide enrichment (FACE) in global change research: a review. Adv Ecol Res 28:1–55
Michelangeli FA (2000) Species composition and species-area relationships in vegetation isolates on the summit of a sandstone mountain in southern Venezuela. J Trop Ecol 16:69–82
Mutke J, Kier G, Braun G, Schultz C, Barthlott W (2002) Patterns of African vascular plant diversity – a GIS based analysis. Syst Geogr Pl 71 (in press)
Myers N, Mittermeier RA, Mittermeier CG, da Fonseca GAB, Kent J (2000) Biodiversity hotspots for conservation priorities. Nature 403:853–858
Neilson RP, Prentice IC, Smith B, Kittel T, Viner D (1998) Simulated changes in vegetation distribution under global warming. In: Watson RT, Zinyowera MC, Moss RH (eds) The regional impacts of climate change. Cambridge University Press, Cambridge, pp 439–456
Nepstad DC, Verissimo A, Alencar A, Nobre C, Lima E, Lefebvre P, Schlesinger P, Potter C, Moutinho P, Mendoza E, Cochrane M, Brooks V (1999) Large-scale impoverishment of Amazonian forests by logging and fire. Nature 398:505–508
Niklaus PA, Leadley PW, Schmid B, Körner C (2001) A long-term field study on biodiversity × elevated CO_2 interactions in grassland. Ecol Monogr 71:341–356
O'Brien E, Field R, Whittaker RJ (2000) Climatic gradients in woody plant (tree and shrub) diversity: water-energy dynamics, residual variation, and topography. Oikos 89:588–600
Oechel WC, Vourlitis GL, Hastings SJ, Zulueta RC, Hinzman L, Kane D (2000) Acclimation of ecosystem CO_2 exchange in the Alaskan Arctic in response to decadal climate warming. Nature 406:978–981
Partzsch M (2001) Die Porphyrkuppenlandschaft des unteren Saaletals – Vergleich der Vegetation in Abhängigkeit von Flächengröße und Genese der Porphyrkuppen. Bot Jahrb Syst 123:1–45
Phoenix GK, Gwynn-Jones D, Callaghan TV, Sleep D, Lee JA (2001) Effects of global change on a sub-Arctic heath: effects of enhanced UV-B radiation and increased summer precipitation. J Ecol 89:256–267
Porembski S, Barthlott W (eds) (2000) Inselbergs: biotic diversity of isolated rock outcrops in tropical and temperate regions. Springer, Berlin Heidelberg New York
Prance GT (2001) Discovering the plant world. Taxon 50: 345–359

Prieur-Richard A-H, Lavorel S (2000) Invasions: the perspective of diverse plant communities. Aust Ecol 25:1–7

Putz FE (2001) Tropical forest management and conservation of biodiversity: an overview. Conserv Biol 15:7–20

Putz FE, Dykstra DP, Heinrich R (2000) Why poor logging practices persist in the tropics. Conserv Biol 14:951–956

Qian H, Ricklefs RE (2000) Large-scale processes and the Asian bias in species diversity of temperate plants. Nature 407:180–182

Rao M, Terborgh J, Nuñez P (2001) Increased herbivory in forest isolates: implications for plant community structure and composition. Conserv Biol 15:624–633

Reich PB, Knops J, Tilman D et al. (2001) Plant diversity enhances ecosystem responses to elevated CO_2 and nitrogen deposition. Nature 410:809–812

Rejmánek M (2000) Invasive plants: approaches and predictions. Aust Ecol 25:497–506

Richardson DM, Pysek P, Rejmánek M, Barbour MG, Panetta FD, West CJ (2000) Naturalization and invasion of alien plants: concepts and definitions. Div Distrib 6: 93–107

Saatchi S, Agosti D, Alger K, Delabie J, Musinsky J (2001) Examining fragmentation and loss of primary forest in the southern Bahian Atlantic forest of Brazil with radar imagery. Conserv Biol 15:867–875

Sakai AK, Allendorf FW, Holt JS et al. (2001) The population biology of invasive species. Annu Rev Ecol Syst 32:305–332

Sala OE, Chapin FS III, Armesto JJ et al. (2000) Global biodiversity scenarios for the year 2100. Science 287:1770–1774

Saxe H, Cannell MGR, Johnsen Ø, Ryan MG, Vourlitis G (2001) Tree and forest functioning in response to global warming. New Phytol 149:369–400

Schimel DS, House JI, Hibbard KA et al. (2001) Recent patterns and mechanisms of carbon exchange by terrestrial ecosystems. Nature 414:169–172

Schläpfer F, Schmid B (1999) Ecosystem effects of biodiversity – a classification of hypotheses and cross-system exploration of empirical results. Ecol Appl 9:893–912

Schmiedel U, Jürgens N (1999) Community structure on unusual habitat islands: quartz-fields in the Succulent Karoo, South Africa. Plant Ecol 142:57–69

Schwartz MW, Brigham CA, Hoeksema JD, Lyons KG, Mills MH, van Mantgem PJ (2000) Linking biodiversity to ecosystem function: implications for conservation ecology. Oecologia 122:297–305

Smith SD, Huxman TE, Zitzer SF et al. (2000) Elevated CO_2 increases productivity and invasive species success in an arid ecosystem. Nature 408:79–82

Spehn EM, Joshi J, Schmid B, Diemer M, Körner C (2000) Above-ground resource use increases with plant species richness in experimental grassland ecosystems. Funct Ecol 14:326–337

Steininger MK, Tucker CJ, Ersts P, Killeen TJ, Villegas Z, Hecht SB (2001) Clearance and fragmentation of tropical deciduous forest in the Tierras Bajas, Santa Cruz, Bolivia. Conserv Biol 15:856–866

Sturm M, Racine C, Tape K (2001) Increasing shrub abundance in the Arctic. Nature 411:546–547

Tabarelli M, Mantovani W, Peres CA (1999) Effects of habitat fragmentation on plant guild structure in the montane forest of southeastern Brazil. Biol Conserv 91:119–127

Tilman D (1999) The ecological consequences of changes in biodiversity: a search for general principles. Ecology 80:1455–1474

Tilman D (2000) Causes, consequences and ethics of biodiversity. Nature 405:208–211

Tilman D, Fargione J, Wolff B, D'Antonio C, Dobson A, Howarth R, Schindler D, Schlesinger WH, Simberloff D, Swackhamer D (2001) Forecasting agriculturally driven global environmental change. Science 292:281–284

Valentini R, Matteucci G, Dolman AJ et al. (2000) Respiration as the main determinant of carbon balance in European forests. Nature 404:861–865

Van der Putten WH, Mortimer SR, Hedlund K et al. (2000) Plant species diversity as a driver of early succession in abandoned fields: a multi-site approach. Oecologia 124:91–99

Vitousek PM, Mooney HA, Lubchenko J, Melillo JM (1997) Human domination of Earth's ecosystems. Science 277:494–499

Walker S, Steffen W, Canadell J, Ingram J (1999) The terrestrial biosphere and global change. Cambridge University Press, Cambridge

Wardle DA, Huston MA, Grime JP et al. (2000) Biodiversity and ecosystem function: an issue in ecology. Bull Ecol Soc Am 81:235–239

WBGU, Wissenschaftlicher Beirat der Bundesregierung Globale Umweltveränderungen (2000) Welt im Wandel: Erhaltung und nachhaltige Nutzung der Biosphäre. Jahresgutachten 1999. Springer, Berlin Heidelberg New York

Whittaker RJ, Field R (2000) Tree species richness modelling: an approach of global applicability. Oikos 89:399–402

Williamson M (1999) Invasions. Ecography 22:5–12

Wilsey BJ, Potvin C (2000) Biodiversity and ecosystem functioning: importance of species evenness in an old field. Ecology 81:887–892

Wolf AT (2001) Conservation of endemic plants in serpentine landscapes. Biol Conserv 100:35–44

Wolf AT, Harrison SP (2001) Effects of habitat size and patch isolation on reproductive success of the serpentine Morning Glory. Conserv Biol 15:111–121

Wolf AT, Harrison SP, Hamrick JL (2000) Influence of habitat patchiness on genetic diversity and spatial structure of a serpentine endemic plant. Conserv Biol 14:454–463

Yao J, Holt RD, Rich PM, Marshall WS (1999) Woody plant colonization in an experimentally fragmented landscape. Ecography 22:715–728

Zabel J, Tscharntke T (1998) Does fragmentation of *Urtica* habitats affect phytophagous and predatory insects differentially? Oecologia 116:419–425

Zschokke S, Dolt C, Rusterholz H-P, Oggier P, Braschler B, Thommen GH, Lüdin E, Erhardt A, Baur B (2000) Short-term responses of plants and invertebrates to experimental small-scale grassland fragmentation. Oecologia 125:559–572

Prof. Dr. Stefan Porembski
Universität Rostock
Institut für Biodiversitätsforschung
Allgemeine und Spezielle Botanik
Wismarsche Strasse 8
18051 Rostock, Germany

e-mail: stefan.porembski@biologie.uni-rostock.de

Subject Index

Progress in Botany, Vol. 64